Bacterial Physiology and Metabolism

Bacterial Physiology and Metabolism

Editor: Cherilyn Gray

AMERICAN
MEDICAL PUBLISHERS
www.americanmedicalpublishers.com

AMERICAN
MEDICAL PUBLISHERS
www.americanmedicalpublishers.com

Cataloging-in-Publication Data

Bacterial physiology and metabolism / edited by Cherilyn Gray.
p. cm.
Includes bibliographical references and index.
ISBN 978-1-63927-911-1
1. Bacteria--Physiology. 2. Microbial metabolism. 3. Metabolism.
4. Bacteria. I. Gray, Cherilyn.
QR88 .B33 2023
576.113 3--dc23

American Medical Publishers,
41 Flatbush Avenue,
1st Floor, New York,
NY 11217, USA

ISBN 978-1-63927-911-1 (Hardback)

Contents

Preface

Over the recent decade, advancements and applications have progressed exponentially. This has led to the increased interest in this field and projects are being conducted to enhance knowledge. The main objective of this book is to present some of the critical challenges and provide insights into possible solutions. This book will answer the varied questions that arise in the field and also provide an increased scope for furthering studies.

Bacterial physiology refers to a discipline concerning the life supporting processes and functions of the bacteria, which facilitate the growth and reproduction of the bacterial cells. Bacterial metabolism is a significant component of bacterial physiology, as it provides the means through which nutrients and energy is obtained by the bacteria for its survival and reproduction. Bacterial metabolism is focused on chemical diversity of dissimilation reactions and substrate oxidations, which create energy in the bacteria. Enzymatic reactions in bacterial cells are regulated by two models, which involve the regulation of enzyme activity and regulation of enzyme synthesis. The regulation of enzyme activity is involved in the modulation of biosynthetic pathways. The regulation of enzyme synthesis focuses on catabolite repression and enzyme induction, which control the degradative pathways. This book provides comprehensive insights into bacterial physiology and metabolism. It aims to serve as a resource guide for students and experts alike.

I hope that this book, with its visionary approach, will be a valuable addition and will promote interest among readers. Each of the authors has provided their extraordinary competence in their specific fields by providing different perspectives as they come from diverse nations and regions. I thank them for their contributions.

Editor

1

A Replisome's journey through the bacterial chromosome

*Thomas R. Beattie and Rodrigo Reyes-Lamothe**

Department of Biology, McGill University, Montreal, QC, Canada

***Correspondence:**
Rodrigo Reyes-Lamothe, Department of Biology, McGill University, 3649 Sir William Osler, Montreal, QC H3G 0B1, Canada rodrigo.reyes@mcgill.ca

Genome duplication requires the coordinated activity of a multi-component machine, the replisome. In contrast to the background of metabolic diversity across the bacterial domain, the composition and architecture of the bacterial replisome seem to have suffered few changes during evolution. This immutability underlines the replisome's efficiency in copying the genome. It also highlights the success of various strategies inherent to the replisome for responding to stress and avoiding problems during critical stages of DNA synthesis. Here we summarize current understanding of bacterial replisome architecture and highlight the known variations in different bacterial taxa. We then look at the mechanisms in place to ensure that the bacterial replisome is assembled appropriately on DNA, kept together during elongation, and disassembled upon termination. We put forward the idea that the architecture of the replisome may be more flexible that previously thought and speculate on elements of the replisome that maintain its stability to ensure a safe journey from origin to terminus.

Keywords: DNA replication, replisome, bacteria, chromosome, evolution, DNA polymerase, *Escherichia coli*, *Bacillus subtilis*

The Architecture of the Bacterial Replisome

Replication of chromosomal DNA is fundamental to the propagation of all bacterial species. This essential process faces many complex mechanistic challenges. Parental DNA must be unwound, and the resulting template strands replicated simultaneously with great efficiency and accuracy. Furthermore, because these two template strands have opposing polarity, the two new DNA strands are synthesized differently, with one strand—the leading strand—synthesized continuously, and the other—the lagging strand—synthesized as a series of short Okazaki fragments. A multi-protein complex known as the replisome has evolved to coordinate all the core enzymatic activities required for these coupled processes into a single molecular machine.

Replisome Structure in *E. coli*

Replisome structure is currently best understood in the Gram-negative bacterium *Escherichia coli*. Decades of genetic and biochemical research have enabled all the essential components of the replisome in this organism to be identified, and a series of protein-protein interactions which link them into a single entity to be mapped. Indeed, it is possible to reconstitute a fully functional replisome *in vitro* from purified *E. coli* proteins (Wu et al., 1992). At the core of the replisome is the DnaB homohexameric helicase, which encircles single-stranded DNA (ssDNA) on the lagging strand and unwinds the parental DNA duplex. Copying of the resulting template strands is performed by DNA polymerase III (Pol III), which consists of a catalytic subunit α, a proofreading exonuclease subunit ε, and a poorly conserved non-essential subunit θ. The exact function of θ is still unclear, but it may moderately stimulate the exonuclease activity of ε (Slater et al., 1994). To achieve the processivity needed to synthesize the entire chromosome, Pol III associates with the dimeric β sliding

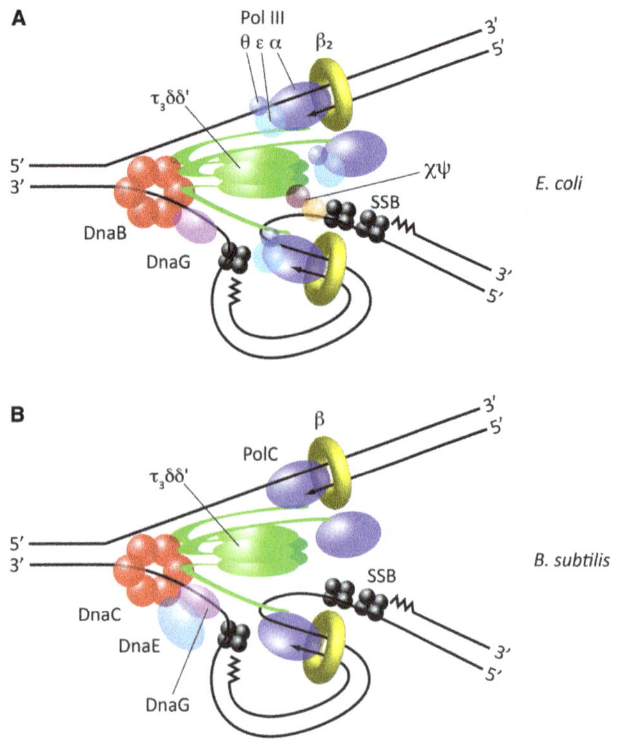

FIGURE 1 | Replisome architecture in bacteria. (A) Architecture of the *E. coli* replisome, derived from *in vitro* studies and direct observation *in vivo*. **(B)** Architecture of the *B. subtilis* replisome, predominantly derived from *in vitro* reconstitution studies.

clamp, which is assembled around DNA by the pentameric $\tau_3\delta\delta'$ clamp loader complex. Crucially, both DnaB and Pol III α directly interact with the C-terminal domain of τ, and thus the clamp loader additionally provides an architectural function, physically coupling template unwinding with DNA synthesis (**Figure 1A**).

It was previously assumed that only two Pol III complexes were present within the replisome, divided between the leading and lagging strands. However, elegant *in vitro* reconstitution experiments surprisingly demonstrated that the *E. coli* replisome can incorporate three Pol III complexes, multimerised by the trimeric τ within the clamp loader (McInerney et al., 2007). More recently, an unprecedented level of detail has been obtained by directly visualizing individual components of the replisome in living *E. coli* cells by fluorescence microscopy (Reyes-Lamothe et al., 2010). This work has confirmed the presence of three active polymerases within a single replisome. Further *in vitro* studies have suggested the third replicative polymerase may serve as a backup to support efficient synthesis of the lagging strand (Georgescu et al., 2011).

A trimeric polymerase architecture was previously unanticipated because the clamp loader was thought to consist of $\tau_2\gamma\delta\delta'$, γ being an alternative C-terminally truncated form of τ which is unable bind to Pol III and derives from a frameshift during translation of the dnaX gene. The role of γ is now unclear; like *E. coli*, clamp loader complexes from a number of other bacterial species appear to possess a τ-only structure (Bruck and O'Donnell, 2000; Bruck et al., 2002, 2005; Bullard et al., 2002; Jarvis et al., 2005a; Sanders et al., 2010). However, although the generation of γ by a frameshift mechanism appears to be limited to enterobacteria (Blinkova et al., 1997), γ-like subunits have been detected in *Thermus thermophilus* and *Caulobacter crescentus*, and furthermore these arise from distinct mechanisms for truncating the dnaX gene product—transcriptional slippage and controlled proteolysis, respectively (Yurieva et al., 1997; Vass and Chien, 2013). This hints at convergent evolution for generating γ, suggesting that in at least some bacterial species a clamp loader containing this truncated product may play a specialized role. It has been suggested for example that γ may be important in dealing with DNA damage (Reyes-Lamothe et al., 2010; Vass and Chien, 2013), but its precise role remains to be elucidated.

Multimerisation of Pol III within the *E. coli* replisome presents a potential topological problem as polymerases must track in opposite directions on the leading and lagging strands, but yet are arranged symmetrically through their interaction with the clamp loader. The replisome has evolved to ensure that DNA unwinding and synthesis proceed uninterrupted by allowing the transient accumulation of ssDNA loops between Pol III and the helicase on the lagging strand. These loops are rapidly bound by the tetrameric ssDNA binding protein, SSB, which protects them from damage. The cyclical assembly, growth and disassembly of these loops, which accompanies the synthesis of each Okazaki fragment, has been compared to the slide of a trombone (trombone model; Sinha et al., 1980). The presence of such trombone loops has been directly observed in simpler bacteriophage replisomes by electron microscopy and single-molecule biophysical studies (Park et al., 1998; Hamdan et al., 2009).

Initiation of each Okazaki fragment requires the synthesis of an RNA primer, which is subsequently extended by Pol III following loading of the β clamp onto the primer-template junction. In *E. coli* this is performed by the DnaG primase, which interacts directly with DnaB, thus facilitating primer deposition on the emerging lagging strand template. However, biochemical studies suggest that DnaG is not a constitutive component of the replisome, but rather associates with DnaB transiently; indeed it has been suggested that each primer is synthesized by a new DnaG molecule recruited from solution (Wu et al., 1992). In contrast, biophysical measurements on bacteriophage T4 replisomes have shown that primase behavior is stochastic, sometimes dissociating from the helicase following primer synthesis, but sometimes remaining tightly bound (Manosas et al., 2009). Unlike other replisome components, it has so far not been possible to observe DnaG *in vivo* by microscopy. It will be interesting in the future to directly determine the nature of the DnaG-DnaB interaction in its native environment of the living cell.

Since the predicted number of Okazaki fragments generated during chromosomal replication in *E. coli* exceeds the number of β clamps available in the cell, these clamps must be reused throughout replication (Leu et al., 2000). While some *in vitro* data suggests that clamps may be immediately recycled with Pol III between successive Okazaki fragments (Tanner et al., 2011), live cell microscopy data from both *E. coli* and *Bacillus subtilis* has demonstrated that clamps accumulate on DNA behind the replisome (Su'etsugu and Errington, 2011; Moolman et al., 2014).

TABLE 1 | A comparison of replication protein nomenclature in *E. coli* and *B. subtilis*.

Protein function	*E. coli*	*B. subtilis*
Origin activator	DnaA	DnaA
Helicase loader	DnaC	DnaI
Helicase	DnaB	DnaC
Primase	DnaG	DnaG
Clamp loader	$\tau_3\delta\delta'\chi\psi$	$\tau_3\delta\delta'$
Primary replicative polymerase	Pol III ($\alpha\varepsilon\theta$)	PolC
Accessory replicative polymerase	–	DnaE
ssDNA binding protein	SSB	SSB
Sliding clamp	β_2	β_2

This therefore supports an alternative model whereby clamps are reloaded from a pool left on the lagging strand. Such an accumulation of β clamps may serve as an important landing platform on DNA for the recruitment of DNA modifying enzymes following the passage of the replisome. Indeed, it has even been proposed that it is for this function, rather than their role as processivity factors, that clamp molecules originally evolved (Georgescu et al., 2015).

Replisome Structure in *B. subtilis*

To date, the replisome has only been characterized in detail in one organism other than *E. coli*: the low-GC Gram-positive bacterium *Bacillus subtilis* (**Figure 1B** and **Table 1**). A fully functional replisome with physiological rates of coupled leading and lagging strand synthesis has been reconstituted *in vitro* from purified *B. subtilis* proteins, defining its minimal protein requirements (Sanders et al., 2010). This study demonstrated a conserved core replisome structure with *E. coli*, consisting of a homohexameric helicase (DnaC) which interacts with both a distributive primase (DnaG) and a pentameric $\tau_3\delta\delta'$ clamp loader complex. Replication is also similarly dependent on a dimeric β clamp and SSB.

However, the *B. subtilis* replisome also exhibits some notable differences compared to *E. coli*. Two distinct C-type DNA polymerases—DnaE and PolC—are required for chromosomal replication in this organism (Dervyn et al., 2001), a feature common to many species of Gram-positive bacteria (Timinskas et al., 2014). DnaE is related to *E. coli* Pol III α, while PolC is significantly different in domain organization and additionally possesses an intrinsic exonuclease activity. Initial genetic analysis suggested that the activities of PolC and DnaE may be divided between the leading and lagging strands respectively (Dervyn et al., 2001), but more recent biochemical analysis suggests that DnaE may in fact purely be used to extend RNA primers a short distance on the lagging strand, before PolC rapidly displaces it to synthesize the majority of DNA on both lagging and leading strands (Sanders et al., 2010). This division of function may be enforced by differences in the mode of interaction of the two polymerases with other components of the replisome; while PolC is linked to the helicase via τ, as in *E. coli* (Bruck and O'Donnell, 2000), it has been reported that DnaE interacts directly with the helicase and DnaG, forming a "primosome" complex (Rannou et al., 2013).

A mechanism of primer extension through the sequential action of two distinct DNA polymerases is also used in eukaryotic organisms. However, it has been noted that this mechanism is inherently wasteful, as the short stretch of error-prone DNA synthesized by the first DNA polymerase is entirely removed later in replication anyway (Forterre, 2013). This is also likely to be the case in *B. subtilis* since DnaE lacks a proofreading activity, although it is unclear which polymerase replaces the DNA synthesized by DnaE; it is possible that DNA Pol I fulfils this function as it removes RNA primers prior to Okazaki fragment ligation. Since the respective proteins involved in lagging strand synthesis in eukaryotes and bacteria are non-orthologous this dual polymerase mechanism likely arose through convergent evolution. At present however, it is still unclear what selective advantage, if any, DnaE-PolC coordination on the lagging strand provides.

While the core replisome components have been identified in *B. subtilis*, their relative stoichiometry within the complex has yet to be fully determined. In particular it is unknown how many polymerases are typically present within the *B. subtilis* replisome. The role of PolC on both leading and lagging strands suggests that at least two molecules of this polymerase must be simultaneously present within the replisome. Furthermore, given the interaction between PolC and the trimeric τ of the clamp loader it is possible that three molecules of PolC may be present, similar to *E. coli* Pol III. Rapid displacement of non-proofreading DnaE by PolC on the lagging strand is presumably important for genome integrity, and therefore it may be speculated that maintaining a high relative local concentration of PolC within the replisome would be a mechanism for achieving this. It should be noted that the interaction between τ and PolC is significantly weaker than the equivalent interaction in *E. coli* (Bruck and O'Donnell, 2000). It is currently unclear whether this represents a physiological property of the *B. subtilis* replisome with implications for its architecture or is due to the absence of other stabilizing components that have yet to be identified. Further characterisation of the *B. subtilis* replisome both *in vitro* and *in vivo* will be important for determining the precise structure of the replisome in this organism and how its multiple polymerases are efficiently coordinated.

Notably, an analysis of polymerase distribution across bacterial genomes has demonstrated that a third major combination of replicative polymerases is found in some organisms: PolC together with a DnaE distinct from that of both *E. coli* and *B. subtilis* (Timinskas et al., 2014). The DNA replication machinery has yet to be examined in any of these species, and it will be interesting in the future to determine if their replisomes employ similar or different strategies to those already described for coordinating multiple DNA polymerases.

Replisome Structure in Other Bacterial Organisms

Despite the extreme diversity present across the bacterial kingdom in many areas of cell biology, a relatively high degree of conservation has been noted for many DNA replication proteins, highlighting the central importance of this cellular process. However, it is not clear whether the assembly of these proteins into a higher order replisome complex also follows a conserved architecture. Beyond *E. coli* and *B. subtilis* this question has not been extensively addressed.

Replisome subassemblies have been successfully reconstituted using purified proteins from *Aquifex aeolicus* (Bruck et al., 2002), *Streptococcus pyogenes* (Bruck and O'Donnell, 2000), *Staphylococcus aureus* (Bruck et al., 2005), *Thermus thermophilus* (Bullard et al., 2002), and *Pseudomonas aeruginosa* (Jarvis et al., 2005a). These studies demonstrate that the core replisome structure of a $\tau_3\delta\delta'$ clamp loader linked to the catalytic subunit of the replicative polymerase is likely to be a conserved feature among bacteria. In the case of *T. thermophilus* the clamp loader can also incorporate γ, although the same caveats should apply as in *E. coli*. However, the organisms examined to date still constitute a tiny sample of total bacterial diversity. Furthermore, it is still unclear if the mode of interaction between this replisome core and other components such as the helicase, primase, and additional polymerases is similarly conserved.

For example, two additional clamp loader subunits, χ and ψ, are present in *E. coli*. While not essential, χ and ψ are required for normal growth. These two components form a tight dimer and interact with both τ and SSB, resulting in both the stabilization of the clamp loader complex and the replisome overall (Olson et al., 1995; Marceau et al., 2011). The $\chi\psi$ subassembly has also been proposed to play a role in promoting the access of Pol III to newly synthesized primers (Yuzhakov et al., 1999). However, sequence analysis suggests these two subunits are only present in proteobacteria, and it is unclear what, if anything, fulfils their replisome function in other bacterial lineages. It should be noted that in *P. aeruginosa* a ψ subunit was identified through its co-purification with χ despite it being unidentifiable on the basis of sequence homology alone (Jarvis et al., 2005b). This raises the possibility that highly diverged functional homologs of these subunits may indeed exist in other organisms where they have not yet been detected, and highlights the importance of empirical research to describing replisome composition across the bacterial kingdom.

In vivo characterisation of replisome architecture is still lacking outside of *E. coli*. Tools are gradually being developed to study the DNA replication machinery in living cells of other species, such as *Helicobacter pylori* (Sharma et al., 2014), *Mycobacterium smegmatis* (Santi and McKinney, 2015; Trojanowski et al., 2015), and *B. subtilis* (Lemon and Grossman, 1998), and it is hoped that in the future determining replisome stoichiometries *in vivo* will enable a better assessment of replisome structure conservation.

Mechanisms of Replisome Assembly

Assembly of the bacterial replisome is a tightly regulated process. DNA replication is normally initiated at a specific origin locus by the assembly of just two replisomes, each of which synthesizes half of the circular chromosome. It is important to restrict replisome assembly beyond this specific initiation event, to prevent over-replication of DNA and chromosome instability. Different mechanisms exist to ensure specificity in replisome assembly, but they appear to converge at the level of controlling the loading of the DnaB helicase onto ssDNA (Fang et al., 1999). Once this loading step has been overcome, DnaB can serve as the platform upon which the rest of the replisome is assembled, through its direct interactions with the primase and clamp loader. Interestingly, a similar strategy is observed in eukaryotic organisms, despite the use of completely non-orthologous proteins, whereby loading and activation of the replicative helicase Mcm2-7 is also the limiting step in replisome assembly (Yeeles et al., 2015).

Replisome Assembly During Initiation

Initiation of replication is best understood in *E. coli*, where replication originates from the *oriC* locus. *oriC* is recognized and melted through sequence-specific binding of the AAA+ protein, DnaA (Leonard and Grimwade, 2011). The prevailing current model proposes that DnaA oligomerizes into a helical filament, around which bound *oriC* DNA is wrapped (Erzberger et al., 2006). This destabilizes the neighboring duplex unwinding element (DUE), leading to DNA melting as the DnaA filament extends onto transiently exposed ssDNA (Duderstadt et al., 2011). The resulting bubble of SSB-coated ssDNA is a substrate for the assembly of two DnaB helicases (Fang et al., 1999). Crucially, however, DnaB loading onto such structures is not spontaneous; it is restricted to the *oriC* locus through a dependence on the helicase loader protein, DnaC (**Figure 2A**).

DnaC has been shown to interact with both DnaA and DnaB, and thus serves as link to recruit DnaB to melted origins (Mott et al., 2008). Furthermore, it appears to play an active role in loading the helicase. Structural analysis of DnaC has shown that this protein resembles DnaA and similarly adopts a helical filament when bound to ssDNA (Mott et al., 2008). More recently, it has been demonstrated that this spiral configuration enables DnaC to break open the DnaB hexameric ring as the two proteins interact, allowing ssDNA to enter the central chamber of the helicase (Arias-Palomo et al., 2013). Association of DnaG primase with DnaB triggers the release of DnaC and thus activates the helicase, beginning the process of replisome assembly (Davey et al., 2002; Makowska-Grzyska and Kaguni, 2010). Interestingly, it has been shown that while two DnaB helicases must be loaded at *oriC* (Fang et al., 1999), the first of these is always deposited on the lower strand (Weigel and Seitz, 2002). A model for asymmetric loading of DnaB has thus been proposed, whereby DnaC-DnaB on one strand is loaded via the interaction of DnaC with DnaA, while on the other strand DnaC-DnaB is loaded via a direct interaction observed between DnaB and DnaA (Mott et al., 2008). Such a loading mechanism is attractive because it accounts for the need for the two helicases—and their associated replisomes—to proceed bidirectionally from *oriC*.

This helicase loading pathway may not be well conserved across bacterial organisms. *B. subtilis* for example possesses orthologs of DnaA, DnaC (called DnaI) and DnaB (called DnaC) but also requires two additional proteins for helicase loading that have no orthologs in *E. coli*: DnaB, and DnaD (Smits et al., 2010). The precise role of these additional proteins is still unclear. Furthermore, despite being related to *E. coli* DnaC, the *B. subtilis*

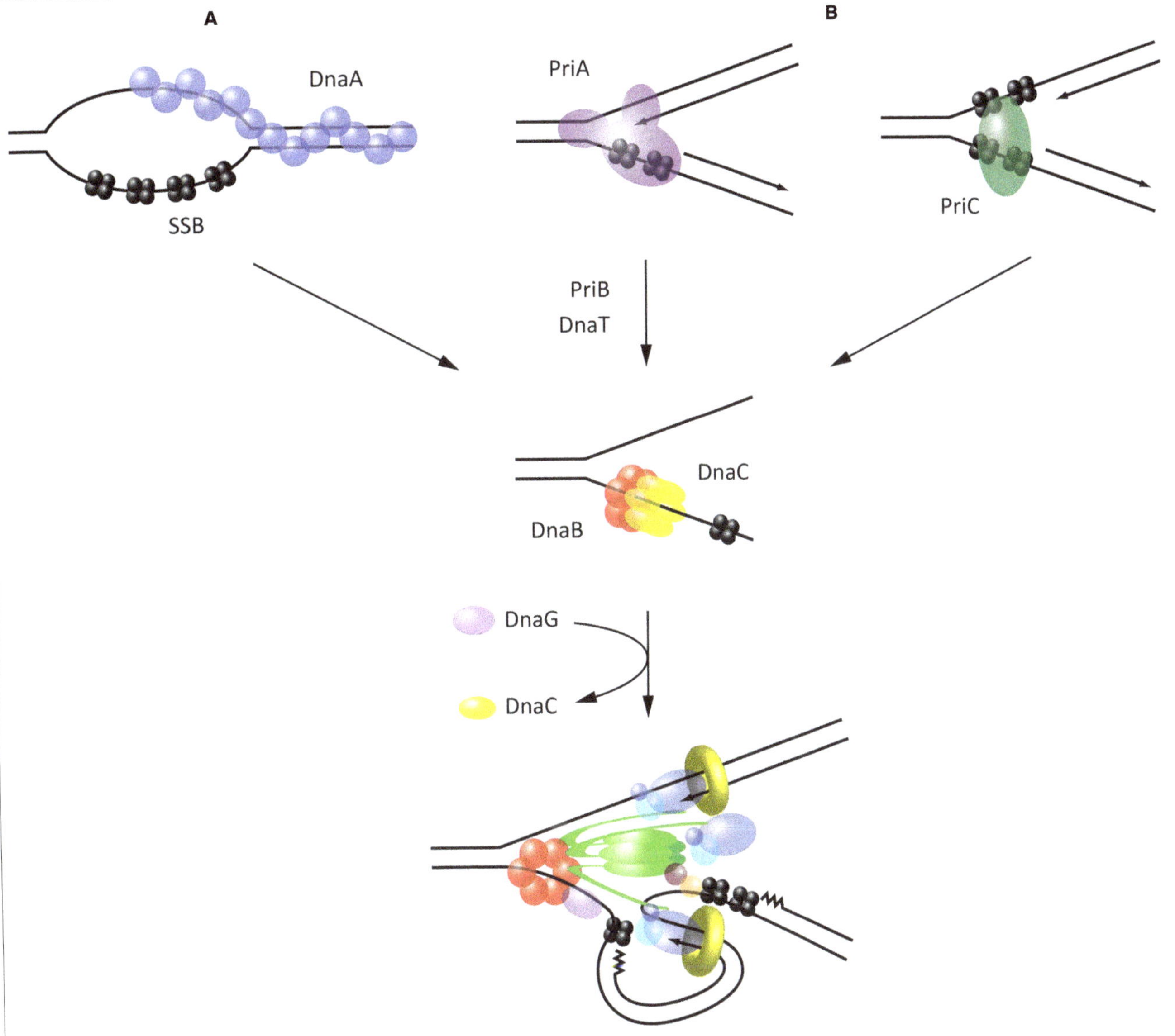

FIGURE 2 | Mechanisms of helicase loading leading to replisome assembly in *E. coli*. (A) Recognition and melting of the *oriC* locus during initiation by DnaA. **(B)** Recognition of abandoned fork structures during replisome reloading by PriA and PriC. All pathways converge on the loading of the replicative helicase DnaB, which acts as an assembly platform for the remaining replisome components.

helicase loader DnaI may employ a different mechanism to load the helicase on DNA; structural and biochemical data suggest that rather than acting as a ring-opener, DnaI may act as a template for the assembly of the hexameric helicase around ssDNA from individual monomers (Velten et al., 2003; Liu et al., 2013). It has been proposed from sequence analysis that some bacterial species including *A. aeolicus* posses a helicase loader distinct from that present in either *E. coli* or *B. subtilis* (Robinson et al., 2012). However, a low-resolution structural comparison suggests that *E. coli* and *A. aeolicus* DnaC form similar spiral oligomers and therefore may employ the same mechanism for opening the DnaB hexameric ring (Arias-Palomo et al., 2013).

Helicase loading may differ even further in other bacterial organisms, since many species lack an identifiable DnaC homolog. For example, helicases from *Pseudomonas* species can be loaded *in vitro* in the absence of a loader protein, in contrast to *E. coli* (Caspi et al., 2001). Similarly, the helicase from *H. pylori* is sufficient to complement *E. coli* lacking DnaC, suggesting it is functionally loaded onto DNA without a loader (Soni et al., 2005). Interestingly, this correlates with the ability of this helicase to uniquely form a double hexamer (Stelter et al., 2012). Thus it may be speculated that in some bacterial species at least, the replicative helicase possesses structural adaptations that confer intrinsic self-loading ability. It is unclear how site-specific loading would be maintained in such cases.

Replisome Assembly Following Fork Collapse

It is now apparent that replication fork structures can frequently break down during chromosomal replication, leading to collapse of the replisome. This is especially the case under stress conditions, when lesions in DNA become more prevalent, but is also likely to be the case during unperturbed replication in healthy cells (Maisnier-Patin et al., 2001). Thus to ensure complete replication of the chromosome, there is a requirement for systems that can efficiently reload replisomes at sites where they have collapsed. Crucially, these mechanisms must be specific enough to ensure that replisomes aren't simply loaded indiscriminately onto any ssDNA structure.

In *E. coli*, this reloading process is restricted to abandoned replication forks through the activity of two proteins, PriA and PriC, which recognize specific DNA structures (**Figure 2B**). PriA possesses a modular arrangement of DNA-binding domains which allow it to specifically bind the three arms of forked DNA structures, with a preference for those possessing a fully extended leading strand (Heller and Marians, 2005a; Bhattacharyya et al., 2014). PriC provides complementary activity by recognizing fork structures lacking a leading strand; these can arise when the leading strand polymerase stalls at a lesion and becomes uncoupled from the continuing lagging strand polymerase (Heller and Marians, 2005a). PriC probably recognizes such fork structures through an interaction with SSB (Wessel et al., 2013). In some cases the presence of SSB may rely on prior unwinding of the lagging strand by Rep or even PriA (Heller and Marians, 2005b).

Supported by genetic analysis, reconstitution of helicase reloading *in vitro* has demonstrated that these different modes of substrate recognition lead to two distinct pathways of helicase reloading. PriA substrate recognition leads to the sequential recruitment of PriB, DnaT and finally the DnaC-DnaB complex for loading (Lopper et al., 2007). Alternatively, substrate recognition by PriC leads directly to the recruitment of DnaC-DnaB (Heller and Marians, 2006). Notably, while the components of these pathways have been characterized extensively, the precise mechanisms that lead to the recruitment of DnaC and loading of the helicase specifically onto the lagging strand, are still unknown. Intriguingly, DnaT has been suggested from structural studies to form a helical filament on DNA (Liu et al., 2014), raising the possibility that it could act analogously to DnaA in recruiting DnaC. The mode of interaction between PriC and DnaC-DnaB has yet to be characterized, although high-throughput studies have identified a potential interaction between PriC and DnaB (Butland et al., 2005).

Notably, many bacteria lack an ortholog of *E. coli* PriC and thus it is unclear what, if anything, fulfils its function in these species. Furthermore, while PriA does seem to be relatively well-conserved, its mode of action may not be; reconstitution of helicase reloading in *B. subtilis* suggests that PriA acts to recruit the helicase through the DnaD, DnaB and DnaI proteins used during initiation, rather than a specialized set of proteins as in *E. coli* (Manhart and McHenry, 2013). The exact mechanism of replisome assembly both at origins of replication and following replication fork collapse may therefore differ significantly between bacterial species. Nonetheless, it appears that these different pathways converge on the limiting step of loading of the replicative helicase, which serves as the core replisome component upon which other replication proteins are assembled.

Dynamics and Stability of the Replisome

Looking at the replisome in action must surely be like looking at a beehive. Like bees performing their collection duties, multiple replisome components are continuously arriving and leaving with every cycle of synthesis on the lagging strand. One new copy of primase and the β clamp dimer is needed to start every Okazaki fragment, and multiple copies of SSB are recruited and displaced as a result of DNA melting by helicase and synthesis by Pol III respectively (Wu et al., 1992).

A long-standing debate on the coordination of replisome activities has been over how Pol III, evolved for high processivity, frequently dissociates from the lagging strand to ensure efficient cycles of Okazaki fragment synthesis. The collision model hypothesizes that Pol III dissociates upon encountering the previous Okazaki fragment. It has been suggested that Pol III senses elimination of template ssDNA as it approaches the end of each Okazaki fragment, which weakens its affinity for the β clamp through an unknown mechanism, thus decreasing the stability of the polymerase on DNA (Georgescu et al., 2009). In contrast, the signaling model proposes that a specific signal intrinsic to the replisome triggers Pol III dissociation following synthesis of each new primer (Wu et al., 1992). This is supported by the fact that Pol III dissociation rates following collision *in vitro* are insufficient to support the rate of synthesis required during chromosomal replication—at least in the context of a dimeric Pol III replisome (Dohrmann et al., 2011). Recent single-molecule data has suggested that the source of this dissociation signal may be the accumulation of topological stress at the replication fork as the physically coupled polymerases track around helical DNA during synthesis (Kurth et al., 2013). Although their relative importance is still contested, some *in vitro* experiments have suggested that in fact both mechanisms operate redundantly to maximize the efficiency of Okazaki fragment synthesis (Li and Marians, 2000; Hamdan et al., 2009).

In contrast with this dynamic picture, the replisome also needs to be a very stable assembly to accomplish replication of the long chromosomal DNA molecules. In *E. coli*, chromosome duplication takes the replisome over forty minutes of continuous work even at synthesis rates of 1 kb per second (Kubitschek and Newman, 1978; Kornberg and Baker, 1992). The importance of keeping the replisome active and in one piece is demonstrated by the observation that the majority of double strand breaks, which are potentially lethal lesions to the cell, occur during DNA replication (Cox et al., 2000; Pennington and Rosenberg, 2007) often as a result of extended pauses in replication fork progression (Michel et al., 2004). Judging by relatively infrequent replisome collapse events—once every five generations in *E. coli* (Maisnier-Patin et al., 2001)—high stability seems to be inherent to the replisome.

Stability may be the result of the extended lifetime of some of the subassemblies of the active replisome. The idea of tight binding between its components is intimately linked to the

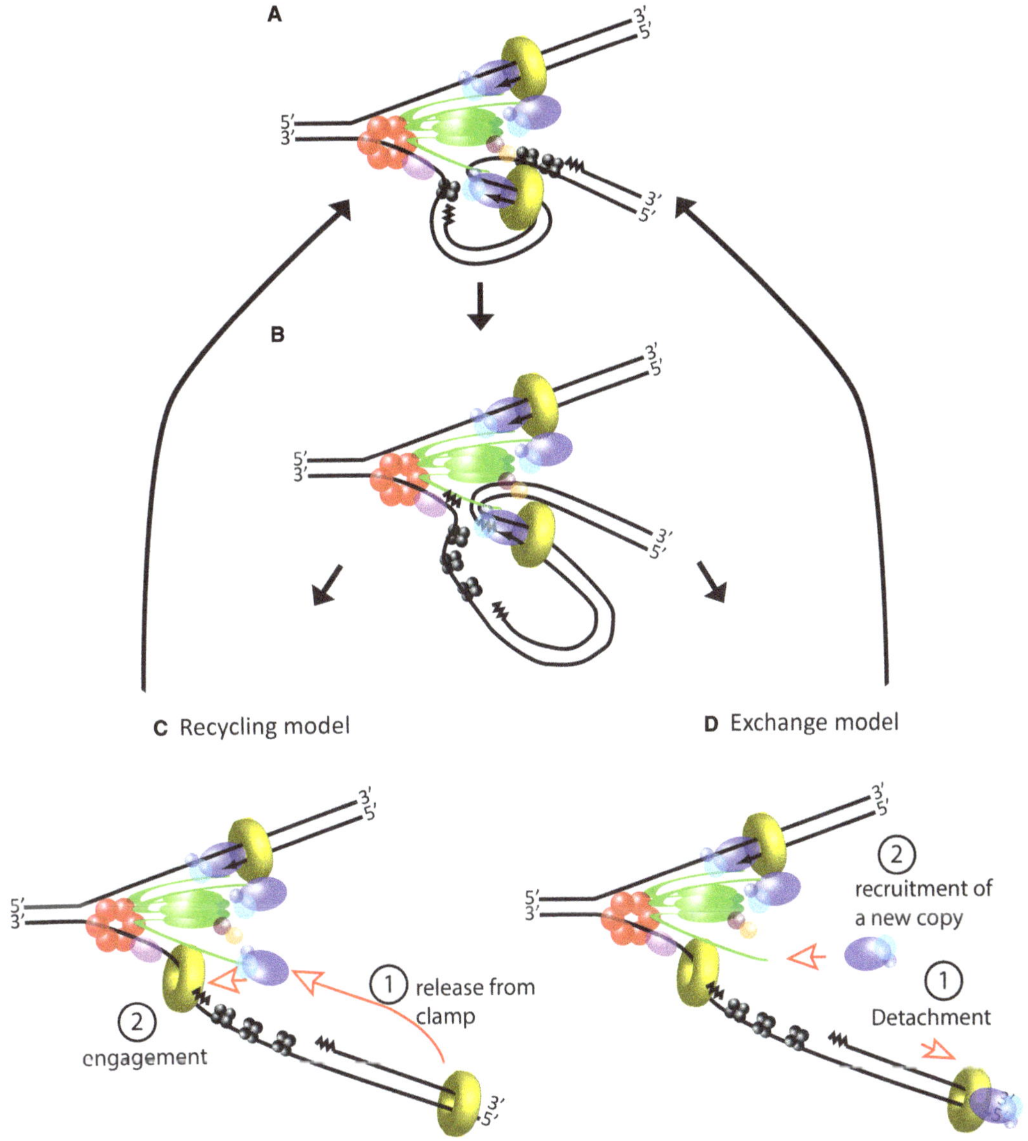

FIGURE 3 | Usage of DNA polymerase during lagging strand synthesis. (A) Schematic of the *E. coli* replisome during the elongation step of an Okazaki fragment. **(B)** Lagging strand polymerase meets the RNA primer of the previous Okazaki fragment and stops synthesis. **(C)** Current model of events following completion of an Okazaki fragment. DNA polymerase is released from the β clamp (step 1) and the same molecule rebinds to a new β clamp to start the next Okazaki fragment (step 2). **(D)** An alternative model based on evidence from T4 and T7 replisomes. After completing the Okazaki fragment, the DNA polymerase detaches from the rest of the replisome (step 1). A new molecule of DNA polymerase is recruited to the replisome (step 2) and engages in the synthesis of a new Okazaki fragment. In this tentative model, a local pool of "spare" polymerases may facilitate their exchange and additional components may exchange along with the polymerase (not depicted).

trombone model of DNA replication. At its conception, this model arose to explain synchronous action of polymerases on both strands, and postulated a physical coupling between polymerases at the leading and lagging strands during synthesis. Importantly, it also suggested that the same copy of Pol III is recycled for multiple rounds of synthesis at the lagging strand (Sinha et al., 1980). Validation of this model came from experimental evidence showing that the replisome in *E. coli* can contain two polymerases (Maki et al., 1988; Onrust et al., 1995), and that they synthesize the two strands simultaneously (Yuzhakov et al., 1996). Long residence times of core replisome components have been demonstrated by ensemble and single-molecule *in vitro* studies of the *E. coli* replisome that show processive DNA synthesis in the absence of any free helicase, clamp loader or Pol III in the buffer (Li and Marians, 2000; Yao et al., 2009). From this data it is inferred that helicase, clamp loader and polymerases form a tight structure in the replisome, so that the same molecules of these proteins may be used over long periods, potentially even the whole replication event (**Figures 3A–C**).

However, this model of the replisome contrasts with evidence from T4 and T7 phages, which shows that the replicative polymerase is frequently replaced during fork progression (Yang et al., 2004; Johnson et al., 2007). Although the number and nature of their components differ, both of these systems have a similar architecture to the replisome of *E. coli*, indeed the trombone model was originally proposed for T4. Similarly to *E. coli*, coordination between leading and lagging strand synthesis has been extensively validated in these systems, and their replisomes can also carry out DNA synthesis in the absence of excess polymerase in the buffer (Debyser et al., 1994; Yang et al., 2004). Nevertheless, polymerase exchange can be detected shortly after addition of mutant polymerase to a reaction where the replisome is engaged in DNA synthesis (Yang et al., 2004; Johnson et al., 2007). How can we reconcile frequent polymerase turnover with efficient recycling on the lagging strand? At least in the case of the T7 replisome, which has been studied using elegant single-molecule studies, this seems to be achieved by creating a local pool of "spare" polymerases available for fast switching. The replisome in this phage consists of only four proteins: gp5, the DNA polymerase; gp4, which acts as helicase and primase; gp2.5, a ssDNA binding protein; and *E. coli*'s thioredoxin which serves as processivity factor for the polymerase (serving the same function as the β clamp in the bacterial replisome). The gp5 polymerase interacts directly with gp4, but importantly, these proteins have two modes of interaction resulting in tight or loose association, respectively (Hamdan et al., 2005; Kulczyk et al., 2012). The tight interaction has a greater contribution to the processivity of the polymerase, but the weak interaction additionally permits the presence of up to six polymerases at the replication fork (Loparo et al., 2011; Geertsema et al., 2014).

The dynamics of phage replisomes are relevant to *E. coli* since they both show different behaviors depending on the concentration of their components in solution. Studies performed *in vitro* in the absence of excess components show the *E. coli* replisome can function without exchange of its subunits, but it is currently unclear if a different behavior will be observed inside the cell (**Figure 3D**), where the diffusing pool of replisome components is one-to-two orders of magnitude higher than the number of active molecules (Leu et al., 2000; Reyes-Lamothe et al., 2010). It is easy to imagine how the capacity to detach polymerases from the replisome would be advantageous in order to rapidly respond to lesions and roadblocks on DNA. Multiple studies have shown that after encountering an obstacle, polymerase can unbind from DNA and subsequently re-engage via alternative mechanisms, leaving a gap in the double strand (Pomerantz and O'Donnell, 2008, 2010; Yeeles and Marians, 2011). The high processivity of polymerase bound to the β clamp conflicts with the observed "hopping" over obstacles on DNA, but complete detachment from the replisome would reconcile the long residence of polymerase on DNA with progression of the replication fork, potentially providing greater flexibility to the replisome.

In this scenario, replisome components otherwise thought to be stable are actually moving parts that actively exchange; one wonders what would then help maintain the stability of the replisome as an assembly. Reiterating its central role in the control of DNA replication, the most likely candidate would be DnaB helicase. Indeed, measurements on the processivity of the replisome show that helicase is its most stable component, although multiple contacts to DNA from engaged polymerases seem to also help increase the processivity of the replisome (Yao et al., 2009). Whatever strategy the replisome may use, work still needs to be done to shed light on the interplay between its dynamics and stability.

Disassembly of the Replisome

A key mechanistic challenge faces the bacterial replisome during termination of DNA replication. The circular nature of the bacterial chromosome dictates that a pair of replisomes that initiate from a single origin of replication will eventually converge on each other in a head-to-head orientation. Positive supercoiling accumulates between the two replisomes as they converge, but the activity of DNA gyrase, which normally removes positive supercoils, becomes limited by the decreasing amount of template DNA available. Instead, supercoils may diffuse behind the replisomes, forming precatenanes between newly replicated DNA; in *E. coli* these must be resolved by Topoisomerase IV for chromosome segregation to occur (Wang et al., 2008). Alternatively, it has been shown *in vitro* that a combination of the 3′-5′ helicase RecQ and Topoisomerase III is sufficient to directly resolve topological stress ahead of converging replication forks, in the presence of SSB (Suski and Marians, 2008).

In at least some bacteria, the progression of replisomes is modulated to ensure that their convergence is restricted to a specific terminus region of the chromosome. This is best characterized in *E. coli*, where the Tus protein binds tightly to specific DNA sequences in the terminus, designated *ter* sites, and halts progression of the replisome. Crucially this effect is dependent on the orientation of *ter* sites, such that replisomes can enter but not exit the terminus region (**Figure 4**). The exact mechanism by which Tus-*ter* is able to block the replisome is still a matter of debate. Biochemical and structural studies have demonstrated that DnaB-catalyzed unwinding of *ter* DNA in the blocking orientation results in flipping of a specific cytosine into a binding pocket in Tus, strengthening its interaction with DNA and blocking further helicase progression. In contrast, unwinding from the opposing direction does not trigger this base flipping and results in displacement of the Tus protein from DNA (Neylon et al., 2000; Mulcair et al., 2006). However, an alternative model proposes that Tus mediates its effect on the replisome through direct protein-protein interactions with DnaB (Mulugu et al., 2001). This argument is supported by the fact that Tus can arrest DnaB translocation on dsDNA *in vitro* independently of *ter* unwinding (Bastia et al., 2008). However, given that DnaB within the replisome translocates on ssDNA, the relevance of this observation is unclear.

Regardless of the precise mechanism by which Tus operates, its regulation of the replisome does not appear to play a direct role in the process of termination, since it can be deleted from *E. coli* without producing a detectable phenotype (Roecklein et al., 1991). Furthermore, the Tus-*ter* system is poorly conserved among bacteria; while *B. subtilis* possesses a functional homolog of Tus (RTP) the two proteins lack any significant sequence or structural homology. The role of these replisome

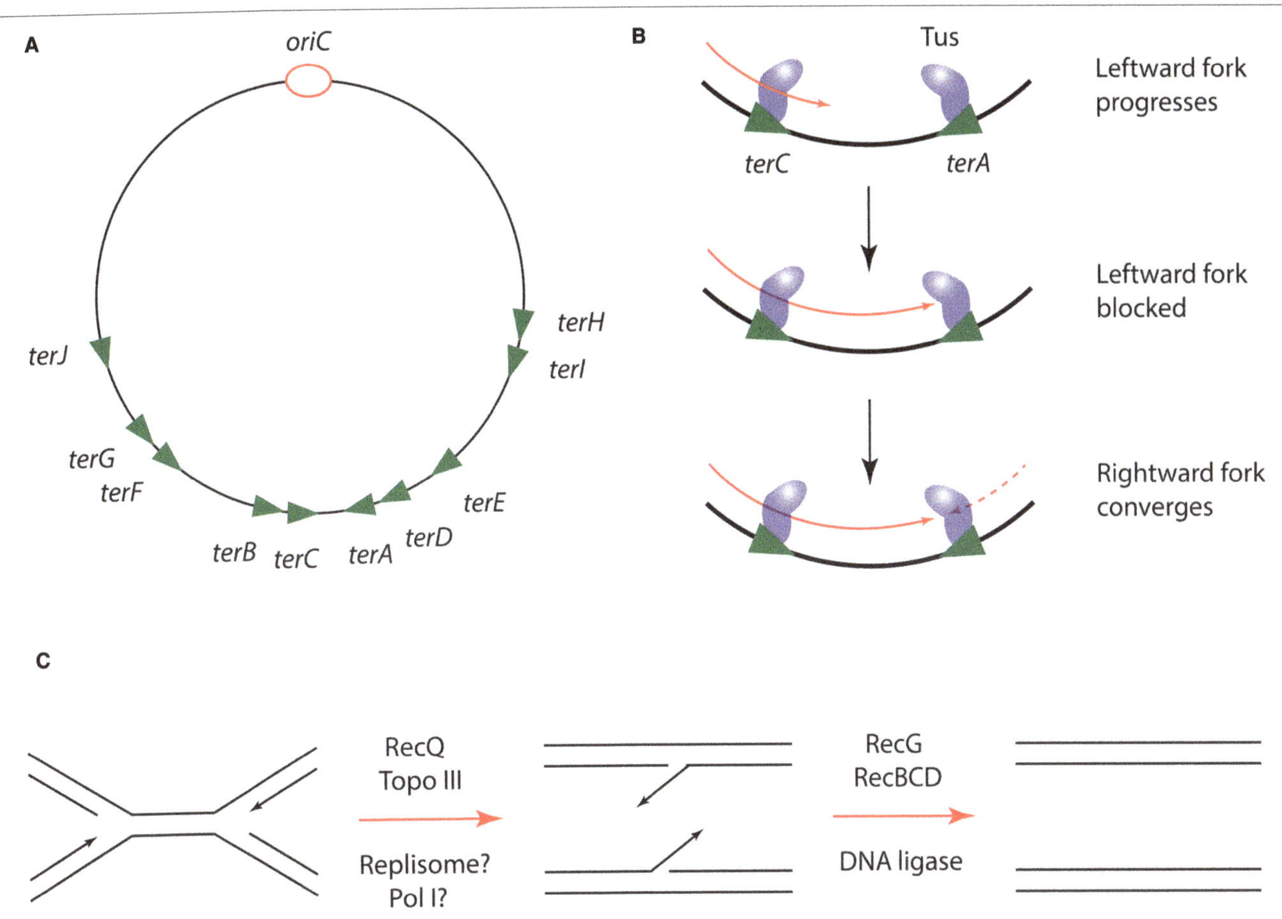

FIGURE 4 | Mechanism of replication termination in *E. coli*. (A) Schematic of the *E. coli* chromosome showing the approximate location of *ter* sites relative to *oriC*. Arrowheads indicate the permitted direction of replisome progression. Adapted from Duggin et al., 2008. **(B)** Mechanism of replisome trapping in the terminus region by Tus-*ter*. Just two *ter* sites are shown for clarity but the same mechanism can operate at any *ter* site. **(C)** Resolution of DNA structures following replisome convergence during termination.

"traps" is thus more likely to be in coordinating termination with other processes localized to the terminus region of the chromosome, such as chromosome dimer resolution by XerCD (Duggin et al., 2008).

Genetic analysis suggests that the actual site of replication termination contains regions of overlapping DNA sequence that must be resolved into single daughter strands through the action of the RecG DNA translocase and the helicase-nuclease RecBCD (Rudolph et al., 2013; Wendel et al., 2014). Crucially however, it is unclear exactly how these termination structures are generated. Some analysis has suggested that DNA Pol I may play a role in replicating DNA at the terminus, but it is unclear if this occurs in coordination with the Pol III machinery or following its disassembly (Markovitz, 2005). Thus in particular, the role of the replisome in the final stages of chromosomal replication, and the fate of the replisome during termination are poorly understood.

Failure to disassemble replisomes in a timely manner is likely to result in genome instability; *in vitro* studies have demonstrated that *E. coli* replisomes can switch strands after converging upon each other, resulting in chromosome over-replication (Hiasa and Marians, 1994). However, it has not been addressed whether replisomes simply dissociate upon converging, or whether their disassembly is an active process. It has recently been shown that disassembly of the replisome in eukaryotic organisms is an active process, triggered by post-translational modification of the replicative helicase, Mcm2-7 (Maric et al., 2014; Moreno et al., 2014). Given the extensive regulation of bacterial helicase loading during initiation and replisome re-loading, it may be speculated that the bacterial replisome is subject to similar mechanisms of regulated disassembly.

Future Perspectives

The information summarized in this work suggests that replisome organization and function follow a common theme with minor variations across bacteria, underlying the importance of DNA replication as a basic metabolic function of the cell. However, more work is needed in order to assess if these conclusions can be extended to all bacteria. Novel technologies in imaging, sequencing and genome editing will likely help us in the study of the replisome at a single-molecule level and *in vivo*, and to extend this analysis to non-classical model organisms. Future research promises a greater understanding of the composition of the replisome; how it is assembled; how it remains assembled; and how it comes apart. Furthermore, work in other bacteria will be especially revealing in understanding how specialized additions to the replisome help adapt organisms to their particular physiological needs. In turn, this information can catalyze the discovery of a set of basic considerations that any replisome must meet in order to achieve fast and efficient genome duplication.

Acknowledgments

This work was supported by the Natural Sciences and Engineering Research Council of Canada (NSERC# 435521-2013), the Canada Research Chairs program and McGill University.

References

Arias-Palomo, E., O'Shea, V. L., Hood, I. V., and Berger, J. M. (2013). The bacterial DnaC helicase loader is a DnaB ring breaker. *Cell* 153, 438–448. doi: 10.1016/j.cell.2013.03.006

Bastia, D., Zzaman, S., Krings, G., Saxena, M., Peng, X., and Greenberg, M. M. (2008). Replication termination mechanism as revealed by Tus-mediated polar arrest of a sliding helicase. *Proc. Natl. Acad. Sci. U.S.A.* 105, 12831–12836. doi: 10.1073/pnas.0805898105

Bhattacharyya, B., George, N. P., Thurmes, T. M., Zhou, R., Jani, N., Wessel, S. R., et al. (2014). Structural mechanisms of PriA-mediated DNA replication restart. *Proc. Natl. Acad. Sci. U.S.A.* 111, 1373–1378. doi: 10.1073/pnas.1318001111

Blinkova, A., Burkart, M. F., Owens, T. D., and Walker, J. R. (1997). Conservation of the *Escherichia coli* dnaX programmed ribosomal frameshift signal in *Salmonella typhimurium*. *J. Bacteriol.* 179, 4438–4442.

Bruck, I., Georgescu, R. E., and O'Donnell, M. (2005). Conserved interactions in the *Staphylococcus aureus* DNA PolC chromosome replication machine. *J. Biol. Chem.* 280, 18152–18162. doi: 10.1074/jbc.M413595200

Bruck, I., and O'Donnell, M. (2000). The DNA replication machine of a gram-positive organism. *J. Biol. Chem.* 275, 28971–28983. doi: 10.1074/jbc.M003565200

Bruck, I., Yuzhakov, A., Yurieva, O., Jeruzalmi, D., Skangalis, M., Kuriyan, J., et al. (2002). Analysis of a multicomponent thermostable DNA polymerase III replicase from an extreme thermophile. *J. Biol. Chem.* 277, 17334–17348. doi: 10.1074/jbc.M110198200

Bullard, J. M., Williams, J. C., Acker, W. K., Jacobi, C., Janjic, N., and McHenry, C. S. (2002). DNA polymerase III holoenzyme from *Thermus thermophilus* identification, expression, purification of components, and use to reconstitute a processive replicase. *J. Biol. Chem.* 277, 13401–13408. doi: 10.1074/jbc.M110833200

Butland, G., Peregrin-Alvarez, J. M., Li, J., Yang, W., Yang, X., Canadien, V., et al. (2005). Interaction network containing conserved and essential protein complexes in *Escherichia coli*. *Nature* 433, 531–537. doi: 10.1038/nature03239

Caspi, R., Pacek, M., Consiglieri, G., Helinski, D. R., Toukdarian, A., and Konieczny, I. (2001). A broad host range replicon with different requirements for replication initiation in three bacterial species. *EMBO J.* 20, 3262–3271. doi: 10.1093/emboj/20.12.3262

Cox, M. M., Goodman, M. F., Kreuzer, K. N., Sherratt, D. J., Sandler, S. J., and Marians, K. J. (2000). The importance of repairing stalled replication forks. *Nature* 404, 37–41. doi: 10.1038/35003501

Davey, M. J., Fang, L., McInerney, P., Georgescu, R. E., and O'Donnell, M. (2002). The DnaC helicase loader is a dual ATP/ADP switch protein. *EMBO J.* 21, 3148–3159. doi: 10.1093/emboj/cdf308

Debyser, Z., Tabor, S., and Richardson, C. C. (1994). Coordination of leading and lagging strand DNA synthesis at the replication fork of bacteriophage T7. *Cell* 77, 157–166. doi: 10.1016/0092-8674(94)90243-7

Dervyn, E., Suski, C., Daniel, R., Bruand, C., Chapuis, J., Errington, J., et al. (2001). Two essential DNA polymerases at the bacterial replication fork. *Science* 294, 1716–1719. doi: 10.1126/science.1066351

Dohrmann, P. R., Manhart, C. M., Downey, C. D., and McHenry, C. S. (2011). The rate of polymerase release upon filling the gap between Okazaki fragments is inadequate to support cycling during lagging strand synthesis. *J. Mol. Biol.* 414, 15–27. doi: 10.1016/j.jmb.2011.09.039

Duderstadt, K. E., Chuang, K., and Berger, J. M. (2011). DNA stretching by bacterial initiators promotes replication origin opening. *Nature* 478, 209–213. doi: 10.1038/nature10455

Duggin, I. G., Wake, R. G., Bell, S. D., and Hill, T. M. (2008). The replication fork trap and termination of chromosome replication. *Mol. Microbiol.* 70, 1323–1333. doi: 10.1111/j.1365-2958.2008.06500.x

Erzberger, J. P., Mott, M. L., and Berger, J. M. (2006). Structural basis for ATP-dependent DnaA assembly and replication-origin remodeling. *Nat. Struct. Mol. Biol.* 13, 676–683. doi: 10.1038/nsmb1115

Fang, L., Davey, M. J., and O'Donnell, M. (1999). Replisome assembly at oriC, the replication origin of *E. coli*, reveals an explanation for initiation sites outside an origin. *Mol. Cell* 4, 541–553. doi: 10.1016/S1097-2765(00)80205-1

Forterre, P. (2013). Why are there so many diverse replication machineries? *J. Mol. Biol.* 425, 4714–4726. doi: 10.1016/j.jmb.2013.09.032

Geertsema, H. J., Kulczyk, A. W., Richardson, C. C., and van Oijen, A. M. (2014). Single-molecule studies of polymerase dynamics and stoichiometry at the bacteriophage T7 replication machinery. *Proc. Natl. Acad. Sci. U.S.A.* 111, 4073–4078. doi: 10.1073/pnas.1402010111

Georgescu, R. E., Kurth, I., and O'Donnell, M. (2011). Single-molecule studies reveal the function of a third polymerase in the replisome. *Nat. Struct. Mol. Biol.* 19, 113–116. doi: 10.1038/nsmb.2179

Georgescu, R. E., Kurth, I., Yao, N. Y., Stewart, J., Yurieva, O., and O'Donnell, M. (2009). Mechanism of polymerase collision release from sliding clamps on the lagging strand. *EMBO J.* 28, 2981–2991. doi: 10.1038/emboj.2009.233

Georgescu, R., Langston, L., and O'Donnell, M. (2015). A proposal: evolution of PCNA's role as a marker of newly replicated DNA. *DNA Repair* 29, 4–15. doi: 10.1016/j.dnarep.2015.01.015

Hamdan, S. M., Loparo, J. J., Takahashi, M., Richardson, C. C., and van Oijen, A. M. (2009). Dynamics of DNA replication loops reveal temporal control of lagging-strand synthesis. *Nature* 457, 336–339. doi: 10.1038/nature07512

Hamdan, S. M., Marintcheva, B., Cook, T., Lee, S. J., Tabor, S., and Richardson, C. C. (2005). A unique loop in T7 DNA polymerase mediates the binding of helicase-primase, DNA binding protein, and processivity factor. *Proc. Natl. Acad. Sci. U.S.A.* 102, 5096–5101. doi: 10.1073/pnas.0501637102

Heller, R. C., and Marians, K. J. (2005a). The disposition of nascent strands at stalled replication forks dictates the pathway of replisome loading during restart. *Mol. Cell* 17, 733–743. doi: 10.1016/j.molcel.2005.01.019

Heller, R. C., and Marians, K. J. (2005b). Unwinding of the nascent lagging strand by Rep and PriA enables the direct restart of stalled replication forks. *J. Biol. Chem.* 280, 34143–34151. doi: 10.1074/jbc.M507224200

Heller, R. C., and Marians, K. J. (2006). Replication fork reactivation downstream of a blocked nascent leading strand. *Nature* 439, 557–562. doi: 10.1038/nature04329

Hiasa, H., and Marians, K. J. (1994). Tus prevents overreplication of oriC plasmid DNA. *J. Biol. Chem.* 269, 26959–26968.

Jarvis, T. C., Beaudry, A. A., Bullard, J. M., Janjic, N., and McHenry, C. S. (2005a). Reconstitution of a minimal DNA replicase from *Pseudomonas aeruginosa* and stimulation by non-cognate auxiliary factors. *J. Biol. Chem.* 280, 7890–7900. doi: 10.1074/jbc.M412263200

Jarvis, T. C., Beaudry, A. A., Bullard, J. M., Ochsner, U., Dallmann, H. G., and McHenry, C. S. (2005b). Discovery and characterization of the cryptic psi subunit of the pseudomonad DNA replicase. *J. Biol. Chem.* 280, 40465–40473. doi: 10.1074/jbc.M508310200

Johnson, D. E., Takahashi, M., Hamdan, S. M., Lee, S. J., and Richardson, C. C. (2007). Exchange of DNA polymerases at the replication fork of bacteriophage T7. *Proc. Natl. Acad. Sci. U.S.A.* 104, 5312–5317. doi: 10.1073/pnas.0701062104

Kornberg, A., and Baker, T. (1992). *DNA Replication*. New York, NY: Freeman.

Kubitschek, H. E., and Newman, C. N. (1978). Chromosome replication during the division cycle in slowly growing, steady-state cultures of three *Escherichia coli* B/r strains. *J. Bacteriol.* 136, 179–190.

Kulczyk, A. W., Akabayov, B., Lee, S. J., Bostina, M., Berkowitz, S. A., and Richardson, C. C. (2012). An interaction between DNA polymerase and helicase is essential for the high processivity of the bacteriophage T7 replisome. *J. Biol. Chem.* 287, 39050–39060. doi: 10.1074/jbc.M112.410647

Kurth, I., Georgescu, R. E., and O'Donnell, M. E. (2013). A solution to release twisted DNA during chromosome replication by coupled DNA polymerases. *Nature* 496, 119–122. doi: 10.1038/nature11988

Lemon, K. P., and Grossman, A. D. (1998). Localization of bacterial DNA polymerase: evidence for a factory model of replication. *Science* 282, 1516–1519. doi: 10.1126/science.282.5393.1516

Leonard, A. C., and Grimwade, J. E. (2011). Regulation of DnaA assembly and activity: taking directions from the genome. *Annu. Rev. Microbiol.* 65, 19–35. doi: 10.1146/annurev-micro-090110-102934

Leu, F. P., Hingorani, M. M., Turner, J., and O'Donnell, M. (2000). The δ subunit of DNA polymerase III holoenzyme serves as a sliding clamp unloader in *Escherichia coli*. *J. Biol. Chem.* 275, 34609–34618. doi: 10.1074/jbc.M005495200

Li, X., and Marians, K. J. (2000). Two distinct triggers for cycling of the lagging strand polymerase at the replication fork. *J. Biol. Chem.* 275, 34757–34765. doi: 10.1074/jbc.M006556200

Liu, B., Eliason, W. K., and Steitz, T. A. (2013). Structure of a helicase-helicase loader complex reveals insights into the mechanism of bacterial primosome assembly. *Nat. Commun.* 4, 2495. doi: 10.1038/ncomms3495

Liu, Z., Chen, P., Wang, X., Cai, G., Niu, L., Teng, M., et al. (2014). Crystal structure of DnaT84-153-dT10 ssDNA complex reveals a novel single-stranded DNA binding mode. *Nucleic Acids Res.* 42, 9470–9483. doi: 10.1093/nar/gku633

Loparo, J. J., Kulczyk, A. W., Richardson, C. C., and van Oijen, A. M. (2011). Simultaneous single-molecule measurements of phage T7 replisome composition and function reveal the mechanism of polymerase exchange. *Proc. Natl. Acad. Sci. U.S.A.* 108, 3584–3589. doi: 10.1073/pnas.1018824108

Lopper, M., Boonsombat, R., Sandler, S. J., and Keck, J. L. (2007). A hand-off mechanism for primosome assembly in replication restart. *Mol. Cell* 26, 781–793. doi: 10.1016/j.molcel.2007.05.012

Maisnier-Patin, S., Nordstrom, K., and Dasgupta, S. (2001). Replication arrests during a single round of replication of the *Escherichia coli* chromosome in the absence of DnaC activity. *Mol. Microbiol.* 42, 1371–1382. doi: 10.1046/j.1365-2958.2001.02718.x

Maki, H., Maki, S., and Kornberg, A. (1988). DNA Polymerase III holoenzyme of *Escherichia coli*. IV. The holoenzyme is an asymmetric dimer with twin active sites. *J. Biol. Chem.* 263, 6570–6578.

Makowska-Grzyska, M., and Kaguni, J. M. (2010). Primase directs the release of DnaC from DnaB. *Mol. Cell* 37, 90–101. doi: 10.1016/j.molcel.2009.12.031

Manhart, C. M., and McHenry, C. S. (2013). The PriA replication restart protein blocks replicase access prior to helicase assembly and directs template specificity through its ATPase activity. *J. Biol. Chem.* 288, 3989–3999. doi: 10.1074/jbc.M112.435966

Manosas, M., Spiering, M. M., Zhuang, Z., Benkovic, S. J., and Croquette, V. (2009). Coupling DNA unwinding activity with primer synthesis in the bacteriophage T4 primosome. *Nat. Chem. Biol.* 5, 904–912. doi: 10.1038/nchembio.236

Marceau, A. H., Bahng, S., Massoni, S. C., George, N. P., Sandler, S. J., Marians, K. J., et al. (2011). Structure of the SSB-DNA polymerase III interface and its role in DNA replication. *EMBO J.* 30, 4236–4247. doi: 10.1038/emboj.2011.305

Maric, M., Maculins, T., De Piccoli, G., and Labib, K. (2014). Cdc48 and a ubiquitin ligase drive disassembly of the CMG helicase at the end of DNA replication. *Science* 346, 1253596. doi: 10.1126/science.1253596

Markovitz, A. (2005). A new *in vivo* termination function for DNA polymerase I of *Escherichia coli* K12. *Mol. Microbiol.* 55, 1867–1882. doi: 10.1111/j.1365-2958.2005.04513.x

McInerney, P., Johnson, A., Katz, F., and O'Donnell, M. (2007). Characterization of a triple DNA polymerase replisome. *Mol. Cell* 27, 527–538. doi: 10.1016/j.molcel.2007.06.019

Michel, B., Grompone, G., Flores, M. J., and Bidnenko, V. (2004). Multiple pathways process stalled replication forks. *Proc. Natl. Acad. Sci. U.S.A.* 101, 12783–12788. doi: 10.1073/pnas.0401586101

Moolman, M. C., Krishnan, S. T., Kerssemakers, J. W., van den Berg, A., Tulinski, P., Depken, M., et al. (2014). Slow unloading leads to DNA-bound β2-sliding clamp accumulation in live *Escherichia coli* cells. *Nat. Commun.* 5, 5820. doi: 10.1038/ncomms6820

Moreno, S. P., Bailey, R., Campion, N., Herron, S., and Gambus, A. (2014). Polyubiquitylation drives replisome disassembly at the termination of DNA replication. *Science* 346, 477–481. doi: 10.1126/science.1253585

Mott, M. L., Erzberger, J. P., Coons, M. M., and Berger, J. M. (2008). Structural synergy and molecular crosstalk between bacterial helicase loaders and replication initiators. *Cell* 135, 623–634. doi: 10.1016/j.cell.2008.09.058

Mulcair, M. D., Schaeffer, P. M., Oakley, A. J., Cross, H. F., Neylon, C., Hill, T. M., et al. (2006). A molecular mousetrap determines polarity of termination of DNA replication in *E. coli*. *Cell* 125, 1309–1319. doi: 10.1016/j.cell.2006.04.040

Mulugu, S., Potnis, A., Shamsuzzaman, Taylor, J., Alexander, K., and Bastia, D. (2001). Mechanism of termination of DNA replication of *Escherichia coli* involves helicase-contrahelicase interaction. *Proc. Natl. Acad. Sci. U.S.A.* 98, 9569–9574. doi: 10.1073/pnas.171065898

Neylon, C., Brown, S. E., Kralicek, A. V., Miles, C. S., Love, C. A., and Dixon, N. E. (2000). Interaction of the *Escherichia coli* replication terminator protein (Tus) with DNA: a model derived from DNA-binding studies of mutant proteins by surface plasmon resonance. *Biochemistry* 39, 11989–11999. doi: 10.1021/bi001174w

Olson, M. W., Dallmann, H. G., and McHenry, C. S. (1995). DnaX complex of *Escherichia coli* DNA polymerase III holoenzyme. The χ ψ complex functions by increasing the affinity of τ and γ for δ.δ′ to a physiologically relevant range. *J. Biol. Chem.* 270, 29570–29577. doi: 10.1074/jbc.270.49.29570

Onrust, R., Finkelstein, J., Turner, J., Naktinis, V., and O'Donnell, M. (1995). Assembly of a chromosomal replication machine: two DNA polymerases, a clamp loader, and sliding clamps in one holoenzyme particle. III. Interface between two polymerases and the clamp loader. *J. Biol. Chem.* 270, 13366–13377. doi: 10.1074/jbc.270.22.13366

Park, K., Debyser, Z., Tabor, S., Richardson, C. C., and Griffith, J. D. (1998). Formation of a DNA loop at the replication fork generated by bacteriophage T7 replication proteins. *J. Biol. Chem.* 273, 5260–5270. doi: 10.1074/jbc.273.9.5260

Pennington, J. M., and Rosenberg, S. M. (2007). Spontaneous DNA breakage in single living *Escherichia coli* cells. *Nat. Genet.* 39, 797–802. doi: 10.1038/ng2051

Pomerantz, R. T., and O'Donnell, M. (2008). The replisome uses mRNA as a primer after colliding with RNA polymerase. *Nature* 456, 762–766. doi: 10.1038/nature07527

Pomerantz, R. T., and O'Donnell, M. (2010). Direct restart of a replication fork stalled by a head-on RNA polymerase. *Science* 327, 590–592. doi: 10.1126/science.1179595

Rannou, O., Le Chatelier, E., Larson, M. A., Nouri, H., Dalmais, B., Laughton, C., et al. (2013). Functional interplay of DnaE polymerase, DnaG primase and DnaC helicase within a ternary complex, and primase to polymerase hand-off during lagging strand DNA replication in *Bacillus subtilis*. *Nucleic Acids Res.* 41, 5303–5320. doi: 10.1093/nar/gkt207

Reyes-Lamothe, R., Sherratt, D. J., and Leake, M. C. (2010). Stoichiometry and architecture of active DNA replication machinery in *Escherichia coli*. *Science* 328, 498–501. doi: 10.1126/science.1185757

Robinson, A., Causer, R. J., and Dixon, N. E. (2012). Architecture and conservation of the bacterial DNA replication machinery, an underexploited drug target. *Curr. Drug Targets* 13, 352–372. doi: 10.2174/138945012799424598

Roecklein, B., Pelletier, A., and Kuempel, P. (1991). The tus gene of *Escherichia coli*: autoregulation, analysis of flanking sequences and identification of a complementary system in *Salmonella typhimurium*. *Res. Microbiol.* 142, 169–175. doi: 10.1016/0923-2508(91)90026-7

Rudolph, C. J., Upton, A. L., Stockum, A., Nieduszynski, C. A., and Lloyd, R. G. (2013). Avoiding chromosome pathology when replication forks collide. *Nature* 500, 608–611. doi: 10.1038/nature12312

Sanders, G. M., Dallmann, H. G., and McHenry, C. S. (2010). Reconstitution of the B. subtilis replisome with 13 proteins including two distinct replicases. *Mol. Cell* 37, 273–281. doi: 10.1016/j.molcel.2009.12.025

Santi, I., and McKinney, J. D. (2015). Chromosome organization and replisome dynamics in *Mycobacterium smegmatis*. *MBio* 6, e01999-01914. doi: 10.1128/mBio.01999-14

Sharma, A., Kamran, M., Verma, V., Dasgupta, S., and Dhar, S. K. (2014). Intracellular locations of replication proteins and the origin of replication during chromosome duplication in the slowly growing human pathogen *Helicobacter pylori*. *J. Bacteriol.* 196, 999–1011. doi: 10.1128/JB.01198-13

Sinha, N. K., Morris, C. F., and Alberts, B. M. (1980). Efficient *in vitro* replication of double-stranded DNA templates by a purified T4 bacteriophage replication system. *J. Biol. Chem.* 255, 4290–4293.

Slater, S. C., Lifsics, M. R., O'Donnell, M., and Maurer, R. (1994). holE, the gene coding for the θ subunit of DNA polymerase III of *Escherichia coli*: characterization of a *holE* mutant and comparison with a *dnaQ* (ϵ-subunit) mutant. *J. Bacteriol.* 176, 815–821.

Smits, W. K., Goranov, A. I., and Grossman, A. D. (2010). Ordered association of helicase loader proteins with the *Bacillus subtilis* origin of replication *in vivo*. *Mol. Microbiol.* 75, 452–461. doi: 10.1111/j.1365-2958.2009.06999.x

Soni, R. K., Mehra, P., Mukhopadhyay, G., and Dhar, S. K. (2005). *Helicobacter pylori* DnaB helicase can bypass *Escherichia coli* DnaC function *in vivo*. *Biochem. J.* 389, 541–548. doi: 10.1042/BJ20050062

Stelter, M., Gutsche, I., Kapp, U., Bazin, A., Bajic, G., Goret, G., et al. (2012). Architecture of a dodecameric bacterial replicative helicase. *Structure* 20, 554–564. doi: 10.1016/j.str.2012.01.020

Su'etsugu, M., and Errington, J. (2011). The replicase sliding clamp dynamically accumulates behind progressing replication forks in *Bacillus subtilis* cells. *Mol. Cell* 41, 720–732. doi: 10.1016/j.molcel.2011.02.024

Suski, C., and Marians, K. J. (2008). Resolution of converging replication forks by RecQ and topoisomerase III. *Mol. Cell* 30, 779–789. doi: 10.1016/j.molcel.2008.04.020

Tanner, N. A., Tolun, G., Loparo, J. J., Jergic, S., Griffith, J. D., Dixon, N. E., et al. (2011). *E. coli* DNA replication in the absence of free β clamps. *EMBO J.* 30, 1830–1840. doi: 10.1038/emboj.2011.84

Timinskas, K., Balvociute, M., Timinskas, A., and Venclovas, C. (2014). Comprehensive analysis of DNA polymerase III alpha subunits and their homologs in bacterial genomes. *Nucleic Acids Res.* 42, 1393–1413. doi: 10.1093/nar/gkt900

Trojanowski, D., Ginda, K., Pioro, M., Holowka, J., Skut, P., Jakimowicz, D., et al. (2015). Choreography of the *Mycobacterium* replication machinery during the cell cycle. *MBio* 6, e02125-02114. doi: 10.1128/mBio.02125-14

Vass, R. H., and Chien, P. (2013). Critical clamp loader processing by an essential AAA+ protease in Caulobacter crescentus. *Proc. Natl. Acad. Sci. U.S.A.* 110, 18138–18143. doi: 10.1073/pnas.1311302110

Velten, M., McGovern, S., Marsin, S., Ehrlich, S. D., Noirot, P., and Polard, P. (2003). A two-protein strategy for the functional loading of a cellular replicative DNA helicase. *Mol. Cell* 11, 1009–1020. doi: 10.1016/S1097-2765(03)00130-8

Wang, X., Reyes-Lamothe, R., and Sherratt, D. J. (2008). Modulation of *Escherichia coli* sister chromosome cohesion by topoisomerase IV. *Genes Dev.* 22, 2426–2433. doi: 10.1101/gad.487508

Weigel, C., and Seitz, H. (2002). Strand-specific loading of DnaB helicase by DnaA to a substrate mimicking unwound oriC. *Mol. Microbiol.* 46, 1149–1156. doi: 10.1046/j.1365-2958.2002.03232.x

Wendel, B. M., Courcelle, C. T., and Courcelle, J. (2014). Completion of DNA replication in *Escherichia coli*. *Proc. Natl. Acad. Sci. U.S.A.* 111, 16454–16459. doi: 10.1073/pnas.1415025111

Wessel, S. R., Marceau, A. H., Massoni, S. C., Zhou, R., Ha, T., Sandler, S. J., et al. (2013). PriC-mediated DNA replication restart requires PriC complex formation with the single-stranded DNA-binding protein. *J. Biol. Chem.* 288, 17569–17578. doi: 10.1074/jbc.M113.478156

Wu, C. A., Zechner, E. L., and Marians, K. J. (1992). Coordinated leading- and lagging-strand synthesis at the *Escherichia coli* DNA replication fork. I. Multiple effectors act to modulate Okazaki fragment size. *J. Biol. Chem.* 267, 4030–4044.

Yang, J., Zhuang, Z., Roccasecca, R. M., Trakselis, M. A., and Benkovic, S. J. (2004). The dynamic processivity of the T4 DNA polymerase during replication. *Proc. Natl. Acad. Sci. U.S.A.* 101, 8289–8294. doi: 10.1073/pnas.0402625101

Yao, N. Y., Georgescu, R. E., Finkelstein, J., and O'Donnell, M. E. (2009). Single-molecule analysis reveals that the lagging strand increases replisome processivity but slows replication fork progression. *Proc. Natl. Acad. Sci. U.S.A.* 106, 13236–13241. doi: 10.1073/pnas.0906157106

Yeeles, J. T., Deegan, T. D., Janska, A., Early, A., and Diffley, J. F. (2015). Regulated eukaryotic DNA replication origin firing with purified proteins. *Nature* 519, 431–435. doi: 10.1038/nature14285

Yeeles, J. T., and Marians, K. J. (2011). The *Escherichia coli* replisome is inherently DNA damage tolerant. *Science* 334, 235–238. doi: 10.1126/science.1209111

Yurieva, O., Skangalis, M., Kuriyan, J., and O'Donnell, M. (1997). *Thermus thermophilis* dnaX homolog encoding γ- and τ-like proteins of the chromosomal replicase. *J. Biol. Chem.* 272, 27131–27139. doi: 10.1074/jbc.272.43.27131

Yuzhakov, A., Kelman, Z., and O'Donnell, M. (1999). Trading places on DNA—a three-point switch underlies primer handoff from primase to the replicative DNA polymerase. *Cell* 96, 153–163. doi: 10.1016/S0092-8674(00)80968-X

Yuzhakov, A., Turner, J., and O'Donnell, M. (1996). Replisome assembly reveals the basis for asymmetric function in leading and lagging strand replication. *Cell* 86, 877–886. doi: 10.1016/S0092-8674(00)80163-4

2

Absence of the min system does not cause major cell division defects in *Agrobacterium tumefaciens*

*Sue A. Flores†, Matthew Howell†, Jeremy J. Daniel, Rebecca Piccolo and Pamela J. B. Brown**

Division of Biological Sciences, University of Missouri, Columbia, MO, United States

***Correspondence:**
Pamela J. B. Brown
brownpb@missouri.edu

In *A. tumefaciens*, the essential FtsZ protein is located at the growth pole before shifting to the mid-cell right before division. Loss of FtsZ causes a halt in cell separation and lysis of cells. To understand how FtsZ polymerization is regulated to properly localize the FtsZ ring at the mid-cell, we have conducted a systematic characterization of the Min system in *A. tumefaciens*. Our findings indicate that the Min system is not required for cell survival. Yet, we find that the deletion of either *minE* or *minCDE* results in a broad cell size distribution, including an increase in the proportion of short and long cells. We observe that the site of constriction is misplaced in the *minE* or *minCDE* deletion strains allowing for short cells to arise from sites of constriction near the cell poles. Remarkably, the short cells are viable and contain DNA. In order to observe chromosome replication and segregation in these strains, YFP-ParB is used as a proxy to track the origin of replication as cells elongate and divide. In the absence of the Min proteins, duplication and segregation of the origin of replication is frequently delayed. Taken together, our data suggest that the Min system contributes to the proper regulation of FtsZ placement and subsequent cell division. Furthermore, the failure to precisely place FtsZ rings at mid-cell in the *min* mutants impacts other cell cycle features including chromosome segregation.

Keywords: *Agrobacterium*, cell division, Min system, FtsZ, chromosome segregation

INTRODUCTION

Most bacteria precisely place the site of cell division at or near mid-cell through proper positioning of FtsZ to initiate divisome assembly. FtsZ forms a ring-like structure at or near the mid-cell along the cytoplasmic surface of the inner membrane (Bi and Lutkenhaus, 1991). FtsZ is stabilized along the membrane by interactions with FtsA leading to the formation of FtsZ filaments (Szwedziak et al., 2014). The FtsZ filaments form a ring-like structure at the future site of cell division and guide other divisome proteins, including peptidoglycan synthases, to the mid-cell (Ma et al., 1996; Sun and Margolin, 1998; Den Blaauwen et al., 1999; Li et al., 2007; Goley et al., 2011). The order of recruitment and essentiality of individual cell division proteins varies across species, but the general mechanism of cell division appears to be broadly conserved (Lutkenhaus and Du, 2017). After assembly of the cell division machinery is complete, peptidoglycan biosynthesis is activated at mid-cell (Addinall et al., 1997; Müller et al., 2007; Möll and Thanbichler, 2009). As the ring constricts, septal peptidoglycan is synthesized inwards to build the new poles of the daughter cells. Septal peptidoglycan synthesis requires the monofunctional PBP3, the SEDS protein FtsW, and the bifunctional PBP1b (Botta and Park, 1981; Bertsche et al., 2006; Cho et al., 2016).

The GTPase-dependent treadmilling motion of the FtsZ filaments drives the movement of the peptidoglycan biosynthesis machinery around the circumference of the cell division site, enabling the synthesis of concentric rings of peptidoglycan to form the septum (Bisson-Filho et al., 2017; Yang et al., 2017).

How is FtsZ properly positioned at mid-cell? There are several well characterized mechanisms to ensure proper positioning of the FtsZ rings (for review see Wu and Errington, 2011; Rowlett and Margolin, 2015). In rod-shaped bacteria such as *Escherichia coli* and *Bacillus subtilis*, the Min system and nucleoid occlusion (NO) prevent FtsZ rings from forming near cell poles or over nuceloids, respectively. In *E. coli*, the MinCDE proteins rapidly oscillate from pole to pole guiding FtsZ to the mid-cell (Hu and Lutkenhaus, 1999; Raskin and de Boer, 1999a,b; Fu et al., 2001; Hale et al., 2001). Since MinC is an inhibitor of FtsZ-ring assembly (Hu and Lutkenhaus, 2000; Justice et al., 2000; Dajkovic et al., 2008), FtsZ rings only form in mid-cell regions where the concentration of MinC is low. MinC forms a complex with MinD, an ATPase which dimerizes and binds to the cell membrane when bound to adenosine triphosphate (ATP) (de Boer et al., 1991, 1992; Hu and Lutkenhaus, 2003). MinE binds to MinD-ATP causing ATP hydrolysis and release of the MinCD complex from the membrane (Hu and Lutkenhaus, 2001). Remarkably, MinE can remain bound to the membrane at the pole to dislodge additional MinCD complexes (Loose et al., 2011; Park et al., 2011, 2017). Meanwhile, MinD-ATP is regenerated and cooperatively binds the membrane at the opposite pole. After removing the MinCD complexes from one cell pole, MinE will travel to the opposite pole. Thus, MinE chases MinCD from pole to pole giving rise to regular oscillations. The oscillation of the Min proteins results in a minimum of FtsZ inhibitory activity at mid-cell (Hale et al., 2001; Bonny et al., 2013).

When the genes encoding the Min proteins in *E. coli* are removed simultaneously or individually, cell division and survival is adversely impacted. In the absence of MinE, the cells usually cannot divide and the cells form long, smooth filaments. The block in cell division occurs because inhibition of FtsZ polymerization by MinC occurs throughout the cell (de Boer et al., 1989; Hu and Lutkenhaus, 2000). In the absence of MinC, or its activator MinD, a broad distribution of cell lengths is observed (de Boer et al., 1989). Both mini cells and long filaments are observed since FtsZ polymerization can occur at the cell poles or near mid-cell leading to asymmetric cell division events. FtsZ polymerization is restricted to poles and mid-cell in the absence of the Min system due to the presence of the nucleoid occlusion protein, SlmA (Bernhardt and de Boer, 2005). The FtsZ inhibitory activity of SlmA is activated by binding specific sites on the DNA near the origin of replication (Cho et al., 2011; Tonthat et al., 2011). Thus, as DNA replication is completed and the origins segregate to the cell poles, a minimal inhibitory zone is formed at mid-cell. SlmA binding to DNA activates its ability to bind the C-terminal tail of FtsZ causing depolymerization of FtsZ filaments (Du and Lutkenhaus, 2014). Under nutrient rich conditions, loss of the Min system and nucleoid occlusion is synthetically lethal; however under nutrient limited conditions the cells continue to grow and divide relatively well (Bernhardt and de Boer, 2005). When both the Min proteins and SlmA are absent, FtsZ ring placement is more accurate than in cells with only SlmA suggesting that other mechanisms contribute to the proper placement of FtsZ-ring in the absence of both Min proteins and SlmA (Bailey et al., 2014; Cambridge et al., 2014).

Indeed, the Min system is not universally distributed among bacteria suggesting the existence of alternative mechanisms of FtsZ positioning. MinCD is present in diverse bacteria, MinE is found in a more restricted range of bacteria, and other bacteria do not contain a Min system (Rothfield et al., 2005). For example, the Caulobacterales clade of alphaproteobacteria do not contain obvious homologs of the Min proteins. Furthermore, in *Caulobacter crescentus*, a bacterial model system within the Caulobacterales, only the final step of cell division takes place after completion of chromosome segregation suggesting that nucleoid occlusion does not occur (Jensen, 2006). Instead, *Caulobacter cresentus* uses at least two distinct mechanisms for regulation of cell division (Thanbichler and Shapiro, 2006; Radhakrishnan et al., 2010; Kiekebusch et al., 2012). MipZ is a distinct member of the MinD/ParA family of ATPases that contribute to spatial organization with bacterial cells (Lutkenhaus, 2012). MipZ forms a bipolar gradient on the nucleoid by binding to DNA sites near the origin of replication and directly interacts with FtsZ, inhibiting filament formation near the cell poles (Thanbichler and Shapiro, 2006; Kiekebusch et al., 2012). KidO is an NAD(H)-binding oxidoreductase that provides temporal regulation of FtsZ-ring assembly (Radhakrishnan et al., 2010). KidO binds FtsZ and prevents premature filament assembly at mid-cell. KidO is proteolytically cleared from the cell during elongation and the initiation of cell division, enabling efficient FtsZ-ring formation at mid-cell. KidO reappears late during cell division and is recruited to the mature divisome where it likely contributes to FtsZ-ring disassembly during constriction. Thus, together MipZ and KidO restrict FtsZ-ring formation to the mid-cell of predivisional cells.

Remarkably, not all alphaproteobacterial species lack a Min system. Among the alphaproteobacteria, the MinCDE proteins are found among the Rhodospirallales, Rhodobacterales, and Rhizobiales clades. The *minCDE* cluster is likely regulated by CtrA, the master cell cycle regulator, in several Rhizobiales species including *Brucella abortus* (Bellefontaine et al., 2002), *Prosthecomicrobium hirschii* (Williams et al., 2016), and *Sinorhizobium meliloti* (Pini et al., 2015). In *S. meliloti*, CtrA negatively regulates *minCD* expression (Pini et al., 2015) and overexpression of MinCD inhibits cell division (Cheng et al., 2007). Together, these observations suggest that the Min system may contribute to the regulation of cell division in the Rhizobiales. Here, we expand our knowledge about the function of the Min system in Rhizobiales by characterizing its contribution to regulation of cell division in *Agrobacterium tumefaciens*. The *A. tumefaciens* genome reveals the presence of an operon predicted to encode the MinCDE proteins, but there is not an obvious nucleoid occlusion system or MipZ homolog (Curtis and Brun, 2014). The MinCDE proteins from *Agrobacterium* share significant sequence similarity with the *E. coli* proteins (percent identities: MinC 31.70%, MinD 61.05%, and MinE 39.29%) suggesting that they may have conserved functions. In this work, we have systematically characterized the role of the Min proteins on cellular morphology, constriction site placement, and chromosome segregation. Our results suggest that the Min system contributes to the regulation of cell division; however, other FtsZ-positioning proteins likely exist in *A. tumefaciens*.

MATERIALS AND METHODS

Bacterial Strains and Culture Conditions

A list of all bacterial strains and plasmids used in this study is provided in **Table 1**. *A. tumefaciens* C58 and derived strains were grown in AT minimal media with 0.5% glucose (ATGN) (Morton and Fuqua, 2012b) without exogenous iron at 28°C with shaking. For *sacB* counterselection during construction of deletion mutants 5% sucrose replaced glucose as the sole carbon source (ATSN). *E. coli* DH5α and S17-1 λ *pir* were routinely cultivated in Luria-Bertani (LB) medium at 37°C with shaking. When appropriate, kanamycin was used at 300 µg/ml for *A. tumefaciens* and 50 µg/ml for *E. coli*. When indicated, IPTG was used as an inducer at a concentration of 1 mM.

Plasmid Construction

PCR was used to amplify ~ 500 bp of flanking sequence upstream (primers 1 and 2) and downstream (primers 3 and 4) of the gene targeted for deletion. Primers used for amplification of regions upstream and downstream of *minC* (Atu 3249), *minD* (Atu3248), *minE* (Atu3247) are shown in **Table 2**. For regions upstream and downstream of the *minCDE* locus, primers Atu 3249 P1-SpeI, Atu 3249 P2, Atu 3247 P3, and Atu 3247 P4-BamHI were used. All PCR reactions contained 10 µM of each primer, 100 ng of genomic DNA purified from wildtype *A. tumefaciens* C58, 10 mM deoxynucleotides (dNTPs; New England Biolabs), 0.5 U Phusion DNA Polymerase (Thermo scientific), 1.5% DMSO, and 5X Phusion GC Buffer (Thermo Scientific). Upstream and downstream DNA fragments were amplified by routine PCR reactions with the following cycling conditions: denaturation 98°C for 30 s, annealing 2° higher than the calculated annealing temperature of the primers for 30s, extension 72°C for 30 s, and final extension was done at 72°C for 5 min. The PCR ran for 30 cycles. PCR products were run on a 0.8% agarose gel by electrophoresis, stained with DNA SafeStain (Midwest Scientific) and gel purified using GeneJet PCR purification kit (Thermo Scientific). A second PCR reaction was done using PCR SOEing (synthesis by overlap extension) to anneal linker sequence from the 500 bp upstream and downstream together as described previously (Merritt et al., 2007). Briefly, purified PCR products were used as both templates and primers for a five-cycle PCR. A final PCR step with primers 1 and 4, using 2 µl of the second-step reaction mix as the template, generates the full-length spliced product. PCR products were then gel purified. PCR products and the pNTPS139 vector were then digested overnight at 37°C using enzymes SpeI and BamHI (New England Biolabs). Digested products were gel purified and ligated together using T4 DNA ligase (New England Biolabs). Ligations were transformed into *E.coli* DH5α competent cells using the suggested manufacturer protocol (Invitrogen). Transformants were plated on LB-agar plates containing kanamycin. Individual colonies from the transformation where then grown up overnight in LB with Kan50 and plasmid extraction was done using GenJet plasmid miniprep kit (Thermo Scientific). Plasmid inserts were verified by sequencing and the plasmids were transformed into *E. coli* S-17 by electroporation.

TABLE 1 | Bacterial strains and plasmids used in this study.

Strain or plasmid	Relevant characteristics	References/Source
PLASMIDS		
pNTPS138/139	Km^r; Suicide vector containing *oriT* and *sacB*	D. Alley
pNTPS139Δ*minC*	Km^r Suc^S; deletion plasmid for *minC*	This study
pNTPS139Δ*minD*	Km^r Suc^S; deletion plasmid for *minD*	This study
pNTPS139Δ*minE*	Km^r Suc^S; deletion plasmid for *minE*	This study
pNTPS139Δ*minCDE*	Km^r Suc^S; deletion plasmid for *minCDE*	This study
pRV-Pvan-FtsZ-GFP	Km^r; replicating plasmid for constitutive expression of FtsZ-GFP	Howell et al., 2017a
pSRKKm-Plac-YFP-ParB	Km^r; replicating plasmid for inducible expression of YFP-ParB	Ehrle et al., 2017
***E. COLI* STRAINS**		
DH5α	Cloning strain	Life Technologies
S17-1	Sm^r; RP4-2, Tc::Mu,Km-Tn7, for plasmid mobilization	Simon et al., 1983
***A. TUMEFACIENS* STRAINS**		
C58	Nopaline type strain; pTiC58; pAtC58	Watson et al., 1975
C58 Δ*minC*	C58 with deletion of *minC*	This study
C58 Δ*minD*	C58 with deletion of *minD*	This study
C58 Δ*minE*	C58 with deletion of *minE*	This study
C58 Δ*minCDE*	C58 with deletion of *minCDE*	This study

Deletion of Target Genes in *A. tumefaciens*

Nonpolar, markerless deletions of the *A. tumefaciens* individual *minC* (Atu 3249), minD (Atu 3248), and minE (Atu 3247) genes and the entire locus were generated using the plasmids pNPTS139Δ*minC*, pNPTS139Δ*minD*, pNPTS139Δ*minE, and* pNTPS139Δ*minCDE* following an established protocol (Morton and Fuqua, 2012a). Deletion of target genes was confirmed by colony PCR using the indicated primer pairs: Δ*minC* (Atu 3249 P1-SpeI and MinCDE P6), Δ*minD* (Atu 3248 P1-SpeI and Atu 3249 P4-BamHI), Δ*minE* (Atu 3247 P1-SpeI and Atu 3248 P4-BamHI), and Δ*minCDE* (MinCDE P5 and Atu 3249 P4-BamHI). PCR products were gel purified and sequence verified to confirm deletion of the target gene.

TABLE 2 | Synthesized DNA primers used in this study.

Primers	Sequence
Atu 3249 P1-SpeI	5′-ACA CGT ACT AGT CAG GCC GAT GCG G–3′
Atu 3249 P2	5′-AAG CTT GGT ACC GAAA TTC GCG AAG CTC G–3′
Atu 3249 P3	5′-GAA TTC GGT ACC AAGCTT CGA ACGCTG G–3′
Atu 3249 P4-BamHI	5′-CGC GCG GGA TCC GCA ATC GAA TTG ACC A–3′
Atu 3248 P1-SpeI	5′-ACA CGT ACT AGT TAT GGC CTG ATG CTG C–3′
Atu 3248 P2	5′-AAG CTT GGT ACC GAA TTC AGG AGG GCT G–3′
Atu 3248 P3	5′-GAA TTC GGT ACC AAG CTT TAC AAC GAC TA–3′
Atu 3248 P4-BamHI	5′-CGC GCG GGA TCC GTT CGC CCG TCG ATG A–3′
Atu 3247 P1-SpeI	5′-ACA CGT ACT AGT GCC GAT CTT GCC GGG C–3′
Atu 3247 P2	5′-AAG CTT GGT ACC GAA TTC CTG CGC GCT–3′
Atu 3247 P3	5′-GAA TTC GGT ACC AAG CTT GAT GCT CAT GC–3′
Atu 3247 P4-BamHI	5′-CGC GCG GGA TCC GAA TGG GTC ATC GCC G–3′
MinCDE P5	5′-CAT GGA ATG CGT GGC GAG CAC GAA TAC G–3′
MinCDE P6	5′-GAA GCC CGC ATG CCA TAGG ATA CGT TGC AG–3′

Growth Curve Analysis

Strains were grown in ATGN until exponential phase was reached, then back diluted to reach an $OD_{600} = 0.1$ in 1 ml of ATGN. 200 μl of culture was placed into 3 wells of a 96-well plate. A BioTek Synergy H1 Hybrid Reader was programed to read the optical density at 600 nm every 5 min after shaking for 1 min. The plate reader was maintained at a temperature of 28°C for a period of 36 h. Growth curve experiments were completed in triplicate and a total of four biological replicates were completed.

Cell Viability Assays

Serial dilutions of *A. tumefaciens* cells were spotted on ATGN plates to assess the viability of the *min* mutants. Exponential cultures ($OD_{600} = 0.4$–0.6) were diluted to OD_{600} 0.05 in ATGN. Cells were then serially diluted and 4 μl of each dilution was spotted onto ATGN plates. Plates were grown for 3 days at 28°C and imaged.

Phase Contrast Microscopy, Cell Length Analysis, and Quantitation of Constriction Position

Cells were grown in ATGN media until exponential phase was reached. A small volume (0.6–0.8 μl) of live cells were then placed onto a 1% agarose ATGN pad as described previously (Brown et al., 2012; Howell et al., 2017b). Phase contrast microscopy was performed with an inverted Nikon Eclipse TiE with a QImaging Rolera em-c^2 1 K EMCCD camera and Nikon Elements Imaging Software. Cell length distributions of 948 cells per strain were determined using the longest medial axis measured using MicrobeJ software (Ducret et al., 2016). Sites of cell constrictions were determined for ~1,000 individual cells for each strain using stacked phase contrast images. Sites of constriction were autodetected using a preset MicrobeJ constriction detection function. To determine the polar orientation of each cell, old poles were identified using the lipophilic dye FM4-64 as previously described (Howell et al., 2017b).

Fluorescence Microscopy

For DNA staining, 1 ml of cells at an $OD_{600} = 0.4$–0.6 were treated with 1 μl of Sytox Orange (Invitrogen) for 5 min in the dark. Cells were pelleted and washed with PBS two times to remove excess dye. Cell pellets were resuspended in PBS and cells were imaged immediately on an agarose pad. Replicating plasmids pRV-Pvan-FtsZ-GFP and pSRKKm-Plac-YFP-ParB were introduced in wildtype cells and *min* mutants via an established electroporation protocol (Morton and Fuqua, 2012a). For imaging of YFP-ParB, exponential phase cells were diluted to $OD_{600} = 0.2$ and were induced with IPTG for 4 h until reaching an $OD_{600} = 0.4$–0.6 and were then imaged on agarose pads. Dual channel images were stacked and cell outlines and YFP-ParB foci were automatically detected using MicrobeJ (Ducret et al., 2016). Demographs depicting YFP-ParB localization were created by capturing the fluorescent intensity along the midline of the longitudinal axis of each cell and ordering these images by cell length. Both cell outlines and identified YFP-ParB foci were manually reviewed to ensure that the plots reflect accurate sites of YFP-ParB foci. For imaging of cells expressing constitutive FtsZ-GFP, cells were grown to $OD_{600} = 0.4$–0.6 and placed on agarose pads. Cells were imaged using phase contrast and epifluorescence microscopy with the appropriate filters. For timelapse microscopy cells were grown on ATGN 1.5% agarose pads with images collected either every 5 or 10 min.

RESULTS

Deletion of *min* Genes Does Not Have a Large Impact on Cell Growth or Viability, but Causes a Broader Cell Length Distribution

Transposons accumulate in the *min* region of *A. tumefaciens* during saturating transposon mutagenesis experiments in *A. tumefaciens* suggesting that the Min system is not required for cell survival (Curtis and Brun, 2014). To verify these results, we constructed markerless deletions in each *min* gene and the entire *min* locus. Growth curves of *min* mutants were indistinguishable from the growth curve of the parent strain (**Figure 1A**). The *min* mutants are viable (**Figure 1B**) with only a slight decrease in viability in Δ*minE*, suggesting that unregulated MinCD is more problematic than loss of the entire *min* locus. Although the *min* mutants are viable, phase contrast images of the cell populations revealed a small but consistent proportion of cells with atypical morphologies, including short cells (**Figure 2A**). Quantitative image analysis was used to determine the cell length distributions of the *min* mutants (**Figure 2B**). While the medians of the cell length distributions of the *min* mutants are actually slightly longer than that of the parent strain, the length of the whiskers is significantly increased suggesting that both short and long cells are accumulating in the *min* mutants. Next, we determined

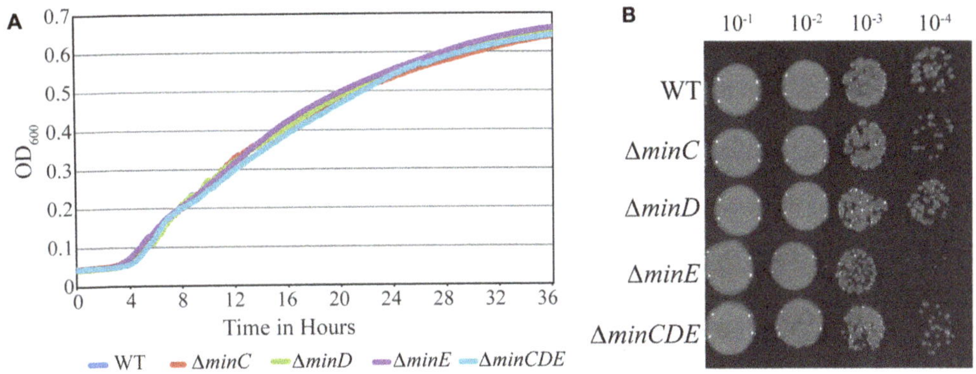

FIGURE 1 | Growth and viability of *min* mutants is not compromised. **(A)** Growth of Wildtype *A. tumefaciens* strain C58 and *min* mutants is monitored over 36 h by observing the increase in optical density. **(B)** Exponentially growing cells were diluted to OD $_{600}$ = 0.05 and serial dilutions were spotted on nutrient rich medium to observe viability.

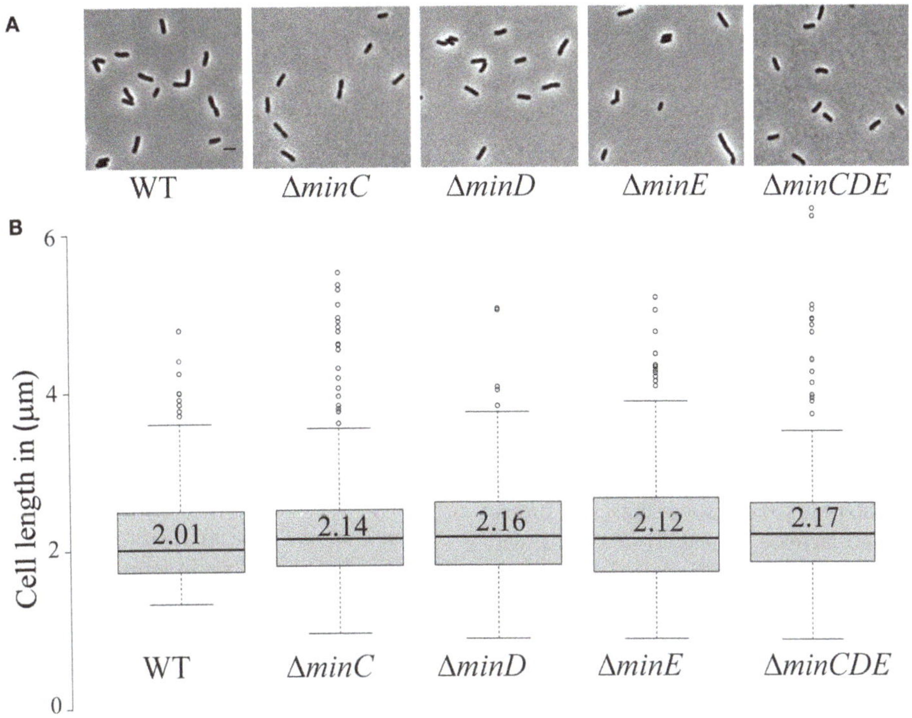

FIGURE 2 | Short cells accumulate in *min* mutants. **(A)** Phase contrast images of representative wildtype and *min* mutant cells. Scale bar = 2 μm. **(B)** Box plots illustrate cell length distributions of wildtype cells and *min* mutants. Medians are shown by the labeled center lines. Box limits indicate the 25th and 75th percentiles as determined by R software; whiskers extend 1.5 times the interquartile range and outliers are represented by dots. Cell lengths measured from 948 cells for each strain.

the percentage of cells with typical cell lengths (defined as 1.5–3.5 μm), cells with branches or bulges, short cells (< 1.5 μm), and cells with visible constrictions (**Figure 3**). Indeed, these observations confirm the presence of a small but reproducible proportion of short cells in the *min* mutant cells (**Figure 3**, panel iii). Furthermore, although the overall proportion of cells with constrictions is not impacted by the loss of individual *min* genes or the entire locus, we observe an increase in cells with obvious asymmetric constrictions or multiple constrictions (**Figure 3**, panel iv).

The Min System Contributes to Precise Placement of Constriction Sites

The previous observations suggest that the Min system may contribute to the proper placement of the site of cell division in *A. tumefaciens*. To better understand the function of the Min system in the establishment of constriction sites, we quantitated the position of constriction sites in wildtype cells ($n = 186$). In this analysis, the true-mid cell is defined as 0 and negative values indicate positions of constrictions closer to the old pole whereas positive values indicate positions of constrictions closer to the new pole. Using the wildtype data set, we define a typical constriction placement to occur near mid-cell with a bias toward the new pole: 95% of constrictions formed between −0.1 and 0.35 μm from the true mid-cell position (**Figure 4A**, left). In addition, constrictions are consistently observed in cells with lengths between 2.5 and 3.5 μm (**Figure 4A**, right). These observations indicate that the site of constriction is established well before the cell has completed elongation. Cells longer than 4 μm in length with constrictions are rarely observed, presumably because the cells have successfully completed cell division.

In the *min* mutants, constrictions form in ~20% of the cell population which is similar to what is observed in wildtype (**Figure 3**); however, the placement of the constrictions is perturbed. The *min* mutants have a broader distribution of constriction placement (**Figure 4A**). The Δ*minE* mutant in particular has lost the ability to maintain the proper bias of constriction site placement near the new pole. In contrast, when the entire *min* locus is absent 82% of constrictions are observed in the typical position and the bias for constriction placement toward the new pole is retained. These observations suggest that the unregulated activity of MinCD results in a more random positioning of constrictions, but proper positioning of constrictions is frequently retained when the *min* locus is entirely deleted. Although constriction sites form more randomly in *min* mutants, we rarely observe constriction sites immediately adjacent to the cell poles (**Figure 4B**) and short cells, rather than mini cells are formed. Since constriction sites are not biased only toward the new pole in *min* mutants, this phenotype cannot be readily explained by the continuation of growth at the new pole and may indicate the existence of another mechanism to protect the poles. Furthermore, an increase in long cells (>4 μm) with visible constrictions (**Figure 4B**) is observed in *min* mutants. This observation suggests that cell division efficiency is adversely impacted in the *min* mutants. It is possible that the perturbations of the Min system result in a delay in cell division or an increase in the frequency of constrictions that do not lead to a productive cell division event.

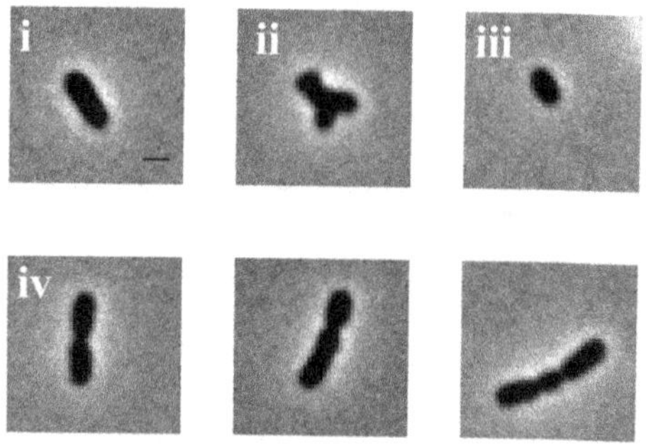

	(i) Rod-shaped	(ii) Branched	(iii) Small <1.5μm	(iv) Constricting
WT	75%	<1%	5%	20%
Δ*minC*	71%	1%	8%	20%
Δ*minD*	70%	1%	8%	21%
Δ*minE*	61%	1%	12%	25%
Δ*minCDE*	71%	1%	8%	20%

FIGURE 3 | Atypical morphologies are observed in *min* mutants. Phase contrast images of representative morphologies. All images taken from the Δ*minE* cell population. Scale bar = 1 μm. **(top)** Quantitation of bacterial morphologies. **(bottom)** The percentage of cells with a rod-shaped morphology (i: 1.5–3.5 μm in length), branched morphology (ii), short cell morphology (iii: <1.5 μm in length), and cells with visible constrictions (iv) are shown in the table. Cell morphologies were categorized from at least 948 cells for each strain.

FtsZ-Rings Form in *min* Mutants

In order to provide additional insights into the position of constriction placement, we introduced a plasmid which constitutively expresses *ftsZ-gfp* at a low level (Howell et al., 2017a) into wildtype and *min* mutant cells. In wildtype cells, FtsZ-rings initially form at an asymmetric position near the new pole and mark the future site of cell division (**Figures 5A,B**, Movie S1). As the cell continues to elongate at the new pole, the FtsZ ring is ultimately positioned near mid-cell and constriction leads to the appearance of a discrete focus of FtsZ in late predivisional cells. In the *min* mutants, FtsZ rings are observed at asymmetric positions, near mid-cell and in some cells multiple FtsZ rings form (**Figure 5A**). Unlike the pattern of FtsZ ring formation in wildtype cells which is very consistent, the pattern of FtsZ ring formation is variable in the *min* mutants. For example, asymmetric FtsZ rings form in positions biased toward either the new or old pole in Δ*minE* mutants (**Figure 5C**, top two panels, Movies S2, S3). In either case, the establishment of an FtsZ ring can lead to a cell division event which creates daughter cells of different cell sizes. These observations are consistent with the broader cell length distribution of the *min* mutants which includes both short and long cells (**Figure 2**) and with the asymmetric positioning of constrictions (**Figure 4**). In addition, Δ*minC*, Δ*minE*, and Δ*minCDE* cells with multiple FtsZ rings are readily observed (**Figure 5A**). In Δ*minE* cells with two FtsZ rings, both sites marked with FtsZ undergo constriction leading to the production of a bow-tie morphology and cell division ultimately occurs at either both or a single site marked by an FtsZ ring (**Figure 5C**, bottom two panels, Movies S4, S5).

Cell Division Is Initiated Prior to Nucleoid Separation

After observing that FtsZ rings are present in wildtype cells without visible constrictions (**Figure 5B**), we next aimed to determine if FtsZ rings form over DNA (**Figure 6**). In elongating wildtype cells, FtsZ is typically observed either as a polar focus or in an asymmetric FtsZ ring, whereas Sytox Orange labeling

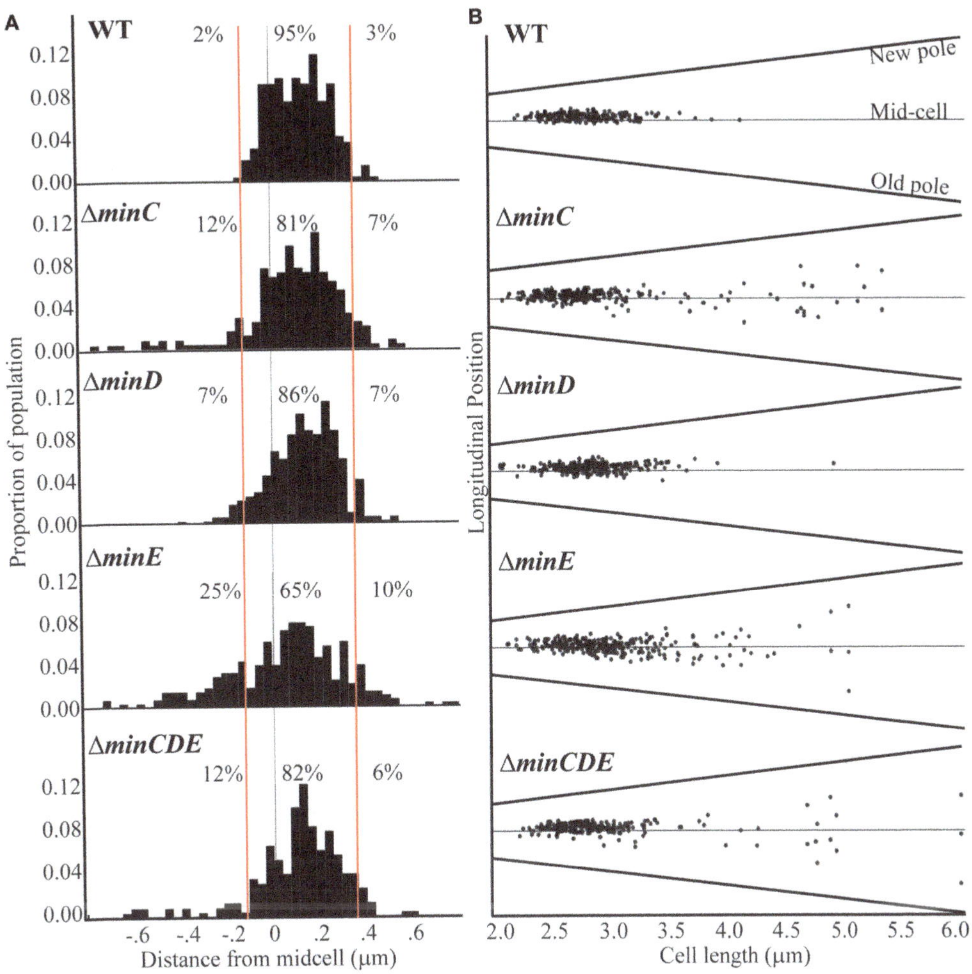

FIGURE 4 | Cell constriction placement in wildtype and *min* mutant cells. **(A)** Histograms of constriction position relative to distance from the midcell (gray line; relative position = 0). The red lines at −0.1 and 0.35 μm mark the region where constrictions typically form in wildtype cells. Negative numbers are closer to the old pole whereas the positive numbers are closer to the new pole. Percentages indicate the proportion of constrictions placed farther than −0.1 μm from mid-cell, placed between −0.1 and 0.35 μm, and placed more than 0.35 μm from mid-cell. **(B)** The longitudinal position of constrictions is plotted against cell length. For this analysis, cells are ordered by cell length and black dots indicate the position of the constriction. Mid-cell is indicated by the center gray line and the slanted lines indicate the position of the new and old pole for each cell. For these analysis, the predivisional cells within a population of ~1,000 cells were analyzed (*n* = 186 for wildtype; *n* = 242 for Δ*minC*, *n* = 262 for Δ*minD*, *n* = 285 for Δ*minE*, and *n* = 225 for Δ*minCDE*).

indicates that the DNA is diffuse (**Figure 6A**, columns 1-2). The DNA remains diffuse in cells with early constrictions (**Figure 6A**, column 3) and only forms separated nucleoids in deeply constricted cells (**Figure 6A**, column 4). These observations are consistent with the possibility that *A. tumefaciens* does not use a nucleoid occlusion mechanism to position FtsZ at mid-cell. Similar results are observed in the *min* mutants, as exemplified by Δ*minE* (**Figure 6B**). Notably, in the Δ*minE* mutant FtsZ rings form over the top of DNA at asymmetric positions and when multiple FtsZ rings are present. Finally, we observed that short cells in the Δ*minE* population typically contain diffuse DNA (**Figure 6B**, right column).

In the wildtype cells and each of the *min* mutants, 90–95% of the cell population (based on observations of ~230 cells/strain) has diffuse DNA. In the *min* mutants, DNA is diffuse in cells without constrictions irrespective of cell shape (shown for Δ*minE* in **Figure 6C**, panel i), including short and branched cells. Two distinct nucleoids are observed in 5–10% of the wildtype and *min* cell populations (shown for Δ*minE* in **Figure 6C**, panel ii). As expected due to the formation of asymmetric sites of cell constriction, the *min* mutants contain a higher proportion of cells with asymmetrically separated nucleoids. In <1% of the Δ*minC and* Δ*minE* cells, the presence of more than 2 nucleoids is observed (**Figure 6C**, panel iii).

Chromosome Segregation Is Delayed in Single *min* Mutants

Since the inefficiency of cell division in the *min* mutants may cause a delay in chromosome segregation, we next tracked the early stages of chromosome segregation by introducing an IPTG

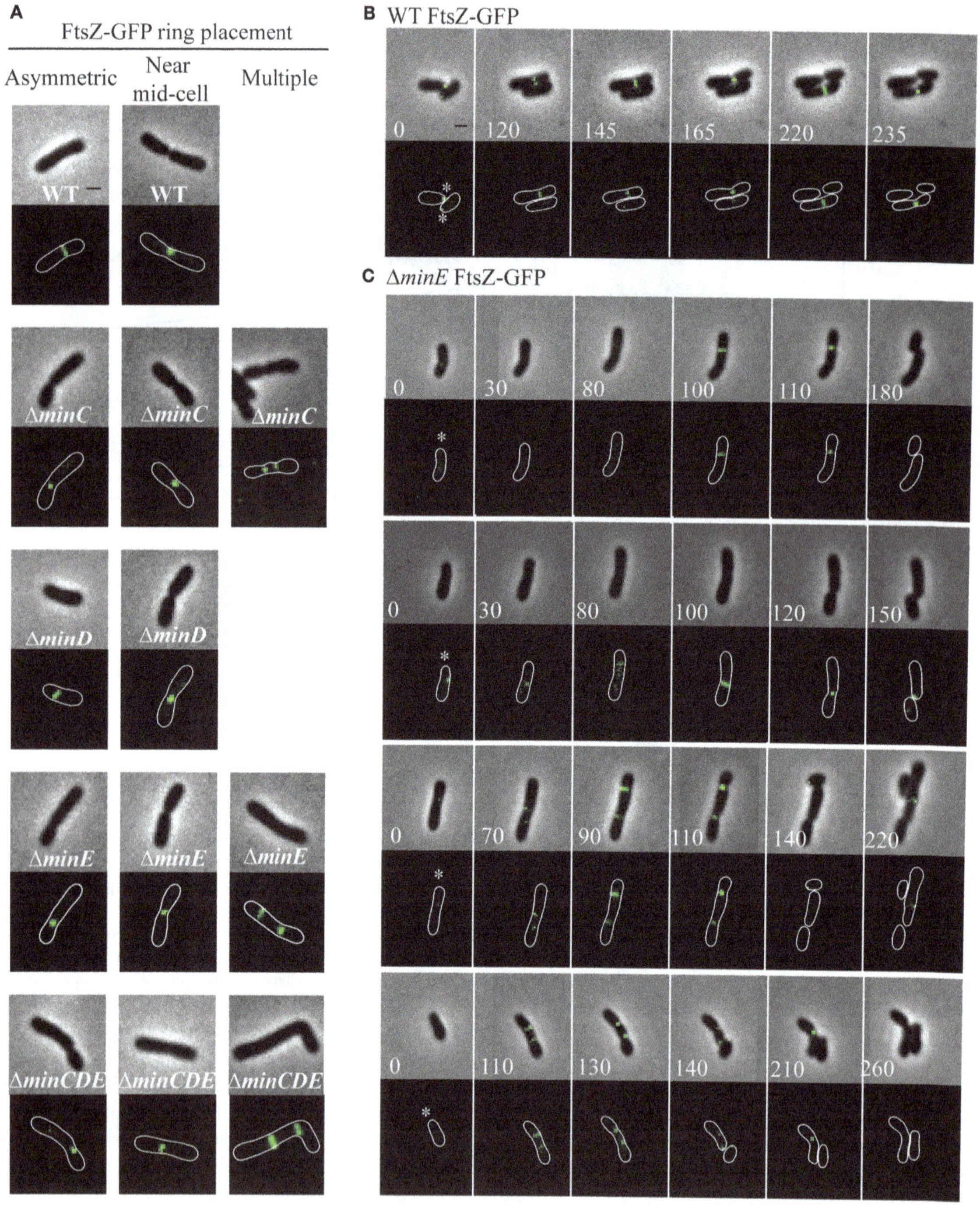

FIGURE 5 | FtsZ-GFP ring position in wildtype and *min* mutant cells. **(A)** Representative images of FtsZ-GFP localization patterns in wildtype and *min* mutants including asymmetric localization, near mid cell localization, and multiple ring formation. The absence of an image indicates that the localization pattern is not observed in the strain. **(B)** Timelapse microscopy showing typical FtsZ-GFP localization patterns in wildtype cells. Asterisks mark the growing poles. **(C)** Timelapse microscopy of FtsZ-GFP localization in Δ*minE* cells. Asterisks mark the growing poles. Images shown in panels **B** and **C** are at the same scale. Scale bars = 1 μm.

inducible plasmid expressing *yfp-parB* (Ehrle et al., 2017) into the *min* mutants. ParB attaches to the *parS* site near the origin of replication, making it suitable to track the movement of newly replicated origin (Ehrle et al., 2017). In *A. tumefaciens* ParB localizes at the old pole and as the cell nears division a second focus appears and rapidly tracks across the cell to the new pole ensuring both cells receive a copy of the chromosome (**Figure 7**; Ehrle et al., 2017). The longitudinal profile of over 500 cells expressing YFP-ParB were aligned in order of cell length to create a demograph depicting the localization of YFP-ParB

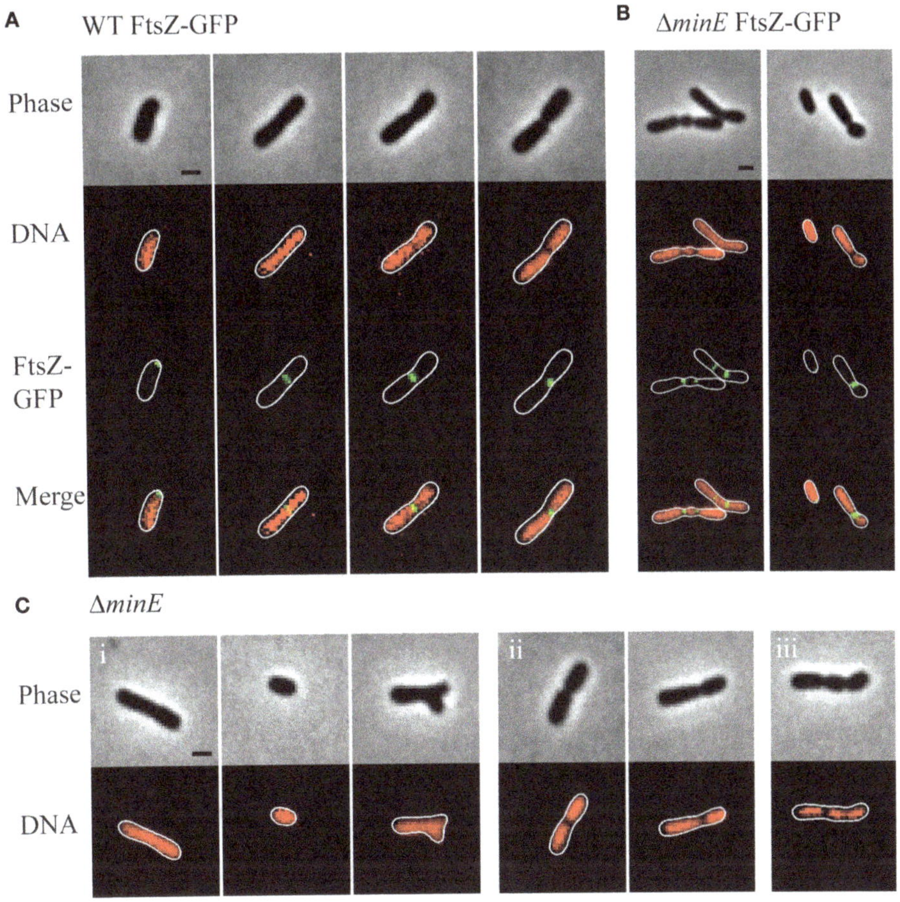

FIGURE 6 | Nucleoid position in wildtype and *min* mutant cells. **(A)** Representative images demonstrating FtsZ-GFP localization in wildtype cells with Sytox Orange stained DNA. **(B)** Representative images of FtsZ-GFP localization in Δ*minE* cells with Sytox Orange stained DNA. **(C)** Representative images from the Δ*minE* mutant population. Sytox Orange was used to label DNA and observe localization patterns including (i) diffuse, (ii) separated nucleoids, and (iii) other patterns. Scale bars = 1 μm.

throughout the cell cycle (**Figure 7A**). In wildtype cells, YFP-ParB is observed in three patterns: first, a single focus is observed in short cells, next we observe a brief transition period where a second focus of YFP-ParB appears and transits along the longitudinal axis of the cell, finally in predivisional cells, both foci are anchored at opposite poles. To examine the YFP-ParB localization pattern at the population level, the positions of the YFP-ParB foci were normalized by cell length and plotted along the cell axis (**Figure 7B**). In wildtype cells, a larger number of foci are observed at the old pole than at the new pole due to the presence of a single focus in short cells and two foci near the old pole at the onset of DNA replication. Notably, very few YFP-ParB foci are observed transiting from old pole to new pole presumably due to the rapid rate of DNA replication and chromosome segregation.

The deletion of single *min* genes results in less consistent patterns of YFP-ParB localization throughout the cell cycle. In the absence of MinC, MinD, or MinE, YFP-ParB is located at a single focus in short cells; however, not all of the foci are found at the old pole (**Figure 7A**). Furthermore, we observe an increase in the number of foci observed between the poles (**Figures 7A,B**). This phenotype is due to a combination of cells which accumulate more than 2 YFP-ParB foci and cells in which the duplicated origin is not efficiently transited from pole to pole. Although MinD has been described as a candidate protein to tether DNA to the membrane during chromosome segregation in *E. coli* (Di Ventura et al., 2013), we do not observe a more severe phenotype in the Δ*minD* mutant compared to the Δ*minC* or Δ*minE* mutants. Furthermore, deletion of the entire *min* locus results in a localization pattern of YFP-ParB which is more similar to wildtype. Since the *min* mutants are viable (**Figure 1**) and even short cells are capable of resuming growth (**Figure 5C**, bottom panel), we infer that chromosome segregation is delayed presumably due to inefficient cell division in the *min* mutants.

DISCUSSION

In this work, we have characterized the impact of the Min system on the cell division of *A. tumefaciens*. Similar to findings in *S. meliloti* (Cheng et al., 2007) and consistent with a saturating

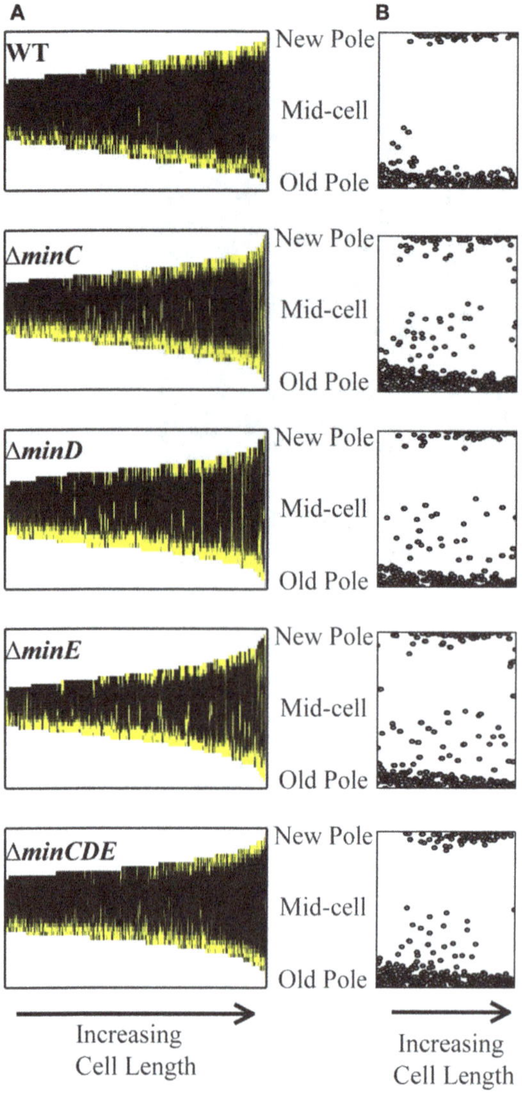

FIGURE 7 | YFP-ParB localization in wildtype and *min* mutant cells. **(A)** Demographs of YFP-ParB localization. Cells are ordered according to cell length and positions of the new pole, mid-cell, and old pole are indicated. **(B)** Position of YFP-ParB foci are plotted relative to longitudinal position in the cells. At least 500 cells were analyzed for each strain.

transposon screen in *A. tumefaciens* suggesting that the *min* genes are not essential (Curtis and Brun, 2014), we have confirmed that the *min* genes are not required for cell viability (**Figure 1**). Quantitative image analysis of cell morphology reveals that cell length distributions and placement of sites of cell constrictions are perturbed when single *min* genes are deleted compared to wildtype cells (**Figures 2–4**). In particular, the absence of MinE leads to the accumulation of both long and short cells suggesting that the placement of the septum is perturbed. Indeed, the sites of cell constriction (**Figure 4**) and localization of FtsZ-GFP is more variable in the Δ*minE* strain (**Figure 5**) suggesting that the misregulation of MinCD is detrimental to efficient cell division. These observations are consistent with the *E. coli* model of the Min system where MinE regulates the activity of the MinCD complex by driving the oscillation of MinC and MinD from pole-to-pole and preventing the establishment of polar FtsZ rings (Lutkenhaus, 2007; Rowlett and Margolin, 2015). In *A. tumefaciens*, the absence of MinE leads to a more random distribution of active of MinCD complexes, allowing the observed misplacement of FtsZ-rings and constriction sites. Remarkably, asymmetric FtsZ-GFP rings are not observed immediately adjacent to the cell poles in *min* mutants (**Figures 5A,C**) and can form over DNA in both wildtype cells and *min* mutants (**Figure 6**). Finally, the absence of the entire Min system has a relatively mild phenotype enabling most cells to divide near mid-cell (**Figures 2, 4**). Together, these results suggest that other mechanisms for proper placement of FtsZ-rings must exist in *A. tumefaciens*.

While the *A. tumefaciens min* mutant phenotypes are generally consistent with *E. coli* Min model, a key component of this model remains to be tested in *A. tumefaciens*. In *E. coli*, the Min proteins oscillate from pole-to-pole producing a local minimum inhibition zone at mid-cell in which FtsZ-rings can form (Raskin and de Boer, 1999a,b; Meinhardt and de Boer, 2001). In contrast, in *B. subtilis* the MinCDJ system does not use protein oscillation. DivIVA binds to regions of the membrane with negative curvature (Lenarcic et al., 2009; Ramamurthi and Losick, 2009). Next, MinJ acts as an adaptor protein and enables the recruitment of MinD and subsequently MinC to sites of DivIVA localization (Bramkamp et al., 2008; Patrick and Kearns, 2008). Initially, it was thought that the MinCD complexes formed a static bipolar gradient which protects the poles from FtsZ ring assembly (Adams and Errington, 2009; Bramkamp and van Baarle, 2009); however, DivIVA is recruited to the nascent division site when constriction is initiated leading to the formation of DivIVA rings on either side of the division site (Eswaramoorthy et al., 2011). MinCDJ complexes are assembled at these DivIVA rings and presumably prevent the formation of additional FtsZ rings near mid-cell. Based on the presence of *minE* and the absence of *divIVA* and *minJ* in the genome of *A. tumefaciens*, we hypothesize that the MinCD proteins should localize at cell poles and oscillate from pole-to-pole and MinE should exhibit a dynamic localization pattern with enrichment at sub-polar region. Our initial efforts to construct C-terminal fluorescent protein fusions to the *A. tumefaciens* Min proteins have been unsuccessful. A rigorous effort will be needed to observe the localization patterns of the Min proteins in *A. tumefaciens* to determine if pole-to-pole oscillation of the Min proteins contributes to the efficient establishment of constriction sites at the proper position.

In addition to cell division defects, single deletion of *min* genes results in delayed chromosome segregation (**Figure 7**). At first glance, this finding appears to be consistent with observations of chromosome partitioning defects in *E. coli min* mutants (Akerlund et al., 1992, 2002; Di Ventura et al., 2013); however, there are notable phenotypic differences. First, in *E. coli*, the production of mini-cells devoid of DNA due to polar cell division events is a hallmark of *min* mutants (de Boer et al., 1989). In contrast, short cells arising from misplacement of constriction sites in *A. tumefaciens min* mutants typically contain DNA. This difference may suggest that *A. tumefaciens* does not employ a nucleoid occlusion system to prevent the establishment of FtsZ rings over unsegregated nucleoids (Wu and Errington, 2004; Bernhardt and de Boer, 2005) and is consistent with the absence of an obvious nucleoid occusion protein in the *A. tumefaciens genome* (Goodner et al., 2001; Wood et al., 2001) and the formation of FtsZ-GFP rings over DNA in unconstricted cells (**Figure 6A**). Second, in *E. coli*, a Δ*minCDE* mutant has a more severe defect in chromosome segregation than a Δ*minC* mutant leading to the hypothesis that MinD may directly contribute to chromosome segregation (Di Ventura et al., 2013). The MinD/ParA family of proteins share an evolutionary history and MinD and ParA function in providing positional information for spatial organization of the FtsZ ring and segregating chromosome, respectively (Lutkenhaus, 2012). *E. coli* MinD can nonspecifically bind chromosomal DNA and may provide polar gradients of DNA tethering sites during chromosome segregation (Di Ventura et al., 2013). If *A. tumefaciens* MinD is capable of binding DNA and tethering chromosomes to the membrane, the predicted random distribution of MinCD complexes in the absence of MinE may explain why the YFP-ParB foci do not rapidly transit from pole-to-pole. Nevertheless, if *A. tumefaciens* MinD is involved in chromosome segregation we would expect to observe segregation defects in the Δ*minD* and Δ*minCDE* strains. Remarkably, the Δ*minC*, Δ*minD*, and Δ*minE* strains exhibit a strikingly similar phenotype with a delay in transition of the YFP-ParB focus from the old pole to the new pole (**Figure 7**). Furthermore, the Δ*minCDE* strain has a less severe phenotype and YFP-ParB is bipolar in the longest cells. Together, these data suggest that the chromosome segregation defect in the *min* mutants likely arises indirectly as a consequence of less efficient cell division.

Remarkably, in some *C. crescentus* cells lacking MipZ, productive cell division events occur resulting in the production of mini cells which contain DNA (Thanbichler and Shapiro, 2006). Although constrictions form over chromosomes that have not completed segregation, most isolated mini cells contain both an origin of replication and a terminus. These observations suggest that cell division is delayed until DNA replication is finished and the complete chromosome is delivered to the mini cell compartment (Thanbichler and Shapiro, 2006). In most bacteria, DNA replication and chromosome segregation occur simultaneously. These processes consist of three major stages: separation and translocation of the duplicated origin, segregation of the bulk chromosome, and separation of the terminus region (Badrinarayanan et al., 2015; Surovtsev and Jacobs-Wagner, 2018). *C. crescentus* uses the widely distributed ParABS system (Livny et al., 2007) to segregate the *ori* region of the chromosome. Briefly, ParB binds to DNA at the *parS* site which is proximal to the origin of replication. Following duplication of *ori*, one of the ParB-bound *ori* regions remains at the old pole and the other is translocated across the cell to the opposite pole following a receding cloud of ParA (Shebelut et al., 2009; Ptacin et al., 2010; Schofield et al., 2010). ParB is anchored to the poles through a direct interaction with the polar organizing protein PopZ (Bowman et al., 2008; Ebersbach et al., 2008). When PopZ is absent, the chromsomes become untethered from the pole and mini cells without DNA are formed (Ebersbach et al., 2008). MipZ not only inhibits FtsZ-ring assembly, it also binds to ParB, protecting the *ori* proximal regions from FtsZ-ring formation (Thanbichler and Shapiro, 2006). Thus, the processes of cell division and *ori* partitioning are tightly coupled through MipZ. Later stages of cell division and chromosome segregation are also coupled through FtsK. In *C. crescentus*, the N-terminus of FtsK contributes to the stability of FtsZ-rings and the C-terminus of FtsK is responsible for clearing the termini from the division plane (Wang et al., 2006).

The observation that the *A. tumefaciens min* mutants produce short cells that contain DNA may suggest that *A. tumefaciens* also couples the processes of cell division and chromosome segregation. Indeed, the deletion of *popZ* in *A. tumefaciens* results in untethered chromosomes and the production of cells devoid of DNA (Ehrle et al., 2017). In some Δ*popZ* cells, DNA appears to be segregated in the wrong direction across the division plane. FtsK functions as a DNA translocase that assists in the completion of cell division by moving DNA across the division plane in the direction of the termini (Besprozvannaya and Burton, 2014; Badrinarayanan et al., 2015). If a terminus is trapped on the wrong side of the division plane in the absence of PopZ, FtsK may pump DNA in the wrong direction leading to the production of cells without DNA. Notably, whereas the deletion of *popZ* leads to the production of mini cells in *C. crescentus* (Ebersbach et al., 2008), the loss of *popZ* results in the production of a broad distribution of cell lengths in *A. tumefaciens* (Howell et al., 2017a) suggesting that the poles are still largely protected from FtsZ-ring formation. Similarly, when the Min system is removed the short cells sometimes arise (**Figure 2**) and although *ori* partitioning appears to be delayed (**Figure 7**), most cells are viable (**Figure 1**) suggesting that even short cells inherit an intact chromosome. Timelapse microscopy of the Δ*minE* mutant illustrates that short cells are capable of resuming growth following cell division (**Figure 5C**, bottom panel). Together, these observations suggest that *A. tumefaciens* must use another FtsZ-positioning mechanism to protect the poles and that the processes of DNA replication, chromosome segregation, and cell division must be coordinated.

How might DNA replication, chromosome segregation, and cell division be properly coordinated in *A. tumefaciens*? In *S. meliloti*, expression of *minC* and *minD* was upregulated during depletion of CtrA, implicating this master cell cycle regulator as a transcriptional repressor of this operon (Pini et al., 2015). Remarkably, *minCD* are the only known cell division genes directly regulated by CtrA; however, introduction of a deletion of *minCDE* into the CtrA depletion strain did not rescue the cell division defect suggesting the Min overexpression is not exclusively responsible for the cell division phenotype. Depletion of CtrA in *A. tumefaciens* leads to a block in cell division (Figueroa-Cuilan et al., 2016) and a putative consensus CtrA binding site (TTAA-N_7-TTAA) is present upstream of *minC* in the *A. tumefaciens* genome. Thus, it is tempting to speculate that the transcription of the *min* genes is under the control of CtrA. Cell-cycle regulation of the Min system may ensure that these proteins are functioning as needed when the cells approach cell division. In *C. crescentus* the expression of *ftsZ* is directly regulated by CtrA (Kelly et al., 1998; Laub et al., 2002) and FtsZ is subject to proteolysis by ClpAP and ClpXP (Williams et al., 2014) leading to cell-cycle variability of FtsZ levels. Even when *ftsZ* is expressed constitutively, it is subject to post-translational control leading to cell cycle variability of FtsZ levels (Williams et al., 2014). The cell cycle variability in FtsZ levels may be a common feature among bacteria with an alphaproteobacterial cell cycle. Thus, in *A. tumefaciens* the cell cycle regulation of the Min system may temporally coordinate the expression levels of the Min and FtsZ proteins. Future studies will be necessary to determine if FtsZ, other divisome components, and the Min proteins are coordinated through cell-cycle regulation in *A. tumefaciens*. Such studies are necessary to better understand how the processes of DNA replication, chromosome segregation, and cell division are coordinated in *A. tumefaciens*.

Overall these results suggest that while the *A. tumefaciens* Min system contributes to the precise positioning of an FtsZ-ring and constriction site near mid-cell, other mechanisms must exist to ensure proper spatial organization during cell division. In *A. tumefaciens*, the phenotype of the Δ*minCDE* strain is milder than that of the individual Δ*minC*, Δ*minD*, or Δ*minE* strains suggesting that this FtsZ positioning system is dispensable for the completion of cell division. There are a number of alternative FtsZ positioning proteins including nucleoid occlusion proteins, MipZ in *C. crescentus* which forms a bipolar gradient and directly inhibits FtsZ-ring assembly near the poles (Thanbichler and Shapiro, 2006; Kiekebusch et al., 2012), and positive regulators which localize to mid-cell and promote FtsZ-ring assembly (Rowlett and Margolin, 2015). Other than the *min* locus, genes encoding candidate FtsZ-positioning proteins cannot be readily identified in the *A. tumefaciens* genome (Goodner et al., 2001; Wood et al., 2001). Thus, further studies of *A. tumefaciens* cell division are likely to reveal novel strategies to ensure proper mid-cell assembly of the divisome.

AUTHOR CONTRIBUTIONS

SF, MH, JD, and PB designed the experiments. SF, MH, JD, and RP conducted the experiments. SF, MH, RP, and PB analyzed the data. All authors contributed to writing and editing of the manuscript.

FUNDING

Research in the Brown lab on *A. tumefaciens* cell growth and division is supported by the National Science Foundation (IOS1557806).

ACKNOWLEDGMENTS

We thank Grant Bowman (University of Wyoming) for the plasmid containing YFP-ParB and members of the Brown lab for feedback during the preparation of this manuscript.

REFERENCES

Adams, D. W., and Errington, J. (2009). Bacterial cell division: assembly, maintenance and disassembly of the Z ring. *Nat. Rev. Microbiol.* 7, 642–653. doi: 10.1038/nrmicro2198

Addinall, S. G., Cao, C., and Lutkenhaus, J. (1997). FtsN, a late recruit to the septum in *Escherichia coli*. *Mol. Microbiol.* 25, 303–309. doi: 10.1046/j.1365-2958.1997.4641833.x

Akerlund, T., Bernander, R., and Nordstrom, K. (1992). Cell division in *Escherichia coli minB* mutants. *Mol. Microbiol.* 6, 2073–2083. doi: 10.1111/j.1365-2958.1992.tb01380.x

Akerlund, T., Gullbrand, B., and Nordstrom, K. (2002). Effects of the Min system on nucleoid segregation in *Escherichia coli*. *Microbiology* 148(Pt 10), 3213–3222. doi: 10.1099/00221287-148-10-3213

Badrinarayanan, A., Le, T. B., and Laub, M. T. (2015). Bacterial chromosome organization and segregation. *Annu. Rev. Cell Dev. Biol.* 31, 171–199. doi: 10.1146/annurev-cellbio-100814-125211

Bailey, M. W., Bisicchia, P., Warren, B. T., Sherratt, D. J., and Mannik, J. (2014). Evidence for divisome localization mechanisms independent of the Min system and SlmA in Escherichia coli. *PLoS Genet.* 10:e1004504. doi: 10.1371/journal.pgen.1004504

Bellefontaine, A. F., Pierreux, C. E., Mertens, P., Vandenhaute, J., Letesson, J. J., and De Bolle, X. (2002). Plasticity of a transcriptional regulation network among alpha-proteobacteria is supported by the identification of CtrA targets in *Brucella abortus*. *Mol. Microbiol.* 43, 945–960. doi: 10.1046/j.1365-2958.2002.02777.x

Bernhardt, T. G., and de Boer, P. A. (2005). SlmA, a nucleoid-associated, FtsZ binding protein required for blocking septal ring assembly over chromosomes in *E. coli*. *Mol. Cell* 18, 555–564. doi: 10.1016/j.molcel.2005.04.012

Bertsche, U., Kast, T., Wolf, B., Fraipont, C., Aarsman, M. E., Kannenberg, K., et al. (2006). Interaction between two murein (peptidoglycan) synthases, PBP3 and PBP1B, in *Escherichia coli*. *Mol. Microbiol.* 61, 675–690. doi: 10.1111/j.1365-2958.2006.05280.x

Besprozvannaya, M., and Burton, B. M. (2014). Do the same traffic rules apply? Directional chromosome segregation by SpoIIIE and FtsK. *Mol. Microbiol.* 93, 599–608. doi: 10.1111/mmi.12708

Bi, E. F., and Lutkenhaus, J. (1991). FtsZ ring structure associated with division in *Escherichia coli*. *Nature* 354, 161–164.

Bisson-Filho, A. W., Hsu, Y. P., Squyres, G. R., Kuru, E., Wu, F., Jukes, C., et al. (2017). Treadmilling by FtsZ filaments drives peptidoglycan synthesis and bacterial cell division. *Science* 355, 739–743. doi: 10.1126/science.aak9973

Bonny, M., Fischer-Friedrich, E., Loose, M., Schwille, P., and Kruse, K. (2013). Membrane binding of MinE allows for a comprehensive description of Min-protein pattern formation. *PLoS Comput. Biol.* 9:e1003347. doi: 10.1371/journal.pcbi.1003347

Botta, G. A., and Park, J. T. (1981). Evidence for involvement of penicillin-binding protein 3 in murein synthesis during septation but not during cell elongation. *J. Bacteriol.* 145, 333–340.

Bowman, G. R., Comolli, L. R., Zhu, J., Eckart, M., Koenig, M., Downing, K. H., et al. (2008). A polymeric protein anchors the chromosomal origin/ParB complex at a bacterial cell pole. *Cell* 134, 945–955. doi: 10.1016/j.cell.2008.07.015

Bramkamp, M., Emmins, R., Weston, L., Donovan, C., Daniel, R. A., and Errington, J. (2008). A novel component of the division-site selection system of *Bacillus subtilis* and a new mode of action for the division inhibitor MinCD. *Mol. Microbiol.* 70, 1556–1569. doi: 10.1111/j.1365-2958.2008.06501.x

Bramkamp, M., and van Baarle, S. (2009). Division site selection in rod-shaped bacteria. *Curr. Opin. Microbiol.* 12, 683–688. doi: 10.1016/j.mib.2009.10.002

Brown, P. J., de Pedro, M. A., Kysela, D. T., Van der Henst, C., Kim, J., De Bolle, X., et al. (2012). Polar growth in the Alphaproteobacterial order Rhizobiales. *Proc. Natl. Acad. Sci. U.S.A.* 109, 1697–1701. doi: 10.1073/pnas.1114476109

Cambridge, J., Blinkova, A., Magnan, D., Bates, D., and Walker, J. R. (2014). A replication-inhibited unsegregated nucleoid at mid-cell blocks Z-ring formation and cell division independently of SOS and the SlmA nucleoid occlusion protein in *Escherichia coli*. *J. Bacteriol.* 196, 36–49. doi: 10.1128/JB.01230-12

Cheng, J., Sibley, C. D., Zaheer, R., and Finan, T. M. (2007). A *Sinorhizobium meliloti minE* mutant has an altered morphology and exhibits defects in legume symbiosis. *Microbiology* 153(Pt 2), 375–387. doi: 10.1099/mic.0.2006/001362-0

Cho, H., McManus, H. R., Dove, S. L., and Bernhardt, T. G. (2011). Nucleoid occlusion factor SlmA is a DNA-activated FtsZ polymerization antagonist. *Proc. Natl. Acad. Sci. U.S.A.* 108, 3773–3778. doi: 10.1073/pnas.1018674108

Cho, H., Wivagg, C. N., Kapoor, M., Barry, Z., Rohs, P. D., Suh, H., et al. (2016). Bacterial cell wall biogenesis is mediated by SEDS and PBP polymerase families functioning semi-autonomously. *Nat. Microbiol.* 1:16172. doi: 10.1038/nmicrobiol.2016.172

Curtis, P. D., and Brun, Y. V. (2014). Identification of essential alphaproteobacterial genes reveals operational variability in conserved developmental and cell cycle systems. *Mol. Microbiol.* 93, 713–735. doi: 10.1111/mmi.12686

Dajkovic, A., Lan, G., Sun, S. X., Wirtz, D., and Lutkenhaus, J. (2008). MinC spatially controls bacterial cytokinesis by antagonizing the scaffolding function of FtsZ. *Curr. Biol.* 18, 235–244. doi: 10.1016/j.cub.2008.01.042

de Boer, P. A., Crossley, R. E., Hand, A. R., and Rothfield, L. I. (1991). The MinD protein is a membrane ATPase required for the correct placement of the *Escherichia coli* division site. *EMBO J.* 10, 4371–4380.

de Boer, P. A., Crossley, R. E., and Rothfield, L. I. (1989). A division inhibitor and a topological specificity factor coded for by the minicell locus determine proper placement of the division septum in *E. coli*. *Cell* 56, 641–649. doi: 10.1016/0092-8674(89)90586-2

de Boer, P. A., Crossley, R. E., and Rothfield, L. I. (1992). Roles of MinC and MinD in the site-specific septation block mediated by the MinCDE system of *Escherichia coli*. *J. Bacteriol.* 174, 63–70. doi: 10.1128/jb.174.1.63-70.1992

Den Blaauwen, T., Buddelmeijer, N., Aarsman, M. E., Hameete, C. M., and Nanninga, N. (1999). Timing of FtsZ assembly in *Escherichia coli*. *J. Bacteriol.* 181, 5167–5175.

Di Ventura, B., Knecht, B., Andreas, H., Godinez, W. J., Fritsche, M., Rohr, K., et al. (2013). Chromosome segregation by the *Escherichia coli* Min system. *Mol. Syst. Biol.* 9:686. doi: 10.1038/msb.2013.44

Du, S., and Lutkenhaus, J. (2014). SlmA antagonism of FtsZ assembly employs a two-pronged mechanism like MinCD. *PLoS Genet.* 10:e1004460. doi: 10.1371/journal.pgen.1004460

Ducret, A., Quardokus, E. M., and Brun, Y. V. (2016). MicrobeJ, a tool for high throughput bacterial cell detection and quantitative analysis. *Nat. Microbiol.* 1:16077. doi: 10.1038/nmicrobiol.2016.77

Ebersbach, G., Briegel, A., Jensen, G. J., and Jacobs-Wagner, C. (2008). A self-associating protein critical for chromosome attachment, division, and polar organization in *Caulobacter*. *Cell* 134, 956–968. doi: 10.1016/j.cell.2008.07.016

Ehrle, H. M., Guidry, J. T., Iacovetto, R., Salisbury, A. K., Sandidge, D. J., and Bowman, G. R. (2017). Polar organizing protein PopZ Is required for chromosome segregation in *Agrobacterium tumefaciens*. *J. Bacteriol.* 199:e00111-17. doi: 10.1128/JB.00111-17

Eswaramoorthy, P., Erb, M. L., Gregory, J. A., Silverman, J., Pogliano, K., Pogliano, J., et al. (2011). Cellular architecture mediates DivIVA ultrastructure and regulates min activity in *Bacillus subtilis*. *MBio* 2:e00257-11. doi: 10.1128/mBio.00257-11

Figueroa-Cuilan, W., Daniel, J. J., Howell, M., Sulaiman, A., and Brown, P. J. (2016). Mini-Tn7 insertion in an artificial *att*Tn7 site enables depletion of the essential master regulator CtrA in the Phytopathogen *Agrobacterium tumefaciens*. *Appl. Environ. Microbiol.* 82, 5015–5025. doi: 10.1128/AEM.01392-16

Fu, X., Shih, Y. L., Zhang, Y., and Rothfield, L. I. (2001). The MinE ring required for proper placement of the division site is a mobile structure that changes its cellular location during the *Escherichia coli* division cycle. *Proc. Natl. Acad. Sci. U.S.A.* 98, 980–985. doi: 10.1073/pnas.98.3.980

Goley, E. D., Yeh, Y. C., Hong, S. H., Fero, M. J., Abeliuk, E., McAdams, H. H., et al. (2011). Assembly of the *Caulobacter* cell division machine. *Mol. Microbiol.* 80, 1680–1698. doi: 10.1111/j.1365-2958.2011.07677.x

Goodner, B., Hinkle, G., Gattung, S., Miller, N., Blanchard, M., Qurollo, B., et al. (2001). Genome sequence of the plant pathogen and biotechnology agent *Agrobacterium tumefaciens* C58. *Science* 294, 2323–2328. doi: 10.1126/science.1066803

Hale, C. A., Meinhardt, H., and de Boer, P. A. (2001). Dynamic localization cycle of the cell division regulator MinE in *Escherichia coli*. *EMBO J.* 20, 1563–1572. doi: 10.1093/emboj/20.7.1563

Howell, M., Aliashkevich, A., Salisbury, A. K., Cava, F., Bowman, G. R., and Brown, P. J. B. (2017a). Absence of the polar organizing protein PopZ results in reduced and asymmetric cell division in *Agrobacterium tumefaciens*. *J. Bacteriol.* 199:e00101-17. doi: 10.1128/JB.00101-17

Howell, M., Daniel, J. J., and Brown, P. J. B. (2017b). Live cell fluorescence microscopy to observe essential processes during microbial cell growth. *J. Vis. Exp.* e56497. doi: 10.3791/56497

Hu, Z., and Lutkenhaus, J. (1999). Topological regulation of cell division in *Escherichia coli* involves rapid pole to pole oscillation of the division inhibitor MinC under the control of MinD and MinE. *Mol. Microbiol.* 34, 82–90. doi: 10.1046/j.1365-2958.1999.01575.x

Hu, Z., and Lutkenhaus, J. (2000). Analysis of MinC reveals two independent domains involved in interaction with MinD and FtsZ. *J. Bacteriol.* 182, 3965–3971. doi: 10.1128/JB.182.14.3965-3971.2000

Hu, Z., and Lutkenhaus, J. (2001). Topological regulation of cell division in *E. coli*. Spatiotemporal oscillation of MinD requires stimulation of its ATPase by MinE and phospholipid. *Mol. Cell* 7, 1337–1343. doi: 10.1016/S1097-2765(01)00273-8

Hu, Z., and Lutkenhaus, J. (2003). A conserved sequence at the C-terminus of MinD is required for binding to the membrane and targeting MinC to the septum. *Mol. Microbiol.* 47, 345–355. doi: 10.1046/j.1365-2958.2003.03321.x

Jensen, R. B. (2006). Coordination between chromosome replication, segregation, and cell division in *Caulobacter crescentus*. *J. Bacteriol.* 188, 2244–2253. doi: 10.1128/JB.188.6.2244-2253.2006

Justice, S. S., Garcia-Lara, J., and Rothfield, L. I. (2000). Cell division inhibitors SulA and MinC/MinD block septum formation at different steps in the assembly of the *Escherichia coli* division machinery. *Mol. Microbiol.* 37, 410–423. doi: 10.1046/j.1365-2958.2000.02007.x

Kelly, A. J., Sackett, M. J., Din, N., Quardokus, E., and Brun, Y. V. (1998). Cell cycle-dependent transcriptional and proteolytic regulation of FtsZ in *Caulobacter*. *Genes Dev.* 12, 880–893. doi: 10.1101/gad.12.6.880

Kiekebusch, D., Michie, K. A., Essen, L. O., Lowe, J., and Thanbichler, M. (2012). Localized dimerization and nucleoid binding drive gradient formation by the bacterial cell division inhibitor MipZ. *Mol. Cell* 46, 245–259. doi: 10.1016/j.molcel.2012.03.004

Laub, M. T., Chen, S. L., Shapiro, L., and McAdams, H. H. (2002). Genes directly controlled by CtrA, a master regulator of the *Caulobacter* cell cycle. *Proc. Natl. Acad. Sci. U.S.A.* 99, 4632–4637. doi: 10.1073/pnas.062065699

Lenarcic, R., Halbedel, S., Visser, L., Shaw, M., Wu, L. J., Errington, J., et al. (2009). Localisation of DivIVA by targeting to negatively curved membranes. *EMBO J.* 28, 2272–2282. doi: 10.1038/emboj.2009.129

Li, Z., Trimble, M. J., Brun, Y. V., and Jensen, G. J. (2007). The structure of FtsZ filaments *in vivo* suggests a force-generating role in cell division. *EMBO J.* 26, 4694–4708. doi: 10.1038/sj.emboj.7601895

Livny, J., Yamaichi, Y., and Waldor, M. K. (2007). Distribution of centromere-like parS sites in bacteria: insights from comparative genomics. *J. Bacteriol.* 189, 8693–8703. doi: 10.1128/JB.01239-07

Loose, M., Fischer-Friedrich, E., Herold, C., Kruse, K., and Schwille, P. (2011). Min protein patterns emerge from rapid rebinding and membrane interaction of MinE. *Nat. Struct. Mol. Biol.* 18, 577–583. doi: 10.1038/nsmb.2037

Lutkenhaus, J. (2007). Assembly dynamics of the bacterial MinCDE system and spatial regulation of the Z ring. *Annu. Rev. Biochem.* 76, 539–562. doi: 10.1146/annurev.biochem.75.103004.142652

Lutkenhaus, J. (2012). The ParA/MinD family puts things in their place. *Trends Microbiol.* 20, 411–418. doi: 10.1016/j.tim.2012.05.002

Lutkenhaus, J., and Du, S. (2017). *E. coli* cell cycle machinery. *Subcell Biochem.* 84, 27–65. doi: 10.1007/978-3-319-53047-5_2

Ma, X., Ehrhardt, D. W., and Margolin, W. (1996). Colocalization of cell division proteins FtsZ and FtsA to cytoskeletal structures in living *Escherichia coli* cells by using green fluorescent protein. *Proc. Natl. Acad. Sci. U.S.A.* 93, 12998–13003. doi: 10.1073/pnas.93.23.12998

Meinhardt, H., and de Boer, P. A. (2001). Pattern formation in *Escherichia coli*: a model for the pole-to-pole oscillations of Min proteins and the localization of the division site. *Proc. Natl. Acad. Sci. U.S.A.* 98, 14202–14207. doi: 10.1073/pnas.251216598

Merritt, P. M., Danhorn, T., and Fuqua, C. (2007). Motility and chemotaxis in *Agrobacterium tumefaciens* surface attachment and biofilm formation. *J. Bacteriol.* 189, 8005–8014. doi: 10.1128/JB.00566-07

Möll, A., and Thanbichler, M. (2009). FtsN-like proteins are conserved components of the cell division machinery in proteobacteria. *Mol. Microbiol.* 72, 1037–1053. doi: 10.1111/j.1365-2958.2009.06706.x

Morton, E. R., and Fuqua, C. (2012a). Genetic manipulation of *Agrobacterium*. *Curr. Protoc. Microbiol.* Chapter 3:Unit 3D.2. doi: 10.1002/9780471729259.mc03d02s25

Morton, E. R., and Fuqua, C. (2012b). Laboratory maintenance of *Agrobacterium*. *Curr. Protoc. Microbiol.* Chapter 1:Unit3D.1. doi: 10.1002/9780471729259.mc03d01s24

Müller, P., Ewers, C., Bertsche, U., Anstett, M., Kallis, T., Breukink, E., et al. (2007). The essential cell division protein FtsN interacts with the murein (peptidoglycan) synthase PBP1B in *Escherichia coli*. *J. Biol. Chem.* 282, 36394–36402. doi: 10.1074/jbc.M706390200

Park, K. T., Villar, M. T., Artigues, A., and Lutkenhaus, J. (2017). MinE conformational dynamics regulate membrane binding, MinD interaction, and Min oscillation. *Proc. Natl. Acad. Sci. U.S.A.* 114, 7497–7504. doi: 10.1073/pnas.1707385114

Park, K. T., Wu, W., Battaile, K. P., Lovell, S., Holyoak, T., and Lutkenhaus, J. (2011). The Min oscillator uses MinD-dependent conformational changes in MinE to spatially regulate cytokinesis. *Cell* 146, 396–407. doi: 10.1016/j.cell.2011.06.042

Patrick, J. E., and Kearns, D. B. (2008). MinJ (YvjD) is a topological determinant of cell division in *Bacillus subtilis*. *Mol. Microbiol.* 70, 1166–1179. doi: 10.1111/j.1365-2958.2008.06469.x

Pini, F., De Nisco, N. J., Ferri, L., Penterman, J., Fioravanti, A., Brilli, M., et al. (2015). Cell cycle control by the master regulator CtrA in *Sinorhizobium meliloti*. *PLoS Genet.* 11:e1005232. doi: 10.1371/journal.pgen.1005232

Ptacin, J. L., Lee, S. F., Garner, E. C., Toro, E., Eckart, M., Comolli, L. R., et al. (2010). A spindle-like apparatus guides bacterial chromosome segregation. *Nat. Cell Biol.* 12, 791–798. doi: 10.1038/ncb2083

Radhakrishnan, S. K., Pritchard, S., and Viollier, P. H. (2010). Coupling prokaryotic cell fate and division control with a bifunctional and oscillating oxidoreductase homolog. *Dev. Cell* 18, 90–101. doi: 10.1016/j.devcel.2009.10.024

Ramamurthi, K. S., and Losick, R. (2009). Negative membrane curvature as a cue for subcellular localization of a bacterial protein. *Proc. Natl. Acad. Sci. U.S.A.* 106, 13541–13545. doi: 10.1073/pnas.0906851106

Raskin, D. M., and de Boer, P. A. (1999a). MinDE-dependent pole-to-pole oscillation of division inhibitor MinC in *Escherichia coli*. *J. Bacteriol.* 181, 6419–6424.

Raskin, D. M., and de Boer, P. A. (1999b). Rapid pole-to-pole oscillation of a protein required for directing division to the middle of *Escherichia coli*. *Proc. Natl. Acad. Sci. U.S.A.* 96, 4971–4976.

Rothfield, L., Taghbalout, A., and Shih, Y. L. (2005). Spatial control of bacterial division-site placement. *Nat. Rev. Microbiol.* 3, 959–968. doi: 10.1038/nrmicro1290

Rowlett, V. W., and Margolin, W. (2015). The Min system and other nucleoid-independent regulators of Z ring positioning. *Front. Microbiol.* 6:478. doi: 10.3389/fmicb.2015.00478

Schofield, W. B., Lim, H. C., and Jacobs-Wagner, C. (2010). Cell cycle coordination and regulation of bacterial chromosome segregation dynamics by polarly localized proteins. *EMBO J.* 29, 3068–3081. doi: 10.1038/emboj.2010.207

Shebelut, C. W., Jensen, R. B., and Gitai, Z. (2009). Growth conditions regulate the requirements for *Caulobacter* chromosome segregation. *J. Bacteriol.* 191, 1097–1100. doi: 10.1128/JB.00862-08

Simon, R., Priefer, U., and Puhler, A. (1983). A broad host range mobilization system for *in vivo* genetic engineering: transposon mutagenesis in gram negative bacteria. *Nat. Biotechnol.* 1, 784–791. doi: 10.1038/nbt1183-784

Sun, Q., and Margolin, W. (1998). FtsZ dynamics during the division cycle of live *Escherichia coli* cells. *J. Bacteriol.* 180, 2050–2056.

Surovtsev, I. V., and Jacobs-Wagner, C. (2018). Subcellular organization: a critical feature of bacterial cell replication. *Cell* 172, 1271–1293. doi: 10.1016/j.cell.2018.01.014

Szwedziak, P., Wang, Q., Bharat, T. A., Tsim, M., and Lowe, J. (2014). Architecture of the ring formed by the tubulin homologue FtsZ in bacterial cell division. *Elife* 3:e04601. doi: 10.7554/eLife.04601

Thanbichler, M., and Shapiro, L. (2006). MipZ, a spatial regulator coordinating chromosome segregation with cell division in *Caulobacter*. *Cell* 126, 147–162. doi: 10.1016/j.cell.2006.05.038

Tonthat, N. K., Arold, S. T., Pickering, B. F., Van Dyke, M. W., Liang, S., Lu, Y., et al. (2011). Molecular mechanism by which the nucleoid occlusion factor, SlmA, keeps cytokinesis in check. *EMBO J.* 30, 154–164. doi: 10.1038/emboj.2010.288

Wang, S. C., West, L., and Shapiro, L. (2006). The bifunctional FtsK protein mediates chromosome partitioning and cell division in *Caulobacter*. *J. Bacteriol.* 188, 1497–1508. doi: 10.1128/JB.188.4.1497-1508.2006

Watson, B., Currier, T. C., Gordon, M. P., Chilton, M. D., and Nester, E. W. (1975). Plasmid required for virulence of *Agrobacterium tumefaciens*. *J. Bacteriol.* 123, 255–264.

Williams, B., Bhat, N., Chien, P., and Shapiro, L. (2014). ClpXP and ClpAP proteolytic activity on divisome substrates is differentially regulated following the *Caulobacter* asymmetric cell division. *Mol. Microbiol.* 93, 853–866. doi: 10.1111/mmi.12698

Williams, M., Hoffman, M. D., Daniel, J. J., Madren, S. M., Dhroso, A., Korkin, D., et al. (2016). Short-stalked *Prosthecomicrobium hirschii* cells have a *Caulobacter*-like cell cycle. *J. Bacteriol.* 198, 1149–1159. doi: 10.1128/JB.00896-15

Wood, D. W., Setubal, J. C., Kaul, R., Monks, D. E., Kitajima, J. P., Okura, V. K., et al. (2001). The genome of the natural genetic engineer *Agrobacterium tumefaciens* C58. *Science* 294, 2317–2323. doi: 10.1126/science.1066804

Wu, L. J., and Errington, J. (2004). Coordination of cell division and chromosome segregation by a nucleoid occlusion protein in *Bacillus subtilis*. *Cell* 117, 915–925. doi: 10.1016/j.cell.2004.06.002

Wu, L. J., and Errington, J. (2011). Nucleoid occlusion and bacterial cell division. *Nat. Rev. Microbiol.* 10, 8–12. doi: 10.1038/nrmicro2671

Yang, X., Lyu, Z., Miguel, A., McQuillen, R., Huang, K. C., and Xiao, J. (2017). GTPase activity-coupled treadmilling of the bacterial tubulin FtsZ organizes septal cell wall synthesis. *Science* 355, 744–747. doi: 10.1126/science.aak9995

3

Segregation of chromosome arms in growing and non-growing *Escherichia coli* cells

Conrad L. Woldringh [1]*, Flemming G. Hansen [2], Norbert O. E. Vischer [1] and Tove Atlung [3]

[1] Bacterial Cell Biology, Faculty of Science, Swammerdam Institute for Life Sciences, University of Amsterdam, Amsterdam, Netherlands, [2] Department of Systems Biology, Technical University of Denmark, Lyngby, Denmark, [3] Department of Science, Systems and Models, Roskilde University, Roskilde, Denmark

***Correspondence:**
Conrad L. Woldringh,
Swammerdam Institute for Life Sciences, University of Amsterdam,
Science Park 904, 1098 XH
Amsterdam, Netherlands
c.woldringh@gmail.com

In slow-growing *Escherichia coli* cells the chromosome is organized with its left (L) and right (R) arms lying separated in opposite halves of the nucleoid and with the origin (O) in-between, giving the pattern L-O-R. During replication one of the arms has to pass the other to obtain the same organization in the daughter cells: L-O-R L-O-R. To determine the movement of arms during segregation six strains were constructed carrying three colored loci: the left and right arms were labeled with red and cyan fluorescent-proteins, respectively, on loci symmetrically positioned at different distances from the central origin, which was labeled with green-fluorescent protein. In non-replicating cells with the predominant spot pattern L-O-R, initiation of replication first resulted in a L-O-O-R pattern, soon changing to O-L-R-O. After replication of the arms the predominant spot patterns were, L-O-R L-O-R, O-R-L R-O-L or O-L-R L-O-R indicating that one or both arms passed an origin and the other arm. To study the driving force for these movements cell growth was inhibited with rifampicin allowing run-off DNA synthesis. Similar spot patterns were obtained in growing and non-growing cells, indicating that the movement of arms is not a growth-sustained process, but may result from DNA synthesis itself. The distances between loci on different arms (LR-distances) and between duplicated loci (LL- or RR-distances) as a function of their distance from the origin, indicate that in slow-growing cells DNA is organized according to the so-called sausage model and not according to the doughnut model.

Keywords: *Escherichia coli*, nucleoid, DNA segregation, chromosome arms (replichores), ordering pattern, rifampicin-treatment, run-off DNA synthesis

Introduction

The chromosome of *Escherichia coli* can be readily visualized in living cells by phase contrast microscopy as a separate and dynamic structure (Mason and Powelson, 1956; Yamaichi and Niki, 2004). We now know that within this nucleoid structure the DNA is confined as a single branched, plectonemic supercoil formed and maintained by topoisomerases (Zechiedrich et al., 2000) and by nucleoid associated proteins (NAPs; Luijsterburg et al., 2006). The primary cause for the phase separation between nucleoid and cytoplasm is the physical phenomenon of excluded-volume interactions between DNA and soluble proteins (Odijk, 1998). Studies of the physical structure of isolated bacterial DNA (Cunha et al., 2001; Pelletier et al., 2012) have indicated that the

DNA segments behave as entropic springs showing diffusive motion within a visco-elastic network (Cunha et al., 2005).

Against this physical background we must consider the process of bacterial segregation and the dynamics of replicated DNA strands collapsed or confined into the nucleoid of living cells. Do these strands become entangled or mixed as is to be expected for such polymer chains? It is now well accepted that DNA daughter strands segregate as they are replicated and that in slow growing *E. coli* cells the two chromosome arms move to different halves of the nucleoid with the origin in-between (Nielsen et al., 2006a; Wang et al., 2006). This organization suggests that the replicated daughter strands do not entangle or mix, but stay separated as Left and Right arms of the chromosome during the entire replication-segregation process (review Jun and Wright, 2010).

Can such an organization and movement be explained without the help of an underlying biological structure as suggested by several authors (Wiggins et al., 2010; Le Chat and Espéli, 2012; Yazdi et al., 2012), a hypothetical structure that, in its turn, has to become organized? This question has recently been considered by Youngren et al. (2014) for the even more complicated situation in fast growing *E. coli* cells undergoing multifork replication. These authors propose that in the wider cells replicating chromosomes are thermodynamically driven into ring polymers in which replicated strands segregate spontaneously by entropic demixing without the help of any additional, biological mechanism.

Nevertheless, it has been proposed that segregation proceeds in growing cells with the help of transcription/transertion processes (Woldringh, 2002). In addition, it was suggested that the demixing process may be sustained by regulatory interactions between transcription factors and target genes that help to self-organize the chromosome into topological domains that do not mix (Fritsche et al., 2011). According to Fisher et al. (2013), the non-intermingling of sister strands occurs in pulses of nucleoid elongation at defined times in the cycle of slowly grown cells while cell elongation continues monotonically. The periodical movements are proposed to depend on the accumulation and release of tethers between sister strands, processes that can be expected to depend on cell growth.

In view of these growth-sustained, active processes proposed to organize and segregate the replicating chromosome we analyzed the positions of fluorescently tagged loci in either growing *E. coli* cells or in rifampicin-treated cells that do not grow but only carry out run-off DNA synthesis. The measurements indicate that separation and migration of loci is very similar in growing and non-growing cells suggesting that the process of segregation continues in the absence of RNA synthesis without cellular or nucleoid elongation. The measurements are discussed in the light of two segregation models proposed in the literature, the so-called doughtnut and sausage model.

Materials and Methods

Strains and Growth Conditions

Strains of *E. coli* MG1655 Δ*lacIZYA* (Nielsen et al., 2006b) carrying three different ParB/*parS* systems were constructed by recombineering. The strains with the three different *parS* sequences were transformed with plasmid pFH4034 which carry three different *parB* gene fusions to three different genes for fluorescent proteins with different colors (green (G), cyan (C), and red (R); see Supplementary Data Figure S1) to give the strains depicted in **Figure 1A**. See for strain construction Supplementary Material and Table S1.

The strains were grown in minimal glycerol medium supplemented with lysine (20 μg/ml) and ampicillin (100 μg/ml). Strain FH4035 has a lysine requirement while the strains FH404056-4060 are *lysA*$^+$. We add 10 μg/ml uracil due to a possible mutation in the *rph* gene which might affect the expression of the *pyrE* gene. Cell growth was monitored with a spectrophotometer at 450 nm. Cells were grown undisturbed at 32°C (doubling time about 150 min) for about 16 h while the OD was kept below 0.5 by periodical dilutions.

Strain FH4035 was grown in a slightly different growth medium at 28°C, with a doubling time of ~180 min. For unknown reasons these cells were shorter than those of the other constructs while showing the same cell diameter (0.7 μm). The smaller cell length leading to smaller segregation distances is evident from the Tables, but in our view does not influence our

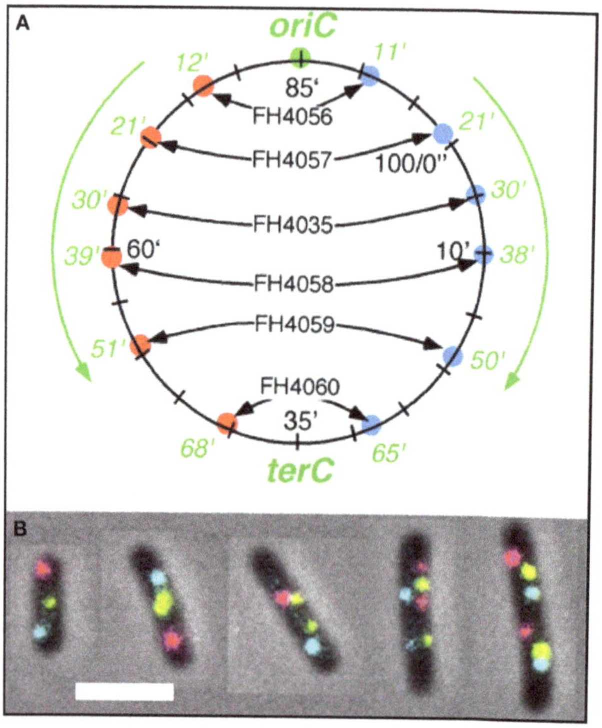

FIGURE 1 | Fluorescent loci and spots. (A) Schematic representation of the chromosome of the 6 strains showing the positions of the 3 probes. Green fluorescent protein labels a locus ~1 kb to the right of the origin (O). The chromosome arms are labeled, respectively, with red (L, left arm) and cyan (R, right arm) fluorescent protein on opposite sites and at different distances from *oriC*. Green numbers indicate the theoretical time when the loci are replicated during progression of the replication forks, based on their distance in kb from *oriC* and on the replication velocity, assuming the replication time to be 75 min (Nielsen et al., 2006a). **(B)** Composite image of the 4 channels (phase contrast, red, green, and cyan filters) of cells of *E. coli* FH4035 showing the most abundant pattern of 3-spot cells (LOR) and 6-spot cells (LOR LOR), as well as some replicative intermediates. Magnification bar 2 μm.

interpretation of patterns and movement of the chromosome arms.

Microscopy

Cell samples (1 ml) were concentrated about 20x by centrifugation, prepared on an agar slab or a poly-lysine coated microscope slide and photographed within 30 min after sampling.

For the constructs FH4056, 4057, 4058, 4059, and 4060 pictures were taken on a Zeiss Axioplan microscope equipped with a Kappa DX 2 camera using an ImageBase capture program. For construct FH4035 an Olympus BX microscope was used in combination with programs ImageJ and MicroManager. Both microscopes had 100 × 1.3 oil immersion lens (PH3) giving a magnification of 15 pixels per μm. Four pictures of each field were obtained using the respective capture programs. Both microscopes had filters for detecting red, cyan, and green fluorescence.

Measurement of Cells and Spots

Image analysis was performed with Coli-Inspector, which is a set of scripts packed in a "project file" running under the ImageJ plugin ObjectJ. After spatial alignment of the 4 channels, cell length and diameter were automatically measured. The positions of the three kinds of spots were automatically measured using different thresholds. Cells in which one kind of spot was lacking were manually discarded. In some cells missing spots could be manually added after better adjusting brightness and contrast of the image. After checking all cells by eye, the spot positions along the long and short axis of the cell, their pattern, the frequency of each pattern and the (half-) distance between duplicated spots were calculated by Coli-Inspector (see: https://sils.fnwi.uva.nl/bcb/oblectj/examples/). By making use of the so-called Qualifier-possibility the distribution of, for instance, cells that show a specific spot pattern can be obtained (cf. **Table 3**).

Results

Analysis of Three- and Four-Spot Patterns in Growing Cells

The segregation of chromosome arms in slowly growing *E. coli* cells (doubling time at 28°C or 32°C Td = 180/150 min), was studied by measuring the cellular positions of three fluorescently labeled loci. The chromosomes contain one locus of green fluorescent-protein near *oriC* (O) and two loci on opposite positions on the chromosome arms at different distances from *oriC* in six constructs. Red fluorescent-protein tags the left arm (L) and cyan fluorescent-protein the right arm (R). See **Figure 1A** for a schematic representation of the chromosome in the different strains. **Figure 1B** shows composite images of the 4 channels for typical cells of *E. coli* FH4035.

In non-replicating cells with only 3 spots the most prominent pattern is LOR (46–78% of cells containing 3 spots; **Table 1A**, column I). This reflects the occurrence of unmixed chromosome arms in the two halves of the nucleoid as originally described by Nielsen et al. (2006b) and Wang et al. (2006). However, in many cells the origin is lying outside the two arms, giving the patterns OLR or ORL (column II in **Table 1A**). This pattern is more prominent in the constructs with loci close to the terminus (strains FH4059 and FH4060), where it increases to 50 per cent of the cells. It does not necessarily mean that the origin lies at the end of the nucleoid as depicted in the cells of **Table 1A**. In these constructs (FH4059 and FH4060), origin proximal loci may well have passed the origin. On average the origin in the 3 spot cells showing the LOR configuration is located in the middle of the cell with a standard deviation of 0.08–0.16 (Figure S2A) in agreement with a previous study (Nielsen et al., 2006a).

We measured the distance between the L and R loci in the 3 spot cells (**Figure 2**, Table S2). The distances were small for the origin proximal loci increasing with increasing distance from the origin, except for the most terminus proximal pair; these two loci are closer together, but not as close as for the origin proximal one's and with larger cell to cell variation (Figure S2). The trends were the same for LOR cells and for OLR/ORL cells, but L-R distances were shorter for the latter cells in all six constructs. Also in 4-spot cells the L-R distances were found to increase for the 4 first constructs in all ordering patterns (see Table S3).

Initiation of DNA replication occurs after a significant increase in cell length (compare lengths indicated in columns I and II with those in columns IIIA and B in **Table 1A**) and results in cells with 4 spots. In general, the duplicated O-spots are found adjacent to each other, either in between the L- and R-arm spots (column III-A) or in rare cases on the outside (column III-B in **Table 1A**). The relatively low percentage of this latter pattern suggests that the origins soon move apart and pass either one or both of the other labeled loci giving the patterns as depicted in column IV and V in **Table 1A**. The relatively high percentage of cells in which the duplicated O-spots occur on the outside of spots on both chromosome arms in some of the constructs (column V in **Table 1A**) shows that some unreplicated loci stay relatively close to the mid of the long axis of the cell. It should be kept in mind that for the constructs with origin proximal loci 4 spot cells are a fairly homogeneous cohort, whereas the 4 spot cells of constructs with origin distal loci are a mixture of cells with different amounts of replicated DNA.

Analysis of the O-O distances (Table S3) as a function of LR locus position shows that soon after initiation the origins are lying close together (strains FH4056 and FH4057), and when later replicated loci are duplicated the origins have moved further apart (strains FH4035 and FH4058). Comparison of the L-R distances in the 3-spot cells (**Figure 2**) with the distances in the 4-spot cells (Table S3) suggests that upon initiation the un-replicated LR loci first move apart (LOOR pattern), but move inward again when the origin passes one (OLOR/OROL) or two (OLRO) arms, causing smaller distances. Table S3 further indicates that while the distance between origins (O-O) increases gradually in the origin-distal constructs, the L-R distances show, after an initial increase, a decrease in the terminal-proximal constructs (FH4059 and FH4060). This could be ascribed to the association of the terminus region with the divisome (Espéli et al., 2012).

Cells with a 5-spot pattern are obtained when a locus on only one of the chromosome arms has duplicated. The relatively low percentage of these cells (last column in **Table 1A**) indicates that the oppositely positioned loci on the two arms are replicated and segregated more or less synchronously, directly

TABLE 1A | Percentages of most abundant patterns in non-replicating cells with 3 spots (column I and II) and in replicating cells with 4 spots that have a replicated origin (O; column III–V).

Strains	Total number cells with spots (mean cell length, μm)	Number cells with 3 spots (% of total population)	Non-replicating cells. % of 3-spot cells (mean cell length, μm)		nr. cells with 4 spots (% of total population)	Replicating cells with only the origin (O) duplicated. % of 4-spot cells (mean cell length, μm)				nr cells with 5 spots (% of total population)
			I LOR 	II OLR/ORL 		III-A LOOR 	III-B OOLR/OORL 	IV OLOR/OLOR 	V OLRO 	
4056[a] 	1376 (3.27)	527 (33)	62 (2.61)	38 (2.56)	63 (5)	35 (3.27)	6 (2.65)	40 (3.04)	19 (2.89)	80 (6)
4057[a] 	1099 (3.30)	487 (44)	74 (2.75)	26 (2.63)	138 (13)	32 (3.21)	9 (3.11)	42 (3.21)	15 (3.17)	56 (5)
4035[b] 	1003 (2.55)	384 (38)	77 (2.07)	23 (2.57)	129 (13)	36 (2.27)	1 (2.31)	36 (2.33)	26 (2.36)	48 (5)
4058[a] 	1166 (3.04)	617 (53)	78 (2.61)	22 (2.54)	191 (16)	44 (3.03)	8 (2.88)	35 (3.14)	13 (3.25)	33 (3)
4059[a] 	705 (3.09)	448 (64)	46 (2.75)	54 (2.80)	56 (8)	43 (3.16)	7 (2.75)	32 (3.48)	20 (3.38)	41 (6)
4060[a] 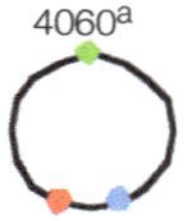	989 (2.98)	306 (31)	50 (2.49)	50 (2.37)	412 (42)	14 (2.92)	11 (2.83)	43 (2.88)	32 (3.17)	85 (9)

The average cell length (μm) in each subpopulation is given within brackets.
[a]*Cells grown at 32°C; doubling time Td = ~150 min.*
[b]*Cells grown at 28°C (Td = 180 min) showing smaller cell lengths (see Materials and Methods).*

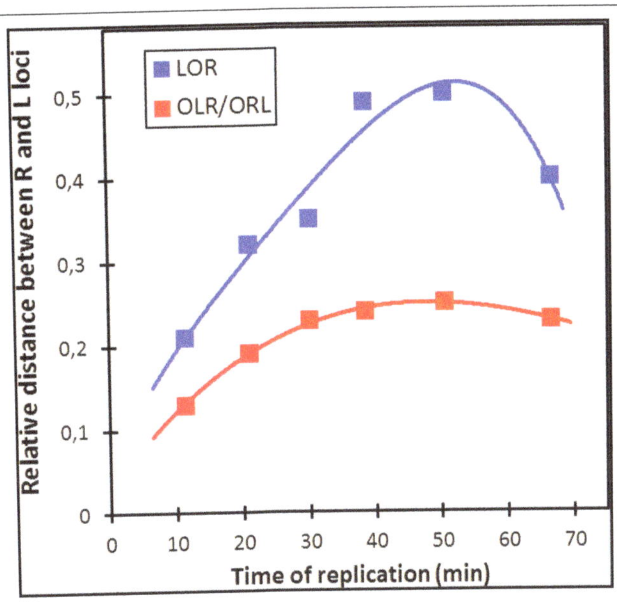

FIGURE 2 | Relative distances between L and R loci for the two different configurations of 3-spot cells. See Table S2 for SD values. For each strain the distances between the L and R spots in the two different configuration types were measured and normalized to the average cell length. The distances are plotted as a function of the time of replication of the loci (see legend to **Figure 1A**).

giving rise to the cells with 6 spots, discussed in the next section.

Analysis of Six-Spot Patterns in Growing Cells

Table 1B presents the cells which have replicated and segregated the three loci thus containing 6 spots. The spot patterns appeared to be very variable. With the 3 spots doubling to 6 spots, a range of 40–70 different patterns were obtained for the six constructs. Except for strain FH4060 the most predominant patterns are those shown in columns VI and VIII of **Table 1B**. The pattern in column VIII will upon division give rise to LOR cells shown in column I of **Table 1A** and the patterns shown in column VI will give rise to the 3-spot patterns shown in column I (LOR) and II (OLR/ORL) of **Table 1A**.

Cells which still have the origins lying outside (ORL LRO) occur in relatively low percentages (column VII in **Table 1B**) and are in all cases the smallest of the 6 spot cells. This suggests that origins which are located at the quarter position relatively early in the cell cycle soon are passed by one or both of the newly replicated loci. Only in construct FH4060, where the loci lie close to the terminus, many of the duplicated L- and R- spots did not pass the origins probably reflecting that these loci, which have the terminus located in between them, remain close for some time after replication. This may reflect a slow movement of loci in the terminal region of the nucleoid up to the division of the cell.

Patterns where the loci on either the Left or Right arm remain adjacent are less frequent (columns IX and X in **Table 1B**). Patterns where loci on both the Left and Right arm remain adjacent (e.g. OLL RRO) are rarely observed (<1% of the cells with 6 spots), indicating a fast separation of duplicated spots along the length axis of the cell, thereby passing loci on the other chromosome arm. However, patterns with adjacent LL or RR spots do occur with an average of 4% in subpopulations of long cells (>3.5 μm). This could indicate that such cells may have experienced an unusual difficulty in replication (and thus segregation) or an extended cohesion time (see Wang et al., 2008), that postponed cell division.

While a minority of the cells show either ajacent loci on the Left or Right arm (columns IX and X in **Table 1B**), cells with all three duplicated loci remaining next to each other (LLOORR) have not been observed. We conclude that both the origin and the loci at different positions on the chromosome arms separate soon after their replication. The observed patterns are well in agreement with movements of chromosome arms as previously described by Nielsen et al. (2006a,b) and Wang et al. (2006).

Segregation Patterns during Run-off DNA Replication in Non-Growing Cells

When cell growth is stopped by inhibiting protein synthesis (Maaløe and Hanawalt, 1961) or RNA synthesis (Lark, 1972), ongoing rounds of DNA replication terminate as is seen by both flow cytometry (Michelsen et al., 2003; Nielsen et al., 2006a), and image cytometry (Huls et al., 1999) of similar samples. To investigate whether segregation of chromosome arms still takes place under such conditions, spot patterns were determined in cells of *E. coli* strain FH4035 treated with 300 μg/ml rifampicin for 210 min at 28°C (cf. Skarstad et al., 1986). If replicated spots would not segregate in rifampicin treated cells we would expect to observe a high percentage of cells with adjacent spots (LLOORR).

For comparison of nucleoid sizes and shapes, cells were also treated with chloramphenicol under the same conditions. **Figure 3** shows the various cell samples fixed with osmium tetroxide and stained with DAPI for visualizing and measuring nucleoids. In contrast to the control cells showing a skewed distribution of nucleoid lengths (average 1.21 μm), the rifampicin-treated cells showed a bimodal distribution with peaks at 0.77 and 1.5 μm (results not shown). It seems probable that these peaks represent 1 and 2 chromosome equivalents.

We performed three independent growth experiments in which cells of strain FH4035 were treated with 300 μg/ml rifampicin for 210 min. These populations were analyzed for spot patterns and the combined results are shown in **Table 2**, together with the data of strain FH4035 given in **Tables 1A,B**, as a reference.

Rifampicin treatment caused a decrease of average cell length from 2.55 to 2.37 μm. This can be attributed to residual divisions, also indicated by the decrease in the percentage of cells with divided nucleoids (from 15 to 9%) and of constricted cells from 14% in the control to 3% in the rifampicin-treated samples (results not shown). Also because of residual division, the percentage of cells with 3 spots increased (3rd column in **Table 2**). In these 3-spot cells the distributions of individual spots in the rifampicin-treated cells are similar to those in the control (results not shown). The nucleoid organization in the 3-spot cells is also very similar to that in the untreated cells, the L-O-R configuration being found in 74% of the rifampicin cells (data not shown) vs. 77% for the control cells (**Table 1A**, column I).

TABLE 1B | Percentages of most abundant patterns of 6 spots in cells with replicating or replicated chromosomes of the different constructs (see Figure 1A).

Strains	Number cells with 6 spots (mean cell length, μm)	% of total population	Percentage of 6-spot patterns					
			VI ORL ROL/OLR LOR	VII OLR LRO	VIII LOR LOR	IX LOR RLO/ROL LRO	X ROL LOR/LOR ROL	Other combi-nations (% of total population)
			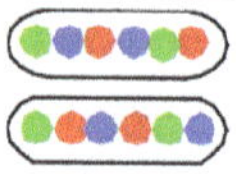	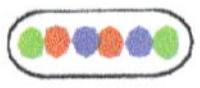		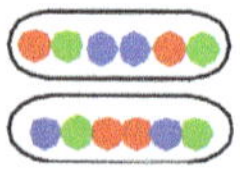	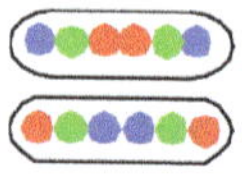	
4056[a] 	603 (3.80)	44	20 (3.85)	5 (3.64)	22 (3.92)	10 (3.67)	9 (3.96)	34 (2)
4057[a] 	318 (4.04)	29	27 (4.11)	4 (3.92)	28 (4.19)	8 (3.99)	8 (4.04)	25 (2)
4035[b] 	412 (3.02)	41	23 (2.89)	7 (2.79)	22 (3.23)	17 (3.02)	15 (3.10)	16 (2)
4058[a] 	243 (3.89)	21	28 (3.82)	7 (3.60)	24 (4.07)	14 (3.88)	10 (4.31)	17 (2)
4059[a] 	120 (3.97)	17	15 (4.05)	5 (3.75)	15 (3.82)	12 (4.16)	7 (4.14)	46 (7)
4060[a] 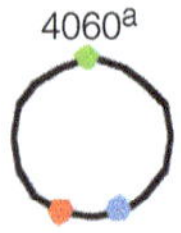	132 (3.85)	13	32 (4.05)	21 (3.66)	11 (4.24)	12 (4.01)	0	24 (2)

The average cell length (μm) in each subpopulation is given within brackets.
[a]Cells grown at 32°C; doubling time Td = ~150 min.
[b]Cells grown at 28°C (Td = 180 min) showing smaller cell lengths (see Materials and Methods).

Due to continued replication, the percentage of cells with 4 spots (only the origin replicated) and 5 spots decreased dramatically causing an increase in the percentage of 6-spot cells (6th column in **Table 2**).

To better compare the patterns in cells with replicated arms (6 spots), **Table 2** shows the same numbering of columns VI–X as used in **Table 1B**. The percentages of cells in **Table 2** showing either one (column VI or IX) or two origin spots (column VII) lying outside of the Left- and Right-arm loci have decreased in comparison to the control cells, whereas the percentage of cells showing the LOR LOR pattern (column VIII) or other patterns with the origin spot between the Left and Right arm (column X) have increased. The percentages in columns IX and X reflect the same trend as in the control. Apparently, the patterns in columns VIII and X can be considered to be a final stage of segregation, while the patterns in column VI, VII, and IX reflect a class of cells that is still in the process of segregation.

In view of (i) the significantly decreased percentage of cells with 4 or 5 spots, (ii) the absence of cells with adjacent LL, OO, RR spots, and (iii) the decrease in percentage of cells showing the origin lying outside of the other loci (columns VI and VII in **Table 2**), we conclude that chromosome movements in non-growing cells occur in a similar way as in growing cells, placing the two arms in different halves of the nucleoid with the origin in between.

Analysis of Segregation Distances between Spot Pairs (LL, RR) and Loci on Different Arms (LR)

The schematic cell images included in **Tables 1, 2** suggest a seemingly ordered, lengthwise distribution of duplicated loci.

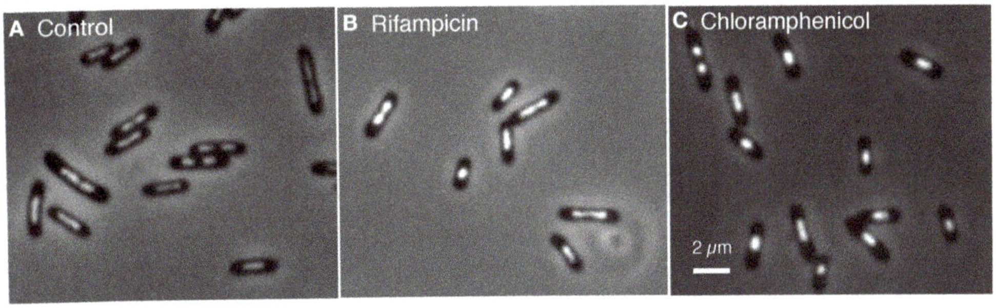

FIGURE 3 | Cells of *E. coli* FH4035, grown in glycerol minimal medium at 28°C, were treated during 100 min with the indicated inhibitors, fixed with 0.1% OsO4 and stained with 10 μg/ml DAPI. (A) In the control cells divided and well separated nucleoids occur in 15% of the cells. **(B)** In rifampicin-treated cells only 9% of the cells show divided nucleoids. **(C)** In chloramphenicol-treated cells the nucleoids (only 3% divided) have a more spherical shape.

TABLE 2 | Percentages of most abundant patterns in cells of strain FH4035, during growth (cf. Tables 1A,B) and after inhibition with rifampicin (300 μg/ml for 210 mi.

Cells	Total nr. cells with spots	% cells with 3 spots	% cells with 4 spots	% cells with 5 spots	% cells with 6 spots	Percentage of patterns in cells with 6 spots					
						VI ROL RLO/LOR LRO	VII ORL RLO	VIII LOR LOR	IX ORL LOR/OLR ROL	X ROL LOR/LOR ROL	Other combinations
						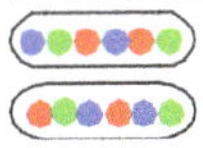	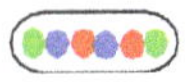		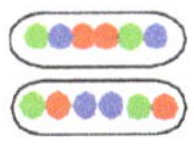	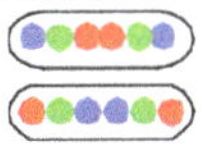	
Control[a]	1003 (2.55)	38	13	5	41	23 (2.89)	7 (2.79)	22 (3.23)	17 (3.02)	15 (3.10)	16
Rifampicin[b]	2065 (2.37)	46	1	1	49	13 (2.69)	2 (2.34)	25 (2.86)	12 (2.72)	20 (2.82)	28

The average cell length (μm) for each subpopulation s given within brackets.
[a] *Strain FH4035; compare with data in* ***Tables 1A,B***.
[b] *Cells of strain FH4035 treated with rifampicin; data from three independent growth experiments have been averaged.*

However, in many cells the two groups of duplicated L-, O-, and R-spots are irregularly positioned along the length axis (see **Figure 1B**). To get a better insight in the positioning and movement of the loci, we determined the distances over which duplicated spots, i.e., spot pairs replicated from the same chromosome arm (L-L and R-R distances), have migrated to opposite cell halves. Depending on the spot pattern and thus the Left/Right arm configuration, the distances between spot pairs will differ. But how will the segregation distance between spot pairs differ in dependence of the distance of the loci from the origin? **Table 3** shows that for the LRLR pattern these distances remain more or less constant in the first 5 constructs. For the RLLR pattern the R-R distances tend to increase, while the L-L distances decrease with their distance from the origin. For the LRRL pattern the opposite behavior is observed (see also Figure S3). For all arm configurations the distances in the last construct (FH4060) are smaller. A possible explanation is that these terminus-proximal loci remain tethered together by the terminus region (see Espéli et al., 2012).

As to be expected, in cells with the RLLR or LRRL configuration, the mean distance between spot pairs on the outer arms is significantly larger than on the inner arms (see **Table 3**). For the LRLR ordering the mean segregation distance between the origins (O-O distance) is slightly larger (1.57 μm) than between the loci on either the Right arm (R-R distance) or the Left arm (L-L distance), which are similar for the first 5 constructs (**Table 3**).

For the rifampicin-treated cells (last row in **Table 3**) it can be seen that all segregation distances are about 0.3 μm smaller than in the growing cells, and cell length is also 0.3 μm smaller. The percentage of cells showing the LRLR ordering pattern has decreased from 57% in the control cells of FH4035, to an average of 52% in the FH4035 cultures treated with rifampicin (**Table 3**). This could indicate a more random segregation with respect to whether either the leading or the lagging strand moves faster (Wang et al., 2005) in growth-inhibited cells.

We conclude that in the majority of cells showing the LRLR configuration, spot pairs segregate over equal distances independent of their distance from the origin (Figure S3). This is in accordance to the movements of chromosome arms as previously described by Nielsen et al. (2006a,b) and Wang et al. (2006). Also in rifampicin-treated cells with the LRLR configuration different spot pairs segregate during run-off DNA synthesis over equal but slightly shorter distances. In contrast, the L-R distances show an increase as a function of the time of replication, except for the latest replicated loci (68 min; **Figure 2**

TABLE 3 | Average distances between spot pairs (LL, OO and RR) in 6-spot cells for the three ordering patterns of Left (L) and Right (R) chromosome arms in the different constructs[a].

Strains	Mean length 6-spot cells (μm)	Number 6-spot cells	LRLR				RLLR				LRRL			
			%	Mean distance spot pairs (μm)			%	Mean distance spot pairs (μm)			%	Mean distance spot pairs (μm)		
				L-L	O-O	R-R		L-L	O-O	R-R		L-L	O-O	R-R
4056[b] 	3.80 ± 0.78	625	65	1.39 ± 0.54	1.43 ± 0.55	1.33 ± 0.53	14	1.12 ± 0.52	1.43 ± 0.57	1.67 ± 0.60	21	1.83 ± 0.6	1.54 ± 0.64	1.18 ± 0.51
4057[b] 	4.04 ± 0.87	326	72	1.49 ± 0.51	1.63 ± 0.54	1.49 ± 0.50	17	1.11 ± 0.44	1.76 ± 0.65	1.91 ± 0.61	12	2.10 ± 0.7	1.63 ± 0.68	1.06 ± 0.51
4035[c] 	3.02 ± 0.49	420	57	1.28 ± 0.44	1.49 ± 0.43	1.29 ± 0.47	17	0.96 ± 0.36	1.57 ± 0.46	1.73 ± 0.53	26	1.68 ± 0.5	1.50 ± 0.42	0.91 ± 0.34
4058[b] 	3.89 ± 0.71	243	68	1.51 ± 0.57	1.77 ± 0.57	1.46 ± 0.55	12	0.85 ± 0.44	1.86 ± 0.65	2.21 ± 0.79	20	2.26 ± 0.5	1.91 ± 0.53	0.98 ± 0.34
4059[b] 	3.89 ± 0.75	123	65	1.47 ± 0.50	1.53 ± 0.47	1.44 ± 0.50	24	0.81 ± 0.25	1.55 ± 0.55	1.97 ± 0.54	11	1.92 ± 0.5	1.69 ± 0.52	0.68 ± 0.32
4060[b] 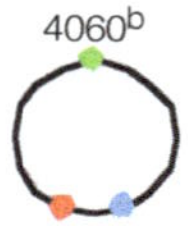	3.85 ± 0.61	131	70	1.01 ± 0.46	1.79 ± 0.51	1.10 ± 0.40	16	0.66 ± 0.29	1.68 ± 0.57	1.41 ± 0.60	15	1.37 ± 0.5	1.83 ± 0.49	0.67 ± 0.39
4035 +rifampicin[d] 	2.73 ± 0.48	1036	52	1.07 ± 0.36	1.12 ± 0.38	1.07 ± 0.35	30	0.64 ± 0.28	1.09 ± 0.37	1.45 ± 0.42	18	1.51 ± 0.4	1.08 ± 0.41	0.72 ± 0.27

[a] *Spot patterns were qualified using the ObjectJ qualification mode, (cf.* ***Table 1B****).*
[b] *Cells grown at 32°C; doubling time Td = ~150 min.*
[c] *Cells grown at 28°C (Td = 180 min) showing smaller cell lengths (see Materials and Methods).*
[d] *The data of three populations treated with rifampicin have been averaged (cf.* ***Table 2****).*

and Tables S2, S3). In the Discussion below we will consider these results in the light of two proposed segregation models, the doughnut- and the sausage-model.

Discussion

Segregation in Non-Growing Cells

The main conclusion from the present study is that segregation of chromosome arms continues in non-growing cells during run-off DNA replication. This result falsifies a previous proposal that the process of transertion drives DNA segregation (Woldringh, 2002) and supports the hypothesis that segregation is merely driven by the process of *de novo* DNA synthesis and accumulation. Ideas of a replication-driven segregation have been proposed previously by Grossman (Lemon and Grossman, 2001), Hansen (Nielsen et al., 2006a,b), Sherratt (Wang et al., 2006), Austin (Nielsen et al., 2007; Youngren et al., 2014), and Wiggins (Wiggins et al., 2010). A direct link between DNA replication and chromosome organization has been demonstrated and emphasized by the group of Sherratt (Liu et al., 2010), who also presented evidence that transertion played no role in *E. coli* chromosome segregation (Wang and Sherratt, 2010).

Doughnut and Sausage Models for Segregation

Two different models have been suggested for explaining observations on the position and movement of fluorescent DNA loci in *E. coli* cells, the doughnut model (Niki et al., 2000) and the sausage model (Wang et al., 2006; Liu et al., 2010).

In the doughnut model newly replicated DNA is deposited in parallel by the replisomes resulting in a separation of the arms in the short axis of the cell (**Figure 4A**, panels 2, 3). This mode of segregation is reminiscent of the situation in *Caulobacter* (Toro and Shapiro, 2010), but has also been observed in fast-growing *E. coli* cells (Youngren et al., 2014).

As depicted for the doughnut model (**Figure 4A**), the replicated origins move to and remain positioned at the polar ends of the developing nucleoid until division (**Figure 4A**, panel 4). In rapidly growing cells this polar position of the origins and the parallel orientation of the chromosome arms in the short axis is maintained during the next round of replication. In these cells replication occurs at their quarter positions (see Figure 7 in Youngren et al., 2014), giving rise to a branched doughnut structure with partial overlap of chromosome arms of different replication cycles. However, in slow-growing cells, replication initiates in the center of the cell (Reyes-Lamothe et al., 2008). The doughnut model therefore requires a reorientation of the origin by a "sliding movement" that places the two chromosome arms in the two halves of the nucleoid along the cell length axis with the origin in the cell center (Niki et al., 2000; see **Figure 4A**, panels 1a,b). The developing nucleoid arms have the earliest replicated loci (colored circles in **Figure 4**) moved further

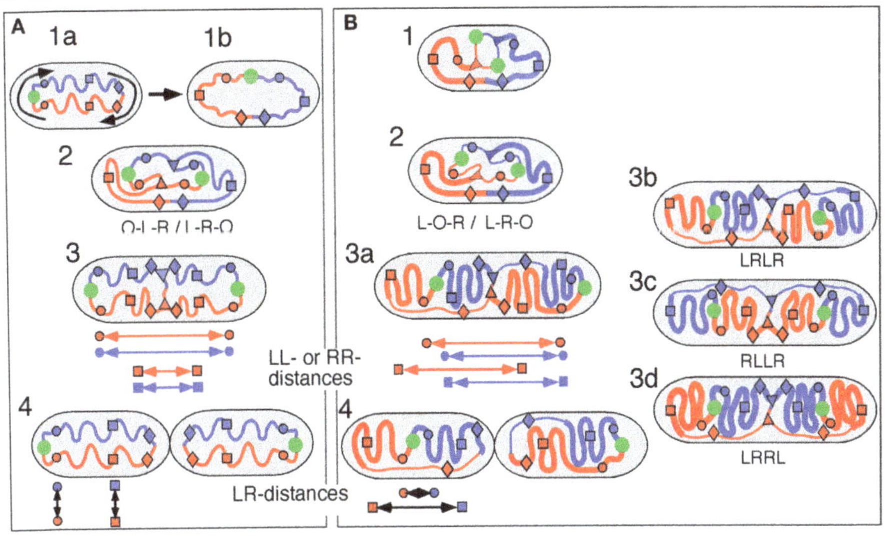

FIGURE 4 | Schematic representation of doughnut and sausage models. Green circle, origin; red and blue circles, origin-proximal loci; red and blue squares, origin-distal loci; red and blue diamonds, terminus-proximal loci; red and blue triangles, replisomes; double arrows indicate distances between loci. **(A)** Doughnut model. The drawings are based on the assumption of parallel, symmetric deposition of the daughter strands leading to a configuration with both origins at the end of the nucleoid (OLR/LRO; panels 3, 4). The circular nucleoid (panel 1a) has to be re-arranged to bring the origin and terminus to the cell center before replication initiates (panel 1b). The model illustrates distances between spot pairs (LL-distances; panel 3) decreasing with their distance from the origin, whereas LR-distances (panel 4) remain constant for origin-proximal and origin-distal loci. **(B)** Sausage model. Replicating chromosome leading to both LOR and LRO configurations (panel 2). The drawings are based on the assumption of an alternating, asymmetric deposition of the daughter strands, which requires stretched regions of replicated DNA to feed the newly developing nucleoids (thin lines). The model illustrates equal distances between spot pairs (LL-distances; panels 3a,b) and an increase in LR-distance for origin-distal loci (panel 4). Cells with the arm configuration LRLR, show LL- and RR-distances that are the same (cf. panels 3a,b); cells with the LRRL configuration show that the LL-distance increases and the RR-distance decreases with distance from the origin (panel 3c); cells with the RLLR configuration, show the LL-distance to decrease and the RR-distance to increase with distance from the origin (panel 3d).

apart than later replicated loci (colored squares in **Figure 4**). This causes a decrease in the distance between replicated loci pairs (LL and RR) with their distance from the origin (**Figure 4A**, panel 3). In contrast, the distances between loci on the parallel arms (L-R-distances) remain constant (**Figure 4A**, panel 4).

In the sausage model, depicted in **Figure 4B**, the replicated origins remain in the center of the nucleoid (**Figure 4B**, panel 2), although a polar positioning can also be observed (**Table 1A**, column II; see also **Figure 4B**, panel 3a). To obtain the alternating orientation of chromosome arms in the long axis, newly replicated DNA of each arm is layered to both inner and outer edges of the newly developing nucleoids. This implies that newly replicated DNA of one arm passes the origin and the other arm, possibly by means of a thread-like structure (**Figure 4B**, panels 2, 3b). In this way, both the earliest and the later replicated loci show similar distances between loci pairs (LL and RR), independent of their distance from the origin (**Figure 4B**, arrows in panel 3a). In contrast, the distances between loci on either arm (LR-distance) increase with their distance from the origin (**Figure 4B**, arrows in panel 4).

Comments on the Doughnut Model

As discussed above, the doughnut model requires a rearrangement of the ring-shaped chromosome before initiation of replication (Niki et al., 2000). A somewhat different reorganization has been proposed by Fisher et al. (2013). In their model, initiation of DNA replication first starts at the polar origin region before one of the daughter strands switches place with unreplicated, parental DNA, placing the terminus in the central part of the nucleoid (see Figure 5B in Fisher et al., 2013). Our data on the 4 spot cells argue against this proposal as we observed very few cells with the OORL configuration even for the most origin proximal RL pair (**Table 1A**).

The high percentage of cells showing the origin outside of the loci on one or two arms (columns II, IIIB and V in **Table 1A** and VI and VII in **Table 1B**) would support the doughnut model (**Figure 4A**). However, an origin lying outside the other loci (OLR-pattern) can still be envisaged to occur from asymmetric, alternating deposition of chromosome arms (see **Figure 4B**, right cell half in panels 2–4). In the 3-spot cells of Figure S2B showing the RLO-pattern, the long axis distributions of L- and R-loci are not fully overlapping as to be expected for the doughnut model (**Figure 4A**, panel 1a).

Comments on the Sausage Model

In the sausage model (Liu et al., 2010; Wiggins et al., 2010) the alternating deposition of replicated DNA to the inner and outer edges of the newly developing nucleoids requires that at least one of the replicated chromosome arms passes the other arm (and the origin). This remote deposition could occur through a thread-like structure (thin lines as depicted in **Figure 4B**). We do not know whether such "feeding threads" exist or how much DNA they contain. They may be revealed by a faster (Brownian) movement of DNA segments through these narrow threads. In the case of a LRLR arm configuration, each replisome forms feeding threads of unequal length resulting in an alternating arm pattern (**Figure 4B**, panel 3b). Presently, indications for the existence of such threads are still lacking.

Concluding Remarks

The results in **Table 3** (see also Figure S3) show that the distances measured between loci pairs (LL and RR) for the different arm configurations stay either constant (LRLR) or show an increase or decrease (LRRL or RLLR) as predicted by the sausage model (**Figure 4B**, panel 3a). Likewise, the distances between loci on the two arms (LR-distances) in the non-replicating (3 spot) cells showing an increase as a function of their distance from the origin (**Figure 3**), conform clearly to the sausage model (**Figure 4B**, panel 4) rather than to the doughnut model (**Figure 4A**, panel 1b). In addition, it should be noted that distributions of spots in the short axis of the present cells did not show the bi-modality as expected for the doughnut model (**Figure 4A**) and as observed in the wide cells of Youngren et al. (2014) (results not shown).

The large variation in the ordering patterns of loci in all constructs suggests that the position and orientation of the nucleoid in the cell is not fixed, so sometimes the origin can lie outside the L and R loci, with a thin thread of DNA extending to the origin close to one pole (**Figure 4B**, panel 3a). This variation is in agreement with the images of nucleoid shape and dynamics obtained by Fisher et al. (2013). Such a flexibility of nucleoid organization and loci positioning is necessary when considering the shape and size changes cells undergo during nutritional shift-up.

At some point during the transition to multifork replication and wider cells, the linear, sausage-like nucleoid will obtain a ring-shaped doughnut structure. Future theoretical work on ring- and linear-polymers will help to understand how, during nutritional shift-up, these two structures move from one state into one another. At present, it only seems clear that in both structures the duplicated origins are separated spontaneously along the cell's long axis by *de novo* DNA synthesis and accumulation and thermodynamic demixing of the newly replicated strands (Youngren et al., 2014).

Segregation can be considered to occur in two stages, (i) the separation of daughter strands or replicated loci as considered here in growing and non-growing cells and (ii) the division of the nucleoid. Because under conditions of inhibited growth nucleoid divisions do not seem to occur (**Figure 3**) it remains possible that an active mechanism like transertion or cell constriction functions in the second stage, the division of the nucleoid in growing cells (see also discussion in Woldringh and Nanninga, 2006).

Acknowledgments

We thank Suckjoon Jun for comments on the manuscript; Alan Leonard, Theo Odijk, Norman Grover, Tanneke den Blaauwen, Arieh Zaritsky, and Mogens Kilstrup for discussions and Jolanda Verheul and Søs Koefoed for practical help. Some of this work was supported by a grant to FH and TA by the Danish National Science Foundation.

References

Cunha, S., Woldringh, C. L., and Odijk, T. H. (2001). Polymer-mediated compaction and internal dynamics of isolated *Escherichia coli* nucleoids. *J. Struct. Biol.* 136, 53–66. doi: 10.1006/jsbi.2001.4420

Cunha, S., Woldringh, C. L., and Odijk, T. H. (2005). Restricted diffusion of DNA segments within the isolated *Escherichia coli* nucleoid. *J. Struct. Biol.* 150, 226–232. doi: 10.1016/j.jsb.2005.02.004

Espéli, O., Borne, R., Dupaigne, P., Thiel, A., Gigant, E., Mercier, R., et al. (2012). A MatP-divisome interaction coordinates chromosome segregation with cell division in *E. coli*. *EMBO J.* 31, 3198–3211. doi: 10.1038/emboj.2012.128

Fisher, J. K., Bourniquel, A., Witz, G., Weiner, B., Prentiss, M., and Kleckner, N. (2013). Four-dimensional imaging of *E. coli* nucleoid organization and dynamics in living cells. *Cell* 153, 882–895. doi: 10.1016/j.cell.2013.04.006

Fritsche, M., Li, S., Heermann, D. W., and Wiggins, P. A. (2011). A model for *Escherichia coli* chromosome packaging supports transcription factor-induced DNA domain formation. *Nucleic Acids Res.* 40, 972–9801-9. doi: 10.1093/nar/gkr779

Huls, P. G., Vischer, N. O. E., and Woldringh, C. L. (1999). Delayed nucleoid segregation in *Escherichia coli*. *Mol. Microbiol.* 33, 959–970. doi: 10.1046/j.1365-2958.1999.01535.x

Jun, S., and Wright, A. (2010). Entropy as the driver of chromosome segregation. *Nat. Rev. Microbiol.* 8, 600–607. doi: 10.1038/nrmicro2391

Lark, K. G. (1972). Evidence for the direct involvement of RNA in the initiation of DNA replication in *Escherichia coli* 15T-. *J. Mol. Biol.* 64, 47–60. doi: 10.1016/0022-2836(72)90320-8

Le Chat, L., and Espéli, O. (2012). Let's get "Fisical" with bacterial nucleoid. *Mol. Microbiol.* 86, 1285–1290. doi: 10.1111/mmi.12073

Lemon, K. P., and Grossman, A. D. (2001). The extrusion-capture model for chromosome partitioning in bacteria. *Genes Dev.* 15, 2031–2041. doi: 10.1101/gad.913301

Liu, X., Wang, X., Reyes-Lamothe, R., and Sherratt, D. (2010). Replication-directed sister chromosome alignment in *Escherichia coli*. *Mol. Microbiol.* 75, 1090–1097. doi: 10.1111/j.1365-2958.2009.06791.x

Luijsterburg, M. S., Noom, M. C., Wuite, G. J., and Dame, R. T. (2006). The architectural role of nucleoid-associated proteins in the organization of the bacterial chromatin: a molecular perspective. *J. Struct. Biol.* 156, 262–272. doi: 10.1016/j.jsb.2006.05.006

Maaløe, O., and Hanawalt, P. C. (1961). Thymine deficiency and the normal DNA replication cycle. I. *J. Mol. Biol.* 3, 144–155. doi: 10.1016/S0022-2836(61)80041-7

Mason, D. J., and Powelson, D. M. (1956). Nuclear division as observed in live bacteria by a new technique. *J. Bacteriol.* 71, 474–479.

Michelsen, O., Teixeira de Mattos, M. J., Jensen, P. R., and Hansen, F. G. (2003). Precise determinations of C and D periods by flow cytometry in *Escherichia coli* K-12 and B/r. *Microbiology* 149, 1001–1010. doi: 10.1099/mic.0.26058-0

Nielsen, H. J., Li, Y., Youngren, B., Hansen, F. G., and Austin, S. J. (2006a). Progressive segregation of the *Escherichia coli* chromosome. *Mol. Microbiol.* 61, 383–393. doi: 10.1111/j.1365-2958.2006.05245.x

Nielsen, H. J., Ottesen, J. R., Youngren, B., Austin, S. J., and Hansen, F. G. (2006b). The *Escherichia coli* chromosome is organized with the left and right chromosome arms in separate cell halves. *Mol. Microbiol.* 62, 331–338. doi: 10.1111/j.1365-2958.2006.05346.x

Nielsen, H. J., Youngren, B., Hansen, F. G., and Austin, S. (2007). Dynamics of *Escherichia coli* chromosome segregation during multifork replication. *J. Bacteriol.* 189, 8660–8666. doi: 10.1128/JB.01212-07

Niki, H., Yamaichi, Y., and Hiraga, S. (2000). Dynamic organization of chromosomal DNA in *Escherichia coli*. *Genes Dev.* 14, 212–223. doi: 10.1101/gad.14.2.212

Odijk, T. (1998). Osmotic compaction of supercoiled DNA into a bacterial nucleoid. *Biophys. Chem.* 73, 23–29. doi: 10.1016/S0301-4622(98)00115-X

Pelletier, J., Halvorsen, K., Ha, B.-Y., Paparcone, R., Sandler, S. J., Woldringh, C. L., et al. (2012). Physical manipulation of the *Escherichia coli* chromosome reveals its soft nature. *Proc. Natl. Acad. Sci. U.S.A.* 109, E2649–E2656. doi: 10.1073/pnas.1208689109

Reyes-Lamothe, R., Possoz, C., Danilova, O., and Sherratt, D. J. (2008). Independent positioning and action of *Escherichia coli* replisomes in live cells. *Cell* 133, 90–102. doi: 10.1016/j.cell.2008.01.044

Skarstad, K., Boye, E., and Steen, H. B. (1986). Timing of initiation of chromosome replication in individual *Escherichia coli* cells. *EMBO J.* 5, 1711–1717.

Toro, E., and Shapiro, L. (2010). Bacterial chromosome organization and segregation. *Cold Spring Harb. Perspect. Biol.* 2:a000349. doi: 10.1101/cshperspect.a000349

Wang, X., Liu, X., Possoz, C., and Sherratt, D. J. (2006). The two *Escherichia coli* chromosome arms locate to separate cell halves. *Genes Dev.* 20, 1727–1731. doi: 10.1101/gad.388406

Wang, X., Possoz, C., and Sherratt, D. J. (2005). Dancing around the divisome: asymmetric chromosome segregation in *Escherichia coli*. *Genes Dev.* 19, 2367–2377. doi: 10.1101/gad.345305

Wang, X., Reyes-Lamothe, R., and Sherratt, D. J. (2008). Modulation of *Escherichia coli* sister chromosome cohesion by topoisomerase IV. *Genes Dev.* 22, 2426–2433. doi: 10.1101/gad.487508

Wang, X., and Sherratt, D. J. (2010). Independent segregation of the two arms of the *Escherichia coli ori* region requires neither RNA synthesis nor MreB dynamics. *J. Bacteriol.* 192, 6143–6153. doi: 10.1128/JB.00861-10

Wiggins, P. A., Cheveralls, K. C., Martin, J. S., Lintner, R., and Kondev, J. (2010). Strong intranucleoid interactions organize the *Escherichia coli* chromosome into a nucleoid filament. *Proc. Natl. Acad. Sci. U.S.A.* 107, 4991–4995. doi: 10.1073/pnas.0912062107

Woldringh, C. L. (2002). The role of co-transcriptional translation and protein translocation (transertion) in bacterial chromosome segregation. *Mol. Microbiol.* 45, 17–29. doi: 10.1046/j.1365-2958.2002.02993.x

Woldringh, C. L., and Nanninga, N. (2006). Structural and physical aspects of bacterial chromosome segregation. *J. Struct. Biol.* 156, 273–283. doi: 10.1016/j.jsb.2006.04.013

Yamaichi, Y., and Niki, H. (2004). *migS*, a cis-acting site that affects bipolar positioning of *oriC* on the *Escherichia coli* chromosome. *EMBO J.* 23, 221–233. doi: 10.1038/sj.emboj.7600028

Yazdi, N. H., Guet, C. C., Johnson, R. C., and Marko, J. F. (2012). Variation of the folding and dynamics of the *Escherichia coli* chromosome with growth conditions. *Mol. Microbiol.* 86, 1318–1333. doi: 10.1111/mmi.12071

Youngren, B., Nielsen, H. J., Jun, S., and Austin, S. (2014). The multifork *Escherichia coli* chromosome is a self-duplicating and self-segregating thermodynamic ring polymer. *Genes Dev.* 28, 71–84. doi: 10.1101/gad.231050.113

Zechiedrich, E. L., Khodursky, A. B., Bachellier, S., Schneider, R., Chen, D., Lilley, D. M. J., et al. (2000). Roles of topoisomerases in maintaining steady-state DNA supercoiling in *Escherichia coli*. *Biol. Chem.* 275, 8103–8113. doi: 10.1074/jbc.275.11.8103

4

Stable regulation of cell cycle events in mycobacteria: Insights from inherently heterogeneous bacterial populations

Michelle M. Logsdon[1,2] *and Bree B. Aldridge*[1,2,3*]

[1] Department of Molecular Biology and Microbiology, Tufts University School of Medicine, Boston, MA, United States,
[2] Department of Molecular Microbiology, Sackler School of Graduate Biomedical Sciences, Tufts University, Boston, MA, United States, [3] Department of Biomedical Engineering, Tufts University School of Engineering, Medford, MA, United States

***Correspondence:**
Bree B. Aldridge
bree.aldridge@tufts.edu

Model bacteria, such as *E. coli* and *B. subtilis*, tightly regulate cell cycle progression to achieve consistent cell size distributions and replication dynamics. Many of the hallmark features of these model bacteria, including lateral cell wall elongation and symmetric growth and division, do not occur in mycobacteria. Instead, mycobacterial growth is characterized by asymmetric polar growth and division. This innate asymmetry creates unequal birth sizes and growth rates for daughter cells with each division, generating a phenotypically heterogeneous population. Although the asymmetric growth patterns of mycobacteria lead to a larger variation in birth size than typically seen in model bacterial populations, the cell size distribution is stable over time. Here, we review the cellular mechanisms of growth, division, and cell cycle progression in mycobacteria in the face of asymmetry and inherent heterogeneity. These processes coalesce to control cell size. Although *Mycobacterium smegmatis* and *Mycobacterium bovis* Bacillus Calmette-Guérin (BCG) utilize a novel model of cell size control, they are similar to previously studied bacteria in that initiation of DNA replication is a key checkpoint for cell division. We compare the regulation of DNA replication initiation and strategies used for cell size homeostasis in mycobacteria and model bacteria. Finally, we review the importance of cellular organization and chromosome segregation relating to the physiology of mycobacteria and consider how new frameworks could be applied across the wide spectrum of bacterial diversity.

Keywords: cell size, mycobacteria, *Mycobacterium smegmatis*, BCG, chromosome organization, cell cycle, asymmetry, cell size control

INTRODUCTION

Mycobacteria are unusual compared to other well-studied bacteria. They grow slowly, are not stained by dyes traditionally used to identify and differentiate bacteria, and have remarkably thick and waxy cell walls (Jankute et al., 2015). Although they have a similar rod shaped morphology as many model bacterial species, recent attention to their basic physiology demonstrates distinct growth modes, subcellular organization, cell cycle timing, division patterns, and size control. Due to these differences, much of the knowledge of basic cell cycle and growth processes developed in well-studied bacteria cannot be directly translated to mycobacteria. These models, however, provide valuable context for studying how the basic physiology of mycobacteria fits into the spectrum of previously studied microorganisms.

Mycobacterial infection remains a major threat to global health. Tuberculosis, caused by infection with *Mycobacterium tuberculosis*, is currently responsible for more deaths than any other single infectious agent (WHO, 2017). Key to mycobacterial virulence and drug tolerance is the ability to tightly control growth. Despite the importance of *M. tuberculosis* growth patterns contributing to infection dynamics, the basic lifecycle properties of mycobacteria are just beginning to be characterized. Until recently, mycobacteria were considered too slow growing and difficult to study to serve as an informative subject for basic cell physiology investigations. Division times of mycobacterial species vary wildly, and all are much longer than the model bacterium *E. coli*, who doubles every 20 min in rich medium. On the extreme end, *Mycobacterium leprae*, the causative agent of leprosy, doubles only once every 14 days and cannot be cultured *in vitro*. *M. tuberculosis* and Bacillus Calmette-Guérin (BCG) are slow growing, culturable mycobacteria with division times of 18–24 h in rich medium. *M. tuberculosis* is an airborne infectious organism that requires a high level of containment during experiments and can only be manipulated within a biosafety level 3 facility (BL3). BCG is often used as a proxy for *M. tuberculosis* in experiments because it is a closely related slow growing mycobacterium but is not pathogenic and does not require a specialized BL3 facility for experiments. However, BCG exposure can cause false positive reactions to the PPD skin test used to monitor exposure to tuberculosis (known as seroconversion, also seen in patients who have received the BCG vaccination) (Cohn, 2001). Therefore BCG can only be manipulated within a biosafety cabinet. *Mycobacterium smegmatis*, a non-pathogenic soil bacterium with a division time of about 3 h, is a commonly used model mycobacteria because it does not cause infection or seroconversion, can be used on a standard BL2 bench top, and allows experiments to be performed on a shorter, more manageable time scale. Much of the growth and division machinery is conserved between *M. smegmatis* and *M. tuberculosis* and the use of *M. smegmatis* as a model organism has allowed the field to progress rapidly in our understanding of the distinct mechanisms of growth and division in mycobacteria (Hett and Rubin, 2008). Additionally, the availability of microfluidic technologies has made basic cell biology studies more accessible, and in the past decade it has become apparent that growth variation within isogenic populations is an intrinsic property of mycobacteria (Aldridge et al., 2012; Kieser and Rubin, 2014). To gain a better understanding of their lifecycle and persistence, it is imperative that we approach mycobacteria as unique and complex organisms.

Differences in physiology between mycobacteria and model bacteria include mechanisms of cell division and growth. A key characteristic of mycobacterial physiology is their striking pattern of asymmetric growth and division (Aldridge et al., 2012; Kieser and Rubin, 2014; Meniche et al., 2014; Manina et al., 2015; Rego et al., 2017). Mycobacteria elongate asymmetrically, preferentially from the old pole (Aldridge et al., 2012; Meniche et al., 2014; Botella et al., 2017; Rego et al., 2017). The new pole experiences a lag in growth before initiating growth partway through the cell cycle (**Figure 1A**) (Aldridge et al., 2012; Botella et al., 2017). The mechanisms controlling initiation or "licensing" of new pole growth are not well understood. In *M. smegmatis* the new pole grows at a slower rate from licensing to division than the old pole, while in *M. tuberculosis*, the rate of new pole growth catches up to the rate of old pole growth preceding division (Botella et al., 2017). Due to differential growth characteristics in cell poles preceding division, sister cells inheriting the mother's old pole have different polar growth characteristics than sister cells inheriting the new pole (Aldridge et al., 2012; Kieser and Rubin, 2014). The "accelerator" sister cell inherits the mother's old pole and is larger and faster growing while the "alternator" sister inherits the mother's new pole and is smaller and slower growing (**Figure 1A**) (Aldridge et al., 2012). This review will address our current understanding of mycobacterial cell biology regarding the generation of asymmetry among closely related cells, impacts of asymmetry on population structure, coordination of innate asymmetry and key cell cycle events, and regulation of cell size in the face of differential growth and size characteristics in mycobacteria.

THE BASIS OF ASYMMETRY

The molecular mechanisms underlying asymmetric growth and division in mycobacteria are just beginning to be elucidated. Many growth and division factors localize in specific patterns to facilitate temporal regulation of growth from each pole. Most well studied bacteria, including *E. coli*, *B. subtilis*, and *C. crescentus* elongate laterally along the length of the cell wall using actin like protein MreB (Daniel and Errington, 2003; Takacs et al., 2010; Dominguez-Escobar et al., 2011; Garner et al., 2011; Wang et al., 2012; White and Gober, 2012; Kysela et al., 2013; Errington, 2015). Mycobacteria, on the other hand, elongate primarily from subpolar regions adjacent to cell poles where the coiled-coil protein Wag31 (also called DivIVA) serves as a scaffold for the elongation complex (Kang et al., 2008; Meniche et al., 2014). Wag31 is targeted to the cell pole through recognition of membrane curvature, where it anchors peptidoglycan, arabinogalactan, and mycolic acid synthesizing enzymes (MurG, GlfT2, and Pks13, respectively) (Meniche et al., 2014). Wag31 preferentially localizes to the old cell pole, consistent with the observation that the old pole serves as the primary site of cell elongation throughout the mycobacterial cell cycle (**Figure 1A**) (Kang et al., 2008; Meniche et al., 2014). Wag31 moves to the new pole at septation to prepare for eventual new pole elongation (**Figure 1A**) (Kang et al., 2008; Santi et al., 2013). Several of the proteins anchored by Wag31, including arabinoglactan synthesizing protein GlfT2, specifically associate with a specialized membrane domain called the pure membrane free of cell wall components (PMf) (Hayashi et al., 2016). GlfT2 is localized to growing old poles, indicating that lipid biosynthetic reactions required for cell envelope synthesis are targeted to regions of active cell growth (Hayashi et al., 2016). Wag31 interacts with the cell wall associated membrane fraction and not the PMf (Hayashi et al., 2016). It has yet to be determined how the distinct PMf and cell wall associated membrane fractions

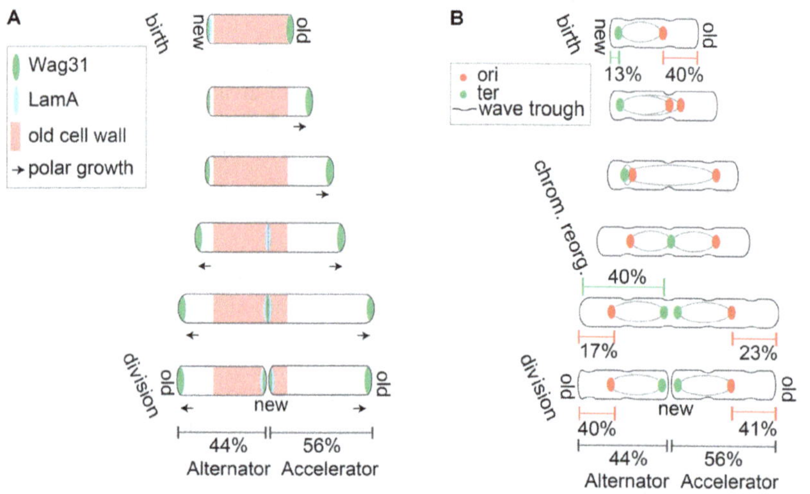

FIGURE 1 | Asymmetric polar growth and division in mycobacteria. **(A)** Mycobacteria grow primarily from subpolar regions (marked with arrows), and elongate preferentially from the old pole. Asymmetric polar growth is directed by elongation and divisome components Wag31 (green) and LamA (blue). At birth, LamA actively inhibits growth from the new pole while Wag31 localizes to the old pole and organizes the elongation complex to promote growth at the old pole. Later in the cell cycle LamA inhibition of new pole growth is relieved, while Wag31 accumulates and promotes growth from the new pole. Finally, LamA and Wag31 accumulate at the septum region where Wag31 assembles the elongation complex while LamA inhibits growth from this site that becomes the new pole after division. **(B)** Mycobacteria divide asymmetrically and organize their chromosomes asymmetrically within the cell. Asymmetric division is partially organized by surface wave troughs inherited from previous generations. Cells divide at the centermost wave trough along the cell body. At birth, the ori region of the chromosome (red) is positioned midcell, but closer to the old pole, and the ter (green) is positioned near the new pole. The chromosome reorganizes part way through the cell cycle. Ori's partition and travel to asymmetric quarter cell positions and the terminus translocates to an asymmetric midcell location that corresponds with surface wave trough and septum placement. At division, proportional chromosome localization is reestablished in daughter cells despite the differential sizes of accelerator and alternator sisters.

work together to promote cell growth. Regardless, it is clear that the localization of a metabolically active membrane domain, as well as scaffold protein Wag31, provides a means of targeting cell growth preferentially to cell poles.

Many mycobacterial growth studies employ fluorescently tagged Wag31 as an elongation and early cell division marker because of its important role in organizing cell elongation complexes (Santi et al., 2013; Meniche et al., 2014; Santi and McKinney, 2015; Botella et al., 2017). Wag31-fluorescent protein fusions form clear and bright fluorescent bands at sites of cell growth and septation. However, the function of Wag31 is very sensitive to fluorescent protein fusion and overexpression. Adding a GFP tag to the C-terminus of native Wag31 leads to aberrant localization of the Wag31-GFP band at the new pole, causing the new pole to become the primary site of growth (Meniche et al., 2014). The C-terminal tagged Wag31-GFP also caused a 40% decrease in cell elongation rate, increased cell width, and an abnormal curved shape in cells (Meniche et al., 2014). However, a merodiploid strain with an unaltered native copy of Wag31 and a fluorescent protein tagged copy of Wag31 integrated elsewhere on the chromosome localized as expected to the old pole and restored the wild type cell shape phenotype (Meniche et al., 2014).

A recent study used super-resolution microscopy and fluorescent D-amino acids (FDAAs) to stain a *M. smegmatis* reporter strain expressing an episomal mCherry tagged Wag31 (Botella et al., 2017). FDAAs are a recently developed class of reporters that are incorporated directly into growing peptidoglycan to identify specific regions of elongation along a bacterial cell (Siegrist et al., 2013; Kuru et al., 2015; Botella et al., 2017). In mycobacteria, FDAA staining was used to differentiate growth from old versus new poles. Cells expressing a fluorescently tagged Wag31 elongated more symmetrically between old and new poles compared to wild type *M. smegmatis*, which elongated preferentially from the old pole (Botella et al., 2017). Together, these bodies of work suggest that Wag31 localization and function is particularly sensitive to fluorescent protein tags, because several of the constructs utilized significantly alter cell growth and polarity (Santi et al., 2013; Meniche et al., 2014; Santi and McKinney, 2015; Botella et al., 2017). Adding a fluorescent protein to any important enzyme can alter enzyme activity, localization, and cell physiology (Landgraf et al., 2012), and this seems especially true of Wag31. A loss of cellular growth asymmetry could have major consequences for bacterial population structure and fitness; therefore we must be cautious in interpreting results from studies performed with fluorescently tagged Wag31 in mycobacteria.

Asymmetric polar growth in mycobacteria is created by a protein conserved among mycobacteria but with no homologs in other characterized bacteria. This protein, called LamA (named for Loss of Asymmetry Mutant A), is a member of the mycobacterial division complex where it actively inhibits growth at the new pole (**Figure 1A**) (Rego et al., 2017). LamA interacts with penicillin binding protein PonA1. Deletion of LamA allows Wag31 to be recruited to the new pole more rapidly, indicating that LamA works to delay assembly of the elongation complex at the new pole (Rego et al., 2017). LamA inhibition of new pole growth in WT cells allows the old pole to grow preferentially, creating the characteristic asymmetric polar growth patterns of mycobacteria (**Figure 1A**).

Though identifying functions of proteins Wag31 and LamA provides us with a glimpse of the networks involved in regulating asymmetric growth in mycobacteria, the extent of their interactions with elongation complexes have yet to be fully elucidated. A current understanding of the processes involved in cell elongation complexes is described in an excellent review by Kieser and Rubin (2014). The enzymatic activity of a large number of proteins is required for synthesizing peptidoglycan, arabinogalactan, and mycolic acid layers in mycobacteria. Asymmetry-promoting proteins Wag31 and LamA likely interact with enzyme complexes and membrane domains involved in cell wall and cell membrane synthesis (Meniche et al., 2014; Hayashi et al., 2016). It will be interesting to understand exactly how previously identified cell elongation networks are asymmetrically regulated at each pole and how growth cues such as nutrient status may interface with these pathways.

In addition to growing asymmetrically, mycobacteria also divide asymmetrically and undergo a fast, mechanical v-snapping process of daughter cell separation (Aldridge et al., 2012; Richardson et al., 2016; Zhou et al., 2016). There is evidence to suggest that sites of asymmetric division are not just established at division, but are inherited from previous generations. Atomic force microscopy (AFM) of the mycobacterial cell surface revealed that cell surface wave troughs can determine division site selection (Eskandarian et al., 2017). Repeating surface waveform troughs are inherited over several generations and the total number of troughs scales with cell size (**Figure 1B**) (Eskandarian et al., 2017). Cells divide at the centermost trough and the other troughs are passed down to subsequent generations to be used as a future division site (**Figure 1B**) (Eskandarian et al., 2017). Furthermore, while cell wall synthetic enzymes GlfT2, MurG, and Pks13 show primary localization to subpolar regions where growth occurs, re-examination of this data indicates secondary localization peaks at quarter cell positions (Meniche et al., 2014). It has yet to be investigated if these elongation complex positions co-localize with surface wave troughs, as a potential means of using division troughs to position cell elongation complexes a generation in advance.

Wild type *M. smegmatis* cells divide at the central surface wave trough while cells deficient in chromosome partitioning can divide at off-center wave troughs, indicating a regulatory role for chromosome segregation in division site selection (Eskandarian et al., 2017). Mycobacteria lack homologues of defined nucleoid occlusion systems (Noc proteins) and minicell (Min) proteins used by other bacteria to prevent chromosome splicing during cell division (Wu and Errington, 2011; Monahan et al., 2014; Schumacher, 2017). Despite the lack of known nucleoid occlusion proteins, *M. smegmatis* cells exhibit a clear relationship between asymmetric chromosome positioning and asymmetric division placement (Eskandarian et al., 2017; Logsdon et al., 2017) (**Figure 1B**). Division occurs in the wave trough nearest the local DNA minimum even in mycobacteria deficient in chromosome segregation, demonstrating the existence of a mechanism to prevent damage to the chromosome during cell division (Eskandarian et al., 2017).

Together, these studies support the idea that chromosome dynamics contributes to asymmetric cell division in mycobacteria. The chromosome is a large, essential macromolecule that may play a key role in determining the layout of many aspects of the intracellular space (Campos and Jacobs-Wagner, 2013). Different bacterial species exhibit a variety of chromosome orientations and arrangements, reflecting the diversity of bacterial cell shapes, sizes, and growth modes. For example, in *E. coli* replication is initiated midcell and continues with two replication forks moving independently before terminating midcell (Reyes-Lamothe et al., 2008). This pattern represents an L-ori-R (left-origin-right) orientation. In *C. crescentus*, DNA replication begins near the flagellar old pole during the swarmer to stalked cell transition. The replisome migrates midcell while the newly replicated ori "flips" and travels across the cell from the stalked pole to the newly formed flagellar pole (Bowman et al., 2008). The *C. crescentus* chromosome organization pattern represents an ori-ter-ter-ori (origin-terminus-terminus-origin) orientation. Other bacteria with ori-ter-ter-ori chromosome organizations include *Pseudomonas aeruginosa*, chromosome 1 of *Vibrio cholerae*, and *Myxococcus xanthus* (Fogel and Waldor, 2005; Harms et al., 2013; Vallet-Gely and Boccard, 2013). *B. subtilis*, on the other hand, oscillates between an ori-ter-ter-ori and L-ori-R pattern, depending on cell cycle stage (Wang et al., 2014).

Chromosome organization and partitioning likely direct many cellular processes. These processes include ensuring genetic inheritance, directing cell division, regulating transcription, and targeting transcripts and their encoded proteins to their necessary location within the cell (Montero Llopis et al., 2010; Nevo-Dinur et al., 2011; Govindarajan et al., 2012; Campos and Jacobs-Wagner, 2013). For example, the order of genes along the chromosome is highly conserved in gammaproteobacteria and correlated with the temporal regulation of gene expression (Sobetzko et al., 2012). The most highly conserved aspect of global gene arrangements is distance from either the ori or ter region (Sobetzko et al., 2012). Proximity of a gene to the ori or ter region on the chromosome could contribute to temporal regulation of its expression through gene dosage, which refers to the difference in copy number of genes located near the ori versus the ter while DNA replication is ongoing (Slager and Veening, 2016). Genes in the ori region are replicated first and thus periodically have two copies while genes near the ter only have one copy until the end of replication. In *Vibrio cholerae*, the ori proximal location and expression of ribosome genes is essential for proper growth and host invasion. Moving half of the genes required for ribosome proteins to distal sites along the chromosome leads to significant defects in growth and invasion processes (Soler-Bistue et al., 2015). In mycobacteria, a notable operon positioned near the ori is called the division cell wall (dcw) cluster and contains many genes whose products are required for cell wall synthesis and division, including *ftsZ, ftsQ, ftsW, murC, murG, murD, murX, murF, murE*, and *wag31* (Kang et al., 2008). The functional implications of maintaining this growth and division operon near the ori have yet to be investigated in mycobacteria, but it stands to reason that this position could allow regulation of expression through gene dosing and/or proper localization of important growth and division proteins.

The molecular machinery responsible for chromosome partitioning and placement within most studied bacterial cells (with the notable exception of *E. coli*) is the ParABs system (Lee and Grossman, 2006; Mierzejewska and Jagura-Burdzy, 2012; Iniesta, 2014; Lagage et al., 2016). The ParABs system in mycobacteria consists of two 16 bp centromere-like *parS* sequences located near the ori on the chromosome (Casart et al., 2008). The ParB DNA binding protein specifically recognizes and binds the *parS* sequences, and its activity is negatively regulated through Ser/Thr kinase phosphorylation (Baronian et al., 2015). ParA, a Walker ATPase, interacts with ParB and directs segrosome partitioning (Ginda et al., 2017). The movement of the ori across the cell to the new pole allows the cell to divide into daughter cells with two neatly packaged chromosomes. ParB colocalizes with the nucleoid associated protein responsible for maintaining chromosome compaction, HupB, indicating a role for HupB in ori segregation (Holowka et al., 2017). The ParABs system is also strongly integrated with aspects of the elongation and division machinery in *M. smegmatis*. Knocking out or overproducing ParA impairs cell division, resulting in filamentous and multinucleoidal cells that have failed to divide (Maloney et al., 2009). ParA interacts directly with the polar protein responsible for the organization of growth and division factories, Wag31 (Ginda et al., 2013). Therefore, studying chromosome partitioning could give us new insight into the integration of chromosome organization and cell division processes.

Several groups have characterized the chromosome organization of mycobacteria using live cell microscopy, and multiple models of chromosome subcellular organization have been proposed. Santi and McKinney (2015) used a fluorescent protein tagged ParB/Wag31 dual reporter strain to conclude that the ori is localized midcell before replication and after replication, both ori's segregate to mirror symmetric quarter cell positions to prepare for midcell localization in the two daughter cells. Chromosomal locus attB1, located at 245° on the left lobe, was found initially localized near the new pole before translocating midcell during replication. Based on the midcell position of the ori and new pole position of the left lobe, they concluded the *M. smegmatis* chromosome organization had a R-ori-L configuration, as in *E. coli*. However, the use of a fluorescent protein labeled Wag31 may have affected the growth polarity in these reporter cells (Meniche et al., 2014; Botella et al., 2017).

Recent studies using ParB (Trojanowski et al., 2015; Ginda et al., 2017) or fluorescent reporter operator system (FROS)-ori (Logsdon et al., 2017) reporters independently converged on a model in which the ori is positioned asymmetrically and closer to the new pole at birth (**Figure 1B**). A few differences between the reporters utilized and results generated should be noted. The ParB reporter consistently shows ori splitting and segregation much earlier than the FROS ori reporter (Santi and McKinney, 2015; Trojanowski et al., 2015, 2017; Ginda et al., 2017; Logsdon et al., 2017). Many cells inherited two ParB foci, indicating these origins had already undergone a second round of replication and a second segregation event before division, which was not observed using the FROS-ori reporter (Ginda et al., 2017; Logsdon et al., 2017; Trojanowski et al., 2017). In these cases, the ParB reporter displayed differing frequencies of re-splitting before division between studies ranging from 0% (Trojanowski et al., 2015), to 11% (Trojanowski et al., 2017), to 75% (Ginda et al., 2017). It is not apparent why these reporters produce different segregation dynamics, however, the asymmetric ori positioning was consistent across studies (Trojanowski et al., 2015, 2017; Ginda et al., 2017; Logsdon et al., 2017). Identification of ori positioning, along with the use of a FROS-terminus reporter, allowed for a more comprehensive investigation of subcellular chromosome localization and strongly supports an ori-ter-ter-ori chromosome orientation (Logsdon et al., 2017). Localization data show the terminus positioned near the new pole at birth followed by a rapid translocation event to the midcell partway through the cell cycle (**Figure 1B**). This asymmetric ori-ter-ter-ori chromosome positioning is not unique to mycobacteria, as it has been previously described in *Myxococcus xanthus* (Harms et al., 2013).

How do chromosome organization patterns in mycobacteria contribute to cell division? The observed ori-ter-ter-ori orientation pattern is asymmetric within the cell throughout the entire cell cycle (Ginda et al., 2017; Logsdon et al., 2017). Though the chromosome terminus translocates midcell, it is not located precisely halfway between each pole. Rather, it is slightly shifted toward the new pole, in a manner that corresponds with asymmetric septum positioning at cell division (**Figure 1B**) (Logsdon et al., 2017). In many cells, the terminus shifts midcell but remains as a single focus (even after replication has completed) until 15–30 min before cell division. Just before division, the two terminus foci move slightly apart and the septum forms directly between them.

Terminus translocation midcell could be a very early determinant of the division site. How the terminus finds this location is unclear – there are systems identified for origin localization in mycobacteria and other bacteria but not terminus localization. The terminus and inherited wave troughs may interact to target termini to the future division spot. Additionally, interactions of the terminus with members of the divisome to prevent division over chromosomes have not been characterized. Mycobacteria do not have homologs of any identified nucleoid occlusion systems but it stands to reason that they have an alternate system that only allows septum formation in the DNA deficient space between termini at division. Overall, many studies indicate that asymmetry is an integral part of mycobacterial physiology that sets this genus apart from other well studied, symmetric bacteria.

BEHAVIORAL CONSEQUENCES OF ASYMMETRY

The establishment of systematic asymmetry in mycobacterial cells likely has functional consequences on behaviors of both single cells and populations of mycobacteria. One consequence of cellular asymmetry is the creation of cellular heterogeneity, even between presumably isogenic sister cells. Phenotypic heterogeneity creates subpopulations of mycobacteria from a single progenitor with varying size, growth, and cell cycle properties (Aldridge et al., 2012; Richardson et al., 2016). Deterministic generation of heterogeneity among populations of cells may serve as a bacterial "bet hedging" mechanism, allowing a subpopulation of cells with a particular phenotype to survive environmental stress (Kysela et al., 2013; Richardson et al., 2016). Differentiation of cell phenotypes leading to bet hedging is apparent in the model organism *C. crescentus*. Upon division, two distinct daughter cell types are generated. One daughter cell is immobile and has a polar stalk while the other has a polar flagellum used for swarming motility. The stalked daughter remains attached to its current environment, while the swarmer daughter takes a risk to swim away and colonize a new, potentially preferential environment (Kysela et al., 2013). In this way, if environmental conditions wipe out all cells at the initial site of colonization, half the cells are already searching for a viable site where they can grow and replicate.

The functional consequences of asymmetric growth and division in mycobacteria are perhaps more subtle. Differences in cell size and cell cycle stage at the time of antibiotic treatment contribute to differential antibiotic susceptibility (Aldridge et al., 2012; Richardson et al., 2016). When treated with cell wall acting antibiotics, faster growing accelerator cells are more sensitive to cycloserine and meropenem than their sister alternator cells (Aldridge et al., 2012). Additionally, *M. smegmatis* cells exhibit differential susceptibility to rifampicin treatment. Rifampicin tolerant cells are larger at birth and inherit older growth poles compared to susceptible cells (Richardson et al., 2016). Together, these studies indicate that asymmetric growth and division within a population of mycobacteria allows certain cells to survive an antibiotic stressor while others with different phenotypic characteristics perish.

The role of asymmetric growth and division in *M. tuberculosis* infection is just beginning to be investigated. Recently, Vijay et al. (2017) examined cell size distributions of 158 clinically isolated *M. tuberculosis* strains and observed that *M. tuberculosis* cell sizes are smaller and less variable after growing in rich liquid broth compared to their sizes directly from patient sputum or passaged through J774 mouse macrophages. These measurements suggest that clinical, non-laboratory adapted strains of *M. tuberculosis* may increase phenotypic heterogeneity in response to host stress conditions. Additionally, cell length and heterogeneity increased in clinically isolated multidrug resistant *M. tuberculosis* (MDR-TB) strains when exposed to the first line antibiotics rifampicin and isoniazid (Vijay et al., 2017). Cell size variability further increased in MDR-TB subjected to both macrophage infection and rifampicin treatment (Vijay et al., 2017). This increase in *M. tuberculosis* size and variability suggests a synergistic effect between drug stress and infection. Thus, phenotypic heterogeneity manifested in cell size in addition to defined genetic mutations, may shape *M. tuberculosis* drug and host tolerance during infection.

Manina et al. (2015) also investigated the effect of host and drug stress on *M. tuberculosis* phenotypic heterogeneity in defined conditions with the lab adapted Erdman strain of *M. tuberculosis*. A GFP tagged rRNA reporter was used to demonstrate increases in *M. tuberculosis* cell-to-cell heterogeneity in nutrient limited medium, stationary phase culture, macrophage infection, and mouse lung infection compared to growth in rich medium (Manina et al., 2015). Increased *M. tuberculosis* cellular variability may be a direct response to stressful environmental conditions, although the physiological and molecular response leading to phenotypic changes in *M. tuberculosis* are poorly understood. Increased phenotypic heterogeneity observed under host and drug stress could be mediated through subtle changes in patterns of *M. tuberculosis* asymmetric growth and division. These studies suggest that amplifying phenotypic heterogeneity may allow a subpopulation of *M. tuberculosis* cells to persist within a host during infection and drug treatment. If true, it is tempting to speculate that identifying and targeting particularly tolerant cell subpopulation could clear TB infection more efficiently.

Finally, molecular support for the hypothesis that phenotypic heterogeneity promotes bacterial survival comes from studies using the symmetrically growing LamA mutant. Δ*lamA* *M. smegmatis* and *M. tuberculosis* show less size heterogeneity than wild type cells and are killed more rapidly and uniformly by rifampicin and cell wall acting antibiotics (Rego et al., 2017). This molecular study, together with population level studies, support the hypothesis that asymmetric polar growth actively creates heterogeneity amongst bacterial cells to serve as a bulk survival mechanism, ensuring a subpopulation of *M. tuberculosis* can persist in stressful environments.

CELL CYCLE

To coordinate chromosome duplication with changes in cell size, growth rate, and variability in stressful growth conditions, mycobacteria alter the dynamics of their cell cycle. These changes are essential to adapt to a wide variety of environments and ensure faithful replication and genetic inheritance. The bacterial cell cycle is usually described as three basic phases: B period occurs after division and before DNA replication, C is the period of active DNA replication, and D occurs after DNA replication but before cell division. In *M. smegmatis*, a fraction of cells begin a new round of DNA replication before division and their daughters subsequently skip B period and continue straight into C (Santi et al., 2013; Sukumar et al., 2014; Trojanowski et al., 2015; Richardson et al., 2016; Logsdon et al., 2017). We named this occurrence the E period (Sukumar et al., 2014). Overall, *M. smegmatis* spends the majority of the cell cycle in C period replicating the chromosome while B and D period are relatively short (**Figure 2A**) (Santi et al., 2013; Sukumar et al., 2014; Logsdon et al., 2017; Trojanowski et al., 2017). E period occurs more frequently and lasts longer in cells that were large at birth (**Figure 2A**) (Logsdon et al., 2017).

A similar but distinct overlap in replication cycles was recently reported in *M. smegmatis*. Trojanowski et al. (2017) observed multifork replication in 11% of *M. smegmatis* cells growing in rich medium. Multifork replication occurs when a second (or

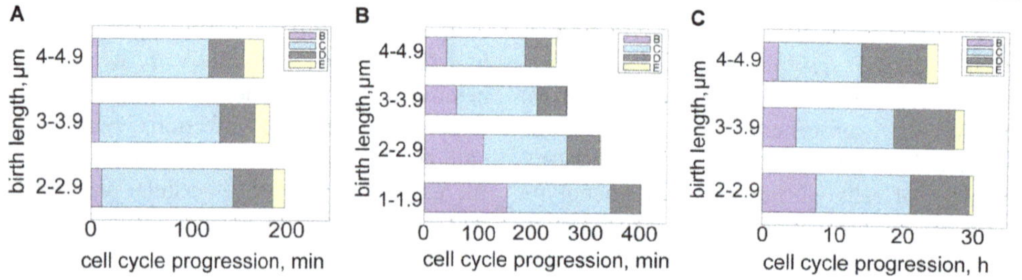

FIGURE 2 | Mycobacterial cell cycle and growth dynamics shift in fast versus slow growth conditions. **(A)** Bar plot of *M. smegmatis* cell cycle progression in rich medium. B period (pre-replication, purple) is short while C (active DNA replication, blue) encompasses most of the cell cycle. D period (post replication, gray) is relatively constant across cell birth sizes but E period (re-initiation, yellow) is associated with cells born larger. **(B)** Bar plot showing *M. smegmatis* cell cycle progression and variability across cell sizes in carbon limited medium. B period is disproportionally extended in carbon limited growth conditions, as smaller cells require more time to initiate DNA replication. E period encompasses less of the cell cycle during carbon limitation. **(C)** Bar plot showing BCG cell cycle progression across cell sizes in rich medium. Duration of B period is associated with cell size, with cells born smaller waiting longer to initiate DNA replication. C and D periods are relatively constant in the BCG cell cycle and E period is infrequent. Data from Logsdon et al. (2017).

third or fourth) round of replication begins before the first has completed (Helmstetter, 1968; Trojanowski et al., 2017). Multifork replication is typically associated with fast growing bacterial species as an adaption to fast growth conditions, in which division time is shorter than the time required to replicate the chromosome. By initiating several rounds of replication before the first has completed, daughter cells inherit an already partially replicated chromosome, preventing the rate of division from outpacing the rate of chromosome duplication. The differing observations of E period and multifork replication in mycobacteria come from studies using different fluorescent DNA replication reporters and imaging/growth conditions, making it difficult to make direct comparisons (Santi et al., 2013; Sukumar et al., 2014; Logsdon et al., 2017; Trojanowski et al., 2017). Even so, both multifork replication and E period are associated with large cells, and the frequency of both re-initiation events decreases under nutrient limitation (Logsdon et al., 2017; Trojanowski et al., 2017).

It is uncommon to observe re-initiation (E period) or multifork replication in slow growing bacteria, particularly a species with a period of DNA replication (C period) shorter than the interdivision time. Mycobacteria therefore break the paradigm in bacterial cell physiology that supposes only fast growing organisms engage in multifork replication. The effects of gene dosage on transcription are multiplied during multifork replication (ori:ter ratios are 4:1 instead of 2:1 or 1:1) and could lead to an overabundance of transcripts from genes located near the ori and set into motion any number of cellular regulatory programs (Slager and Veening, 2016). Ultimately, the effect of gene dosage and transcript abundance could create an additional layer of heterogeneity in the subpopulation of mycobacterial cells experiencing reinitiation. Because *M. smegmatis* cells engaging in multifork replication comprise only a small percent of the total population (11%) many more cells need to be studied to determine the functional utility of this behavior (Trojanowski et al., 2017). Overall, multifork replication may serve a different purpose or evolutionary advantage in slow growing mycobacteria compared to fast growing model bacteria and these cell cycle patterns require further investigation.

Several other distinct cell cycle timing changes occur in mycobacteria under nutrient limitation. Studying cells grown under carbon limitation presents a chance to examine cell cycle dynamics in environmental conditions more similar to those encountered during early infection. When grown in carbon-limited medium, the average *M. smegmatis* interdivision time increases by 2 h (Logsdon et al., 2017). Rather than slowing all periods of the cell cycle equally, B period is extended disproportionally (**Figure 2B**). The proportion of the cell cycle spent in B period is negatively correlated with cell birth size, meaning that small cells require significantly more time before initiating DNA replication than cells born large (**Figure 2B**) (Logsdon et al., 2017). However, post initiation time (C+D+E periods) remains constant and independent of cell size at birth (**Figure 2B**). The extended delay before initiation could be an adaption strategy to early infection, during which there is evidence that mycobacteria slow or halt their growth to avoid host detection (Rohde et al., 2012).

The slow growing mycobacterial species BCG is often used as a non-pathogenic model for another slow growing pathogenic mycobacteria, *M. tuberculosis*. The BCG interdivision time is longer than *M. smegmatis* (and comparable to that of *M. tuberculosis*), spanning from 15 to 20 h with a C period lasting an average of 9.4 h (**Figure 2C**) (Nair et al., 2009; Logsdon et al., 2017). The B period shows an extended duration and negative correlation with cell size at birth while the total time spent initiation to division remains constant across cell size, as in carbon limited *M. smegmatis* (**Figure 2C**) (Logsdon et al., 2017). The very long interdivision and DNA replication times in slow growing species present a long standing puzzle for mycobacteriologists (Ditse et al., 2017). A recent study of the *M. tuberculosis* DNA polymerase DnaE1 showed that the process of nucleotide incorporation is not the rate-limiting step in DNA replication or division times, because recombinant *M. tuberculosis* DnaE1 polymerase works faster *in vitro* than the *E. coli* DNA polymerase PolIIIα (Rock et al., 2015). Therefore there must be additional factors limiting the rate of DNA replication and cell cycle progression in slow growing mycobacteria that should be identified with future work. The necessity of coordinating DNA replication with chromosome partitioning, elongation, division, and metabolic status likely plays a role in determining rates of DNA replication in slow growing mycobacteria.

During growth in nutrient-limited environments, cell cycle timing and cell sizes become more heterogeneous. The mycobacterial cell cycle (particularly in slow growing mycobacteria) may be innately asynchrononus. The length of B period varies significantly based on cell birth size, and daughter birth sizes are intrinsically unequal, thus cell cycle timing is also intrinsically unequal (Aldridge et al., 2012; Logsdon et al., 2017). This asynchrony may provide an additional level of heterogeneity to promote population survival. The cell cycle stage at the start of rifampicin treatment is a strong predictor of bacterial survival on a single cell level (Richardson et al., 2016). It is therefore intriguing that slow growing mycobacteria BCG or carbon limited *M. smegmatis* have a means of generating cells in many stages of the cell cycle within a single colony. These studies suggest that mycobacterial cell cycle dynamics could determine, on a single cell level, which bacteria will survive nutrient or antibiotic pressure. Elucidating the effect of cell cycle dynamics on bacterial stress tolerance and survival will require additional single cell experiments and analysis (Taheri-Araghi et al., 2015b). With continued study focusing on the timing of cell cycle components in individual cells, we can begin to elucidate how mycobacteria regulate their lifecycle to persist in the face of environmental stress.

CELL SIZE CONTROL

It is clear that mycobacteria must manage variability in size, growth, and cell cycle timing in the wide range of conditions potentially encountered during infection. Coordination of these essential cell processes ensures reproducibility of cell sizes in the face of environmental stress. Even under rich growth conditions, asymmetric elongation and division in mycobacteria give rise to increased variability in cell size compared to model bacteria such as *E. coli*, *B. subtilis*, and *C. crescentus*, yet the distribution of this variability is stable over time (Aldridge et al., 2012; Amir, 2014; Campos et al., 2014; Taheri-Araghi et al., 2015a). Cell size homeostasis is an important aspect of the physiology of all living cells because without a mechanism of regulation, cell sizes and physiological properties would rapidly diverge and become unsustainable. Mycobacterial populations maintain stability in cell size variation despite larger sister cells elongating faster and smaller sister cells elongating slower, indicating a strong cellular mechanism of size regulation to prevent differences in size and growth rate from compounding and diverging over time. What mechanism is used by mycobacteria to maintain control of cell sizes when their variability in cell size increases under stress?

In well-studied, model bacterial species the distribution of cell birth sizes is quite narrow. *E. coli* birth size distribution has a coefficient of variation (CV) of 12% and *C. crescentus* maintains a CV of 14%, despite being another asymmetrically dividing bacterial species (Campos et al., 2014; Iyer-Biswas et al., 2014). This tightly controlled population structure has fascinated microbiologists for decades and lead to several attempts to model the factors controlling bacterial cell division and cell size (Schaechter et al., 1958; Koch and Schaechter, 1962; Cooper and Helmstetter, 1968; Donachie, 1968). However, it was not clear how cell size control models, which are well studied in other bacteria, would translate given the physical asymmetry of mycobacteria.

Mathematical models of bacterial cell size control have been introduced beginning in the 1960s (Koch and Schaechter, 1962; Cooper and Helmstetter, 1968; Donachie, 1968; Amir, 2014; Campos et al., 2014; Ho and Amir, 2015; Taheri-Araghi et al., 2015a; Harris and Theriot, 2016; Sauls et al., 2016; Soifer et al., 2016; Wallden et al., 2016; Zheng et al., 2016; Jun et al., 2018). To date, most models of bacterial cell size control have been built on the assumption that bacteria grow exponentially at a single-cell level (Schaechter et al., 1958). However, time resolution of imaging in *M. smegmatis* is limited due to phototoxic effects, such that single cell traces cannot distinguish between linear and exponential modes of growth. Though growth data on a single-cell level have been reported as either an averaged growth rate (length/time) or an exponential growth constant (1/time) depending on the analysis type, consensus whether mycobacterial growth per cell is linear or exponential is lacking. (Aldridge et al., 2012; Santi et al., 2013; Wakamoto et al., 2013; Logsdon et al., 2017; Priestman et al., 2017). There are other advanced techniques used to definitively determine the mode of mass increase within a bacterial cell over time (Godin et al., 2010; Cermak et al., 2016), however, these have not been applied to mycobacteria.

Recent studies have taken advantage of growth measurements from many cells to compare against expected relationships between generation time and cell size of exponentially or linearly growing bacteria. Studies using bulk measurements and single cell traces with model selection are in agreement with exponential growth expectations compared to a linear growth model (Logsdon et al., 2017; Priestman et al., 2017). Interestingly, both found that the fraction of *M. smegmatis* cells with the longest interdivision times deviate from exponential growth and appear more linear (Logsdon et al., 2017; Priestman et al., 2017). It would be interesting to determine if another cellular growth mechanism in these slowly dividing cells affects apparent linear versus exponential growth. Perhaps very large cells inherited old growth poles or experienced a lag in growth and could have distinct stress responses compared to their exponentially growing counterparts (Kysela et al., 2013).

Historically, models of cell size control used measurements of size or time to relate events within the cell cycle of exponentially growing cells ("timer" or "sizer" models) (**Figures 3A,B**) (Koch and Schaechter, 1962; Robert et al., 2014; Willis and Huang, 2017). Time increments measure the minutes or hours between events, (e.g., cells grow 100 min before division), while size measurements record the absolute length at an event (e.g., cells divide once reaching 7 μm in length) (**Figures 3A,B**). Models of cell size control used one or both types of measurements simultaneously to describe bacterial cell division. Model bacteria were largely considered to follow a sizer model as timer models are very sensitive to slight differences in exponential growth

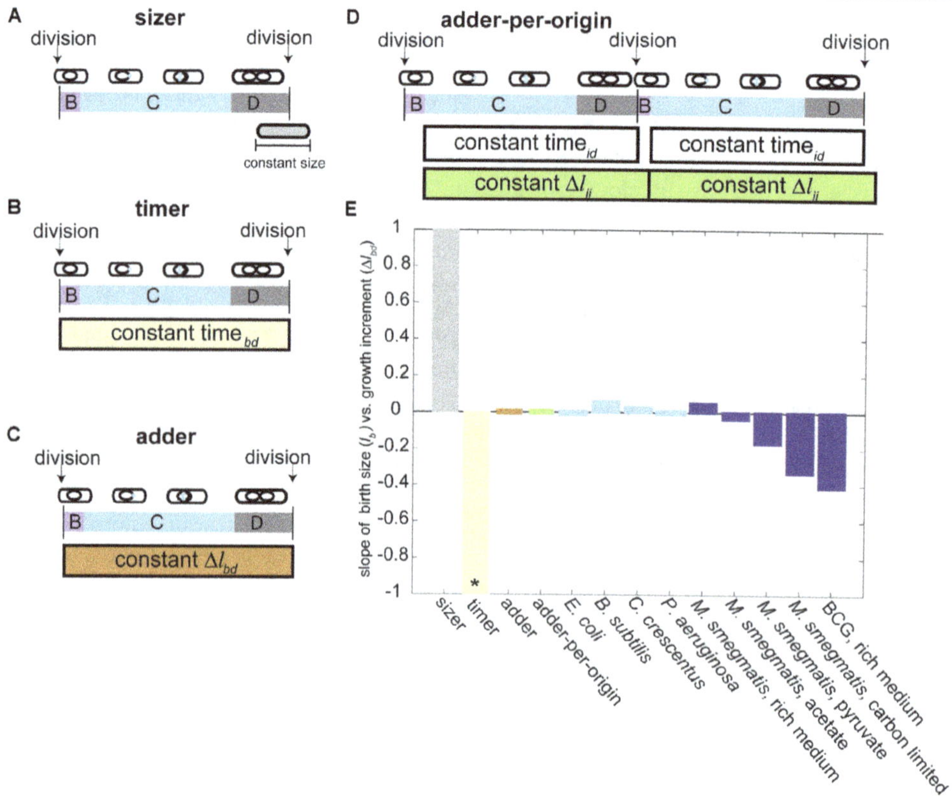

FIGURE 3 | Models of bacterial cell size control. **(A)** In the sizer model, cells divide once they reach a constant size (e.g., cells divide once reaching 7 μm in length). **(B)** In the timer model, cells divide a constant time after birth (e.g., cells divide 100 min after birth). **(C)** In the adder model, cells divide after adding a constant growth increment from birth (e.g., cells divide after adding 3 μm to their birth length). **(D)** In the adder-per-origin model, cells add a constant increment from initiation to initiation and then divide a constant time after initiation (e.g., cells divide 100 min after initiation, and grow 3 μm between initiation events). **(E)** Bar plot depicting the slope of cell birth length (Δl_b) versus birth to division adder increment (Δl_{bd}) for several size control models and bacterial populations. Model predictions for the slope of Δl_b vs. Δl_{bd} are shown on the left of the graph. The sizer model (left, gray bar) predicts a slope of 1, because cells born smaller must grow more to reach the required size for division (Koch and Schaechter, 1962; Sauls et al., 2016). The timer model (yellow bar) predicts a slope of −1. In a population of cells where growth rate is exponentially correlated with cell size, larger cells are expected to grow more than smaller cells in the same amount of time, giving a negative slope (Robert et al., 2014; Sauls et al., 2016; Willis and Huang, 2017). The adder model supposes that all cells add a constant growth increment from birth to division regardless of birth size and thus gives a slope of 0 (Taheri-Araghi et al., 2015a). The adder-per-origin model posits that all cells grow a constant length from initiation to initiation, and then divide a constant time after initiation. However, a constant adder from initiation to initiation mathematically reduces to an adder correlation from birth to division, giving a slope of zero (Amir, 2014). Light blue bars represent l_b vs. Δl_{bd} slope measurements from populations of model bacteria. *E. coli* and *B. subtilis* from Taheri-Araghi et al. (2015a), *C. crescentus* from Campos et al. (2014), *P. aeruginosa* from Deforet et al. (2015). Dark blue bars represent l_b vs. Δl_{bd} slope measurements from mycobacterial populations. *M. smegmatis* rich media, carbon limited, and BCG from Logsdon et al. (2017). *M. smegmatis* acetate and pyruvate from Priestman et al. (2017). *The timer model was shown in several studies to be incompatible with generation of stable bacterial cell size homeostasis (Trucco and Bell, 1970; Robert et al., 2014).

rates among cells and thus do not reach balanced growth and size homeostasis (Trucco and Bell, 1970; Robert et al., 2014). A key sizer model prediction is that birth size plotted against birth to division elongation is negatively correlated, because smaller cells must grow more to reach the absolute cell size for division (**Figure 3E**) (Koch and Schaechter, 1962; Sauls et al., 2016). Alternatively, a non-correlation or negative correlation between these two parameters had been interpreted as a time-based mechanism (**Figure 3E**) (Sveiczer et al., 1996; Robert et al., 2014; Willis and Huang, 2017). Therefore, when we first observed no correlation between *M. smegmatis* birth size and elongation, we concluded that mycobacteria use a time-based mechanism to control cell division (Aldridge et al., 2012).

Soon after, a new framework for understanding cell size control was introduced centered on the amount of growth rather than absolute size or elapsed time. Growth increments measure the change in length between two events (e.g., cells divide after adding 4 μm to their original length). The new model is referred to as an incremental or adder model of cell size control and postulates that every cell grows a constant length between birth and division, regardless of birth size (**Figure 3C**) (Campos et al., 2014; Taheri-Araghi et al., 2015a). With this in mind, a non-correlation between birth size and birth to division elongation (as seen in *M. smegmatis*) could be reinterpreted as arising from an adder (**Figure 3E**). Several studies adapted this model to predict that a constant change in cell length (Δl_{bd}), surface area, or volume from birth to division regulates cell size

(Amir, 2014; Campos et al., 2014; Deforet et al., 2015; Taheri-Araghi et al., 2015a; Harris and Theriot, 2016; Sauls et al., 2016; Soifer et al., 2016). In all cases, homeostasis is achieved because average birth size converges to the length of the increment added at each generation (Jun and Taheri-Araghi, 2015). Adding a constant length increment compensates for aberrant birth sizes by correcting outlier cells back to the population mean with exponential decay at each round of division (Taheri-Araghi et al., 2015a).

Thus, until recently, it was thought that mycobacteria used an adder mechanism of cell size control. Data from several labs corroborated the constant growth increment from birth to division in populations of *M. smegmatis* in rich medium (**Figure 3E**) (Aldridge et al., 2012; Santi et al., 2013; Logsdon et al., 2017; Priestman et al., 2017). Some of these data were produced several years before the mathematical framework for the adder model of cell size control was introduced, and have been retrospectively interpreted through this lens (Aldridge et al., 2012; Santi et al., 2013; Sauls et al., 2016). However, as recent work has shown, this adder behavior is not consistent across growth conditions (**Figure 3E**).

Several studies tested the growth and division of mycobacteria in sub-optimal carbon conditions because intracellular infection is thought to be carbon limited, with cholesterol serving as a key carbon source for *M. tuberculosis* during persistent infection (Pandey and Sassetti, 2008). Cell growth measurements of *M. smegmatis* cells grown with carbon restricted medium or with the cholesterol by-products acetate or pyruvate as the primary carbon source significantly deviated from expected adder correlations (**Figure 3E**). When cells are grown in acetate as a primary carbon source, accelerators continue to follow the adder principle but alternators are not consistent with an adder, showing a significantly negative correlation between birth length and growth birth to division (**Figure 3E**) (Priestman et al., 2017). Both accelerator and alternator cell populations grown in pyruvate and carbon limited medium deviate from the adder model and show a significantly negative correlation between birth length and growth birth to division (**Figure 3E**) (Logsdon et al., 2017; Priestman et al., 2017). BCG cells growing in rich medium also deviate from the adder model expectation (**Figure 3E**) (Logsdon et al., 2017). In other bacteria that follow the adder model, the constant growth correlation is maintained across many nutrient rich or poor growth media. Instead of shifting the correlation between birth length and total elongation, species like *E. coli* and *B. subtilis* change the value of the length added to cell size to allow cell sizes to adapt to nutrient poor or rich media (Taheri-Araghi et al., 2015a). Therefore, the negative correlation between birth length and birth to division growth indicates that the adder model cannot capture size control in the slow growing mycobacterium species BCG or in *M. smegmatis* growing in nutrient poor medium (**Figure 3E**). It is quite remarkable that the model mycobacterium *M. smegmatis* appeared to be an adder in standard, rich-growth conditions. This apparent (but not true) adder behavior also emphasizes the importance of mathematical modeling and nutritional perturbations in studying bacterial cell size control.

Because the correlation between birth size and growth shifts to become negative in nutrient limited growth conditions (Logsdon et al., 2017; Priestman et al., 2017), this *M. smegmatis* birth to division adder correlation in rich medium presents as an emergent property rather than the actual mechanism of size control (**Figure 3E**). To look beyond simple birth to division models of size control, many studies evaluate the relationship between cell size and intermediate cell cycle events (Cooper and Helmstetter, 1968; Donachie, 1968; Hill et al., 2012; Adiciptaningrum et al., 2015; Ho and Amir, 2015; Soifer et al., 2016; Wallden et al., 2016; Zheng et al., 2016; Banerjee et al., 2017; Osella et al., 2017). Several studies investigating the quantitative basis of cell cycle and size control showed that an initiation-to-initiation adder correlation mathematically reduces to an adder correlation from birth to division (**Figure 3E**) (Amir, 2014; Ho and Amir, 2015). On this basis, it was proposed that cell size control in bacteria is simply a consequence of a mechanism evolved to control the number of replication forks in bacteria (Amir, 2017). These observations served as the backbone of the adder-per-origin model, in which cells add a constant increment from initiation to initiation and then divide a constant time after initiation (**Figure 3D**). The adder-per-origin model has the appealing property of regulating cell size based on the growth rate and number of origins in the cell (particularly in bacteria that use multifork replication). Ori number must have an exponential dependence on growth rate, to ensure that cell growth does not outpace DNA replication (Schaechter et al., 1958; Ho and Amir, 2015). Therefore, an initiation dependent model could regulate mycobacterial cells with different growth rates (e.g., accelerator and alternator sister cells) and cells with multiple initiations per cell cycle (e.g., cells with an E period or multifork replication).

We recently developed an initiation-dependent model of cell size control that describes the specific growth and division properties in mycobacteria. These properties include asymmetric polar growth and division, shifting lengths of B and E period during slow growth, and increased variability in cell size during slow growth (Logsdon et al., 2017). The new model is named the parallel adder model because it is based on a regulatory mechanism in which cells add two constant growth increments simultaneously. One growth increment spans between two DNA replication initiation events (Δl_{ii}) while the other increment spans initiation to division (Δl_{id}) (**Figure 4A**). Together, these parallel increments allow coupled regulation of cell division and cell cycle dynamics. When directly tested against previously established models of cell size control, including adder and adder-per-origin models, the parallel adder model was significantly better fit to 13 key cell cycle and cell size measurements in *M. smegmatis* and BCG (Logsdon et al., 2017). The primary difference between initiation-to-initiation adder models in mycobacteria (parallel adder) versus previously studied *E. coli* (adder-per-origin) is that *E. coli* cell division occurs a constant time after initiation, while mycobacteria divide after a constant growth increment from initiation to division (Amir, 2014; Ho and Amir, 2015; Logsdon et al., 2017). However, in both cases initiation of DNA replication provides the key checkpoint controlling cell division and initiation in the next generation.

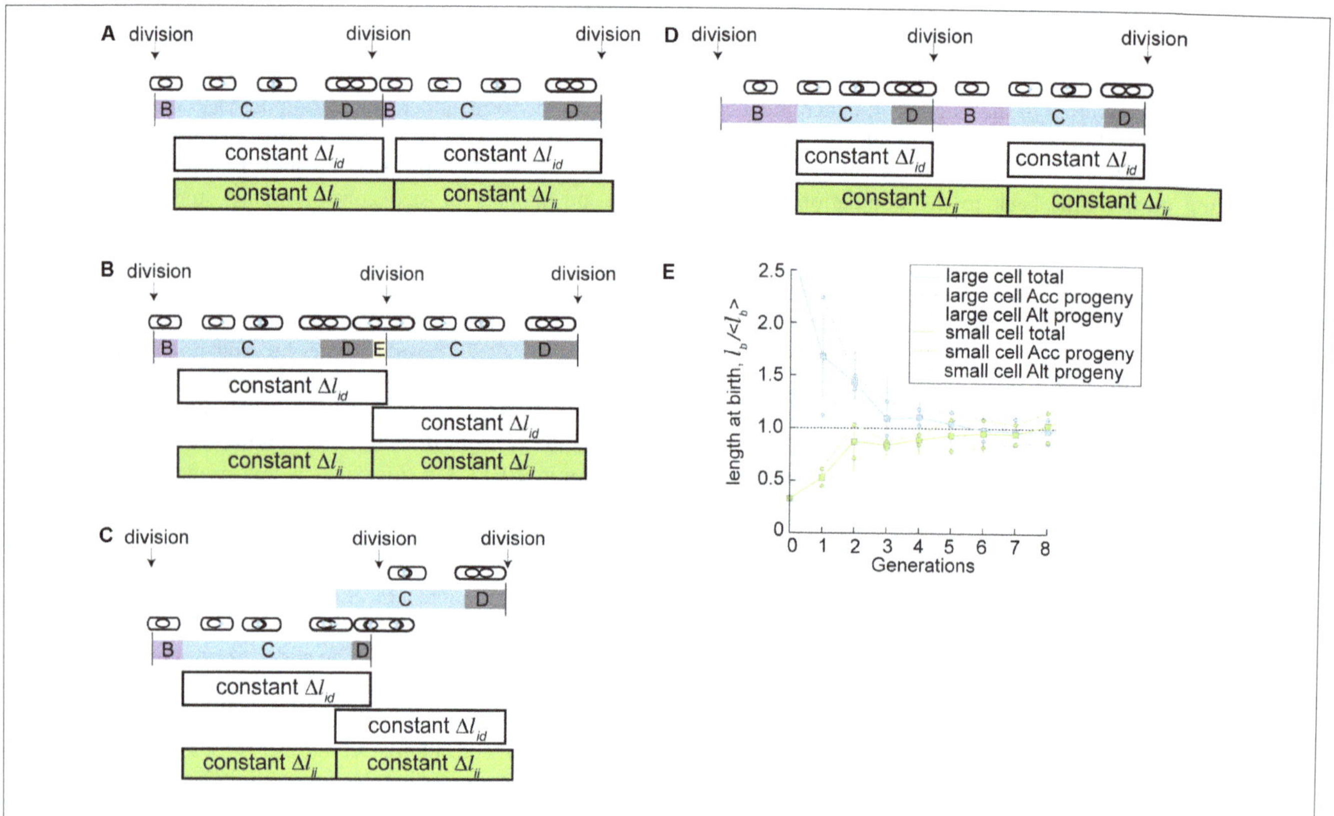

FIGURE 4 | A parallel adder model describes mycobacterial cell size control in fast and slow growth conditions. In the parallel adder model of cell size control, cells grow a constant length between two initiations (Δl_{ii}, green bars), and simultaneously grow a constant length between initiation and division (Δl_{id}, white bars). The mean of these two growth increments represents the average birth size. Initiation of DNA replication occurs at the beginning of C period (blue). **(A)** Parallel adder growth increments (Δl_{id} and Δl_{ii}) describe *M. smegmatis* cell cycle and division dynamics during fast growth. **(B)** The similar length of *M. smegmatis* parallel adder growth increments (Δl_{id} and Δl_{ii}) in fast growth conditions can cause stochastic overlaps in initiation and division, leading to reinitiation during D period, called E period (shown in yellow). The parallel adder accommodates stochastic re-initiation events including E period and allows recovery of population cell cycle dynamics. **(C)** The similar length of *M. smegmatis* parallel adder growth increments (Δl_{id} and Δl_{ii}) in fast growth conditions can cause stochastic overlaps in initiation and division, leading to reinitiation during C period, called multifork replication (shown by overlapping C periods). The parallel adder accommodates stochastic re-initiation events and allows recovery of population cell cycle dynamics. **(D)** The parallel adder model shifts relative lengths of Δl_{id} and Δl_{ii} growth increments during slow growth conditions (carbon limited *M. smegmatis* or BCG in rich medium) to accommodate differences in cell size, growth rate, and cell cycle dynamics during slow growth. However, constant Δl_{id} and Δl_{ii} increments pattern is maintained. **(E)** The parallel adder model leads to cell size convergence in accelerator and alternator subpopulations, despite innate asymmetry. Plot showing parallel adder simulations for progeny of hypothetical cells born extremely large (blue) or small (green). The average cell birth length with SEM bars over eight generations is plotted for a hypothetical cell (accelerator) born 2.53x the population average and a hypothetical cell (alternator) born 0.33x the population average. Average sizes of accelerator and alternator cell progeny from each hypothetical progenitor cell are also plotted with SEM bars over seven generations. Figure **(E)** is adapted from Logsdon et al. (2017).

Importantly, the increments of the parallel adder model shift to allow cells to optimize total size in a variety of growth environments (Logsdon et al., 2017). Parallel adder increments can adjust in total length and in relation to one another, allowing shifts in cell cycle timing and variability, most notably in B period and E period. In rich growth conditions, the Δl_{ii} and Δl_{id} growth increments are similar in length (Logsdon et al., 2017). This necessitates a shorter pre-initiation period (B period) and can often cause an overlap of initiation periods before division, meaning that the mother cell experiences E period (**Figure 4B**) or multifork replication (**Figure 4C**) and daughter cells entirely skip B. However, in BCG and carbon limited *M. smegmatis* cells, Δl_{ii} increments are much longer than Δl_{id}, leading to longer B periods and lowered frequency of both E period and multifork replication (**Figure 4D**). In this way, cell cycles are adjusted to create increased variability in timing under nutrient stress or in slow growing mycobacteria.

With a parallel adder, size convergence occurs by the same principle as other adder models, e.g., if cells add a constant length regardless of their initial size, cells smaller than average will increase in size and cells larger than average will decrease in size until they reach the population average (which is equal to the size of the growth increment added) over several generations (**Figure 4E**) (Amir, 2014; Taheri-Araghi et al., 2015a). The parallel adder regulates birth size by setting the cell length at birth equal to the average of the Δl_{ii} and Δl_{id} increments. Δl_{ii} and Δl_{id} measurements are specific for accelerator and alternator subtypes, and so accelerator and alternator cells converge to distinct average cell sizes (Logsdon et al., 2017). This creates increased variability in mycobacterial size compared to other studied bacteria, and reflects the distinct size, growth, and adder increment properties measured in accelerator vs. alternator cells. As an example, an accelerator cell will add an accelerator specific Δl_{ii} increment as well as an accelerator specific Δl_{id} increment. Averaging the lengths of those two growth increments allows calculation of the birth size of the average accelerator progeny. Accelerator specific increments are larger than alternator specific increments, causing the average accelerator cell size to be larger than that of the larger alternator cell size (**Figure 4E**). Another well characterized asymmetrically dividing bacterial species, *C. crescentus*, exhibits an asymmetric incremental growth pattern that is opposite that of *M. smegmatis*. *C. crescentus* large (stalked) cells add a smaller increment and small (swarmer) cells add a larger increment to compensate for asymmetric birth sizes, leading to less variable birth and division sizes (Campos et al., 2014). Though *M. smegmatis* does not have this mechanism of compensating for asymmetric birth sizes, both accelerator and alternator cell types can recover from aberrant large or small birth sizes to reach homeostasis with the parallel adder model (**Figure 4E**). Therefore, the total population average can also reach homeostasis, albeit slightly more variable than most bacteria (Logsdon et al., 2017).

In the parallel adder model, initiation of DNA replication is the key checkpoint regulating cell division and size control under many conditions. Additional support for an initiation dependent model comes from a recent study identifying a novel essential component of the mycobacterial replication machinery called Rv0004. Rv0004 interacts with helicase loader DnaB and affects the interaction of DnaA and DnaB (Mann et al., 2017). When Rv0004 is depleted from cells growing in rich medium, variability in *M. smegmatis* birth size more than doubles, from a CV of 18% to a CV of 40% (Mann et al., 2017). This increased variability indicates that the process of initiation is molecularly coupled to division, as interrupting replisome assembly disrupts division and cell size regulation. It is possible though that there are additional "backup" mechanisms regulating cell division in the absence of initiation, as many of these cells still do divide, albeit with less size consistency than WT cells (Mann et al., 2017). Additional genetic or chemical inhibition of cell cycle processes will allow us to tease apart regulatory mechanisms of cell division and size control.

There are many additional and important questions regarding size control in mycobacteria to be addressed with future research. While the parallel adder describes mycobacterial growth under all conditions studied thus far (*M. smegmatis* rich growth, *M. smegmatis* carbon limitation, and BCG rich growth), it is possible that other growth conditions may reveal inconsistencies with the model and require further assessment or model modification. Additionally, it has yet to be determined whether any of the developed models of cell size control apply to *M. tuberculosis*. Cell size control models could provide an important tool for probing *M. tuberculosis* population dynamics during infection. Though correlations between cell size and antibiotic tolerance hold under controlled growth conditions, it is not known if they apply to variable and stressful conditions present in a host (Lin et al., 2014; Cadena et al., 2017; Martin et al., 2017). Cell size and cell cycle dynamics are crucial to understand during infection because of previously observed correlations with stress tolerance (Aldridge et al., 2012; Manina et al., 2015; Richardson et al., 2016; Vijay et al., 2017).

In summary, mycobacteria actively create population heterogeneity through growth, division, and cell cycle patterns. The extent of variability is partially determined by growth conditions and environmental stress experienced by mycobacterial cells. Differences in cell size within a population of mycobacteria increase under a wide variety of stressors. *M. smegmatis* and BCG coordinate basic cell cycle properties with their increased variation under stress using a unique model of cell size control. The parallel adder model accommodates inherent variation in cell cycle and growth that is likely important for population survival. Models of the cell cycle and size control could be particularly useful in predicting bacterial population dynamics during infection leading to the development of persistent subpopulations. Understanding the increase in cell-to-cell variability observed in *M. tuberculosis* cells under stress could aid in identifying and targeting the molecular pathways responsible for population heterogeneity and stress tolerance.

Despite the irregularities of mycobacteria compared to other well-studied rod shaped organisms, mycobacterial studies may provide a means of untangling many unanswered questions about bacterial physiology. Mycobacteria actively create exaggerated heterogeneity among closely related cells, allowing the resolution of bacterial population structure with a level of detail never before observed. We can assess functional differences created between (presumably) genetically identical sister cells, and easily identify phenotypic characteristics that allow survival under stress. Mycobacteria do not follow commonly observed patterns of growth, division, and size control, forcing researchers to look beyond established mechanisms developed with data from other bacteria. It became necessary to develop a novel model to account for innate asymmetry and pre-determined cell-to-cell variation observed in mycobacteria. The development of models to describe asymmetric mycobacterial cell properties may provide frameworks through which to understand other asymmetric or pole growing bacteria. In fact, when considering the diversity of bacterial cell sizes and shapes, ranging from filaments to spirochetes to cocci, and spanning sizes of 0.3–750 μm, we cannot help but wonder if mycobacteria are not so unusual after all.

AUTHOR CONTRIBUTIONS

ML and BA conceived of the review topic and structure, wrote the manuscript, and edited the manuscript.

FUNDING

This work was supported by an NIH Director's New Innovator Award 1DP2LM011952-01.

ACKNOWLEDGMENTS

We thank Ariel Amir and Po-Yi Ho for many helpful discussions on bacterial cell size control.

REFERENCES

Adiciptaningrum, A., Osella, M., Moolman, M. C., Cosentino Lagomarsino, M., and Tans, S. J. (2015). Stochasticity and homeostasis in the *E. coli* replication and division cycle. *Sci. Rep.* 5:18261. doi: 10.1038/srep18261

Aldridge, B. B., Fernandez-Suarez, M., Heller, D., Ambravaneswaran, V., Irimia, D., Toner, M., et al. (2012). Asymmetry and aging of mycobacterial cells lead to variable growth and antibiotic susceptibility. *Science* 335, 100–104. doi: 10.1126/science.1216166

Amir, A. (2014). Cell size regulation in bacteria. *Phys. Rev. Lett.* 112:208102. doi: 10.1103/PhysRevLett.112.208102

Amir, A. (2017). Is cell size a spandrel? *eLife* 6:e22186. doi: 10.7554/eLife.22186

Banerjee, S., Lo, K., Daddysman, M. K., Selewa, A., Kuntz, T., Dinner, A. R., et al. (2017). Biphasic growth dynamics control cell division in *Caulobacter crescentus*. *Nat. Microbiol.* 2:17116. doi: 10.1038/nmicrobiol.2017.116

Baronian, G., Ginda, K., Berry, L., Cohen-Gonsaud, M., Zakrzewska-Czerwinska, J., Jakimowicz, D., et al. (2015). Phosphorylation of *Mycobacterium tuberculosis* ParB participates in regulating the ParABS chromosome segregation system. *PLoS One* 10:e0119907. doi: 10.1371/journal.pone.0119907

Botella, H., Yang, G., Ouerfelli, O., Ehrt, S., Nathan, C. F., and Vaubourgeix, J. (2017). Distinct spatiotemporal dynamics of peptidoglycan synthesis between *Mycobacterium smegmatis* and *Mycobacterium tuberculosis*. *mBio* 8:e01183-17. doi: 10.1128/mBio.01183-17

Bowman, G. R., Comolli, L. R., Zhu, J., Eckart, M., Koenig, M., Downing, K. H., et al. (2008). A polymeric protein anchors the chromosomal origin/ParB complex at a bacterial cell pole. *Cell* 134, 945–955. doi: 10.1016/j.cell.2008.07.015

Cadena, A. M., Fortune, S. M., and Flynn, J. L. (2017). Heterogeneity in tuberculosis. *Nat. Rev. Immunol.* 17, 691–702. doi: 10.1038/nri.2017.69

Campos, M., and Jacobs-Wagner, C. (2013). Cellular organization of the transfer of genetic information. *Curr. Opin. Microbiol.* 16, 171–176. doi: 10.1016/j.mib.2013.01.007

Campos, M., Surovtsev, I. V., Kato, S., Paintdakhi, A., Beltran, B., Ebmeier, S. E., et al. (2014). A constant size extension drives bacterial cell size homeostasis. *Cell* 159, 1433–1446. doi: 10.1016/j.cell.2014.11.022

Casart, Y., Gamero, E., Rivera-Gutierrez, S., González-y-Merchand, J. A., and Salazar, L. (2008). *par* genes in *Mycobacterium bovis* and *Mycobacterium smegmatis* are arranged in an operon transcribed from "SigGC" promoters. *BMC Microbiol.* 8:51. doi: 10.1186/1471-2180-8-51

Cermak, N., Olcum, S., Delgado, F. F., Wasserman, S. C., Payer, K. R., A Murakami, M., et al. (2016). High-throughput measurement of single-cell growth rates using serial microfluidic mass sensor arrays. *Nat. Biotechnol.* 34, 1052–1059. doi: 10.1038/nbt.3666

Cohn, D. L. (2001). The effect of BCG vaccination on tuberculin skin testing. Does it matter? *Am. J. Respir. Crit. Care Med.* 164, 915–916. doi: 10.1164/ajrccm.164.6.2107090c

Cooper, S., and Helmstetter, C. E. (1968). Chromosome replication and the division cycle of *Escherichia coli* B/r. *J. Mol. Biol.* 31, 519–540. doi: 10.1016/0022-2836(68)90425-7

Daniel, R. A., and Errington, J. (2003). Control of cell morphogenesis in bacteria: two distinct ways to make a rod-shaped cell. *Cell* 113, 767–776. doi: 10.1016/S0092-8674(03)00421-5

Deforet, M., Van Ditmarsch, D., and Xavier, J. B. (2015). Cell-size homeostasis and the incremental rule in a bacterial pathogen. *Biophys. J.* 109, 521–528. doi: 10.1016/j.bpj.2015.07.002

Ditse, Z., Lamers, M. H., and Warner, D. F. (2017). DNA replication in *Mycobacterium tuberculosis*. *Microbiol. Spectr.* 5:TBTB2-0027-2016.

Dominguez-Escobar, J., Chastanet, A., Crevenna, A. H., Fromion, V., Wedlich-Soldner, R., and Carballido-Lopez, R. (2011). Processive movement of MreB-associated cell wall biosynthetic complexes in bacteria. *Science* 333, 225–228. doi: 10.1126/science.1203466

Donachie, W. D. (1968). Relationship between cell size and time of initiation of DNA replication. *Nature* 219, 1077–1079. doi: 10.1038/2191077a0

Errington, J. (2015). Bacterial morphogenesis and the enigmatic MreB helix. *Nat. Rev. Microbiol.* 13, 241–248. doi: 10.1038/nrmicro3398

Eskandarian, H. A., Odermatt, P. D., Ven, J. X. Y., Hannebelle, M. T. M., Nievergelt, A. P., Dhar, N., et al. (2017). Division site selection linked to inherited cell surface wave troughs in mycobacteria. *Nat. Microbiol.* 2:17094. doi: 10.1038/nmicrobiol.2017.94

Fogel, M. A., and Waldor, M. K. (2005). Distinct segregation dynamics of the two *Vibrio cholerae* chromosomes. *Mol. Microbiol.* 55, 125–136. doi: 10.1111/j.1365-2958.2004.04379.x

Garner, E. C., Bernard, R., Wang, W., Zhuang, X., Rudner, D. Z., and Mitchison, T. (2011). Coupled, circumferential motions of the cell wall synthesis machinery and MreB filaments in *B. subtilis*. *Science* 333, 222–225. doi: 10.1126/science.1203285

Ginda, K., Bezulska, M., Ziolkiewicz, M., Dziadek, J., Zakrzewska-Czerwinska, J., and Jakimowicz, D. (2013). ParA of *Mycobacterium smegmatis* co-ordinates chromosome segregation with the cell cycle and interacts with the polar growth determinant DivIVA. *Mol. Microbiol.* 87, 998–1012. doi: 10.1111/mmi.12146

Ginda, K., Santi, I., Bousbaine, D., Zakrzewska-Czerwinska, J., Jakimowicz, D., and Mckinney, J. (2017). The studies of ParA and ParB dynamics reveal asymmetry of chromosome segregation in mycobacteria. *Mol. Microbiol.* 105, 453–468. doi: 10.1111/mmi.13712

Godin, M., Delgado, F. F., Son, S., Grover, W. H., Bryan, A. K., Tzur, A., et al. (2010). Using buoyant mass to measure the growth of single cells. *Nat. Methods* 7, 387–390. doi: 10.1038/nmeth.1452

Govindarajan, S., Nevo-Dinur, K., and Amster-Choder, O. (2012). Compartmentalization and spatiotemporal organization of macromolecules in bacteria. *FEMS Microbiol. Rev.* 36, 1005–1022. doi: 10.1111/j.1574-6976.2012.00348.x

Harms, A., Treuner-Lange, A., Schumacher, D., and Sogaard-Andersen, L. (2013). Tracking of chromosome and replisome dynamics in *Myxococcus xanthus* reveals a novel chromosome arrangement. *PLoS Genet.* 9:e1003802. doi: 10.1371/journal.pgen.1003802

Harris, L. K., and Theriot, J. A. (2016). Relative rates of surface and volume synthesis set bacterial cell size. *Cell* 165, 1479–1492. doi: 10.1016/j.cell.2016.05.045

Hayashi, J. M., Luo, C. Y., Mayfield, J. A., Hsu, T., Fukuda, T., Walfield, A. L., et al. (2016). Spatially distinct and metabolically active membrane domain in mycobacteria. *Proc. Natl. Acad. Sci. U.S.A.* 113, 5400–5405. doi: 10.1073/pnas.1525165113

Helmstetter, C. E. (1968). DNA synthesis during the division cycle of rapidly growing *Escherichia coli* B/r. *J. Mol. Biol.* 31, 507–518. doi: 10.1016/0022-2836(68)90424-5

Hett, E. C., and Rubin, E. J. (2008). Bacterial growth and cell division: a mycobacterial perspective. *Microbiol. Mol. Biol. Rev.* 72, 126–156. doi: 10.1128/MMBR.00028-07

Hill, N. S., Kadoya, R., Chattoraj, D. K., and Levin, P. A. (2012). Cell size and the initiation of DNA replication in bacteria. *PLoS Genet.* 8:e1002549. doi: 10.1371/journal.pgen.1002549

Ho, P. Y., and Amir, A. (2015). Simultaneous regulation of cell size and chromosome replication in bacteria. *Front. Microbiol.* 6:662. doi: 10.3389/fmicb.2015.00662

Holowka, J., Trojanowski, D., Ginda, K., Wojtas, B., Gielniewski, B., Jakimowicz, D., et al. (2017). HupB is a bacterial nucleoid-associated protein with an indispensable eukaryotic-like tail. *mBio* 8:e01272-17. doi: 10.1128/mBio.01272-17

Iniesta, A. A. (2014). ParABS system in chromosome partitioning in the bacterium *Myxococcus xanthus*. *PLoS One* 9:e86897. doi: 10.1371/journal.pone.0086897

Iyer-Biswas, S., Wright, C. S., Henry, J. T., Lo, K., Burov, S., Lin, Y., et al. (2014). Scaling laws governing stochastic growth and division of single bacterial cells. *Proc. Natl. Acad. Sci. U.S.A.* 111, 15912–15917. doi: 10.1073/pnas.1403232111

Jankute, M., Cox, J. A., Harrison, J., and Besra, G. S. (2015). Assembly of the Mycobacterial Cell Wall. *Annu. Rev. Microbiol.* 69, 405–423. doi: 10.1146/annurev-micro-091014-104121

Jun, S., Si, F., Pugatch, R., and Scott, M. (2018). Fundamental principles in bacterial physiology - history, recent progress, and the future with focus on cell size control: a review. *Rep. Prog. Phys.* doi: 10.1088/1361-6633/aaa628 [Epub ahead of print].

Jun, S., and Taheri-Araghi, S. (2015). Cell-size maintenance: universal strategy revealed. *Trends Microbiol.* 23, 4–6. doi: 10.1016/j.tim.2014.12.001

Kang, C. M., Nyayapathy, S., Lee, J. Y., Suh, J. W., and Husson, R. N. (2008). Wag31, a homologue of the cell division protein DivIVA, regulates growth, morphology and polar cell wall synthesis in mycobacteria. *Microbiology* 154, 725–735. doi: 10.1099/mic.0.2007/014076-0

Kieser, K. J., and Rubin, E. J. (2014). How sisters grow apart: mycobacterial growth and division. *Nat. Rev. Microbiol.* 12, 550–562. doi: 10.1038/nrmicro3299

Koch, A. L., and Schaechter, M. (1962). A model for statistics of the cell division process. *J. Gen. Microbiol.* 29, 435–454. doi: 10.1099/00221287-29-3-435

Kuru, E., Tekkam, S., Hall, E., Brun, Y. V., and Van Nieuwenhze, M. S. (2015). Synthesis of fluorescent D-amino acids and their use for probing peptidoglycan synthesis and bacterial growth in situ. *Nat. Protoc.* 10, 33–52. doi: 10.1038/nprot.2014.197

Kysela, D. T., Brown, P. J., Huang, K. C., and Brun, Y. V. (2013). Biological consequences and advantages of asymmetric bacterial growth. *Annu. Rev. Microbiol.* 67, 417–435. doi: 10.1146/annurev-micro-092412-155622

Lagage, V., Boccard, F., and Vallet-Gely, I. (2016). Regional control of chromosome segregation in *Pseudomonas aeruginosa*. *PLoS Genet.* 12:e1006428. doi: 10.1371/journal.pgen.1006428

Landgraf, D., Okumus, B., Chien, P., Baker, T. A., and Paulsson, J. (2012). Segregation of molecules at cell division reveals native protein localization. *Nat. Methods* 9, 480–482. doi: 10.1038/nmeth.1955

Lee, P. S., and Grossman, A. D. (2006). The chromosome partitioning proteins Soj (ParA) and Spo0J (ParB) contribute to accurate chromosome partitioning, separation of replicated sister origins, and regulation of replication initiation in *Bacillus subtilis*. *Mol. Microbiol.* 60, 853–869. doi: 10.1111/j.1365-2958.2006.05140.x

Lin, P. L., Ford, C. B., Coleman, M. T., Myers, A. J., Gawande, R., Ioerger, T., et al. (2014). Sterilization of granulomas is common in active and latent tuberculosis despite within-host variability in bacterial killing. *Nat. Med.* 20, 75–79. doi: 10.1038/nm.3412

Logsdon, M. M., Ho, P. Y., Papavinasasundaram, K., Richardson, K., Cokol, M., Sassetti, C. M., et al. (2017). A parallel adder coordinates mycobacterial cell-cycle progression and cell-size homeostasis in the context of asymmetric growth and organization. *Curr. Biol.* 27, 3367.e7–3374.e7. doi: 10.1016/j.cub.2017.09.046

Maloney, E., Madiraju, M., and Rajagopalan, M. (2009). Overproduction and localization of *Mycobacterium tuberculosis* ParA and ParB proteins. *Tuberculosis* 89(Suppl. 1), S65–S69. doi: 10.1016/S1472-9792(09)70015-0

Manina, G., Dhar, N., and Mckinney, J. D. (2015). Stress and host immunity amplify *Mycobacterium tuberculosis* phenotypic heterogeneity and induce nongrowing metabolically active forms. *Cell Host Microbe* 17, 32–46. doi: 10.1016/j.chom.2014.11.016

Mann, K. M., Huang, D. L., Hooppaw, A. J., Logsdon, M. M., Richardson, K., Lee, H. J., et al. (2017). Rv0004 is a new essential member of the mycobacterial DNA replication machinery. *PLoS Genet.* 13:e1007115. doi: 10.1371/journal.pgen.1007115

Martin, C. J., Cadena, A. M., Leung, V. W., Lin, P. L., Maiello, P., Hicks, N., et al. (2017). Digitally barcoding *Mycobacterium tuberculosis* reveals in vivo infection dynamics in the macaque model of tuberculosis. *mBio* 8:e00312-17 doi: 10.1128/mBio.00312-17

Meniche, X., Otten, R., Siegrist, M. S., Baer, C. E., Murphy, K. C., Bertozzi, C. R., et al. (2014). Subpolar addition of new cell wall is directed by DivIVA in mycobacteria. *Proc. Natl. Acad. Sci. U.S.A.* 111, E3243–E3251. doi: 10.1073/pnas.1402158111

Mierzejewska, J., and Jagura-Burdzy, G. (2012). Prokaryotic ParA-ParB-parS system links bacterial chromosome segregation with the cell cycle. *Plasmid* 67, 1–14. doi: 10.1016/j.plasmid.2011.08.003

Monahan, L. G., Liew, A. T., Bottomley, A. L., and Harry, E. J. (2014). Division site positioning in bacteria: one size does not fit all. *Front. Microbiol.* 5:19. doi: 10.3389/fmicb.2014.00019

Montero Llopis, P., Jackson, A. F., Sliusarenko, O., Surovtsev, I., Heinritz, J., Emonet, T., et al. (2010). Spatial organization of the flow of genetic information in bacteria. *Nature* 466, 77–81. doi: 10.1038/nature09152

Nair, N., Dziedzic, R., Greendyke, R., Muniruzzaman, S., Rajagopalan, M., and Madiraju, M. V. (2009). Synchronous replication initiation in novel *Mycobacterium tuberculosis* dnaA cold-sensitive mutants. *Mol. Microbiol.* 71, 291–304. doi: 10.1111/j.1365-2958.2008.06523.x

Nevo-Dinur, K., Nussbaum-Shochat, A., Ben-Yehuda, S., and Amster-Choder, O. (2011). Translation-independent localization of mRNA in *E. coli*. *Science* 331, 1081–1084. doi: 10.1126/science.1195691

Osella, M., Tans, S. J., and Cosentino Lagomarsino, M. (2017). Step by step, cell by cell: quantification of the bacterial cell cycle. *Trends Microbiol.* 25, 250–256. doi: 10.1016/j.tim.2016.12.005

Pandey, A. K., and Sassetti, C. M. (2008). Mycobacterial persistence requires the utilization of host cholesterol. *Proc. Natl. Acad. Sci. U.S.A.* 105, 4376–4380. doi: 10.1073/pnas.0711159105

Priestman, M., Thomas, P., Robertson, B. D., and Shahrezaei, V. (2017). Mycobacteria modify their cell size control under sub-optimal carbon sources. *Front. Cell Dev. Biol.* 5:64. doi: 10.3389/fcell.2017.00064

Rego, E. H., Audette, R. E., and Rubin, E. J. (2017). Deletion of a mycobacterial divisome factor collapses single-cell phenotypic heterogeneity. *Nature* 546, 153–157. doi: 10.1038/nature22361

Reyes-Lamothe, R., Possoz, C., Danilova, O., and Sherratt, D. J. (2008). Independent positioning and action of *Escherichia coli* replisomes in live cells. *Cell* 133, 90–102. doi: 10.1016/j.cell.2008.01.044

Richardson, K., Bennion, O. T., Tan, S. M., Hoang, A. N., Cokol, M., and Aldridge, B. B. (2016). Temporal and intrinsic factors of rifampicin tolerance in mycobacteria. *Proc. Natl. Acad. Sci. U.S.A.* 113, 8302–8307. doi: 10.1073/pnas.1600372113

Robert, L., Hoffmann, M., Krell, N., Aymerich, S., Robert, J., and Doumic, M. (2014). Division in *Escherichia coli* is triggered by a size-sensing rather than a timing mechanism. *BMC Biol.* 12:17. doi: 10.1186/1741-7007-12-17

Rock, J. M., Lang, U. F., Chase, M. R., Ford, C. B., Gerrick, E. R., Gawande, R., et al. (2015). DNA replication fidelity in *Mycobacterium tuberculosis* is mediated by an ancestral prokaryotic proofreader. *Nat. Genet.* 47, 677–681. doi: 10.1038/ng.3269

Rohde, K. H., Veiga, D. F., Caldwell, S., Balazsi, G., and Russell, D. G. (2012). Linking the transcriptional profiles and the physiological states of *Mycobacterium tuberculosis* during an extended intracellular infection. *PLoS Pathog.* 8:e1002769. doi: 10.1371/journal.ppat.1002769

Santi, I., Dhar, N., Bousbaine, D., Wakamoto, Y., and Mckinney, J. D. (2013). Single-cell dynamics of the chromosome replication and cell division cycles in mycobacteria. *Nat. Commun.* 4:2470. doi: 10.1038/ncomms3470

Santi, I., and McKinney, J. D. (2015). Chromosome organization and replisome dynamics in *Mycobacterium smegmatis*. *mBio* 6:e01999-14. doi: 10.1128/mBio.01999-14

Sauls, J. T., Li, D., and Jun, S. (2016). Adder and a coarse-grained approach to cell size homeostasis in bacteria. *Curr. Opin. Cell Biol.* 38, 38–44. doi: 10.1016/j.ceb.2016.02.004

Schaechter, M., Maaloe, O., and Kjeldgaard, N. O. (1958). Dependency on medium and temperature of cell size and chemical composition during balanced grown of *Salmonella typhimurium*. *J. Gen. Microbiol.* 19, 592–606. doi: 10.1099/00221287-19-3-592

Schumacher, M. A. (2017). Bacterial nucleoid occlusion: multiple mechanisms for preventing chromosome bisection during cell division. *Subcell. Biochem.* 84, 267–298. doi: 10.1007/978-3-319-53047-5_9

Siegrist, M. S., Whiteside, S., Jewett, J. C., Aditham, A., Cava, F., and Bertozzi, C. R. (2013). (D)-Amino acid chemical reporters reveal peptidoglycan dynamics of an intracellular pathogen. *ACS Chem. Biol.* 8, 500–505. doi: 10.1021/cb3004995

Slager, J., and Veening, J. W. (2016). Hard-wired control of bacterial processes by chromosomal gene location. *Trends Microbiol.* 24, 788–800. doi: 10.1016/j.tim.2016.06.003

Sobetzko, P., Travers, A., and Muskhelishvili, G. (2012). Gene order and chromosome dynamics coordinate spatiotemporal gene expression during the bacterial growth cycle. *Proc. Natl. Acad. Sci. U.S.A.* 109, E42–E50. doi: 10.1073/pnas.1108229109

Soifer, I., Robert, L., and Amir, A. (2016). Single-cell analysis of growth in budding yeast and bacteria reveals a common size regulation strategy. *Curr. Biol.* 26, 356–361. doi: 10.1016/j.cub.2015.11.067

Soler-Bistue, A., Mondotte, J. A., Bland, M. J., Val, M. E., Saleh, M. C., and Mazel, D. (2015). Genomic location of the major ribosomal protein gene locus determines *Vibrio cholerae* global growth and infectivity. *PLoS Genet.* 11:e1005156. doi: 10.1371/journal.pgen.1005156

Sukumar, N., Tan, S., Aldridge, B. B., and Russell, D. G. (2014). Exploitation of *Mycobacterium tuberculosis* reporter strains to probe the impact of vaccination at sites of infection. *PLoS Pathog.* 10:e1004394. doi: 10.1371/journal.ppat.1004394

Sveiczer, A., Novak, B., and Mitchison, J. M. (1996). The size control of fission yeast revisited. *J. Cell Sci.* 109(Pt 12), 2947–2957.

Taheri-Araghi, S., Bradde, S., Sauls, J. T., Hill, N. S., Levin, P. A., Paulsson, J., et al. (2015a). Cell-size control and homeostasis in bacteria. *Curr. Biol.* 25, 385–391. doi: 10.1016/j.cub.2014.12.009

Taheri-Araghi, S., Brown, S. D., Sauls, J. T., Mcintosh, D. B., and Jun, S. (2015b). Single-cell physiology. *Annu. Rev. Biophys.* 44, 123–142. doi: 10.1146/annurev-biophys-060414-034236

Takacs, C. N., Poggio, S., Charbon, G., Pucheault, M., Vollmer, W., and Jacobs-Wagner, C. (2010). MreB drives de novo rod morphogenesis in *Caulobacter crescentus* via remodeling of the cell wall. *J. Bacteriol.* 192, 1671–1684. doi: 10.1128/JB.01311-09

Trojanowski, D., Ginda, K., Pioro, M., Holowka, J., Skut, P., Jakimowicz, D., et al. (2015). Choreography of the *Mycobacterium* replication machinery during the cell cycle. *mBio* 6:e02125-14. doi: 10.1128/mBio.02125-14

Trojanowski, D., Holowka, J., Ginda, K., Jakimowicz, D., and Zakrzewska-Czerwinska, J. (2017). Multifork chromosome replication in slow-growing bacteria. *Sci. Rep.* 7:43836. doi: 10.1038/srep43836

Trucco, E., and Bell, G. I. (1970). A note on the dispersionless growth law for single cells. *Bull. Math. Biophys.* 32, 475–483. doi: 10.1007/BF02476766

Vallet-Gely, I., and Boccard, F. (2013). Chromosomal organization and segregation in *Pseudomonas aeruginosa*. *PLoS Genet.* 9:e1003492. doi: 10.1371/journal.pgen.1003492

Vijay, S., Vinh, D. N., Hai, H. T., Ha, V. T. N., Dung, V. T. M., Dinh, T. D., et al. (2017). Influence of stress and antibiotic resistance on cell-length distribution in *Mycobacterium tuberculosis* clinical isolates. *Front. Microbiol.* 8:2296. doi: 10.3389/fmicb.2017.02296

Wakamoto, Y., Dhar, N., Chait, R., Schneider, K., Signorino-Gelo, F., Leibler, S., et al. (2013). Dynamic persistence of antibiotic-stressed mycobacteria. *Science* 339, 91–95. doi: 10.1126/science.1229858

Wallden, M., Fange, D., Lundius, E. G., Baltekin, O., and Elf, J. (2016). The synchronization of replication and division cycles in individual *E. coli* cells. *Cell* 166, 729–739. doi: 10.1016/j.cell.2016.06.052

Wang, S., Furchtgott, L., Huang, K. C., and Shaevitz, J. W. (2012). Helical insertion of peptidoglycan produces chiral ordering of the bacterial cell wall. *Proc. Natl. Acad. Sci. U.S.A.* 109, E595–E604. doi: 10.1073/pnas.1117132109

Wang, X., Montero Llopis, P., and Rudner, D. Z. (2014). *Bacillus subtilis* chromosome organization oscillates between two distinct patterns. *Proc. Natl. Acad. Sci. U.S.A.* 111, 12877–12882. doi: 10.1073/pnas.1407461111

White, C. L., and Gober, J. W. (2012). MreB: pilot or passenger of cell wall synthesis? *Trends Microbiol.* 20, 74–79. doi: 10.1016/j.tim.2011.11.004

WHO (2017). *Global Tuberculosis Report 2017.* Geneva: World Health Organization.

Willis, L., and Huang, K. C. (2017). Sizing up the bacterial cell cycle. *Nat. Rev. Microbiol.* 15, 606–620. doi: 10.1038/nrmicro.2017.79

Wu, L. J., and Errington, J. (2011). Nucleoid occlusion and bacterial cell division. *Nat. Rev. Microbiol.* 10, 8–12. doi: 10.1038/nrmicro2671

Zheng, H., Ho, P. Y., Jiang, M., Tang, B., Liu, W., Li, D., et al. (2016). Interrogating the *Escherichia coli* cell cycle by cell dimension perturbations. *Proc. Natl. Acad. Sci. U.S.A.* 113, 15000–15005. doi: 10.1073/pnas.1617932114

Zhou, X., Halladin, D. K., and Theriot, J. A. (2016). Fast mechanically driven daughter cell separation is widespread in actinobacteria. *mBio* 7:e00952-16. doi: 10.1128/mBio.00952-16

5

The dynamic nature and territory of transcriptional machinery in the bacterial chromosome

Ding J. Jin, Cedric Cagliero, Carmen M. Martin, Jerome Izard and Yan N. Zhou*

Transcription Control Section, Gene Regulation and Chromosome Biology Laboratory, National Cancer Institute, National Institutes of Health, Frederick, MD, USA

***Correspondence:**
*Ding J. Jin,
Transcription Control Section, Gene Regulation and Chromosome Biology Laboratory, National Cancer Institute,
PO Box B, Frederick, MD 21702, USA
jind@mail.nih.gov*

Our knowledge of the regulation of genes involved in bacterial growth and stress responses is extensive; however, we have only recently begun to understand how environmental cues influence the dynamic, three-dimensional distribution of RNA polymerase (RNAP) in *Escherichia coli* on the level of single cell, using wide-field fluorescence microscopy and state-of-the-art imaging techniques. Live-cell imaging using either an agarose-embedding procedure or a microfluidic system further underscores the dynamic nature of the distribution of RNAP in response to changes in the environment and highlights the challenges in the study. A general agreement between live-cell and fixed-cell images has validated the formaldehyde-fixing procedure, which is a technical breakthrough in the study of the cell biology of RNAP. In this review we use a systems biology perspective to summarize the advances in the cell biology of RNAP in *E. coli*, including the discoveries of the bacterial nucleolus, the spatial compartmentalization of the transcription machinery at the periphery of the nucleoid, and the segregation of the chromosome territories for the two major cellular functions of transcription and replication in fast growing cells. Our understanding of the coupling of transcription and bacterial chromosome (or nucleoid) structure is also summarized. Using *E. coli* as a simple model system, co-imaging of RNAP with DNA and other factors during growth and stress responses will continue to be a useful tool for studying bacterial growth and adaptation in changing environment.

Keywords: RNA polymerase, bacterial nucleolus, replisome, chromosome territories, growth rate regulation, stress responses, superresolution imaging, *E. coli*

Features of E. Coli Genome That Are Important for Cell Growth and Chromosome Replication

Genome, Gene, and Growth Rate Regulation

Escherichia coli cells, such as the prototype K-12 strain MG1655, are small rod-shaped, gram-negative bacteria. The *E. coli* genome contains ~4.6 million base pairs (bp). If fully stretched, a single *E. coli* genomic DNA is ~1600 μm long, ~1000-fold longer than the length of the cell; therefore the genome must be fully compacted to fit into a cell. The genome encodes 4453 genes, which are organized into about 2390 operons (Blattner et al., 1997; Riley et al., 2006). Not all genes are equal in terms of growth (or growth rate) regulation. *E. coli* genes can be broadly categorized in two functional classes: growth-promoting genes, represented by ribosomal RNA (rRNA) operons (for simplicity hereafter called *rrn*) and other genes.

The bacterial growth rate is determined by the growth medium (Kjeldgaard et al., 1958; Schaechter et al., 1958). Growth-promoting genes are few in number and mainly involved the synthesis of the translationary machinery, primarily *rrn* that encodes different species of rRNA and tRNA. Synthesis of rRNA is a rate-limiting step for the production of ribosomes (Gausing, 1980), as ribosomes are assembled onto nascent rRNAs. The number of ribosomes in the cell is proportional to the growth rate, which is needed to meet the demand for protein synthesis (Bremer and Dennis, 1996; Keener and Nomura, 1996). Because of the important role of rRNA synthesis in growth-rate regulation, the regulation of *rrn* has been extensively studied (Condon et al., 1995; Gourse et al., 1996; Wagner, 2002; Paul et al., 2004; Potrykus et al., 2011; Jin et al., 2012; Ross et al., 2013). The promoters of *rrn* are the most actively transcribed, accounting for >80% of total RNA synthesis in cells growing in nutrient-rich media (Bremer and Dennis, 1996), but become marginal under poor growth conditions or by the treatment of serine hydroxamate (SHX), a serine analog that triggers amino acid starvation (Tosa and Pizer, 1971) and induces the stringent response (Cashel et al., 1996). In addition, transcription of *rrn* is also regulated by an antitermination system containing NusA and NusB as well as other factors. While NusA binds to RNAP *in vitro* (Greenblatt and Li, 1981), NusB does not. NusB is thought to bind *in vivo* to the BoxA RNA sequences of nascent rRNA molecules and is also involved in rRNA processing (Torres et al., 2004; Bubunenko et al., 2013). Another difference between the two functional classes of genes is their respective genomic DNA content (see below): while the seven *rrn* operons (each ~5.5 kb in length) represent only ~1% of genomic DNA, other genes represent 99% of the genome.

Bacterial Growth and Chromosome Replication

The *E. coli* genetic map is shown in **Figure 1A**. The chromosome is a circular DNA molecule with a specific origin of chromosome replication (*oriC*). After initiation, DNA replication proceeds bidirectionally as a pair of replication forks toward the terminus region (*ter*). As an integral component of replisomes (Yao and O'Donnell, 2010), single-stranded DNA binding protein (SSB) (Reyes-Lamothe et al., 2008; Marceau et al., 2011) coats the single-stranded DNA at the replication forks and interacts with the DNA polymerase III holoenzyme (O'Donnell, 2006). SeqA protein polymerizes with the nascent hemimethylated DNA at or near the DNA replication forks (Slater et al., 1995; Yamazoe et al., 2005; Waldminghaus et al., 2012).

Several features related to the *E. coli* lifestyle and the location of growth-promoting genes in the genome are important with respect to bacterial growth and chromosome replication.

First, the cell size and the copy number of the bacterial chromosome in a cell are sensitive to growth rate (Jin et al., 2013) (**Figure 1B**). The combined time required to complete a round of replication and subsequently chromosome segregation and cell division varies from ~70 to 150 min, depending on growth conditions (Stokke et al., 2012). Consequently, there are 1.5 genome equivalents (displaying one nucleoid surrounded by ribosomes) in a small slow-growing cell (doubling time 135 min) in M63 nutrient-poor media, containing glycerol as the carbon source at 30°C. When a cell's doubling time is shorter than the combined time required for replication and segregation, multiple initiations from *oriC* occur in preceding generations, manifesting "multi-fork replication" to ensure that at least one completed genome is passed onto each of the two newly formed daughter cells (Cooper and Helmstetter, 1968). For example, in a cell growing in nutrient-rich Lennox Broth (LB) at 37°C with a doubling time of ~20 min, the cell is large and there are up

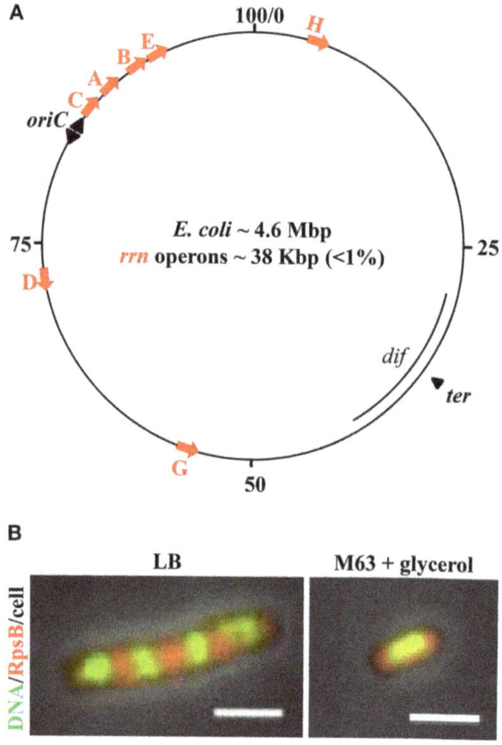

	LB	M63 + glycerol
Doubling time	20min	135min
Genome equivalent	5.0	1.5
Copy of *rrn*	50	12

FIGURE 1 | *E. coli* chromosome map and features important for growth rate regulation and genome replication. (A) The genetic map of *E. coli* K-12. Positions of the *rrn* operons, along with *oriC*, and *ter* regions, are indicated. The red arrows represent each of the seven *rrn* operons, all of which are localized in the *oriC* proximal half of the chromosome; four are near the *oriC*, from which replication initiates and elongates bidirectionally, as indicated by black arrowheads. Transcription of *rrn* occurs in the same direction as replication. **(B)** Cell size and the number of genome equivalents and the *rrn* gene copies in a cell are sensitive to growth rate, as determined by growth medium. Images are overlays of nucleoid (green)/ribosome (red)/cell. Exponential-phase cells were prepared for imaging as described (Cabrera and Jin, 2003b). Ribosomes are tracked by 30S ribosomal subunit protein S2 fused with mCherry at the carboxy terminus (*rpsB-mCherry*). The nucleoid (DNA) is stained with Hoescht 33342. The cell growing in nutrient-rich LB (37°C) is large and has four apparent nascent nucleoids surrounded by ribosomes. The cell growing in minimal M63 + glycerol (30°C) is small and has only one apparent nucleoid surrounded with ribosomes. The scale bar represents 2 μm.

to five genome equivalents (displaying four nascent nucleoids surrounded by plentiful of ribosomes), and 8–16 copies of the *oriC* (Nielsen et al., 2007).

The second feature is the location of the seven *rrn* operons. All seven are located in the *oriC* half of the genome, and four of these are close to *oriC*. Because of the unique location of *rrn* relative to *oriC*, the gene dosage of *rrn* magnifies as the growth rate increases. For example, it can be calculated, based on the growth rate and the location of the *rrn* in the genetic map (Condon et al., 1993; Bremer and Dennis, 1996), that when cells are growing in LB at 30°C, with a doubling time of 45 min, the copy number for *rrn* is 18 per cell with more than two genome equivalents; however, with a doubling time of 20 min (LB, 37°C), the copy number of *rrn* approaches 50 per cell. This feature is particularly significant, as it dictates the differential allocation of RNAP between *rrn* and other genes genome-wide in response to changes in the environment, as described below. In addition, the transcription of *rrn* and most of the components of the translational machinery is in the same direction as that of chromosome replication (Rocha and Danchin, 2003; Lin et al., 2010). This co-directionality has likely evolved to minimize replication–transcription conflicts in bacteria (Merrikh et al., 2012).

Differential Allocation of RNA Polymerase between *rrn* and Other Genes Genome-Wide in Response to a Changing Environment: Evidence from Genetic and Genomic Studies

Competition for Limited RNAP between *rrn* and Other Genes Genome-Wide in Fast-Growing Cells

It becomes evident that RNAP is limiting in the cell for the simultaneous active expression of *rrn* and the broad transcription of other genes genome-wide. This notion was first suggested by the study of a class of RNAP mutants that altered growth rate regulation: these mutants grew slowly in LB because of reduced transcription of *rrn*, as if they were partially starved for amino acids, even in nutrient-rich medium. In contrast, the expression of some other genes that are activated only during starvation in wild type was elevated in those mutants grown in LB (Zhou and Jin, 1997, 1998). Combined with the biochemical studies of these RNAP mutants, it was proposed that RNAP is limiting in the cell for simultaneous active transcription of *rrn* and other genes involved in the stress response and thus RNAP redistributes from *rrn* to other genes in the genome during the stringent response induced by amino acid starvation (Zhou and Jin, 1998). Analysis of these RNAP mutants motivated the initiation of research on the cell biology of RNAP as described below.

RNAP is primarily allocated to the active transcription of *rrn* in cells growing in nutrient-rich media. Micrographs of chromatin spreads of fast-growing cells revealed that RNAP molecules are packed during *rrn* synthesis, at an estimated 65 RNAP molecules per *rrn* (French and Miller, 1989). The estimated number of RNAP molecules in a cell varies by quantitative Western blot analysis (Ishihama, 2000; Grigorova et al., 2006; Piper et al., 2009), but direct counting of single molecules using photoactivated localization microscopy (PALM) yields an independent measurement of ~5000 RNAP molecules in a fast-growing cell (Bakshi et al., 2012; Endesfelder et al., 2013). Consequently, in a fast-growing cell with a doubling time of 20 min (LB, 37°C), active transcription of the estimated 50 copies of *rrn* will consume 3250 RNAP molecules, indicating that ~65% of the total RNAP (assuming ~5000 molecules per cell) is allocated for rRNA synthesis. To use an analogy, in fast-growing cells, most of the bacterial RNAP functions as eukaryotic Pol I/III activities to promote growth. In eukaryotic cells Pol I is responsible for the synthesis of rRNA at the nucleolus (O'Sullivan et al., 2013) and Pol III makes small RNAs including tRNAs. Consequently, only a small fraction of *E. coli* RNAP engages in transcription of other genes in the remaining ~99% of the genome, consistent with the finding that most regions of the chromatin spreads appeared to be "naked," or devoid of RNAP (French and Miller, 1989).

Redistribution of RNAP from *rrn* to Other Genes Genome-Wide during Stress Responses

When fast-growing cells are exposed to nutrient limitation/starvation or to stresses, the immediate response of cells is to "turn off" *rrn* expression, leading to the decrease and/or arrest of growth, and concomitantly to "turn on" other genes involved in the response. The "feast or famine" or "thrive or survive" lifestyle not only reflects the amazing ability of *E. coli* to rapidly respond and/or adapt to changing environments but also represents the two ends of the spectrum in bacterial growth rate regulation. Global transcriptional profiling of cells grown on different carbon sources has indicated that there is an apparent inverse correlation between the growth potential of the environment and the number of genes expressed systematically genome-wide (Liu et al., 2005). The poorer the quality of carbon source used, the larger the number of genes that are expressed genome-wide. Similarly, when fast-growing cells are subject to amino acid starvation with the addition of SHX, leading to the stringent response (Cashel et al., 1996), transcription of *rrn* becomes minimal and the cells' growth arrests, whereas the number of other genes expressed across the genome increases systematically (Durfee et al., 2008; Traxler et al., 2008). To use another analogy, in cells undergoing stress responses, most of the bacterial RNAP functions as eukaryotic Pol II activities for bulk mRNA synthesis at the expense of the *rrn* expression.

This apparent inverse relationship between the expression of *rrn* and the number of other genes expressed genome-wide in response to changes in the environment indicates that growth regulation and stress responses can be explained by genome-wide competition between *rrn* and other genes for limited RNAP in the cell (Jin et al., 2012). The differential allocation of RNAP between the two functional classes of genes in response to changes in the environment also argues against the notion that there are a significant number of "inactive" RNAP molecules in the cell (Bremer and Dennis, 1996).

Extensive studies with different approaches have pointed to the special role of rRNA synthesis in both cell growth and the competition for the limited RNAP in the transcription of other genes genome-wide in response to environmental cues. These studies have provided an intellectual foundation for understanding the cell biology of RNAP in cells under different growth conditions.

Imaging the Dynamic Distribution of RNAP in Cells Responding Rapidly to Environmental Cues: Challenge and a Breakthrough Solution

The Distribution of RNAP in Fast-Growing Cells Undergoes Continuous Change during Sampling Preparation and Live-Cell Imaging

Shortly after green fluorescent protein (GFP) technology was introduced into bacteria (Gordon et al., 1997; Lewis et al., 2000; Margolin, 2000), an *E. coli* chromosomal *rpoC-gfp* fusion as the RNAP-GFP reporter was constructed to study the cell biology of RNAP in fast-growing cells and in cells undergoing an amino acid starvation–induced stress response (Cabrera and Jin, 2003a,b). Since then, many derivatives of the RNAP-GFP reporter have been constructed in *E. coli* to study the distribution of RNAP under different conditions using wide-field microscopy and other cutting-edge imaging techniques (Cabrera et al., 2009; Bakshi et al., 2012; Cagliero and Jin, 2013; Endesfelder et al., 2013; Jin et al., 2013).

It soon became evident that the distribution of RNAP is extremely sensitive to physiological perturbations, including those caused by the sampling and imaging processes; such a property has not been reported for other GFP reporters, such as the *parS*-ParB-GFP system (Nielsen et al., 2007) and the SSB-GFP fusion used in tracking the replisome (Reyes-Lamothe et al., 2008). Initially, no difference was observed in the RNAP distribution in living cells under different growth conditions. **Figure 2A** shows a set of images of living cells growing in a rapidly shaking flask (LB, 37°C), a condition in which most RNAP molecules are engaged in active rRNA synthesis. These cells were imaged after sampling from the flask and then agarose-embedded on coverslips at room temperature (25°C). The pattern of RNAP distribution changes depending on the length of time between sampling preparation and imaging. For example, RNAP foci could be observed in some cells after a short time (3 min); however, as the length of time increases (6 min), only few cells have RNAP foci, and RNAP is homogeneously distributed in the nucleoid in the cell population after 12 min. Note that nucleoid structure also changes in parallel and becomes expanded as time increases during the imaging process, indicating also the dynamic nature of the nucleoid. Experimentally, the time elapsed between sampling and image acquisition is usually longer than 10 min. The same distribution pattern was also observed in fast-growing cells treated either with rifampicin (**Figure 3A**, living cell), or with SHX (**Figure 3B**, living cell), which inhibit *rrn* transcription. RNAP is distributed homogeneously in the nucleoid regardless of growth conditions and status of *rrn* synthesis in living cells under the conditions used.

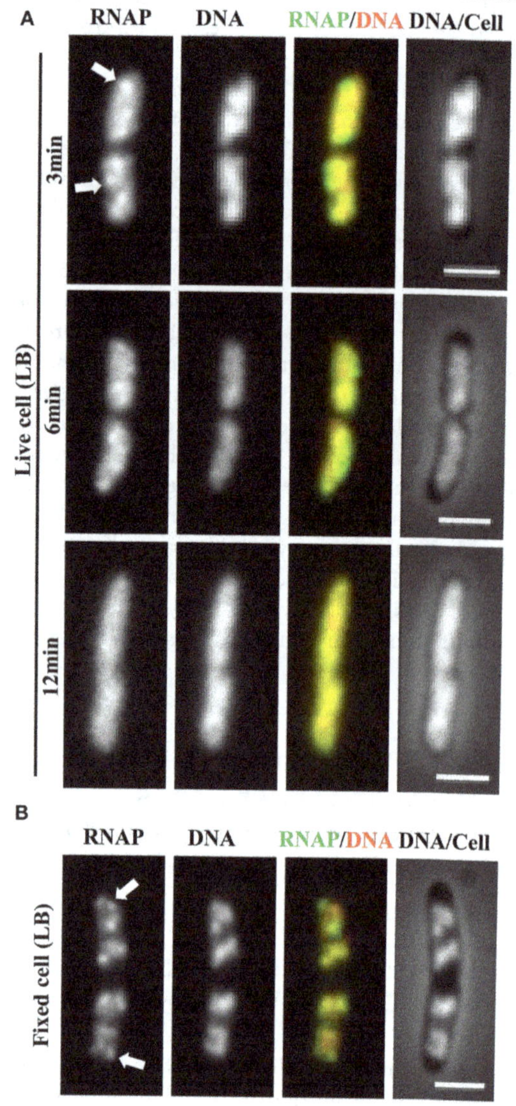

FIGURE 2 | Challenge of imaging RNAP in fast-growing living cells and a breakthrough technique using a formaldehyde-fixed cell procedure. **(A)** Rapidly disappearing RNAP foci during sampling and imaging of fast-growing living cells (LB, 37°C). Exponential-phase cells (*rpoC-venus*) in a rapidly shaking flask in a water bath were sampled and agarose embedded on coverslips for imaging at different time intervals as indicated. The sampling and preparation were performed as described (Cabrera and Jin, 2003b), except that there was no formaldehyde treatment. The nucleoid (DNA) was stained with Hoescht 33342. **(B)** Same as in **(A)**, except that the cells were fixed with formaldehyde immediately after sampling. RNAP foci were stable in the formaldehyde-fixed cells for a few hr before imaging. Foci are indicated by arrows. The scale bar represents 2 μm.

Using Formaldehyde-Fixed Cells Revolutionizes the Imaging of RNAP

Realizing that fast-growing cells in a rapidly shaking flask quickly adapt to slow-growth or growth arrest on coverslips during

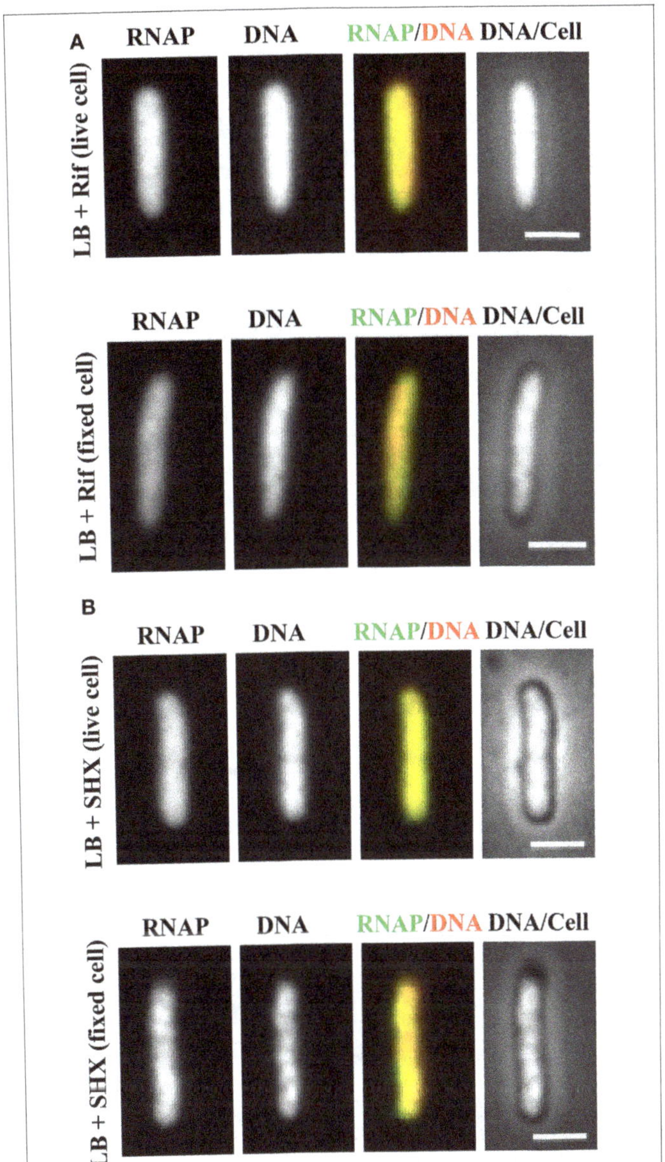

FIGURE 3 | Inhibition of *rrn* expression leads to homogeneous distribution of RNAP across the nucleoid in both living and fixed cells. Fast-growing cells (LB, 37°C) were treated with either rifampicin (100 μg/ml) for 15 min **(A)** or with SHX (500 μg/ml) for 30 min **(B)**. Cells were sampled and agarose embedded on coverslips. Live-cell imaging (live cell); Fixed-cell imaging (fixed cell), cells were fixed by formaldehyde. The scale bar represents 2 μm.

the sampling procedure, which usually takes between 10 and 15 min, and that sequential imaging would capture only the adapted states of RNAP, thereby missing the "true" state of RNAP in fast-growing cells, it became essential to develop a new sampling procedure. To this end, a formaldehyde-fixed cell procedure was developed, in which cells under different physiological conditions are immediately fixed or cross-linked with formaldehyde to "freeze" RNAP and subcellular structures before embedding them in agarose on coverslips for imaging. Formaldehyde is a cell-permeable, small four-atom molecule that penetrates cells rapidly and cross-links stably bound proteins to DNA (Schmiedeberg et al., 2009). This breakthrough technique has enabled the capture of dynamic states of RNAP in *E. coli* as snapshots in changing environments; the results have been shown to be consistent with the genetics and physiology of *E. coli* cells (Cabrera and Jin, 2003b). For example, RNAP forms prominent foci (**Figure 2B**, fixed cell) at the clustering of *rrn*, resembling a bacterial nucleolus in fast-growing cells (LB, 37°C) (more below). In contrast, there are no RNAP foci in cells treated with either rifampicin (**Figure 3A**, fixed cell) or SHX (**Figure 3B**, fixed cell); instead, RNAP distributes homogeneously across the nucleoid. These images have provided the first biological evidence, at a single-cell level, of the redistribution of RNAP from the clustering of *rrn* in fast-growing cells to a broad genome-wide transcription (or RNAP binding) in cells undergoing stresses. Intriguingly, changes in the distribution of RNAP accompany changes in nucleoid structure (see **Figures 2, 3**), revealing the role of RNAP and transcription in the organization of the bacterial chromosome (Jin and Cabrera, 2006; Jin et al., 2013) (more below). On a cautionary note, the dynamic nature of the nucleoid, which responds to changes in the environment and sampling and imaging processes, should be taken into consideration when studying cell cycles using live-cell imaging.

Live-Cell Imaging Using Continuous-Flow Microfluidics Has Further Validated the Use of Formaldehyde in the Study of the Cell Biology of RNAP

Recently, continuous-flow microfluidics has been introduced for live cell imaging of *E. coli* (Wang et al., 2010). This technique has the advantage of enabling continuous live-cell imaging of RNAP in fast-growing and/or in changing environments without sampling interruptions. Time-lapse images from a set of such experiments are shown in **Figure 4**. RNAP foci are evident in fast-growing living cells in LB (RNAP and RNAP/nucleoid overlay), and the foci disappear shortly (~3 min) after the culture is downshifted from LB (at time 0) to nutrient-poor, minimal medium, M9 + glycerol; the process is reversible, as RNAP foci reappear after the culture is upshifted back to LB (RNAP and RNAP/Nucleoid overlay). Treatment of fast-growing cells in LB with SHX for 30 min (**Figure 5**) causes the RNAP foci to disappear (RNAP and RNAP/Nucleoid overlay) almost completely. In addition, RNAP foci are preserved after formaldehyde treatment of fast-growing cells in microfluidics (**Figure 6A**), whereas RNAP maintains a homogenous distribution pattern after the formaldehyde treatment of cells that were starved for amino acid by the SHX treatment (**Figure 6B**). Together, these findings demonstrated that formaldehyde does not cause "artificial" perturbations in the organization of RNAP and DNA in cells, as has been shown in ChIP-chip assays (Davis et al., 2011), and that the distribution of RNAP from fixed-cell images (Cabrera and Jin, 2003b, 2006; Cabrera et al., 2009; Cagliero and Jin, 2013; Endesfelder et al., 2013; Jin et al., 2013), reflects the true dynamic states of RNAP in living cells, thus validating the use of formaldehyde in the study of the cell biology of RNAP.

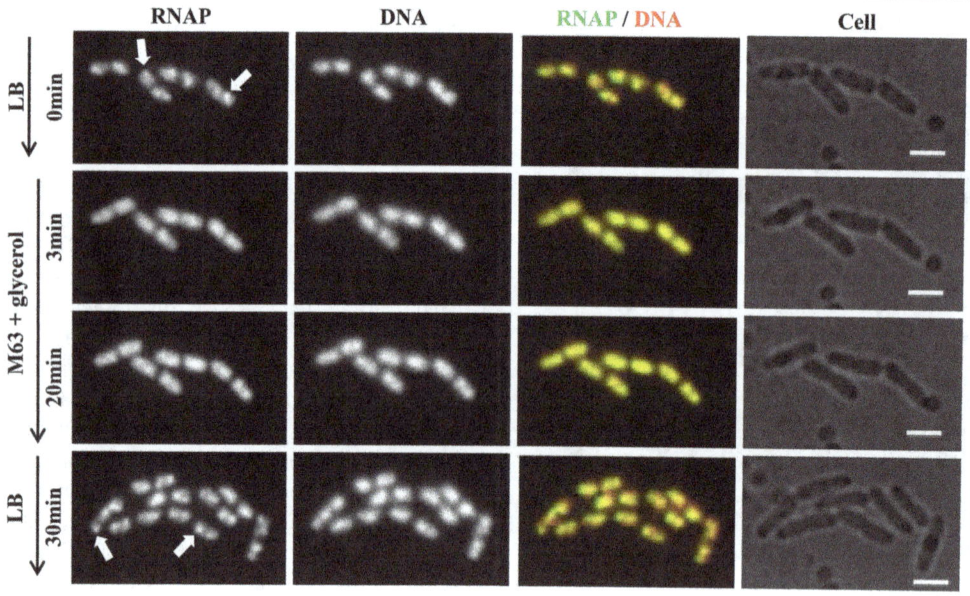

FIGURE 4 | RNAP foci are evident in fast-growing cells and dynamic in response to nutrient downshift and upshift using live-cell imaging with continuous-flow microfluidics. Cells (*rpoC-venus, hupA-mCherry*) were growing in a microfluidic device controlled by the CellASIC ONIX microfluidic perfusion system (EMD Millipore) with continuous flow of LB. After exponential-phase cells in LB were imaged (LB, 0 min), LB was replaced with M63 + glycerol, and time-lapse images were taken, two of which, at 3 min and 20 min, are shown. At 20 min, the minimal medium was replaced with LB, and 30 min later, the image shown was taken. The experiments with microfluidics were performed at 30°C because LB has high background autofluorescence at 37°C. It took about 1 min to complete a medium change in the system. Hu-mCherry was used as a proxy for DNA staining because the microfluidic device used has high autofluorescence in the range of 460–488 nm, which interferes with the detection of DNA-bind dye DAPI or Hoescht 33342. The scale bar represents 2 μm. RNAP foci are indicated by arrows. Note the changes in the cells' position and size during the imaging process.

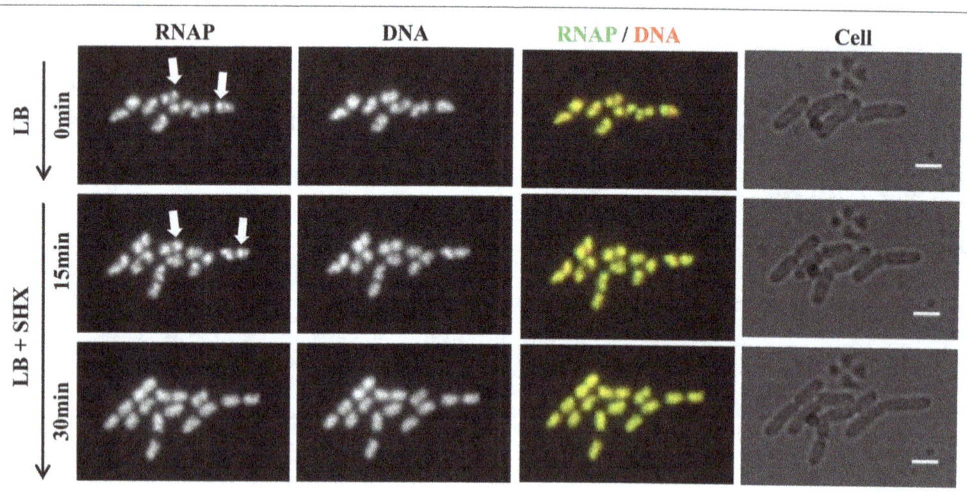

FIGURE 5 | Live-cell imaging using microfluidics confirms that RNAP foci disappear in cells treated with SHX, causing amino acid starvation. Cells (*rpoC-venus, hupA-mCherry*) were grown in LB using the CellASIC ONIX microfluidic system, as described in **Figure 4**. After exponential-phase cells in LB were imaged (LB, 0 min), LB was replaced with LB + SHX (500 μg/ml), and cells were imaged at different intervals after the SHX treatment. Images taken at 15 and 30 min after the addition of SHX are shown. The scale bar represents 2 μm. RNAP foci are indicated by arrows. Note the changes in the cells' position and size during the imaging process.

Imaging of RNAP in formaldehyde-fixed cells has significant advantages over live-cell imaging. Living cells, particularly fast-growing cells, undergo high metabolic activities, such as DNA replication and cell division, which constantly generate internal motions and may also change cells' positions in microfluidics during the imaging process (**Figures 4, 5**). Also, because LB has high background autofluorescence, the quality of RNAP live-cell images, particularly in fast-growing cells in microfluidics,

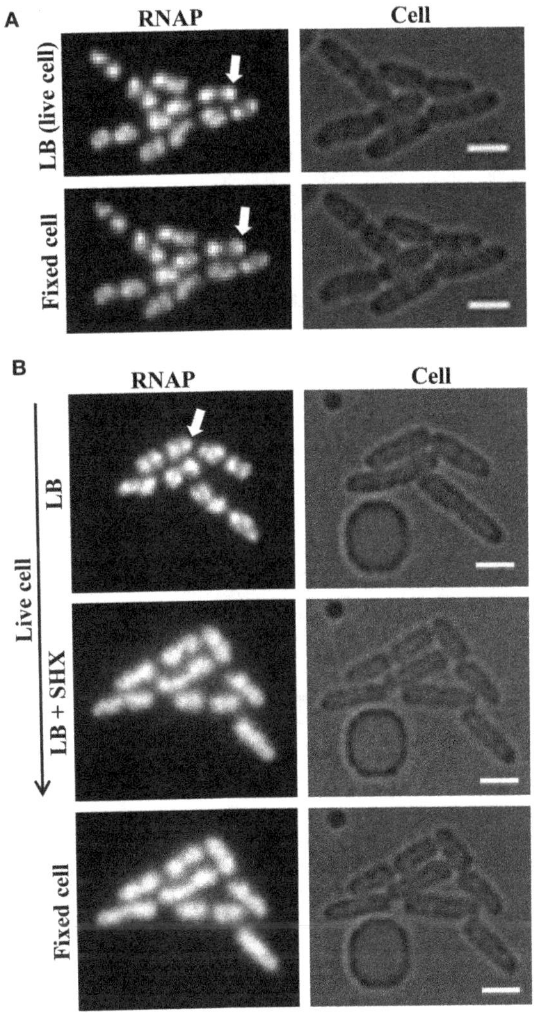

FIGURE 6 | Formaldehyde freezes the dynamic states of RNAP in living cells. (A) RNAP foci in fast-growing cells are preserved by formaldehyde treatment. Live-cell imaging with continuous-flow microfluidics was performed as described in **Figure 4**. After exponential-phase cells in LB were imaged (LB, live cell), LB was replaced with LB + formaldehyde (final concentration 3.7% v/v) for 10 min to fix the cells, followed by flushing with M63 + glycerol for 10 min before imaging (fixed cell). **(B)** Homogenous distribution of RNAP across the nucleoid is maintained in amino acid-starved cells by formaldehyde treatment. The early steps of live-cell imaging, including LB and LB + SHX for 30 min to induce amino acid starvation, were performed as described in **Figure 5**. The subsequent formaldehyde treatment steps were performed as described in **(A)** for fixed-cell imaging. For simplicity, only images of RNAP and cells are shown. RNAP foci are indicated by arrows. The scale bar represents 2 μm.

is rather poor. In addition, for some studies, the physiology of cells in a continuous-flow microfluidic device may be different from that of cells in shaking flasks in a water bath. In contrast, formaldehyde-fixed cells are immobile, and this characteristic is critical and mandatory for acquiring sharp superresolution co-images of RNAP and DNA, as demonstrated below.

It is noteworthy that the formaldehyde-fixed cell procedure has also been widely used in ChIP-chip assays to study the distribution of *E. coli* RNAP genome-wide in one dimension (Grainger et al., 2005; Herring et al., 2005; Davis et al., 2011) and in various chromosome conformation capture assays to probe chromosome organization in living cells (Dekker et al., 2002; Dostie et al., 2006; Umbarger et al., 2011; Cagliero et al., 2013). It is expected that the formaldehyde-fixed cell procedure will also be useful for the capture of dynamic states of other cellular machineries in *E. coli*.

Chromosome Territories in *E. Coli*: A Landscape of Transcription Machinery and Replisome

Significance of the Study of the Cell Biology of RNAP

Significant progress has been made in imaging the distribution of RNAP in *E. coli* in response to changing environments and these studies have shed new light on the differential allocation of RNAP in during bacterial growth and stress responses. Findings from the images of RNAP at the single-cell level complement other genomic analyses of transcription and RNAP binding, including gene arrays and ChIP-chip assays, which are used in population studies and in which the data involved in rRNA synthesis or RNAP binding at the *rrn* regions are purposely deleted to "simplify" the analyses. Consequently, these analyses have missed a significant part of the whole story in the lifestyle of *E. coli*, as transcription of *rrn* plays a key role in growth rate regulation. Specifically, as described below, recent co-imaging of RNAP with DNA and components of the replisome in a single cell has revealed some important features related to the spatial organization of the transcription machinery as well as the landscape of the transcription machinery and replisome (Cagliero et al., 2014). Features of the chromosome territories in bacteria are analogous to those of eukaryotes (Cremer et al., 2006; Meaburn and Misteli, 2007), indicating that *E. coli* is a useful simple model system to study chromosome biology in response to changes in the environment.

Bacterial Nucleolus

In fast-growing cells, most RNAP molecules are concentrated, forming foci at the clustering of *rrn*, which resembles the eukaryotic nucleolus. The presence of the bacterial nucleolus is inferred from the finding using wide-field microscopy, that the number of RNAP foci is significantly lower than the number of copies of *rrn* in a fast-growing cell (Cabrera and Jin, 2003b). Subsequent super-resolution images have further confirmed this finding (Endesfelder et al., 2013; Cagliero et al., 2014). For example, super-resolution structured illumination microscopy (SIM) of RNAP and DNA (Cagliero et al., 2014) (**Figure 7**) reveals that the median number of RNAP foci is eight per cell in fast-growing cells (LB at 37°C, doubling time 20 min) (**Figure 7C**). Compared with the estimated average of 50 *rrn* copies, then, each RNAP focus occurs, on average, at approximately six copies of *rrn*. The resolution (~140 nm lateral and ~300 nm axial for a typical Venus fluorophore) of SIM would be as effective as PALM in detecting small transcription foci in fast-growing cells. Indeed, analysis of PALM images

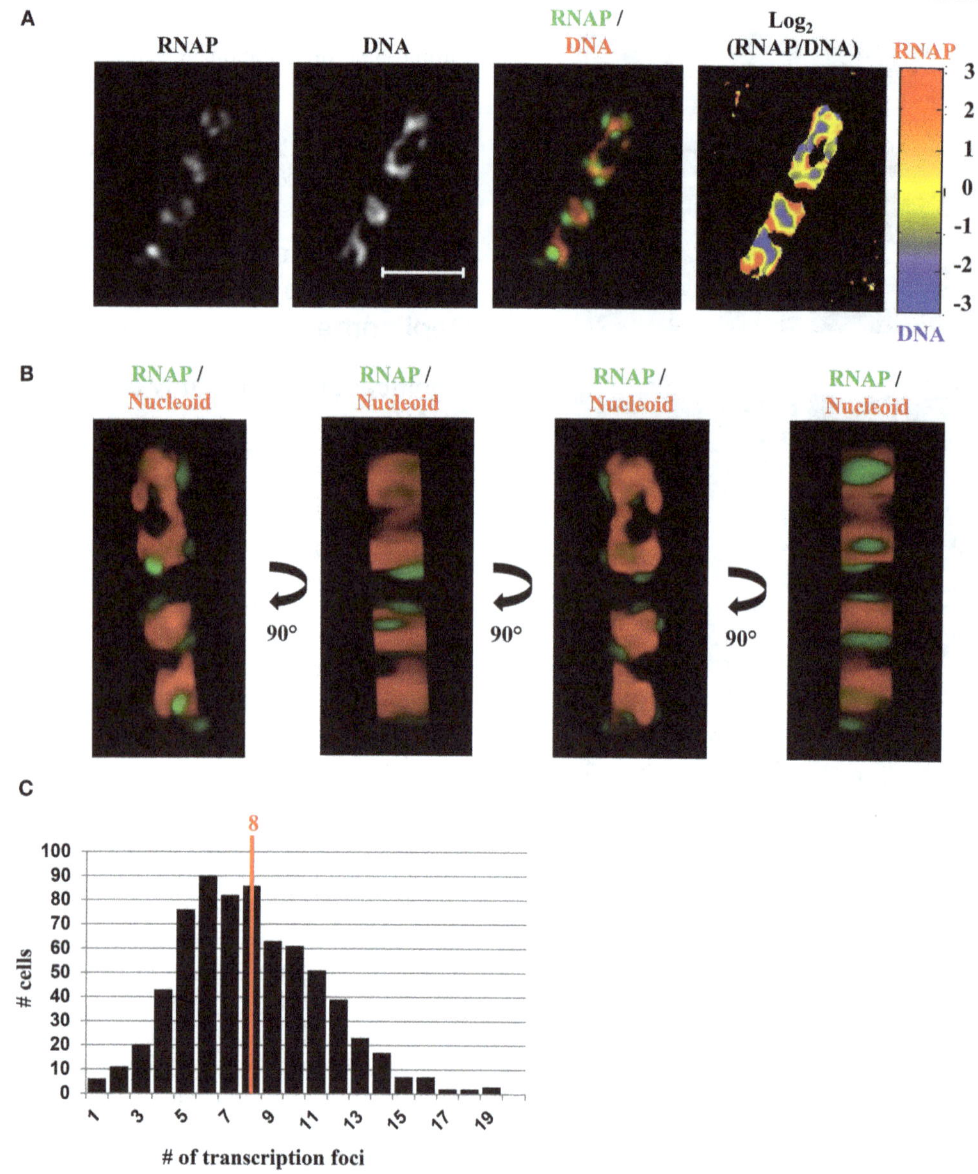

FIGURE 7 | SIM co-imaging of RNAP and DNA reveals spatial compartmentalization of transcription foci in fast-growing cells. **(A)** Images of RNAP, DNA (nucleoid), and an overlay of RNAP (green) and DNA (red) from a representative fast-growing *E. coli* cell (LB, 37°C). The scale bar represents 2 μm. The $\log_2$(RNAP/DNA) plot (heat map) is a quantitative representation of the relationship between RNAP and DNA, which is represented by a color scale bar with values ranging from −3 to 3. Note that regions enriched in RNAP up to eightfold over DNA are at the periphery of the nucleoid (red foci), and regions enriched in DNA up to eightfold over RNAP are in the center of the nucleoid (blue regions). **(B)** 3D representation of the overlay image of RNAP (green) and nucleoid (DNA) (red) in the cell, as shown in **(A)**. Compared with the first and the third panels (x–y axis), the second and fourth panels reflect the limited Z-stacking due to the short axis of the cell. **(C)** Histogram showing the distribution of apparent RNAP-Venus foci in fast-growing cells. The red line indicates the median number of transcription foci in the population of cells. Modified and adapted from Cagliero et al. (2014).

reveals that even small clusters (foci) of 70 RNAPs form a sphere of ~160 nm in diameter (Endesfelder et al., 2013), which is larger than the resolution detectable by SIM. Whether *rrn* from different locations in the chromosome are clustered with RNAP foci remain to be determined experimentally.

There are up to four nascent nucleoids in a fast-growing cell; thus, on average, there are two RNAP foci per nascent nucleoid in the cell. The size of RNAP foci in fast-growing cells varies, probably reflecting the dynamic nature of the foci, as they are likely undergoing disassembly and reassembly when replication forks pass through the *rrn* region. The size variation of RNAP foci is consistent with the super-resolution images, using PALM, of both large clusters of RNAP (up to 800 molecules), which are likely engaging in rRNA synthesis from multiple *rrn* operons, and small clusters of RNAP (~70 molecules), which are attributed to the transcription of a single *rrn* operon (Endesfelder et al.,

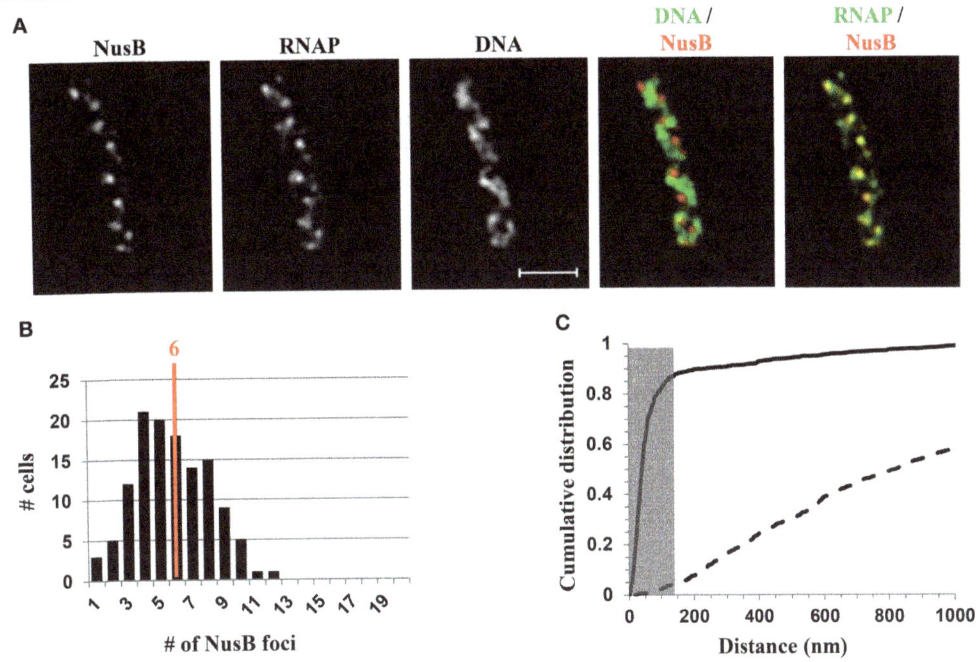

FIGURE 8 | Nascent rRNA-binding protein NusB forms foci and colocalizes with transcription foci in fast-growing cells. (A) Images of NusB, RNAP, DNA (nucleoid), overlay of NusB (red) and DNA (green), and overlay of RNAP (green) and NusB (red) from a representative fast-growing *E. coli* cell, as described in the legend to **Figure 7**. NusB foci are at the periphery of the nucleoid (separate red and green colors on the RNAP/NusB overlay) and the NusB signals perfectly colocalize with RNAP signals (overall yellow color on the RNAP/NusB overlay). **(B)** The distribution of apparent NusB-mCherry foci in fast-growing cells. The red line in the histogram indicates the median number of NusB foci in the population of cells. Note that the median number of NusB foci is close to that of RNAP foci in fast-growing cells. **(C)** Cumulative distribution of the distances between NusB foci and their closest RNAP foci in the population of cells. (—) NusB-mCherry RNAP-Venus, and (- - -) NusB-mCherry RNAP-Venus random. The gray rectangle represents the colocalization area (≤140 nm), as the theoretical SIM microscope resolution for the mCherry is 140 nm. 87.1% of the NusB foci are within 140 nm of the closest transcription foci. Adapted from Cagliero et al. (2014).

2013). As each nascent nucleoid has RNAP foci in place, each of the two newly formed daughter cells will "inherit" the functional transcription machinery without the need for *de novo* synthesis and/or assembly. Because transcription foci at the bacterial nucleolus are critical for growth rate regulation in *E. coli* (Jin et al., 2012), this feature can explain, in part, why newly formed daughter cells maintain the same fast growth rate as the parental cells.

Compartmentalization of RNAP Foci at the Periphery of the Nucleoid

A significant advantage of co-imaging RNAP with DNA is that it enables the determination of the spatial relationship between the transcription machinery and DNA. In a fast-growing cell, RNAP foci appear to be located at the periphery or surroundings of the nucleoid (RNAP/DNA overlay), either in two-dimensional SIM pictures (**Figure 7A**) or reconstituted three-dimensional SIM images (**Figure 7B**). The quantitative heat map, as a normalized Log_2(RNAP/DNA) density plot, further illustrates an inverse relationship between the intensities of RNAP and DNA (**Figure 7A**), indicating that RNAP foci at the clustering of *rrn* are low in DNA density or at loops at the surface of the nucleoid. Such a spatial organization has logistical advantages, including coupling of rRNA synthesis with rRNA processing and ribosome assembly, as well as reducing the traffic jams with other cellular functions.

RNAP foci are likely to be networking hubs for transcription of growth-promoting genes, because genome conformation capture assays have shown that those genes, which are down-regulated during amino acid starvation, are in a highly interactive environment and are present in large clusters (Cagliero et al., 2013). The spatial compartmentalization of RNAP foci suggests that transcription foci could be used as RNAP pools to effectively interact with DNA loops containing non-*rrn* growth-promoting genes located a distance away by hopping three-dimensionally, rather than by traveling linearly or laterally (Wang et al., 2013). Such an organization would reduce the transcription traffic jam and allow for maximum use of limited RNAP in the cell (Jin et al., 2012). It would be interesting to identify other non-*rrn* genes and RNAs as well as proteins in these potential transcription hubs. Other subcellular hyperstructures in bacteria have been described (Norris et al., 2007).

Components of *rrn* Antitermination and rRNA Processing Systems Are Colocalized with RNAP Foci

NusA and NusB are involved in *rrn* antitermination and rRNA processing systems. SIM co-imaging of RNAP-Venus with DNA and NusA-mCherry or NusB-mCherry in fast-growing cells has demonstrated that, (i) like RNAP, NusA, or NusB forms foci at the periphery of the nucleoid, and the median number of NusA or NusB foci per cell is similar to that of the RNAP foci, and (ii) the NusA or NusB foci are co-localized with RNAP foci (Cagliero et al., 2014). For example, **Figure 8** shows the SIM images of NusB and its spatial relationship with DNA (DNA/NusB overlay) and with RNAP (RNAP/NusB) in a typical fast-growing cell (LB at 37°C) (**Figure 8A**). There were six NusB foci per cell on average in a population of fast-growing cells (**Figure 8B**), a value that is close to that of RNAP foci. The cumulative distribution of NusB foci and RNAP foci from a population of fast-growing cells has confirmed that most of NusB foci (>87%) are colocalized with RNAP foci at the clustering of *rrn* or bacterial nucleolus (**Figure 8C**), demonstrating that rRNA synthesis and processing are intimately coupled in space. It remains to be determined what other components are associated with transcription foci in fast-growing cells.

Spatial Segregation of Transcription Machinery and Replisome

Unlike a eukaryotic cell that has defined phases in the cell cycle (S, G2, M, and G1), all processes–such as transcription, replication, and chromosome segregation, are intimately entangled in a rapidly growing *E. coli* cell. Therefore, maximum expression of growth-promoting genes and multiple genome replications are occurring concurrently in a fast-growing cell. A long-standing interest in the field relates to understanding how the two major cellular functions, transcription and replication, maintain harmony to avoid conflicts between the two processes (Merrikh et al., 2012), particularly in fast-growing cells.

Recent SIM co-images of RNAP-Venus with DNA as well as with SeqA-mCherry or SSB-mCherry serving as proxies for replisomes have revealed chromosome landscapes for the two major functions of transcription and replication, which could explain why they remain in harmony in fast-growing cells. The major transcription machinery and replisome are mostly located in different chromosome territories or spatially segregated in the nucleoid (Cagliero et al., 2014). For example, **Figure 9A** shows the SIM images of SeqA and its spatial relationship with DNA (DNA/SeqA overlay) and with RNAP (RNAP/SeqA) in a typical fast-growing cell (LB at 37°C). Similar to RNAP foci and foci of NusA/NusB, SeqA foci also are located at the periphery of the nucleoid (DNA/SeqA overlay), indicating that the replisome is also compartmentalized in regions low in density of DNA or at DNA loops. Image analyses of populations of fast-growing cells showed that on average, each cell contains 10 SeqA foci (**Figure 9B**). In contrast to NusA/NusB, the cumulative distribution of SeqA foci and RNAP foci in the population of fast-growing cells has shown that most (~80%) of the SeqA foci

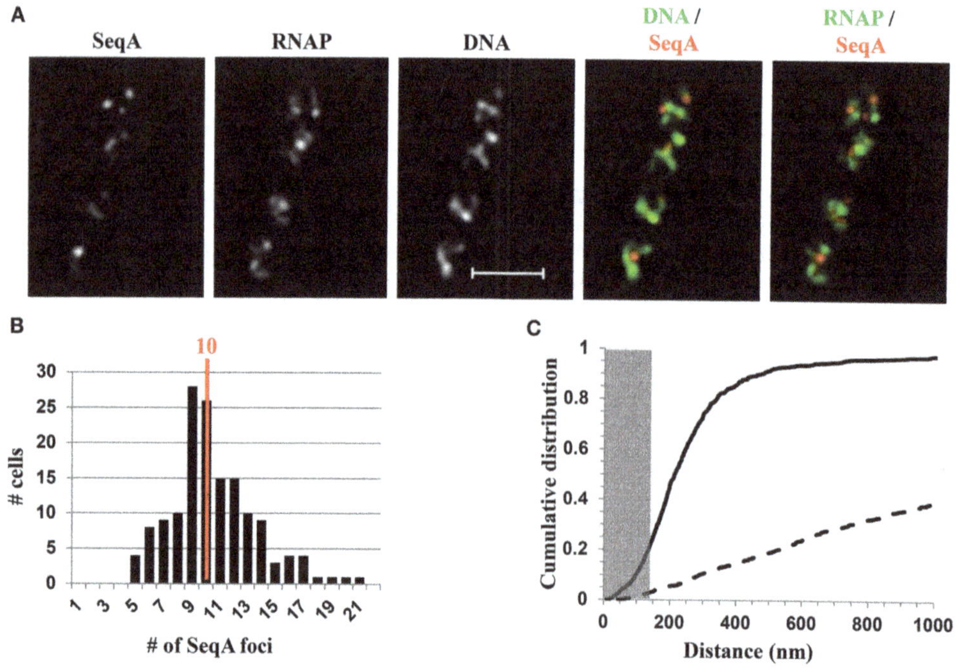

FIGURE 9 | Spatial segregation of transcription foci and replication forks tracked by SeqA in fast-growing cells. (A) Images of SeqA, RNAP, DNA (nucleoid), overlays of SeqA (red) and DNA (green), and RNAP (green) and SeqA (red) from a representative fast-growing *E. coli* cell, as described in the legend to **Figure 7**. SeqA foci and transcription foci are largely located at different positions (red and green colors on the overlay of SeqA and RNAP). Note also that most of the SeqA foci appear to be separated from high intensities of DNA signals in the nucleoids (red and green colors on the overlay of SeqA and DNA). **(B)** The distribution of apparent SeqA-mCherry foci in a population of fast-growing cells. The red line in the histogram indicates the median number of SeqA foci in these cells. **(C)** Cumulative distribution of the distances between SeqA foci and their closest RNAP foci in the population of cells. (—) SeqA-mCherry RNAP-Venus, and (- - -) SeqA-mCherry RNAP-Venus random. The gray rectangle represents the colocalization area (≤140 nm). Only 21.5% of the SeqA foci are within 140 nm of the closest transcription foci. Adapted from Cagliero et al. (2014).

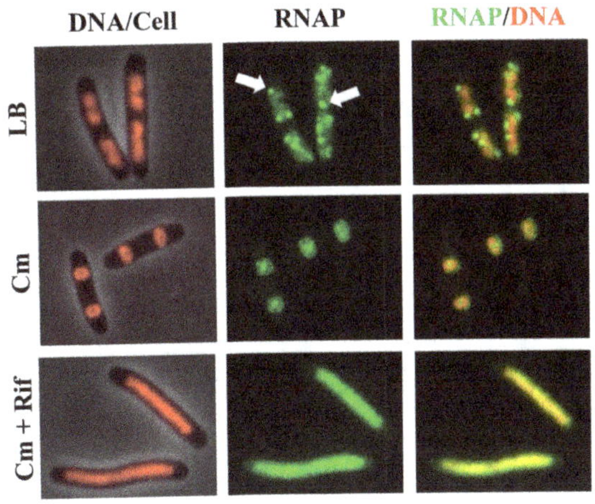

FIGURE 10 | Active *rrn* expression is required to condense the nucleoid in the presence of chloramphenicol. Images of overlay of DNA (nucleoid) and cell, RNAP, and overlay of RNAP and DNA are shown. DNA in the DNA/cell overlay is shown in red (left column), RNAP is shown in green (middle column), and the RNAP (green) and DNA (red) overlay is shown (right column). The panels on the top row show rapidly growing cells (*rpoC-gfp*) in LB, at 30°C, and RNAP foci are indicated by arrows; panels in the center row show cells treated with chloramphenicol (100 μg/ml) for 100 min; panels in the bottom row show cells sequentially treated with chloramphenicol (100 μg/ml, 10 min) and rifampicin (100 μg/ml, 90 min). Modified and adapted from Cabrera et al. (2009).

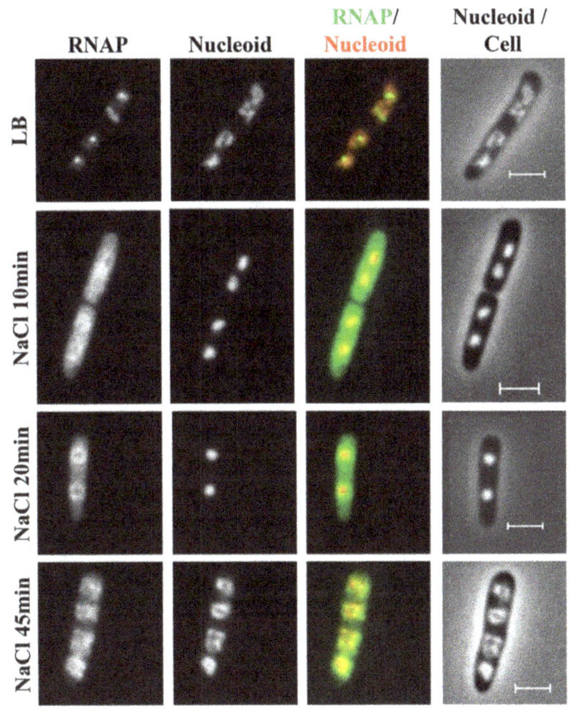

FIGURE 11 | Dynamic interaction of RNAP and DNA during the osmotic stress response. Images of RNAP, nucleoid, overlays of RNAP (green) and nucleoid (red), and nucleoid and cell (*rpoC-venus*) before LB or after osmotic shock (NaCl 10 min to NaCl 45 min). The RNAP rapidly dissociates from the nucleoid, and the released RNAP occupies the entire cytoplasmic space shortly after osmotic shock (NaCl 10 min). The nucleoid becomes hyper-condensed at the same time. Subsequently, the freed RNAP gradually re-associates with the genome by first forming a ring at the periphery of the hyper-condensed nucleoid (NaCl 20 min to NaCl 30 min) before penetrating the interior (NaCl 45 min). The scale bar represents 2 μm. Adapted from Cagliero and Jin (2013).

are not colocalized with the RNAP foci, i.e., the two cellular functions are mostly segregated in space (**Figure 9C**). The low co-localization frequency of SeqA foci and RNAP foci suggests transient overlapping of transcription and replication of *rrn* regions. It is conceivable that during replication of *rrn* operons, RNAP foci are somehow disassembled, allowing replication forks to pass through the region, followed by reassembly of transcription machinery at the *rrn* clusters. Development of fast, super-resolution, time-lapse, live-cell imaging techniques will be necessary to address the dynamic interaction and segregation of the two active cellular functions in fast-growing cells.

Coupling the Distribution of RNAP to the Organization of the Bacterial Nucleoid

Active rRNA Synthesis at the Clustering of *rrn* Causes Nucleoid Compaction

Co-imaging of RNAP and DNA in cells undergoing physiological perturbations has also revealed that global changes in the distribution of RNAP accompany alterations in nucleoid structure, indicating an important role of RNAP and transcription in the organization of the bacterial chromosome (Jin and Cabrera, 2006; Jin et al., 2013). The changes are stress-dependent and in most cases studied, as shown in **Figures 2, 3**, redistribution of RNAP from a few prominent foci at the clustering of *rrn* to the nucleoid homogeneously leads to nucleoid expansion. In an effort to determine whether active rRNA synthesis is required to condense the nucleoid, the effects of the antibiotics chloramphenicol and rifampin as well as two mutations that decrease rRNA synthesis, on the nucleoid structure have been reexamined (Cabrera et al., 2009). The nucleoids are condensed in fast-growing cells (LB) treated with the translation inhibitor chloramphenicol (Woldringh et al., 1995; van Helvoort et al., 1996); however, if cells are treated sequentially with chloramphenicol first and then rifampicin, the nucleoids become expanded and RNAP foci disappear (**Figure 10**). Moreover, the nucleoids remain expanded in the above-described RNAP mutant cells defective in the transcription of *rrn* in LB when treated with chloramphenicol. Similarly, even with chloramphenicol treatment, the nucleoids remain expanded in the Δ6*rrn* cells that have only one copy of *rrn* remaining in the chromosome. Together, these results support the role of active rRNA synthesis from the clustering of *rrn* in nucleoid compaction in fast-growing cells. Because transcription and supercoiling are coupled (Liu and Wang, 1987), it is possible that transcription-induced supercoiling at the clustering of *rrn* (Jin et al., 2012) contributes to the nucleoid compaction in fast-growing cells.

The mechanisms underlying the apparent coupling of the distribution of RNAP with the organization of the bacterial chromosome remain to be determined. Multiple factors are likely

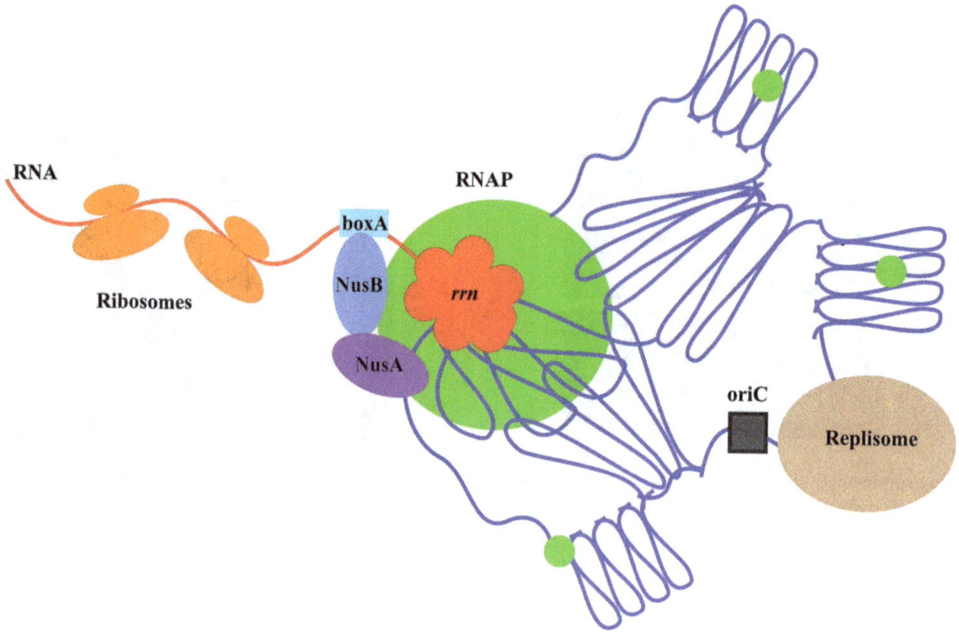

FIGURE 12 | Model illustrating the chromosome territories of the major transcription machinery and replisome in a fast-growing cell. The *E. coli* chromosome is represented by the blue lines folded in loops, the *oriC* by a black square, and the RNAP molecules by green circles. For simplicity, only one of the prominent transcription foci and the replisome are shown. RNAP foci at the clustering of *rrn* (red), resembling a bacterial nucleolus, are spatially organized at the periphery of the nucleoid for compartmentalization. Foci of NusA and NusB co-localize with RNAP foci, indicating the coupling of the synthesis and processing of rRNA, and, possibly, ribosomal assembly. Compartmentalization of the replisome is also shown, but the two major cellular functions of transcription and replication are largely spatially segregated. See text for the details and advantages of the landscapes of transcription machinery and replisome in fast-growing cells.

to be responsible for the organization of the bacterial nucleoid (Woldringh et al., 1995; Cagliero et al., 2013; Jin et al., 2013). Other small nucleoid-associated proteins (NAPs) or "histone-like" proteins, such as FIS, H-NS, HU, and IHF, play important architectural roles and are transcription factors (Luijsterburg et al., 2006; Browning et al., 2010; Dillon and Dorman, 2010; Liu et al., 2010; Pul and Wagner, 2012). The distribution of RNAP will likely to be affected by these NAPs and vice versa; together, they determine the organization of the nucleoid.

Temporal Changes in the Distribution of RNAP and the Nucleoid Structure in Response to Osmotic Stress

During hyperosmotic stress response, the changes in RNAP distribution and nucleoid compaction are temporally manifested (Cagliero and Jin, 2013). Initially, significant numbers of RNAP molecules dissociate from the nucleoid into the cytoplasmic space because of the transient accumulation of the cytoplasmic K^+, and, concomitantly, the nucleoid becomes hyper-condensed (**Figure 11**). Subsequently, when the cytoplasmic K^+ levels decrease during the osmoadaptation phase, the free RNAP re-associates with DNA and initially forms a ring at the periphery of the nucleoid, and the nucleoid gradually expands to a size approaching that prior to the salt shock. The ring of RNAP surrounding the hyper-condensed nucleoid during the early osmoadaptation phase (NaCl 20 min) is thought to indicate the location of DNA loops for the expression of responsive genes. The causes of temporal changes of the nucleoid structure during the osmotic stress response are not clear.

Summary

Co-imaging of RNAP with DNA as well as with other proteins in cells under different growth conditions has been an important tool and a new approach in the long-standing journey to understanding growth rate regulation and stress responses in *E. coli*. Significant progress has been made, including new concepts and findings from these studies, particularly in fast-growing cells (**Figure 12**). It is expected that more will be learned using wide-field and state-of-the-art imaging systems in future studies. In addition, as a simple model system, *E. coli* will have many advantages in seeking fundamental knowledge of the chromosome biology including chromosome territories, a frontier in biology research.

Acknowledgments

This research was supported by the Intramural Research Program of the NIH, National Cancer Institute, Center for Cancer Research. DJJ would like to thank the four reviewers and the editor for critical reading of the manuscript and productive interactions in the reviewing process.

References

Bakshi, S., Siryaporn, A., Goulian, M., and Weisshaar, J. C. (2012). Superresolution imaging of ribosomes and RNA polymerase in live *Escherichia coli* cells. *Mol. Microbiol.* 85, 21–38. doi: 10.1111/j.1365-2958.2012.08081.x

Blattner, F. R., Plunkett, G. III, Bloch, C. A., Perna, N. T., Burland, V., Riley, M., et al. (1997). The complete genome sequence of *Escherichia coli* K-12. *Science* 277, 1453–1474. doi: 10.1126/science.277.5331.1453

Bremer, H., and Dennis, P. P. (1996). "Modulation of chemical composition and other parameters of the cell by growth rate," in *Escherichia coli and Salmonella*, eds F. C. Neidhardt, R. Curtiss III, J. L. Ingraham, E. C. C. Lin, K. B. Low, B. Magasanik, W. S. Reznikoff, M. Riley, M. Schaechter, and H. E. Umbarger (Washington, DC: ASM Press), 1553–1569.

Browning, D. F., Grainger, D. C., and Busby, S. J. (2010). Effects of nucleoid-associated proteins on bacterial chromosome structure and gene expression. *Curr. Opin. Microbiol.* 13, 773–780. doi: 10.1016/j.mib.2010.09.013

Bubunenko, M., Court, D. L., Al Refaii, A., Saxena, S., Korepanov, A., Friedman, D. I., et al. (2013). Nus transcription elongation factors and RNase III modulate small ribosome subunit biogenesis in *Escherichia coli*. *Mol. Microbiol.* 87, 382–393. doi: 10.1111/mmi.12105

Cabrera, J. E., Cagliero, C., Quan, S., Squires, C. L., and Jin, D. J. (2009). Active transcription of rRNA operons condenses the nucleoid in *Escherichia coli*: examining the effect of transcription on nucleoid structure in the absence of transertion. *J. Bacteriol.* 191, 4180–4185. doi: 10.1128/JB.01707-08

Cabrera, J. E., and Jin, D. J. (2003a). Construction, purification, and characterization of *Escherichia coli* RNA polymerases tagged with different fluorescent proteins. *Methods Enzymol.* 370, 3–10. doi: 10.1016/S0076-6879(03)70001-9

Cabrera, J. E., and Jin, D. J. (2003b). The distribution of RNA polymerase in *Escherichia coli* is dynamic and sensitive to environmental cues. *Mol. Microbiol.* 50, 1493–1505. doi: 10.1046/j.1365-2958.2003.03805.x

Cabrera, J. E., and Jin, D. J. (2006). Active transcription of rRNA operons is a driving force for the distribution of RNA polymerase in bacteria: effect of extrachromosomal copies of rrnB on the *in vivo* localization of RNA polymerase. *J. Bacteriol.* 188, 4007–4014. doi: 10.1128/JB.01893-05

Cagliero, C., Grand, R. S., Jones, M. B., Jin, D. J., and O'Sullivan, J. M. (2013). Genome conformation capture reveals that the *Escherichia coli* chromosome is organized by replication and transcription. *Nucleic Acids Res.* 41, 6058–6071. doi: 10.1093/nar/gkt325

Cagliero, C., and Jin, D. J. (2013). Dissociation and re-association of RNA polymerase with DNA during osmotic stress response in *Escherichia coli*. *Nucleic Acids Res.* 41, 315–326. doi: 10.1093/nar/gks988

Cagliero, C., Zhou, Y. N., and Jin, D. J. (2014). Spatial organization of transcription machinery and its segregation from the replisome in fast-growing bacterial cells. *Nucleic Acids Res.* 42, 13696–13705. doi: 10.1093/nar/gku1103

Cashel, M., Gentry, D. R., Hernandez, V. J., and Vinella, D. (1996). "The stringent response," in *Escherichia coli and Salmonella*, eds F. C. Neidhardt, R. Curtiss III, J. L. Ingraham, E. C. C. Lin, K. B. Low, B. Magasanik, W. S. Reznikoff, M. Riley, M. Schaechter, and H. E. Umbarger (Washington, DC: ASM Press), 1458–1496.

Condon, C., French, S., Squires, C., and Squires, C. L. (1993). Depletion of functional ribosomal RNA operons in *Escherichia coli* causes increased expression of the remaining intact copies. *EMBO J.* 12, 4305–4315.

Condon, C., Squires, C., and Squires, C. L. (1995). Control of rRNA transcription in *Escherichia coli*. *Microbiol. Rev.* 59, 623–645.

Cooper, S., and Helmstetter, C. E. (1968). Chromosome replication and the division cycle of *Escherichia coli* B/r. *J. Mol. Biol.* 31, 519–540. doi: 10.1016/0022-2836(68)90425-7

Cremer, T., Cremer, M., Dietzel, S., Muller, S., Solovei, I., and Fakan, S. (2006). Chromosome territories–a functional nuclear landscape. *Curr. Opin. Cell Biol.* 18, 307–316. doi: 10.1016/j.ceb.2006.04.007

Davis, S. E., Mooney, R. A., Kanin, E. I., Grass, J., Landick, R., and Ansari, A. Z. (2011). Mapping *E. coli* RNA polymerase and associated transcription factors and identifying promoters genome-wide. *Methods Enzymol.* 498, 449–471. doi: 10.1016/B978-0-12-385120-8.00020-6

Dekker, J., Rippe, K., Dekker, M., and Kleckner, N. (2002). Capturing chromosome conformation. *Science* 295, 1306–1311. doi: 10.1126/science.1067799

Dillon, S. C., and Dorman, C. J. (2010). Bacterial nucleoid-associated proteins, nucleoid structure and gene expression. *Nat. Rev. Microbiol.* 8, 185–195. doi: 10.1038/nrmicro2261

Dostie, J., Richmond, T. A., Arnaout, R. A., Selzer, R. R., Lee, W. L., Honan, T. A., et al. (2006). Chromosome Conformation Capture Carbon Copy (5C): a massively parallel solution for mapping interactions between genomic elements. *Genome Res.* 16, 1299–1309. doi: 10.1101/gr.5571506

Durfee, T., Hansen, A. M., Zhi, H., Blattner, F. R., and Jin, D. J. (2008). Transcription profiling of the stringent response in *Escherichia coli*. *J. Bacteriol.* 190, 1084–1096. doi: 10.1128/JB.01092-07

Endesfelder, U., Finan, K., Holden, S. J., Cook, P. R., Kapanidis, A. N., and Heilemann, M. (2013). Multiscale spatial organization of RNA polymerase in *Escherichia coli*. *Biophys. J.* 105, 172–181. doi: 10.1016/j.bpj.2013.05.048

French, S. L., and Miller, O. L. Jr. (1989). Transcription mapping of the *Escherichia coli* chromosome by electron microscopy. *J. Bacteriol.* 171, 4207–4216.

Gausing, K. (1980). "Regulation of ribosome production in *Escherichia coli*," in *Ribosomes: Structure, Function and Genetics*, eds G. Chambliss, G. R. Craven, J. Davies, K. Davis, and M. Nomura (Baltimore, MD: University Park Press), 693–718.

Gordon, G. S., Sitnikov, D., Webb, C. D., Teleman, A., Straight, A., Losick, R., et al. (1997). Chromosome and low copy plasmid segregation in *E. coli*: visual evidence for distinct mechanisms. *Cell* 90, 1113–1121. doi: 10.1016/S0092-8674(00)80377-3

Gourse, R. L., Gaal, T., Bartlett, M. S., Appleman, J. A., and Ross, W. (1996). rRNA transcription and growth rate-dependent regulation of ribosome synthesis in *Escherichia coli*. *Annu. Rev. Microbiol.* 50, 645–677. doi: 10.1146/annurev.micro.50.1.645

Grainger, D. C., Hurd, D., Harrison, M., Holdstock, J., and Busby, S. J. (2005). Studies of the distribution of *Escherichia coli* cAMP-receptor protein and RNA polymerase along the *E. coli* chromosome. *Proc. Natl. Acad. Sci. U.S.A.* 102, 17693–17698. doi: 10.1073/pnas.0506687102

Greenblatt, J., and Li, J. (1981). Interaction of the sigma factor and the nusA gene protein of *E. coli* with RNA polymerase in the initiation-termination cycle of transcription. *Cell* 24, 421–428. doi: 10.1016/0092-8674(81)90332-9

Grigorova, I. L., Phleger, N. J., Mutalik, V. K., and Gross, C. A. (2006). Insights into transcriptional regulation and sigma competition from an equilibrium model of RNA polymerase binding to DNA. *Proc. Natl. Acad. Sci. U.S.A.* 103, 5332–5337. doi: 10.1073/pnas.0600828103

Herring, C. D., Raffaelle, M., Allen, T. E., Kanin, E. I., Landick, R., Ansari, A. Z., et al. (2005). Immobilization of *Escherichia coli* RNA polymerase and location of binding sites by use of chromatin immunoprecipitation and microarrays. *J. Bacteriol.* 187, 6166–6174. doi: 10.1128/JB.187.17.6166-6174.2005

Ishihama, A. (2000). Functional modulation of *Escherichia coli* RNA polymerase. *Annu. Rev. Microbiol.* 54, 499–518. doi: 10.1146/annurev.micro.54.1.499

Jin, D. J., and Cabrera, J. E. (2006). Coupling the distribution of RNA polymerase to global gene regulation and the dynamic structure of the bacterial nucleoid in *Escherichia coli*. *J. Struct. Biol.* 156, 284–291. doi: 10.1016/j.jsb.2006.07.005

Jin, D. J., Cagliero, C., and Zhou, Y. N. (2012). Growth rate regulation in *Escherichia coli*. *FEMS Microbiol. Rev.* 36, 269–287. doi: 10.1111/j.1574-6976.2011.00279.x

Jin, D. J., Cagliero, C., and Zhou, Y. N. (2013). Role of RNA polymerase and transcription in the organization of the bacterial nucleoid. *Chem. Rev.* 113, 8662–8682. doi: 10.1021/cr4001429

Keener, J., and Nomura, M. (1996). "Regulation of ribosome synthesis," in *Escherichia coli and Salmonella*, ed F. C. Neidhardt (Washington, DC: ASM Press), 1417–1431.

Kjeldgaard, N. O., Maaloe, O., and Schaechter, M. (1958). The transition between different physiological states during balanced growth of *Salmonella typhimurium*. *J. Gen. Microbiol.* 19, 607–616. doi: 10.1099/00221287-19-3-607

Lewis, P. J., Thaker, S. D., and Errington, J. (2000). Compartmentalization of transcription and translation in Bacillus subtilis. *EMBO J.* 19, 710–718. doi: 10.1093/emboj/19.4.710

Lin, Y., Gao, F., and Zhang, C. T. (2010). Functionality of essential genes drives gene strand-bias in bacterial genomes. *Biochem. Biophys. Res. Commun.* 396, 472–476. doi: 10.1016/j.bbrc.2010.04.119

Liu, L. F., and Wang, J. C. (1987). Supercoiling of the DNA template during transcription. *Proc. Natl. Acad. Sci. U.S.A.* 84, 7024–7027. doi: 10.1073/pnas.84.20.7024

Liu, M., Durfee, T., Cabrera, J. E., Zhao, K., Jin, D. J., and Blattner, F. R. (2005). Global transcriptional programs reveal a carbon source foraging strategy by *Escherichia coli*. *J. Biol. Chem.* 280, 15921–15927. doi: 10.1074/jbc.M414050200

Liu, Y., Chen, H., Kenney, L. J., and Yan, J. (2010). A divalent switch drives H-NS/DNA-binding conformations between stiffening and bridging modes. *Genes Dev.* 24, 339–344. doi: 10.1101/gad.1883510

Luijsterburg, M. S., Noom, M. C., Wuite, G. J., and Dame, R. T. (2006). The architectural role of nucleoid-associated proteins in the organization of bacterial chromatin: a molecular perspective. *J. Struct. Biol.* 156, 262–272. doi: 10.1016/j.jsb.2006.05.006

Marceau, A. H., Bahng, S., Massoni, S. C., George, N. P., Sandler, S. J., Marians, K. J., et al. (2011). Structure of the SSB-DNA polymerase III interface and its role in DNA replication. *EMBO J.* 30, 4236–4247. doi: 10.1038/emboj.2011.305

Margolin, W. (2000). Green fluorescent protein as a reporter for macromolecular localization in bacterial cells. *Methods* 20, 62–72. doi: 10.1006/meth.1999.0906

Meaburn, K. J., and Misteli, T. (2007). Cell biology: chromosome territories. *Nature* 445, 379–781. doi: 10.1038/445379a

Merrikh, H., Zhang, Y., Grossman, A. D., and Wang, J. D. (2012). Replication-transcription conflicts in bacteria. *Nat. Rev. Microbiol.* 10, 449–458. doi: 10.1038/nrmicro2800

Nielsen, H. J., Youngren, B., Hansen, F. G., and Austin, S. (2007). Dynamics of *Escherichia coli* chromosome segregation during multifork replication. *J. Bacteriol.* 189, 8660–8666. doi: 10.1128/JB.01212-07

Norris, V., den Blaauwen, T., Cabin-Flaman, A., Doi, R. H., Harshey, R., Janniere, L., et al. (2007). Functional taxonomy of bacterial hyperstructures. *Microbiol. Mol. Biol. Rev.* 71, 230–253. doi: 10.1128/MMBR.00035-06

O'Donnell, M. (2006). Replisome architecture and dynamics in *Escherichia coli*. *J. Biol. Chem.* 281, 10653–10656. doi: 10.1074/jbc.R500028200

O'Sullivan, J. M., Pai, D. A., Cridge, A. G., Engelke, D. R., and Ganley, A. R. (2013). The nucleolus: a raft adrift in the nuclear sea or the keystone in nuclear structure? *Biomol. Concepts* 4, 277–286. doi: 10.1515/bmc-2012-0043

Paul, B. J., Ross, W., Gaal, T., and Gourse, R. L. (2004). rRNA transcription in *Escherichia coli*. *Annu. Rev. Genet.* 38, 749–770. doi: 10.1146/annurev.genet.38.072902.091347

Piper, S. E., Mitchell, J. E., Lee, D. J., and Busby, S. J. (2009). A global view of *Escherichia coli* Rsd protein and its interactions. *Mol. Biosyst.* 5, 1943–1947. doi: 10.1039/b904955j

Potrykus, K., Murphy, H., Philippe, N., and Cashel, M. (2011). ppGpp is the major source of growth rate control in *E. coli*. *Environ. Microbiol.* 13, 563–575. doi: 10.1111/j.1462-2920.2010.02357.x

Pul, U., and Wagner, R. (2012). "Nucleoid-associated proteins: structural properties," in *Bacterial Chromatin*, eds R. T. Dame and C. J. Dorman (New York, NY: Springer), 149–173.

Reyes-Lamothe, R., Possoz, C., Danilova, O., and Sherratt, D. J. (2008). Independent positioning and action of *Escherichia coli* replisomes in live cells. *Cell* 133, 90–102. doi: 10.1016/j.cell.2008.01.044

Riley, M., Abe, T., Arnaud, M. B., Berlyn, M. K., Blattner, F. R., Chaudhuri, R. R., et al. (2006). *Escherichia coli* K-12: a cooperatively developed annotation snapshot–2005. *Nucleic Acids Res.* 34, 1–9. doi: 10.1093/nar/gkj405

Rocha, E. P., and Danchin, A. (2003). Gene essentiality determines chromosome organisation in bacteria. *Nucleic Acids Res.* 31, 6570–6577. doi: 10.1093/nar/gkg859

Ross, W., Vrentas, C. E., Sanchez-Vazquez, P., Gaal, T., and Gourse, R. L. (2013). The magic spot: a ppGpp binding site on *E. coli* RNA polymerase responsible for regulation of transcription initiation. *Mol. Cell* 50, 420–429. doi: 10.1016/j.molcel.2013.03.021

Schaechter, M., Maaloe, O., and Kjeldgaard, N. O. (1958). Dependency on medium and temperature of cell size and chemical composition during balanced grown of *Salmonella typhimurium*. *J. Gen. Microbiol.* 19, 592–606. doi: 10.1099/00221287-19-3-592

Schmiedeberg, L., Skene, P., Deaton, A., and Bird, A. (2009). A temporal threshold for formaldehyde crosslinking and fixation. *PLoS ONE* 4:e4636. doi: 10.1371/journal.pone.0004636

Slater, S., Wold, S., Lu, M., Boye, E., Skarstad, K., and Kleckner, N. (1995). *E. coli* SeqA protein binds oriC in two different methyl-modulated reactions appropriate to its roles in DNA replication initiation and origin sequestration. *Cell* 82, 927–936. doi: 10.1016/0092-8674(95)90272-4

Stokke, C., Flatten, I., and Skarstad, K. (2012). An easy-to-use simulation program demonstrates variations in bacterial cell cycle parameters depending on medium and temperature. *PLoS ONE* 7:e30981. doi: 10.1371/journal.pone.0030981

Torres, M., Balada, J. M., Zellars, M., Squires, C., and Squires, C. L. (2004). *In vivo* effect of NusB and NusG on rRNA transcription antitermination. *J. Bacteriol.* 186, 1304–1310. doi: 10.1128/JB.186.5.1304-1310.2004

Tosa, T., and Pizer, L. I. (1971). Biochemical bases for the antimetabolite action of L-serine hydroxamate. *J. Bacteriol.* 106, 972–982.

Traxler, M. F., Summers, S. M., Nguyen, H. T., Zacharia, V. M., Hightower, G. A., Smith, J. T., et al. (2008). The global, ppGpp-mediated stringent response to amino acid starvation in *Escherichia coli*. *Mol. Microbiol.* 68, 1128–1148. doi: 10.1111/j.1365-2958.2008.06229.x

Umbarger, M. A., Toro, E., Wright, M. A., Porreca, G. J., Bau, D., Hong, S. H., et al. (2011). The three-dimensional architecture of a bacterial genome and its alteration by genetic perturbation. *Mol. Cell* 44, 252–264. doi: 10.1016/j.molcel.2011.09.010

van Helvoort, J. M., Kool, J., and Woldringh, C. L. (1996). Chloramphenicol causes fusion of separated nucleoids in *Escherichia coli* K-12 cells and filaments. *J. Bacteriol.* 178, 4289–4293.

Wagner, R. (2002). Regulation of ribosomal RNA synthesis in *E. coli*: effects of the global regulator guanosine tetraphosphate (ppGpp). *J. Mol. Microbiol. Biotechnol.* 4, 331–340.

Waldminghaus, T., Weigel, C., and Skarstad, K. (2012). Replication fork movement and methylation govern SeqA binding to the *Escherichia coli* chromosome. *Nucleic Acids Res.* 40, 5465–5476. doi: 10.1093/nar/gks187

Wang, F., Redding, S., Finkelstein, I. J., Gorman, J., Reichman, D. R., and Greene, E. C. (2013). The promoter-search mechanism of *Escherichia coli* RNA polymerase is dominated by three-dimensional diffusion. *Nat. Struct. Mol. Biol.* 20, 174–181. doi: 10.1038/nsmb.2472

Wang, P., Robert, L., Pelletier, J., Dang, W. L., Taddei, F., Wright, A., et al. (2010). Robust growth of *Escherichia coli*. *Curr. Biol.* 20, 1099–1103. doi: 10.1016/j.cub.2010.04.045

Woldringh, C. L., Jensen, P. R., and Westerhoff, H. V. (1995). Structure and partitioning of bacterial DNA: determined by a balance of compaction and expansion forces? *FEMS Microbiol. Lett.* 131, 235–242. doi: 10.1111/j.1574-6968.1995.tb07782.x

Yamazoe, M., Adachi, S., Kanaya, S., Ohsumi, K., and Hiraga, S. (2005). Sequential binding of SeqA protein to nascent DNA segments at replication forks in synchronized cultures of *Escherichia coli*. *Mol. Microbiol.* 55, 289–298. doi: 10.1111/j.1365-2958.2004.04389.x

Yao, N. Y., and O'Donnell, M. (2010). SnapShot: the replisome. *Cell* 141, 1088, 1088.e1081. doi: 10.1016/j.cell.2010.05.042

Zhou, Y. N., and Jin, D. J. (1997). RNA polymerase beta mutations have reduced sigma70 synthesis leading to a hyper-temperature-sensitive phenotype of a sigma70 mutant. *J. Bacteriol.* 179, 4292–4298.

Zhou, Y. N., and Jin, D. J. (1998). The rpoB mutants destabilizing initiation complexes at stringently controlled promoters behave like "stringent" RNA polymerases in *Escherichia coli*. *Proc. Natl. Acad. Sci. U.S.A.* 95, 2908–2913. doi: 10.1073/pnas.95.6.2908

6

Analysis of factors limiting bacterial growth in PDMS mother machine devices

Da Yang [1], *Anna D. Jennings* [1], *Evalynn Borrego* [2], *Scott T. Retterer* [3] *and Jaan Männik* [1*]

[1] *Department of Physics and Astronomy, The University of Tennessee, Knoxville, TN, United States,* [2] *Department of Mechanical, Aerospace and Biomedical Engineering, University of Tennessee, Knoxville, TN, United States,* [3] *Oak Ridge National Laboratory, Center for Nanophase Materials Sciences, Oak Ridge, TN, United States*

***Correspondence:**
Jaan Männik
jmannik@utk.edu

The microfluidic mother machine platform has attracted much interest for its potential in studies of bacterial physiology, cellular organization, and cell mechanics. Despite numerous experiments and development of dedicated analysis software, differences in bacterial growth and morphology in narrow mother machine channels compared to typical liquid media conditions have not been systematically characterized. Here we determine changes in *E. coli* growth rates and cell dimensions in different sized dead-end microfluidic channels using high resolution optical microscopy. We find that *E. coli* adapt to the confined channel environment by becoming narrower and longer compared to the same strain grown in liquid culture. Cell dimensions decrease as the channel length increases and width decreases. These changes are accompanied by increases in doubling times in agreement with the universal growth law. In channels 100 μm and longer, cell doublings can completely stop as a result of frictional forces that oppose cell elongation. Before complete cessation of elongation, mechanical stresses lead to substantial deformation of cells and changes in their morphology. Our work shows that mechanical forces rather than nutrient limitation are the main growth limiting factor for bacterial growth in long and narrow channels.

Keywords: mother machine, nutrient shielding, mechanics of cell growth, peptidoglycan synthesis, cell wall, microfluidics

INTRODUCTION

High resolution optical microscopy is the most widespread method to study bacterial cellular organization and physiology at the single cell level. Most early studies were carried out using fixed cells that were attached to microscope slides. However, both fixation and attachment of cells to slides alter subcellular organization and can lead to imaging artifacts. As a less invasive preparation method, thin agarose pads sandwiched between microscope cover slide and coverslip have been adopted by numerous groups (e.g., in recent reports, Bailey et al., 2014; Adiciptaningrum et al., 2015). Since thin layer of agarose is prone to drying, thicker layers of agarose in Petri dishes with coverslip bottoms can be used (Männik et al., 2017). The dishes and pads allow imaging live cells over several doublings (typically 4–5). Longer imaging is hampered because individual cells start to overlap. Moreover, cells in the interior of the colony experience a different growth environment than the cells at the periphery and consequently grow at different rates. It is unclear for how long exactly, if at all, steady-state growth conditions can be maintained on the pads.

For reproducible quantitative studies, steady-state cell growth is necessary. To be able to grow cells in steady conditions both the physical and chemical environment of cells needs to remain the same over time. Moreover, cells should not overlap as they grow. In practical terms this means that the colony size has to be kept fixed despite exponential growth in cell numbers over time. Different microfluidic platforms have been developed over the past decade to achieve these requirements (Hol and Dekker, 2014). The developed devices either trap cells in narrow channels comparable to bacterial cross-sectional diameter (Wang et al., 2010; Moffitt et al., 2012; Long et al., 2013) or hold them in shallow chambers where bacteria are confined to a single layer (Männik et al., 2012; Ullman et al., 2013). In the latter case bacteria grow packed side-by-side and quantitative analysis of individual cells is more complicated. In addition to providing a steady growth environment, microfluidics can also be used to administer different chemical (Baltekin et al., 2017; Kaiser et al., 2018) and physical stimuli (Yang et al., 2015) to the cells *in situ* while they are imaged under the microscope.

Of these various designs the most wide-spread has been the so-called mother machine platform (Wang et al., 2010) where cells grow in short (10–25 μm long) dead-end channels (**Figure 1A**). The advantage of dead-end channels relative to channels where both ends are open is longer retention time of cells. Pressure fluctuations are more likely to drive cells out from the channels that have both ends open. In mother machine design, all the cells in the channels are clones of the mother cell that resides in the dead-end side of the channel. The size of the colony is maintained fixed in time because flow in the main channel flushes away extra cells that grow out from the dead-end channels. The same flow also maintains a constant media environment in the growth channels by replenishing nutrients and removing metabolic waste products. Both exchanges are thought to occur via diffusion (Wang et al., 2010). Diffusion may set a limit for nutrient availability for cells at the dead-end side of the channel. To increase diffusion rate a design with shallow reservoirs surrounding the dead-end channels has been implemented (Norman et al., 2013; Cabeen et al., 2017). These reservoirs allow diffusion of nutrients from the main channel but are shallow enough to prevent cells from populating them. Faster exchange of media can also be achieved by diverting some flow past the cells via small opening on the "dead-end" side of the channel. While allowing flow of medium, the opening needs to be made small enough to prevent cells from passing through. Such channels have been recently fabricated and tested (Baltekin et al., 2017; Jennings, 2017). However, the fabrication of these devices is rather challenging, especially for smaller size bacteria, such as *E. coli* growing in poor medium, because the opening in the dead-end side needs to be made no more than about 300 nm wide to prevent cells from squeezing through the deformable openings (Männik et al., 2009; Jennings, 2017).

Mother machine platform has been used to study cell aging (Wang et al., 2010), cell cycle control (Taheri-Araghi et al., 2015), and effects of mechanical forces on cell wall growth (Amir et al., 2014). The devices have been also used in studies of gene regulation (Norman et al., 2013; Cabeen et al., 2017; Kaiser et al., 2018) and antibiotic resistance (Baltekin et al., 2017). New open-source computational platforms have been specifically developed to segment and analyze cells in mother machine platform (Sachs et al., 2016; Kaiser et al., 2018). Despite such widespread interest, bacterial growth in narrow dead-end channels has not been systematically compared to their growth in typical liquid media conditions. Nor has it been determined what phenotypic differences appear in cells that grow in such microfluidic devices. Here, we analyze nutritional and mechanical growth limitations to clonal *E. coli* cultures in microfluidic dead-end channels of various widths and lengths (**Figure 1A**). We find that *E. coli* adapt to the confined channel environment by becoming significantly narrower and longer than the same bacteria in liquid cultures. While the aspect ratio is affected, the cell volume remains approximately the same for cells growing in short channels compared to those growing in

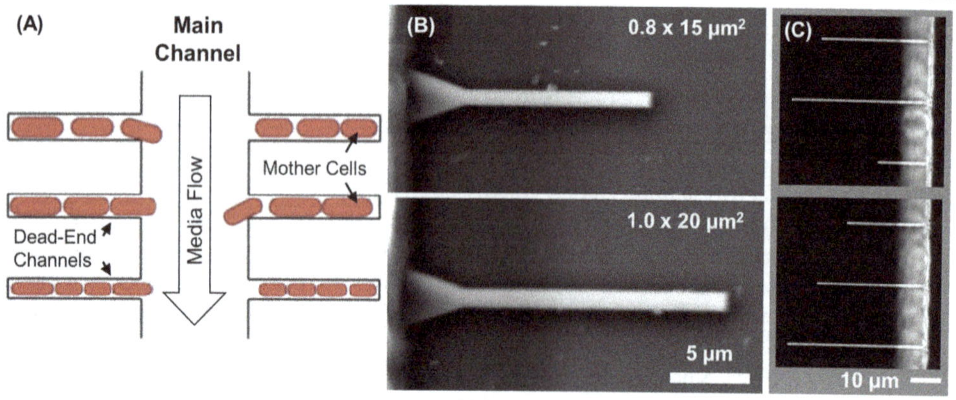

FIGURE 1 | Design of microfluidic chip. **(A)** Schematics showing the mother machine channel layout. Cells grow in dead-end channels. Nutrients diffuse to cells from the main channel where a constant flow is maintained. This flow also removes metabolic waste products and flushes away extra cells. Growth of the mother cell at the end of the channel is studied as a function of channel width and length. **(B)** SEM images of the silicon mold showing patterns of two different size channels. **(C)** Channels form a completed PDMS device imaged using phase contrast microscopy. Channel lengths in the two images vary from 20 to 50 μm. There is a total of 150 channels of each length on a single chip.

the same media conditions in liquid cultures. By increasing the length of the channels, the growth speed and cell volume both decrease until cell growth completely stops in longer channels. We assign the complete cessation of growth to high levels of mechanical stress resulting from colony growth rather than from nutrient limitations. Interestingly, the stress in these 1D colonies can reach levels sufficient to deform cells and to cause them growing into irregular shapes.

MATERIALS AND METHODS

Microfluidic Device Design and Fabrication

The fluidic circuitry in each chip consists of the main channel for media supply and waste product removal and 600 dead-end channels connected to the main channel (**Figure 1A**) following the typical mother machine layout (Wang et al., 2010). The designed length of the dead-end channels varies between 15 and 200 μm, and width from 0.6 to 1.0 μm. The fabrication of microfluidic devices is based on soft-lithography of polydimethylsiloxane (PDMS) elastomers (Weibel et al., 2007).

The channels in PDMS elastomers are created using 4″ silicon (Si) wafer molds. The fabrication of the Si molds follows the process described earlier (Yang et al., 2015). Briefly, the patterns of dead-end channels are defined by e-beam lithography using a JEOL JBX-9300FS electron beam lithography system (JEOL, Japan) with ZEP520A, a positive tone e-beam resist (ZEON Chemical, Japan). After e-beam writing and resist development, a 15 nm chromium layer is deposited. Subsequently, the e-beam resist layer is lifted off using sonication in an acetone bath. A 1.2 μm deep Si etch is carried out in an Oxford Plasmalab 100 inductively coupled plasma reactive ion etching system (Oxford Instruments, MA). The Cr layer acts as a mask for Si etching. The patterns for the larger flow channels are defined using photolithography of SU-8 2015 (MicroChem, MA). The reliefs that result from this step have a typical height of 20 μm. The molds are subsequently silanized in a desiccator using (tridecafluoro-1,1,2,2-tetrahydrooctyl)-1-trichlorosilane (UCT Specialties, CA) for at least 15 min.

PDMS elastomer (Sylgard 184 kit, by Dow Corning, MI) is cast on the mold in a 10:1 base/linker weight ratio. The PDMS is baked at 90°C for 20 min in a convection oven, and then left in the oven for at least two more hours as the oven cools down from 90°C. Individual patterns are cut out, and access holes to the main channels are punched using a biopsy needle. These pieces are subsequently bonded to coverslips. For bonding, #1.5 coverslips are cleaned in isopropyl alcohol (both from Thermo Fisher Scientific, NH) by sonication and then treated in O_2 plasma at 200 mTorr for 70 s. The PDMS elastomer piece and glass coverslips are additionally treated in O_2 plasma for 7s before bonding. After bonding, the chips are left at room temperature at least for 12 h before starting live cell measurements.

The channel heights and widths are measured from the Si mold. For the height measurement, a KLA-Tencor P-6 Stylus (KLA-Tencor Corporation, CA) profilometer is used. The heights of all dead-end channels in different microfluidic chips are within 1.15 ± 0.05 μm. The channels widths are measured using SEM and closely follow their design widths (less than 20 nm differences) (**Figure 1B**). The length of these channels, as measured from the optical images of completed chips (**Figure 1C**), also closely follow their design values.

Bacterial Strain and Culturing

In all measurements, *E. coli* strain AJ5 is used. The strain is created from strain BW25113 (Datsenko and Wanner, 2000) by P1 transduction with lysate from strain FW1401 (Wu et al., 2015). The resulting strain carries a *tagRFP-T* sequence together with kanamycin resistance cassette replacing *leuB*. For an experiment, a colony from LB plate is grown overnight at 28°C with shaking in M9 minimal medium. The medium consists of M9 salts (Teknova, CA), supplemented with 2 mM magnesium sulfate ($MgSO_4$), 0.5% (w/v) glucose (Sigma-Aldrich, MO), 0.2% casamino acids (ACROS Organics, NJ). Liquid media are supplemented with 25 μg/ml kanamycin (Sigma-Aldrich, MO) for selection.

Microfluidic Chip Experiment

Before inoculation of cells to a microfluidic chip, 0.1% (w/v) bovine serum albumin (BSA, Sigma-Aldrich, MO) is added to an overnight liquid culture ($OD_{600} > 0.8$). The cells are then concentrated 100 times in the same medium. About 2 μl of the concentrated culture is injected into the microfluidic device using a pipette and left at 28°C. After a satisfactory amount of dead-end channels are filled with at least one cell (requires a minimum of 20 min), tubing will be connected to the device, and the flow of fresh M9 medium is started. The medium that is used is the same as for the overnight culture, but with 0.1% BSA added to prevent cells from sticking to the main channel surfaces. The medium contains no antibiotics. The flow of this medium is maintained at 4.5 μl/min by an NE-1000 Syringe Pump (New Era Pump Systems, NY). The micro-cultures are grown at 28°C for at least 14 h before imaging is started.

Liquid Culture Measurements

To measure doubling times in liquid culture, the cells are grown in M9 medium overnight ($OD_{600} > 0.8$). The composition of the medium is the same as for the microfluidic chip measurements. The culture is then diluted at least 100× to fresh M9 medium. OD of the culture is measured at $\lambda = 600$ nm using a GENESYS 20 Visible Spectrophotometer (Thermo Fisher Scientific, NH) in 30 min intervals outside the incubator. The linear region of the log(OD) vs. t curve is used to determine the doubling time (Supplementary Figure 1).

To measure cell dimensions in liquid culture, cells are grown to mid-log phase ($OD_{600} \sim 0.15$), concentrated 50×, and then spread onto a 2% agarose pad with the same M9 media composition as in other measurements. The cells are imaged within 30 min after spreading to the pad.

Microscopic Imaging

A Nikon Ti-E inverted epifluorescence microscope (Nikon Instruments, Japan) with a 100X (NA = 1.45) oil immersion phase contrast objective (Nikon Instruments, Japan) is used for imaging the bacteria. Images are captured on an iXon DU-897 EMCCD camera (Andor Technology, UK) and recorded using NIS-Elements software (Nikon Instruments, Japan). Fluorophores are excited by a 200 W Hg lamp through an ND4 neutral density filter. A Chroma 41004 filtercube (Chroma Technology Corp., VT) is used when capturing fluorescent images. A motorized stage and a perfect focus system are utilized during time-lapse imaging. About 100 dead-end channels for each channel size is imaged in each measurement for 6-h period or longer (at least 5 doublings).

Image Analysis

Image analysis was carried out using Matlab (MathWorks, MA) scripts based on Matlab Image Analysis Toolbox, Signal Processing Toolbox and DipImage Toolbox. ImageJ was used to prepare individual images for the figures and the SI movies.

Cell Length Measurements

Both cell lengths and widths are measured from fluorescent images of the cytoplasmic tagRFP-T label. First, the raw fluorescent images are transformed to the second derivative images using the DipImage function *laplace_plus_dgg*, which computes the Laplacian and the second derivative in the gradient direction of an image (Verbeek and Vanvliet, 1994). Thresholding of the second derivative image based on zero crossing yields a binary image, which is then eroded and dilated to resolve individual cells and to fill them, respectively. Cell lengths are measured from these binary images. The measurement consists of finding the greatest Feret diameter for each cell using the *measure* function from DipImage.

Cell Width Measurements

Based on the binary cell masks from the previous step, the gray value center of mass coordinates and the coordinates of the cell long axes are determined. Next, a line perpendicular to the long axes of the cell that passes through the cell center is calculated. To improve the determination of the cell width, an additional set of lines are calculated, where the orientation relative to the previous line is varied in small angular steps. For each orientation, an intensity profile from the fluorescent image is determined, and the profile is fitted to a Gaussian. The Gaussian with the smallest width is then found among all the fits. Due to diffraction, variance of the Gaussian is about 0.04 μm smaller than the cytoplasmic diameter of the cell in our setup (Männik et al., 2009). An additional 0.04 μm is added to the cytoplasmic diameter to account for the width of the periplasmic space and outer membrane layer of the cell. The final, calculated width thus corresponds to the outer diameter of the cell.

Cell Volume Measurements

For the cell volume determination, we assume that every cell is a cylinder with two hemispherical caps. The volume, V_c, based on cell length, L_c, and width, W_c, is:

$$V_c = \frac{\pi}{6} W_c^3 + \frac{\pi}{4} W_c^2 (L_c - W_c) \quad (1)$$

Statistical Analysis

Three replicate measurements of cells were performed for each channel size and for liquid cultures. The average cell length, width, volume and doubling time were calculated from the averages of these three measurements. The error bars associated with the above measurements are the standard errors which have been calculated from the averages of the three measurements. For universal growth law plot the error bars were calculated by propagating the random errors for volumes and doubling times. For statistical significance testing Welch's t-test was used. This test is more reliable than Student's t-test when sample sizes are different. To estimate correlations Pearson R was used.

RESULTS

Dependence of Mother Cell Growth on Channel Width

To investigate how the channel width affects the growth rate and cell dimensions we fabricated channels of 0.6, 0.7, 0.8, 0.9, 1.0, and 1.2 μm in width, 1.15 μm in height and of 15 and 20 μm in length on a single chip. The chosen channel widths were expected to be close to the cell diameter in these growth conditions. The majority of the 0.8, 0.9, and 1.0 μm wide channels had stable bacterial populations that filled the

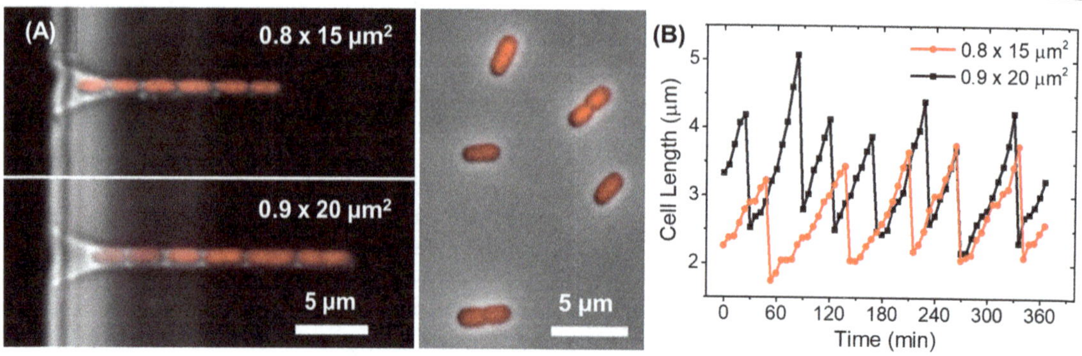

FIGURE 2 | Bacterial growth in channels. **(A)** Composite of phase and fluorescent images of two different size channels filled tagRFP-T labeled *E. coli* (left images). For comparison, composite image of the same strain grown in liquid culture and imaged on agarose pad (right image). The difference in cell width in the two growth environments can be visualized. **(B)** Growth curves of mother cells from channels shown on **(A)**.

whole channel. Cells in these populations grew in single rows (**Figure 2A**, Supplementary Videos 1, 2). In the 1.2 μm wide channels, cells grew in two parallel rows and, as such, these channels were not suitable for analysis. On the other hand, 0.6 μm wide channels were too narrow to support stable colony growth. Although few cells loaded to these channels initially, they all moved out from the channels before imaging started. The same also occurred in the majority of the 0.7 μm wide channels (Supplementary Video 3). However, in some of the 0.7 μm wide channels (8 out of 200), stable populations were present and could be imaged. In both 0.6 and 0.7 μm wide channels the bacteria appeared to be wider than the channel causing deformations to channel walls (Supplementary Figure 2). Note that bacteria from overnight stationary cultures were loaded into channels. These cells are significantly narrower than log phase cells (Männik et al., 2009), and this made their entry to 0.6 and 0.7 μm wide channels possible. As cells started to grow in fresh medium in these channels, their diameter widened beyond the channel width. Higher deformability of open ends of channels may have provided a driving force that pushed the cells out from the channels.

Thus, of all the fabricated channels only those in the range 0.7–1.0 μm were suitable for studying growth of stable 1D bacterial cultures. Cells in these channels were imaged in three independent measurements that each lasted at least 6 h (Supplementary Videos 1, 2, **Figure 2B**). Here, we analyze, in detail, the growth of the mother cell, which is the cell at the dead-end side of the channel. All the cells in the channel are clones of this cell. We use the mother cell for analysis because it is the only cell that strictly grows in steady state conditions. The other cells in the channel move toward the channel entrance during their growth and, because of that movement, we expect both the composition of the growth medium and the mechanical stress that the cells experience to change. Also, to guarantee that mother cells in a channel of a given size grew in comparable conditions we required the channels to be completely full of cells (<1 μm empty space between cells) throughout the entire time-lapse imaging period.

We next compare birth length, width, volume and doubling time measurements of mother cells in differently sized channels to those from cells in liquid cultures. We found cell length (2.12 ± 0.05 μm) to be independent of channel width in the 0.8–1.0 μm range (**Figure 3A**). In 0.7 μm wide channels, where cells appeared to be in contact with channel walls, the cell lengths were significantly (9% difference, $p < 0.02$) smaller than those in other channels. Interestingly, cells were longer in all microfluidic channels than in the test-tube liquid cultures (1.84 μm; $p = 0.19$ for 0.7 μm wide channel, and $p < 10^{-16}$ for all other

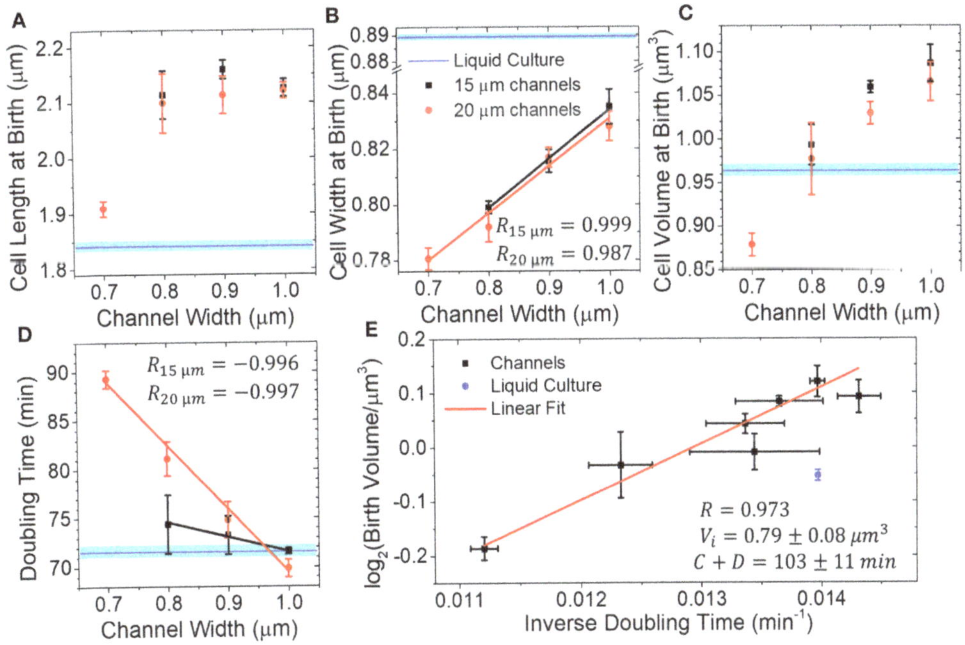

FIGURE 3 | Dimensions of the mother cell and its doubling times in dead-end channels of different widths. **(A–D)** Mother cell length, width, volume and doubling time as a function of channel width in channels of 15 (red circles) and 20 μm (black squares) length. All measurements have been done at 28°C. Data points represent the average of three independent measurements. In each measurement, more than 300 cell births have been analyzed. Error bars are s.e.m. of the three independent measurements. Blue horizontal lines represent measurements of cells from liquid culture tubes. The width of these lines corresponds to s.e.m. All other solid lines are linear fits to the data. **(E)** Logarithm of birth volume, V_b, as a function inverse of doubling time, T_d^{-1}. Solid line shows the fit of the data to the universal growth law, $V_b = \left(\frac{1}{2}\right)V_i \bullet 2^{\frac{C+D}{T_d}}$. In the universal growth law, V_i is cell volume at the initiation of replication, C the time taken to replicate DNA and D the time taken from the end of replication to cell division. Here V_i and $C+D$ are treated as fitting parameters with best fit values $V_i = 0.79 \pm 0.08\ \mu m^3$ and $C+D = 103 \pm 11$ min. The errors are standard error.

channels). At the same time, the coefficients of variation of the cell length at birth distributions were comparable to the values in the liquid culture (all are about 15%) (Supplementary Figure 3). The coefficients of variation determined here are in agreement with previous measurements where values of 12–17% have been reported from agarose pad (Adiciptaningrum et al., 2015) and microfluidic measurements (Campos et al., 2014; Taheri-Araghi et al., 2015).

While cell lengths were longer in microfluidic channels, their widths were significantly smaller (Welch's t-test $p < 10^{-8}$) (**Figure 3B**). Also, coefficients of variation of the cell width distribution were smaller than the ones in liquid culture (4.5%) and decreased as the channel width decreased from 4 to 3% (Supplementary Figure 3). In 0.7 μm wide channels, the cell width was wider than the undeformed channel diameter indicating that channel walls prevented cells from becoming wider. In 0.8 μm wide channels, some cells could still be squeezed by channel walls. In 0.9 and 1.0 μm wide channels, however, the cells were narrower than the channels and not squeezed by the channel walls. Irrespective of squeezing or no squeezing from channel walls, the cell width at birth increased linearly with the channel width (Pearson $R = 0.999$ in 15 μm long and 0.987 in 20 μm long channels). So, in wider channels, the decrease in cell diameter was not linked to a direct mechanical force to cells but must have reflected some form of adaptation of the cells to a confined channel environment.

As expected from the two above measurements, the average mother cell volume also increased with the channel width (**Figure 3C**). Comparison of microfluidic and the liquid culture cells showed that the differences in their lengths and widths, to some degree, compensated each other in cell volumes. In particular, for the 0.8 μm wide channels, the cell volumes matched those in liquid culture.

The doubling times also depended on channel width and increased as the channels become narrower (**Figure 3D**). The effect was pronounced for cells in 20 μm long channels ($R = -0.997$, slope= −65 min/μm), but was less significant in 15 μm channels ($R = -0.996$, slope = −15 min/μm). In the latter case, the doubling times appeared almost indistinguishable from cells grown in liquid cultures despite changes in cell shape. The observed increase in cell doubling times with the decrease of its dimensions is expected. According to the universal growth law, cell volumes at birth depend exponentially on growth rate, which is taken here as an inverse of doubling time (Schaechter et al., 1958; Willis and Huang, 2017). Such correlations are indeed evident in our data (**Figure 3E**). The universal growth law emerged from studies where cells grew unrestricted, and the growth rate was determined by the type of carbon source rather than by its abundance (Schaechter et al., 1958). In later studies, growth rate was also examined in growth limiting conditions and the increase in cell size and growth rate were found to be correlated in most but not all conditions (Shehata and Marr, 1971). Accordingly, we interpreted the exponential decrease in mother cell volume as a function of doubling time to result from nutrient limiting conditions in channels ends which would arise from nutrient shielding in the channel.

Dependence of Mother Cell Growth on Channel Length

To further investigate this hypothesis, we studied how the dimensions and doubling times of mother cell depended on the channel length. We expected that doubling times should increase and cell dimensions should decrease as the length of the channel increases. For these studies, we used microfluidic chips with a fixed channel width of 0.9 μm but varied channel length from about 20–50 μm (19, 29, 39, and 49 μm). We combined results from these measurements with the earlier ones in 15 and 20 μm long channels. Although the 15 and 20 μm long channels had the same designed width as the longer channels, the actual widths of longer channels appeared slightly smaller than 0.9 μm, which likely explains the differences in growth rates of the two data sets.

We found that the mother cell length, width, and volume all decreased in longer channels (**Figures 4A–C**). The decrease was more pronounced in cell length and volume (17 and 33%) than that in cell width (7% change). This contrasts earlier measurements in **Figure 3**, where the cell width showed a larger variation while the length remained approximately constant. The decrease in cell length and volume was in good approximation proportional to channel length ($R = -0.992$ for both cases), while the decrease of cell width showed lower correlations ($R = -0.832$). The doubling time also increased linearly as a function of channel length (**Figure 4D**). In 50 μm long channels, the doubling time was 44% longer than in 15 μm long channels. The increase in doubling time and decrease in cell volumes were, in good approximation ($R = 0.999$), related by the universal growth law (**Figure 4E**). The fit parameters, which determine the cell volume at initiation per replication origin (V_i), and the sum of C and D periods (Cooper and Helmstetter, 1968), matched those found from measurements where the channel width was varied (cf. **Figures 3E, 4E**). The consistency of these two datasets indicates that growth limitation likely had the same origin in the measurements where channel width and length were varied.

Mechanical Impediments to Cell Growth in 1D Cultures

In the above measurements, the doubling time increased linearly as the function of channel length. We hypothesized that if the channel length increased even further then the doubling time of the mother cell would show a non-linear increase and perhaps cell growth would completely stop as the nutrient levels deplete at the end of the channel. To test if doubling time of mother cells as a function of channel length would deviate from linear at longer channel lengths, we fabricated a new batch of microfluidic chips with channel lengths of 20, 50, 100, 150, and 200 μm.

Unexpectedly, during the initial passivation step of the channel walls with bovine serum albumin (BSA), which preceded cell loading, we observed an accumulation of this protein to the channel ends in 100 μm and longer channels in phase contrast images. The effect was completely missing in channels 50 μm and shorter on the same chip even for long incubation times (>12 h). When the same M9 growth medium was used without BSA, no material accumulated to the ends of any channel. Aggregation of BSA was so strong that it completely excluded

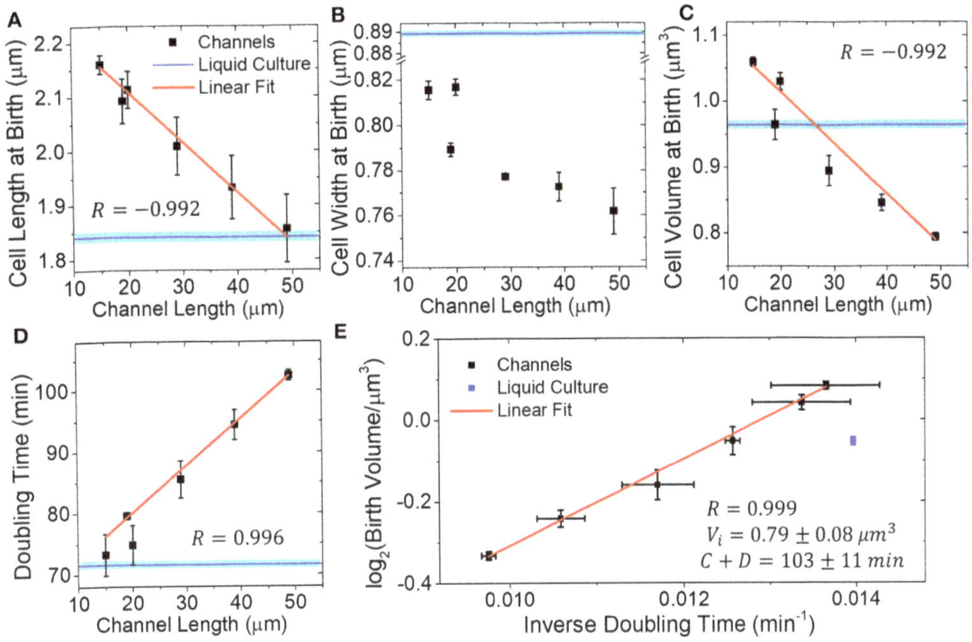

FIGURE 4 | Dimensions and doubling times of mother cells in channels of different lengths. **(A–D)** Mother cell length, width, and volume at birth, and doubling time as a function of channel length. All channels are 0.9 μm wide. Each data point is an average of three independent measurements. In each measurement, more than 300 cell births have been analyzed. Error bars are s.e.m. of the three independent measurements. Blue horizontal lines correspond to measurements of these quantities from liquid culture cells. Solid lines are linear fits to data. **(E)** Logarithm of birth volume, V_b, as a function of inverse doubling time, T_d^{-1}, and it fits to the universal growth law, $V_b = \left(\frac{1}{2}\right)V_i \bullet 2^{\frac{C+D}{T_d}}$.

cells from the channel ends. Further investigation showed that the effect was not specific to BSA because if the channels were left with LB medium that did not contain any BSA, some material still accumulated to the channel ends. Similar effects have been observed before and were explained by water diffusing into PDMS (Randall and Doyle, 2005). The effect depends sensitively on channel length as will be discussed in more details later. Although BSA prevented cells from occupying the ends of 100 μm and longer channels, cells could also be cultivated without BSA. In this case, there was some aggregation of cells near the entrances of the channels. The omission of BSA thus increased sticking of cells to the channel walls, as expected. However, the growth rate of cells in 20–50 μm long channels without BSA passivation were not different from these with BSA (Supplementary Figure 4).

The growth of colonies in 100 μm and longer channels differed significantly from those in shorter channels (≤50 μm). These differences were present irrespective of the presence or absence of BSA in the growth medium. Attachments of cells to channel walls combined with their continued growth lead to considerable pressure buildup, which was evidenced by the widening of the channel in phase contrast images (**Figures 5A–C**). In deformed regions of the channel, the cells grow in multiple rows or were tilted relative to the direction of the channel. Based on our earlier studies (Yang et al., 2015), stress analysis of such channels has shown that pressures in the 0.2 MPa range are needed to widen channels at their midplane by few hundred nanometers. Similar or even larger pressures must have been present in the broadened regions of the channel. From time to time the pressure buildup resulted in a sudden release of cells from the channels as the force resulting from cell growth exceeded some critical value needed to break adhesive contacts between cells and channel walls (**Figure 5A**). However, in some channels, these contacts were likely stronger or differently distributed, and no pressure release occurred during the 12 h observation period.

The pressure buildup had pronounced effects on cell growth and morphology (**Figure 5C**, Supplementary Videos 4, 5). As the pressure increased, cell elongation and division rates slowed down (**Figure 5D**), and the cells started to broaden (**Figure 5E**). For cells that grew tilted relative to the channel axes in the initial stages of pressure buildup, the broadening was not uniform along the cell length due to uneven pushing from other cells (Supplementary Videos 6, 7). The cell regions in contact with the other cells expanded less than the regions where cells were able to expand toward channel walls without these contacts. As a result, the cells acquired irregular pear-shaped morphologies that significantly differed from the regular rod shape (**Figure 5C**). Some mass growth, albeit at a much slower rate than in liquid cultures, still occurred in these conditions (**Figure 5F**). The mass growth in these conditions could be predominantly attributed to cell broadening rather than to elongation. Upon release of pressure, the cell shape immediately returned to the usual elongated rod-shaped morphology. This

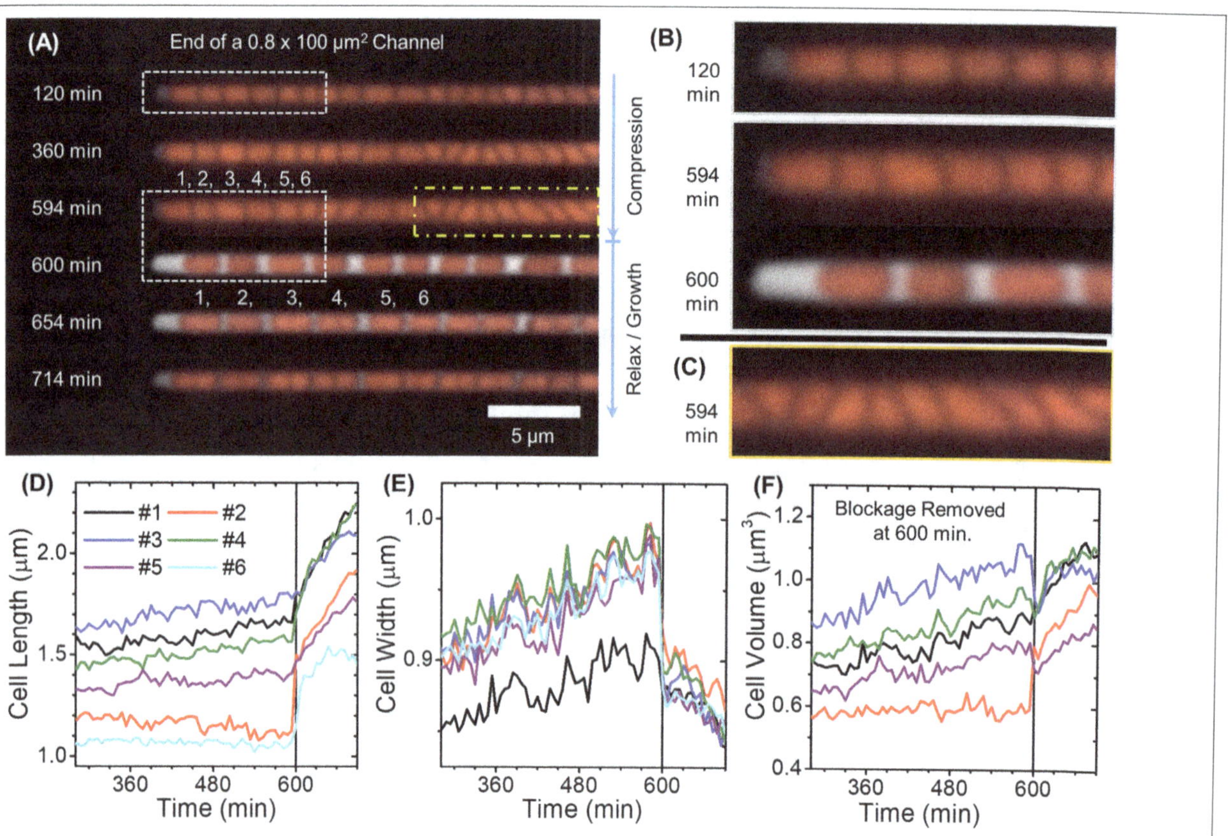

FIGURE 5 | Cell growth in 100 μm long channels. **(A)** Composite time-lapse images of cells growing in a 100 μm long and 0.8 μm wide channel without BSA passivation. Pressure builds up in the channel until the frame at 594 min. Pressure release occurs between 594 and 600 min after which cell growth and division resume. Individual cells are labeled with numbers 1 to 6. **(B,C)** Boxed regions in **(A)** are magnified. **(D–F)** Cell lengths, widths and volumes as a function of time for the first six cells in the channel as labeled on **(A)**. Solid vertical lines show time when the pressure release occurred.

sudden change was accompanied by a decrease in cell width (≈7%) and increase of cell length (≈10%). Interestingly, cell elongation resumed within minutes after pressure was released (**Figure 5D**) even though the elongation rate had been almost zero during the 9 h compression period. Along with resumption of cell elongation, the cell width, after decreasing abruptly upon release of compression, decreased further. Over the course of 1.5 h, it approached its regular value (0.78 μm). The slowdown of cell growth also appeared in measurements in 100 μm and longer channels when BSA was used (Supplementary Figure 5, Supplementary Video 8). The slowdown was followed by a period of faster growth after each pressure release. However, adhesion of cells to the channel wall was clearly weaker in this case; consequently, smaller pressure buildups occurred. In these conditions, cell morphology remained regular, and broadening could not be detected. Altogether, in 100 μm and longer channels, mechanical hindrances were the growth limiting factor instead of nutrient depletion.

DISCUSSION

Cell Shape Shows Adaptations to Channel Geometry

Here we studied how confinement changes the growth of *E. coli* cells in microfluidic channels of different widths and lengths. In 15 and 20 μm long channels with widths of 0.9 and 1.0 μm, the growth rate and cell volume were comparable and even slightly (5–10%) exceeded the values from liquid cultures measurements. Although the confinement related effects to the growth rate in these conditions were negligible, cell morphology was affected. Cells were significantly longer and narrower than in liquid cultures. What factors could have led to the change of cell morphology? In the narrowest channels with widths of 0.7 μm, the cell width was physically limited by the channel walls. The same mechanical constraint could also have limited cell diameter in 0.8 μm wide channels. However, in 0.9 and 1.0 μm wide channels the cells were not squeezed by channel walls, yet they still maintained a higher length to width ratio compared to liquid culture cells (2.7 vs. 2.1, respectively). Even higher ratios (>3.4) can be inferred from previous measurements by another group using mother machine devices at similar growth conditions (Taheri-Araghi et al., 2015). We hypothesize that cell shape in these channels is narrower and longer because of contacts to channel walls. Although the cell adhesion to channel walls was weak, especially to BSA passivated channel walls, these contacts may have had an effect on peptidoglycan synthesizing machinery. Consistent with this idea, the length to diameter ratio of cells was higher when channel walls were not passivated by BSA compared to that when they were (Supplementary Figure 4). By a common view, peptidoglycan synthesis activity is controlled via MreB and FtsZ polymer scaffolds, which reside in the cytosolic side of the inner membrane (Typas et al., 2012). The data presented here point to the possibility that stresses in the outer envelope may also modulate peptidoglycan synthesis activities directly without being transmitted via MreB and FtsZ scaffolds.

Is Mass Transport Limiting Cell Growth in Channels?

Our measurements showed universal growth law-like dependence between cell volume at birth and growth rate. We interpreted this dependence as arising from nutrient limiting conditions at the channel. We hypothesized that adsorption of nutrients by cells growing between the mother cell and the channel entrance lead to depletion of nutrients at the location of the mother cell. This effect has been also referred to as nutrient shielding (Lavrentovich et al., 2013). Taking the simple geometry of channels and growth of bacteria in single rows without gaps, the nutrient shielding effects in these conditions can be quantitatively analyzed using 1D reaction-diffusion equations. Separate equations can be written for each chemical compound present in the growth medium. Here, we assume that there is just one component in the medium that is growth limiting. Denoting its concentration by c, its diffusion coefficient by D, its absorption coefficient per unit cell surface area by k_{abs}, channel width by W and height by H, and cell radius by R_c a reaction-diffusion equation as a function of distance x from the channel entrance for this component can be written as:

$$\left(W \cdot H - \pi R_c^2\right) \cdot D \frac{d^2 c}{dx^2} = 2\pi R_c k_{abs} \cdot c \tag{2}$$

Here we assumed that the uptake of this nutrient component is a first order process that is far from saturation. If the uptake were saturated (kinetically limited) then the corresponding compound would not be a growth limiting factor. The solution to this equation assuming non-adsorbing boundary condition at channel end is:

$$c\,(x) = \frac{c_0 \cosh\left(\frac{x-L}{\lambda}\right)}{\cosh\left(\frac{L}{\lambda}\right)} \quad \lambda = \sqrt{\frac{\left(W \cdot H - \pi R_c^2\right) \cdot D}{2\pi R_c \cdot k_{abs}}} \tag{3}$$

Here L is length of the channel and c_0 is concentration of this compound at the channel entrance. λ defines a characteristic length scale, which is referred to as the nutrient screening length (Lavrentovich et al., 2013). The screening length depends on the cross-sectional area of the channel, being longer for wider channels. We next assume that growth rate (inverse of doubling time), which is controlled by this compound, follows Michelis-Menten type of kinetics (Shehata and Marr, 1971):

$$T_d^{-1}\,(x) = \frac{T_{d,min}^{-1}}{1 + \frac{K}{c(x)}} \tag{4}$$

Here T_d is doubling time of cells at position x from the channel entrance, $T_{d,min}$ is the doubling time in conditions where the compound is not growth limiting and K determines the concentration above which nutrient uptake rate starts to saturate. In dilute liquid cultures $T_{d,liquid} = T_{d,min}(1+K/c_0)$. Using the latter expression and the solution for the reaction-diffusion equation for $c(x)$ yields:

$$T_d\,(x) = T_{d,min} + \left(T_{d,liquid} - T_{d,min}\right) \cosh\left(\frac{L}{\lambda}\right) / cosh\left(\frac{x-L}{\lambda}\right) \tag{5}$$

Our data concerns mother cells. The doubling time of the mother cell T_d then becomes:

$$T_d = T_{d,min} + \left(T_{d,liquid} - T_{d,min}\right) \cosh\left(\frac{L}{\lambda}\right) \tag{6}$$

Here we assumed the position of the mother cell to be $x \approx L$. The formula predicts that the doubling time in short channels $L \ll \lambda$ is independent of channel length and equals the doubling time in liquid cultures. In channels $L \approx \lambda$ it increases as $T_d \sim L^2$ and in channels $L \gg \lambda$ the increase is exponential $T_d \sim exp(L/\lambda)$. Altogether equation 5 thus predicts a distinctly non-linear relationship between doubling time and channel length. In contrast to this prediction, the data in **Figure 4D** is in good approximation linear (Supplementary Figure 6). The discrepancy between the model and the data also appears when one compares dependence of the doubling time on channel width. The model predicts a smaller variation of doubling time as a function of channel width, in particular for 20 μm long channel (Supplementary Figure 6). It is possible that model treats too simplistically the relationships between concentration of growth-limiting compound and growth rate. For example, some deviations from a single Michelis-Menten type relationship for growth rates were observed in glucose limiting conditions (Shehata and Marr, 1971). However, the underlying relationship between doubling time and channel length should still be distinctly non-linear and as such not consistent with the experimental data. Moreover, there is no obvious compound in our growth medium that can be growth limiting. Concentrations of all media components at channel opening are in the millimolar range and as such exceed several orders of magnitude the growth-limiting concentrations. For example, concentration of glucose in the channel opening is 22 mM. This is about 10^4 higher than its growth-limiting concentration of about 1 μM (Shehata and Marr, 1971). The uptake of glucose by cells is thus completely kinetically limited in the vicinity of channel opening. One can solve the reaction-diffusion equation similar to equation 1 also in kinetically limited regime. The solution shows that the concentration of such nutrients decreases quadratically from the channel entrance as:

$$c = c_0 - c\,(L) = L^2 \frac{I_{\max}}{2\left(WH - \pi R_c^2\right) L_c D} \tag{7}$$

Here I_{max} is kinetically limited nutrient uptake rate per cell per second, and L_c is the average cell length. For glucose $I_{max} = 2 \cdot 10^5$ molecules per cell per second (Natarajan and Srienc, 1999) and D–700 $\mu m^2/s$ (Longsworth, 1953) yields concentration change of 2.7 mM (~10%) at the end of 50 μm long channel that is completely filled with bacteria. Resulting 19 mM concentration of glucose in channel end corresponds still to highly saturating level. The same arguments also apply to the casamino acids, which are present in mM concentrations too. Moreover, their depletion with exception of leucine, which is essential for the strain, would not significantly alter the doubling times. If depletion of leucine would have occurred then this would have led to increase in doubling times approximately given by equation 5. Such dependence, however, would not be consistent with the experimentally observed linear dependence in **Figure 4D** as already argued.

The previous analysis assumed mass transport of nutrients to channels is solely via diffusion. However, accumulation of BSA to the ends of 100 μm and longer channels demonstrates that there is significant influx of water into these dead-end channels because of evaporation/diffusion water into PDMS (Randall and Doyle, 2005). This influx was significant even after channels were filled with water for a 12-h period. The steady uptake rate of water to PDMS has been estimated to be $J = 7{\cdot}10^{-6}$ kg/(m^{2}s) (Randall and Doyle, 2005). This update rate explains why BSA accumulates in 100 μm long channels but not in 50 μm long channels. The steady state concentration profile for the BSA in empty channels can be found from equation that balances its diffusive and convection fluxes:

$$W \cdot H \cdot D_{BSA} \frac{dc_{BSA}}{dx} = \frac{J}{\rho}(2H + W) \cdot (L - x) \cdot c_{BSA}(x) \quad (8)$$

Here ρ is density of water. The solution to this equation is:

$$c_{BSA}(x) = c_{0,BSA} \exp\left(\frac{L^2}{2\sigma^2}\right) \exp\left(-\frac{(L-x)^2}{2\sigma^2}\right) \quad \sigma = \sqrt{\frac{\rho \cdot D_{BSA}}{J} \frac{WH}{2H + W}} \quad (9)$$

Here σ is a characteristic length scale for accumulation, which depends on cross-sectional parameters of the channel but not on its length. For BSA, which diffusion coefficient is 70 μm^2/s, $\sigma = 57\ \mu m$ can be estimated for 0.9 μm wide channels. Increase in BSA concentration in channel end is 1.5 times in 50 μm, 5 times in 100 μm and 500 times in 200 μm long channels. 500 times increase corresponds to almost complete precipitation of BSA as observed in our experiments.

While water permeation to PDMS has strong effect on the BSA distribution in the channel, Equation (8) predicts that it does not have significant effect on distribution of nutrients, signaling molecules and metabolic waste products in the channel. The reason is that diffusion coefficients of these small molecules are about an order of magnitude larger (e.g., $D = 700$ μm^2/s for glucose) than for BSA. Even for 200 μm long empty channel one would expect the concentration of these molecules to be less than twice of that in the channel entrance. This increase does not likely have a significant effect on cell growth.

So far we did not consider adsorption of small molecules by channel walls. In current treatment, the adsorption would be completely analogous to adsorption of these molecules by cells. Consequently, adsorption by channel walls can be accounted by increasing k_{abs} in Equation (2) while the functional form of the equations remains the same. Accounting wall adsorption would therefore not lead to better agreement between experiment and model as inclusion of adsorption would not change functional dependence of doubling time on channel length.

Altogether our analysis shows that mass transport related limitations to cell growth in channels $\leq$ 100 μm are not significant in our experimental conditions despite our initial expectations. However, if some of the essential components in the media are present at less than saturating levels then the doubling times of the cells in the channel ends should become strongly affected. Flow of water into channels because of permeability of PDMS should not affect these conclusions because of fast diffusion of nutrients and waste molecules.

Mechanical Constraints to Cell Growth

Our experiments indicate that instead of nutrient limitations, the mechanical impediments set stronger constraints for bacterial growth in channels. In channels 100 μm and longer, friction forces caused cessation of cell elongation. In shorter channels friction could have also been the main growth limiting factor. In accordance with this assumption, friction leads to growth opposing force on a mother cell, F_f, which increases proportionally to channel length $F_f \sim L$. Measurements of *E. coli* growth in agarose hydrogels have shown that forces opposing cell growth decrease elongation rate approximately linearly as the magnitude of the force increases (Tuson et al., 2012). For small forces, this would mean that increase in doubling time from the bulk value is proportional to the opposing force and thus on channel length $\Delta T_d = T_d - T_{d,liquid} \sim F_f \sim L$ in agreement with the experimental data (**Figure 4D**). Increased friction can also explain why in narrower channels doubling time increased (**Figure 3D**).

Although sticking of individual cells to channels walls was weak in channels longer than 50 μm, the cumulative effect of adhesion become strong enough to prevent cell elongation on the dead-end side of the channels. Interestingly, some residual mass growth still occurred in this situation that resulted in broadening of cells. Presumably due to this residual growth cells in some channels were able to overcome static friction forces and release the pressure that opposed their growth. During the release of pressure, the cells behaved similarly to a compressed elastic rod: their length increased (25%) and diameter decreased (7%). The length increase furthermore indicates that cells did synthesize new cell wall during the compression phase. Taken the previously estimated Young modulus in the range of 50–150 MPa for *E. coli* cell for the envelope thickness of 4 nm (Tuson et al., 2012) 25% of compression corresponds to force on the 0.8 μm wide cell of 0.1–0.4 microNewtons (μN). Here we assumed that turgor pressure, during and few minutes after the compression, was the same and that cell wall compression was elastic. The force of 0.1–0.4 μN then corresponds approximately to the stall force for peptidoglycan synthesis in *E. coli*. For comparison, the stalling force has been estimated to be 11 μN in fission yeast (Minc et al., 2009).

Strikingly, the cells were able to restore their growth rates very rapidly (less than 6 min) after more than 9 h of very limited growth during compression. Here the 6 min estimate corresponds to the frame rate of measurements, but actual lag phase could have been shorter. Very short lag time if any is in contrast with the time needed to restore growth in stationary phase cells upon entering fresh medium. In our measurements, this time has typically been in the range of 0.5–2.0 h depending on the length of time cells spent in stationary state. A period of approximately 30 min appears to be needed to restore transcriptional activity and assemble functional peptidoglycan synthesis complexes. Since the cell growth in channels resumed much faster, the cells must have maintained active peptidoglycan machinery throughout the compression period. The compressed cells in channels were clearly different from stationary phase cells as they had access to nutrients and were metabolically more active. Maintenance of enzymatic machinery, even when no elongation occurred, indicates that mechanical stress opposing cell growth alone is not able to suppress transcriptional peptidoglycan synthesis machinery the way the entry to stationary phase does.

Implications to Bacterial Growth in Microfluidic Devices and in Natural Microenvironments

Our work brings out some differences in the growth of *E. coli* in confined, relative to unconfined, conditions in mother machine platform. We find that for mother machine platform the channels need to be in a rather narrow range of dimensions for not to display growth-limiting effects: for *E. coli* growing in M9 medium with glucose and casamino acids only 15 and 20 μm long and 0.9 and 1.0 μm wide channels did not display these effects. However, even in these channels cell morphology differed from the one in liquid cultures. The different aspect ratio that we observed is not likely significant factor for most experiments, but for those dealing with cell size control and cell wall synthesis, these effects need to be considered. To improve the mother machine design, one can use even shorter (<15 μm) channels. However, this comes with the drawback of losing cells from the channel in long-term experiments. On the other hand, our data and analysis indicate that growth limitation due to nutrient and waste diffusion is minimal even in long channels (>100 μm). The designs that increase mass transport to channels, such as done recently by diverting some fluid flows through the channels (Baltekin et al., 2017; Jennings, 2017) or using shallow side channels that surround the cells (Norman et al., 2013; Cabeen et al., 2017), is not likely to alleviate growth limitations in these channels.

The mechanical constraints resulting from cell adhesion may be a major limiting factor for bacterial growth not only in microfluidic channels but also in natural environments such as soil where small pores and channels with dimensions comparable to bacterial size are present (Ranjard and Richaume, 2001). These constraints may also limit bacterial growth in unconstrained colonies such as ones on agar plates. Although nutrient limitation has been considered as the main growth limiting factor for growth of bacterial cells in the interior of the colony (Jeanson et al., 2015), our results point out the possibility that mechanical constraints have an equally important role in limiting growth only to the outer layers of the colony.

AUTHOR CONTRIBUTIONS

DY and JM contributed conception and design of the study. DY and SR fabricated the devices used in the study. DY, AJ, EB performed the measurements. DY analyzed the data. DY and JM wrote the manuscript. All authors contributed to manuscript revision, read and approved the submitted version.

FUNDING

This work has been supported in part by NSF research grant MCB-1252890.

ACKNOWLEDGMENTS

The authors thank Cees Dekker and Fabai Wu for the generous gift of the strain and Ariel Amir, Maxim Lavrentovich, Jaana Männik, Bryant Walker, Conrad Woldringh, and Arieh Zaritsky for discussions and critical reading the manuscript. Authors acknowledge technical assistance and material support from the Center for Environmental Biotechnology at the University of Tennessee. A part of this research was conducted at the Center for Nanophase Materials Sciences, which is sponsored at Oak Ridge National Laboratory by the Scientific User Facilities Division, Office of Basic Energy Sciences, U.S. Department of Energy.

REFERENCES

Adiciptaningrum, A., Osella, M., Moolman, M. C., Lagomarsino, M. C., and Tans, S. J. (2015). Stochasticity and homeostasis in the *E. coli* replication and division cycle. *Sci. Rep.* 5:18261. doi: 10.1038/srep18261

Amir, A., Babaeipour, F., McIntosh, D. B., Nelson, D. R., and Jun, S. (2014). Bending forces plastically deform growing bacterial cell walls. *Proc. Natl. Acad. Sci. U.S.A.* 111, 5778 5783. doi: 10.1073/pnas.1317497111

Bailey, M. W., Bissichia, P., Warren, B. T., Sherratt, D. J., and Männik, J. (2014). Evidence for divisome localization mechanisms independent of the Min system and SlmA in *Escherichia coli*. *PLoS Genet.* 10:e1004504. doi: 10.1371/journal.pgen.1004504

Baltekin, Ö., Boucharin, A., Tano, E., Andersson, D. I., and Elf, J. (2017). Antibiotic susceptibility testing in less than 30 min using direct single-cell imaging. *Proc. Natl. Acad. Sci. U.S.A.* 114, 9170–9175. doi: 10.1073/pnas.1708558114

Cabeen, M. T., Russell, J. R., Paulsson, J., and Losick, R. (2017). Use of a microfluidic platform to uncover basic features of energy and environmental stress responses in individual cells of *Bacillus subtilis*. *PLoS Genet.* 13:e1006901. doi: 10.1371/journal.pgen.1006901

Campos, M., Surovtsev, I. V., Kato, S., Paintdakhi, A., Beltran, B., Ebmeier, S. E., et al. (2014). A constant size extension drives bacterial cell size homeostasis. *Cell* 159, 1433–1446. doi: 10.1016/j.cell.2014.11.022

Cooper, S., and Helmstetter, C. E. (1968). Chromosome replication and division cycle of *Escherichia coli* B/r. *J. Mol. Biol.* 31:519. doi: 10.1016/0022-2836(68)90425-7

Datsenko, K. A., and Wanner, B. L. (2000). One-step inactivation of chromosomal genes in *Escherichia coli* K-12 using PCR products. *Proc. Natl. Acad. Sci. U.S.A.* 97, 6640–6645. doi: 10.1073/pnas.120163297

Hol, F. J., and Dekker, C. (2014). Zooming in to see the bigger picture: Microfluidic and nanofabrication tools to study bacteria. *Science* 346, 438. doi: 10.1126/science.1251821

Jeanson, S., Floury, J., Gagnaire, V., Lortal, S., and Thierry, A. (2015). Bacterial colonies in solid media and foods: a review on their growth and interactions with the micro-environment. *Front. Microbiol.* 6:1284. doi: 10.3389/fmicb.2015.01284

Jennings, A. D. (2017). *Development of Microfluidic Platforms for Studies of Cellular Organization in Escherichia coli.* Master's Thesis, University of Tennessee.

Kaiser, M., Jug, F., Julou, T., Deshpande, S., Pfohl, T., Silander, O. K., et al. (2018). Monitoring single-cell gene regulation under dynamically controllable conditions with integrated microfluidics and software. *Nat. Commun.* 9:212. doi: 10.1038/s41467-017-02505-0

Lavrentovich, M. O., Koschwanez, J. H., and Nelson, D. R. (2013). Nutrient shielding in clusters of cells. *Phys. Rev. E* 87:062703. doi: 10.1103/PhysRevE.87.062703

Long, Z., Nugent, E., Javer, A., Cicuta, P., Sclavi, B., Lagomarsino, M. C., et al. (2013). Microfluidic chemostat for measuring single cell dynamics in bacteria. *Lab Chip* 13, 947–954. doi: 10.1039/c2lc41196b

Longsworth, L. G. (1953). Diffusion measurements, at 25-degrees, of aqueous solutions of amino acids, peptides and sugars. *J. Am. Chem. Soc.* 75, 5705–5709. doi: 10.1021/ja01118a065

Männik, J., Bailey, M. W., O'Neill, J. C., and Männik, J. (2017). Kinetics of large-scale chromosomal movement during asymmetric cell division in *Escherichia coli*. *PLoS Genet.* 13:e1006638. doi: 10.1371/journal.pgen.1006638

Männik, J., Driessen, R., Galajda, P., Keymer, J. E., and Dekker, C. (2009). Bacterial growth and motility in sub-micron constrictions. *Proc. Natl. Acad. Sci. U.S.A.* 106, 14861–14866. doi: 10.1073/pnas.0907542106

Männik, J., Wu, F., Hol, F. J. H., Bissichia, P., Sherratt, D. J., Keymer, J. E., et al. (2012). Robustness and accuracy of cell division in *Escherichia coli* in diverse cell shapes. *Proc. Natl. Acad. Sci. U.S.A.* 109, 6957–6962. doi: 10.1073/pnas.1120854109

Minc, N., Boudaoud, A., and Chang, F. (2009). Mechanical forces of fission yeast growth. *Curr. Biol.* 19, 1096–1101. doi: 10.1016/j.cub.2009.05.031

Moffitt, J. R., Lee, J. B., and Cluzel, P. (2012). The single-cell chemostat: an agarose-based, microfluidic device for high-throughput, single-cell studies of bacteria and bacterial communities. *Lab Chip* 12, 1487–1494. doi: 10.1039/c2lc00009a

Natarajan, A., and Srienc, F. (1999). Dynamics of glucose uptake by single *Escherichia coli* cells. *Metab. Eng.* 1, 320–333. doi: 10.1006/mben.1999.0125

Norman, T. M., Lord, N. D., Paulsson, J., and Losick, R. (2013). Memory and modularity in cell-fate decision making. *Nature* 503:481. doi: 10.1038/nature12804

Randall, G. C., and Doyle, P. S. (2005). Permeation-driven flow in poly(dimethylsiloxane) microfluidic devices. *Proc. Natl. Acad. Sci. U.S.A.* 102, 10813–10818. doi: 10.1073/pnas.0503287102

Ranjard, L., and Richaume, A. (2001). Quantitative and qualitative microscale distribution of bacteria in soil. *Res. Microbiol.* 152, 707–716. doi: 10.1016/S0923-2508(01)01251-7

Sachs, C. C., Grünberger, A., Helfrich, S., Probst, C., Wiechert, W., Kohlheyer, D., et al. (2016). Image-based single cell profiling: high-throughput processing of mother machine experiments. *PLoS ONE* 11:e0163453. doi: 10.1371/journal.pone.0163453

Schaechter, M., Maaloe, O., and Kjeldgaard, N. O. (1958). Dependency on medium and temperature of cell size and chemical composition during balanced growth of *Salmonella Typhimurium*. *J. Gen. Microbiol.* 19, 592–606. doi: 10.1099/00221287-19-3-592

Shehata, T., and Marr, A. G. (1971). Effect of nutrient concentration on growth of *Escherichia coli*. *J. Bacteriol.* 107, 210.

Taheri-Araghi, S., Bradde, S., Sauls, J. T., Hill, N. S., Levin, P. A., Paulsson, J., et al. (2015). Cell-size control and homeostasis in bacteria. *Curr. Biol.* 25, 385–391. doi: 10.1016/j.cub.2014.12.009

Tuson, H. H., Auer, G. K., Renner, L. D., Hasebe, M., Tropini, C., Salick, M., et al. (2012). Measuring the stiffness of bacterial cells from growth rates in hydrogels of tunable elasticity. *Mol. Microbiol.* 84, 874–891. doi: 10.1111/j.1365-2958.2012.08063.x

Typas, A., Banzhaf, M., Gross, C. A., and Vollmer, W. (2012). From the regulation of peptidoglycan synthesis to bacterial growth and morphology. *Nat. Rev. Microbiol.* 10, 123–136. doi: 10.1038/nrmicro2677

Ullman, G., Wallden, M., Marklund, E. G., Mahmutovic, A., Razinkov, I., and Elf, J. (2013). High-throughput gene expression analysis at the level of single proteins using a microfluidic turbidostat and automated cell tracking. *Philos. Trans. R. Soc. B Biol. Sci.* 368:20120025. doi: 10.1098/rstb.2012.0025

Verbeek, P. W., and Vanvliet, L. J. (1994). On the location error of curved edges in low-pass filtered 2-D and 3-D images. *IEEE Trans. Pattern Anal. Mach. Intell.* 16, 726–733. doi: 10.1109/34.297954

Wang, P., Robert, L., Pelletier, J., Dang, W. L., Taddei, F., Wright, A., et al. (2010). Robust growth of *Escherichia coli*. *Curr. Biol.* 20, 1099–1103. doi: 10.1016/j.cub.2010.04.045

Weibel, D. B., DiLuzio, W. R., and Whitesides, G. M. (2007). Microfabrication meets microbiology. *Nat. Rev. Microbiol.* 5, 209–218. doi: 10.1038/nrmicro1616

Willis, L., and Huang, K. C. (2017). Sizing up the bacterial cell cycle. *Nat. Rev. Microbiol.* 15, 606–620. doi: 10.1038/nrmicro.2017.79

Wu, F., van Rijn, E., van Schie, B. G. C., Keymer, J. E., and Dekker, C. (2015). Multicolor imaging of bacterial nucleoid and division proteins with blue, orange and near-infrared fluorescent proteins. *Front. Microbiol.* 6:607. doi: 10.3389/fmicb.2015.00607

Yang, D., Greer, C. M., Jones, B. P., Jennings, A. D., Retterer, S. T., and Männik, J. (2015). Characterization of small microfluidic valves for studies of mechanical properties of bacteria. *J. Vacuum Sci. Technol. B* 33:06F202. doi: 10.1116/1.4929883

7

Light-dependent governance of cell shape dimensions in cyanobacteria

Beronda L. Montgomery [1,2*]

[1] Department of Energy-Plant Research Laboratory, Michigan State University, East Lansing, MI, USA, [2] Department of Biochemistry and Molecular Biology, Michigan State University, East Lansing, MI, USA

***Correspondence:**
Beronda L. Montgomery,
Department of Energy-Plant Research Laboratory, Michigan State University,
612 Wilson Road, Room 106, East Lansing, MI 48824-1312, USA
montg133@msu.edu

The regulation of cellular dimension is important for the function and survival of cells. Cellular dimensions, such as size and shape, are regulated throughout the life cycle of bacteria and can be adapted in response to environmental changes to fine-tune cellular fitness. Cell size and shape are generally coordinated with cell growth and division. Cytoskeletal regulation of cell shape and cell wall biosynthesis and/or deposition occurs in a range of organisms. Photosynthetic organisms, such as cyanobacteria, particularly exhibit light-dependent regulation of morphogenes and generation of reactive oxygen species and other signals that can impact cellular dimensions. Environmental signals initiate adjustments of cellular dimensions, which may be vitally important for optimizing resource acquisition and utilization or for coupling the cellular dimensions with the regulation of subcellular organization to maintain optimal metabolism. Although the involvement of cytoskeletal components in the regulation of cell shape is widely accepted, the signaling factors that regulate cytoskeletal and other distinct components involved in cell shape control, particularly in response to changes in external light cues, remain to be fully elucidated. In this review, factors impacting the inter-coordination of growth and division, the relationship between the regulation of cellular dimensions and central carbon metabolism, and consideration of the effects of specific environment signals, primarily light, on cell dimensions in cyanobacteria will be discussed. Current knowledge about the molecular bases of the light-dependent regulation of cellular dimensions and cell shape in cyanobacteria will be highlighted.

Keywords: cell division, cellular morphology, cyanobacteria, morphogenes, photomorphogenesis

Introduction

Cyanobacteria and other photosynthetic organisms, which have limited mobility in their environment, are exquisitely sensitive to changes in environmental conditions. Cues that impact growth, development, and metabolism of these organisms include light, nutrients, and other factors such as inter-organismal interactions, including predation. As photosynthetic organisms depend upon light for driving energy-producing photosynthesis, these organisms are finely tuned to perceive changes in the photoenvironment. Properties of light that impact organismal form and function include light quality or prevalent wavelengths (visible and ultraviolet or UV), light intensity, daily cycles of dark/light cues, and interactions of light with other factors such as the availability of nutrients.

Light can positively promote organismal development; however, in excess it can also induce light-associated damage or phototoxicity, particularly in photosynthetic organisms (reviewed by Busch and Montgomery, 2015). Thus, changes in growth, development, or metabolism in response to light can be photomorphogenic or photoprotective in nature. Morphological changes can be linked to the optimization of the utilization of resources to support growth and development in bacterial generally (Young, 2006, 2007). A role for light in regulating cell shape or morphology occurs across many organisms, including prokaryotes (Bennett and Bogorad, 1973; Bordowitz and Montgomery, 2008; Singh and Montgomery, 2011) and eukaryotes (Daniel and Rusch, 1962; Desnos et al., 1996; Gendreau et al., 1997; Marwan, 2001; Nishihama et al., 2015). The specific mechanisms by which light impacts morphology have not been vastly studied; however, in some cases photomorphogenesis is linked to the regulation of cytoskeleton proteins (Putzer et al., 1984; Poetsch et al., 1989; Kadota and Wada, 1992; Dyachok et al., 2011; Lindeboom et al., 2013; Singh and Montgomery, 2014). Notably, the regulation of cell shape is correlated with cytoskeletal components and/or function across a range of organisms (Smith, 2003; Mathur, 2004; Cabeen and Jacobs-Wagner, 2005, 2007; Stewart, 2005; Carballido-López, 2006; Osborn and Rothfield, 2007; Terenna et al., 2008; Margolin, 2009; Ueki and Nishii, 2009).

Changes in cyanobacterial shape specifically can be directly linked to growth and/or development—light-dependent changes in vegetative cell shape (Bennett and Bogorad, 1973) or cellular differentiation, e.g., nitrogen fixation-associated development of heterocysts (Kumar et al., 2010), spore-like akinete formation for surviving harsh or stressful environments (Rippka et al., 1979; Flores and Herrero, 2010; Kaplan-Levy et al., 2010), or motility-associated hormogonium induction (Rippka et al., 1979; Meeks and Elhai, 2002; Wong and Meeks, 2002; Adams and Duggan, 2008). These developmentally induced changes in cellular morphology can be tuned by environmental cues, which include light and/or nutrient availability (Singh and Montgomery, 2011; Vadia and Levin, 2015). Stress-induced changes in cellular morphology also occur frequently in bacterial cells. UV, temperature, salt, osmotic, and biotic stresses have known impacts on the morphology of cyanobacterial cells (reviewed Singh and Montgomery, 2011). Knowledge about the mechanisms by which developmental- or stress-related cues result in morphological changes has only recently begun to emerge. Such mechanisms include light-and photoreceptor-dependent regulation of morphogene expression and cytoskeletal protein accumulation (Pattanaik and Montgomery, 2010; Singh and Montgomery, 2014), regulation of reactive oxygen species (ROS) levels that are associated with morphological determination (Singh and Montgomery, 2012; Singh et al., 2013) and light-driven changes in second messenger homeostasis that have the potential to modulate cell shape or dimensions (Neunuebel and Golden, 2008; Agostoni and Montgomery, 2014). Also noted are nutrient-associated, cell division-related cues (Goclaw-Binder et al., 2012; Osanai et al., 2013), among others.

Light-Modulated Regulators Drive Adaptation of Cellular Morphology in Cyanobacteria

Light quality- or wavelength-dependent changes in physiology and development occur in some cyanobacteria (Bennett and Bogorad, 1973; Bordowitz and Montgomery, 2008). The well-studied, freshwater filamentous cyanobacerium *Fremyella* *diplosiphon* exhibits an acclimatory phenomenon, which is historically known as complementary chromatic adaptation or CCA, during which the photosynthetic pigment content is tuned to the external environment (Bennett and Bogorad, 1973; Tandeau de Marsac, 1991). The organism accumulates red-colored, green wavelength-absorbing pigment phycoerythrin (PE) in green-enriched environments, and conversely blue-green-colored, red wavelength-absorbing pigment phycocyanin (PC) in red-enriched environments (Gutu and Kehoe, 2012). The PE and PC pigments are contained in peripheral light-harvesting complexes called phycobilisomes (PBS). The PBS antennae are associated with the thylakoid membranes generally found at the periphery of cyanobacterial cells (**Figure 1**). As the availability of red vs. green wavelengths vary significantly at difference depths in a water column, CCA-dependent tuning of photosynthetic pigmentation is linked to environmental adaptation and tuning of photosynthetic efficiency in natural contexts (Campbell, 1996; Postius et al., 2001). This process is regulated by a soluble, light-absorbing photoreceptor known as RcaE in *F. diplosiphon* (Kehoe and Grossman, 1996). As a part of CCA, there are noted changes in cellular morphology, in addition to pigmentation. Spherical cells and shorter filaments are characteristic of growth under red light (RL), whereas rectangular cells and long filaments are associated with growth under green light (GL; Bennett and Bogorad, 1973; Bordowitz and Montgomery, 2008). RcaE is the regulator for cell shape and filament morphology changes, in addition to its previously mentioned role in fine-tuning photosynthetic pigment levels in *F. diplosiphon* (Bordowitz and Montgomery, 2008). RcaE mutant cells exhibit a rounded cell shape independent of the external light conditions, in contrast to the light-regulated switch between spherical- and rod-shaped cells observed for wild-type (Bennett and Bogorad, 1973; Bordowitz and Montgomery, 2008).

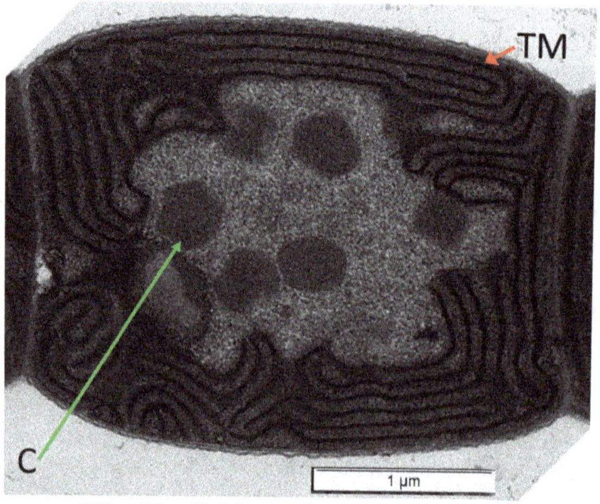

FIGURE 1 | Transmission electron micrograph of wild-type *Fremyella diplosiphon* cell grown in BG-11 medium. Thylakoid membranes (TM; red arrow) are at the periphery of the cell. Carboxysomes (C; green arrow) are found in the center cytoplasmic region of the cell. Bar is 1 micrometer.

Apart from RcaE which is linked to the regulation of pigmentation and morphology in *F. diplosiphon*, another regulator cpeR has also been linked to the regulation of both pigmentation and morphology in this organism (Pattanaik et al., 2011b). CpeR induces synthesis of the photosynthetic pigment PE in GL (Cobley et al., 2002; Seib and Kehoe, 2002; Alvey et al., 2003). However, it was also shown that a Δ*cpeR* mutant in *F. diplosiphon* exhibits an altered cellular morphology compared to wild-type cells (Pattanaik et al., 2011b). Thus, the CpeR regulator, which appears to serve as a transcriptional activator for PE genes (Cobley et al., 2002), also likely controls genes important for photoregulation of cellular morphology. Notably, many other pigment mutants isolated in *F. diplosiphon* impact pigmentation of cells with no measurable impact on cellular morphology (Whitaker et al., 2009, 2011; Bordowitz et al., 2010; Pattanaik et al., 2011a,b).

Light intensity has been shown to impact cellular morphology in cyanobacteria (Bittencourt-Oliveira et al., 2012; Pattanaik et al., 2012; Walters et al., 2013). *Cylindrospermopsis raciborskii* exhibits trichome elongation under reduced light intensity (Bittencourt-Oliveira et al., 2012). In *F. diplosiphon*, lower intensity light results in elongated cells, whereas light of increased intensity results in cells which are more spherical and with reduced length (Pattanaik et al., 2012; Walters et al., 2013). In natural environmental contexts, elongated *F. diplosiphon* cells are prevalent in low light intensities, whereas spherical cells exist in higher intensity regions of the water column (Montgomery, 2008). Thus, it was proposed that cells may regulate cellular morphology in natural contexts under different light intensities to control cellular volume and associated capacity for total photosynthetic membrane content that is correlated with regulation of cellular photosynthetic capacity (Montgomery, 2008; Pattanaik et al., 2012; Walters et al., 2013). As related photoreceptors serve as potent light quality and intensity sensors (e.g., Reed et al., 1994; Rausenberger et al., 2010), it is proposed that RcaE serves as a mediator of light intensity-dependent regulation of morphology in *F. diplosiphon*.

Visible and UV light exposure can result in trichome spiral compression in *Arthrospira platensis* (Wu et al., 2005; Ma and Gao, 2009). The mechanism of this morphological regulation has been implicated as a periplasmic protein of undetermined function (Ma and Gao, 2009). Spiral compression in this organism may serve a role in self-shading that can protect cells against damaging UV exposure (Wu et al., 2005). Thus, the smaller, compressed state under UV has been correlated with reduced surface area that has been hypothesized to serve as a photoprotective state associated with the maintenance of photosynthetic capacity (Singh and Montgomery, 2011). Relatedly, trichome breakage or filament fragmentation under UV light also occurs in *A. platensis* and other cyanobacteria (Gao et al., 2007; Ma and Gao, 2010; Rastogi et al., 2010, 2014). Similar to trichome compression, these responses may be associated with reduced surface area potentially exposed to UV and a potential reduction in UV-associated damage.

Light-Dependent Regulation of Morphogenes is One Mechanism for Regulating Cellular Morphology in Cyanobacteria

Although distinct links between environmental signals and modulation of cellular morphology are clear in cyanobacteria (Singh and Montgomery, 2011), knowledge about the mechanisms by which these changes are mediated is limited. TonB is important for regulating GL-dependent cellular morphology in *F. diplosiphon* (Pattanaik and Montgomery, 2010). Although its exact biochemical function has not been elucidated, the protein has sequence similarity to both iron-associated TonB proteins and a glycine-rich region related to domains in proteins involved in cellular elongation (Sachetto-Martins et al., 2000; Pattanaik and Montgomery, 2010). However, the GL-associated disruption in photoregulation of morphology in a Δ*tonB* mutant in *F. diplosiphon* was shown to be independent of responses to iron limitation (Pattanaik and Montgomery, 2010). This observation, thus, suggests a more direct role for TonB in modulating cell shape in response to GL in this organism.

Recently, photoregulation of morphogenes has emerged as a key mechanism for tuning cellular morphology to light cues in cyanobacteria (Singh and Montgomery, 2014). Morphogene function in the regulation of cellular morphology generally is well known. The bacterial actin MreB is associated with rod-shaped bacteria, and is nearly always absent in spherical bacterial cells (reviewed by Cabeen and Jacobs-Wagner, 2007). MreB has functional homologs in cyanobacteria, both filamentous (Hu et al., 2007; González et al., 2010; Singh and Montgomery, 2014) and unicellular (Koksharova et al., 2007; Savage et al., 2010; Singh and Montgomery, unpublished data), which are associated with rod-shaped cells. Mutation of *mreB* in cyanobacterial systems leads to a loss of shape regulation, thus resulting in the adoption of spherical cells in Δ*mreB* mutants (Hu et al., 2007; Savage et al., 2010). In *F. diplosiphon*, in which rod-shaped cells are prevalent under GL, it appears that photoregulation of *mreB* expression corresponds to the induction of rod-shaped morphology (Singh and Montgomery, 2014).

A well-characterized regulator of MreB is the morphogene BolA (Aldea et al., 1989; Santos et al., 1999). Accumulation of BolA is associated with a spherical cell shape, and imparts this morphology in part through transcriptional downregulation of *mreB* (Aldea et al., 1988, 1989; Santos et al., 1999; Freire et al., 2009). In the first reported functional characterization of a cyanobacterial BolA protein, we recently showed that photoregulation of *bolA* expression in *F. diplosiphon* is correlated with RL-dependent BolA accumulation that is associated with spherical morphology (Singh and Montgomery, 2014). Similar to *E. coli* BolA (Freire et al., 2009), BolA from *F. diplosiphon* binds the promotor of *mreB* to maintain low levels of *mreB* expression in RL (Singh and Montgomery, 2014). Lower levels of BolA are found in GL which are correlated with derepression of *mreB* expression, thereby resulting in the induction of a rod-shaped morphology in *F. diplosiphon* under these conditions (Singh and Montgomery, 2014).

Additional Signals that Impact Cyanobacterial Morphology or Cellular Dimensions—ROS, Second Messengers, and Cell Cycle Factors

Additional factors apart from morphogenes have been correlated with the regulation of cellular morphology in response to environmental cues in cyanobacteria. Some factors induced by environmental cues, including light, that impact cellular dimensions include reactive oxygen species (ROS), second messengers, and cell cycle or division cues.

ROS as a Signal that Impacts Cellular Morphology of Cyanobacteria

Reactive oxygen species are potent cues that can arise from photooxidative stress. When light absorption by the photosynthetic light-harvesting complexes exceeds the need for carbon fixation or the capacity of the photosynthetic electron transfer chains, absorbed light energy can be transferred to other targets resulting in production of ROS (reviewed by Busch and Montgomery, 2015). Generated ROS can then result in cellular photodamage and/or ROS molecules can serve as signals that impact cellular function or development (Busch and Montgomery, 2015). In cyanobacteria, light-generated ROS have been associated with the adoption of particular cellular morphologies. In *F. diplosiphon*, ROS levels are elevated in RL and this light-induced ROS accumulation is associated with spherical morphology (Singh and Montgomery, 2012; Singh et al., 2013). ROS levels are lower in GL and associated with rod-shaped *F. diplosiphon* cells (Singh and Montgomery, 2012; Singh et al., 2013). ROS specifically appear to be the signal impacting morphological adaptation as treatment of cells grown in RL with ROS-scavenging antioxidants reverses the ROS-associated impacts on cellular dimensions (Singh and Montgomery, 2012). In addition to the impact of ROS on cellular morphology, filament length is regulated by photoregulation of ROS. RL-associated elevated ROS accumulation is correlated with shorter filaments, presumably due to filament fragmentation (Singh and Montgomery, 2012). Notably, the RL vs. GL effects on ROS levels are regulated by the RL-responsive RcaE photoreceptor (Singh and Montgomery, 2012), a photoreceptor which regulates pigmentation and the photoregulation of morphology as described above (Kehoe and Grossman, 1996; Bordowitz and Montgomery, 2008). ROS levels are also elevated when light intensity is increased, which also is associated with the induction of a spherical cellular morphology (Walters et al., 2013). ROS have been associated with the regulation of cellular morphology in other systems, including non-photosynthetic organisms such as *Aspergillus nidulans* (Semighini and Harris, 2008) and specific phenotypes such as pollen germination and development, root elongation and cell expansion in plants (Foreman et al., 2003; Carol and Dolan, 2006; Li et al., 2007; Potocký et al., 2007; Müller et al., 2009; Speranza et al., 2012). One function of ROS in this regard is in cell wall loosening (Müller et al., 2009), potentially due to lipid peroxidation (He and Häder, 2002). This ROS-dependent impact could be linked to observed changes in cellular morphology.

The role of UV irradiation in spiral breakage or filament fragmentation in cyanobacteria has been associated with ROS formation (Ma and Gao, 2010; Rastogi et al., 2010, 2014). The accumulation of ROS has been implicated in the induction of lipid oxidation, which ultimately is hypothesized to lead to cellular damage and/or lysis (Ma and Gao, 2010; Rastogi et al., 2010). Thus, as in other systems, ROS molecules have the potential to serve as signaling molecules as proposed for the regulation of cellular morphology in *F. diplosiphon* or as molecules that cause potential damage such as cellular lysis (Busch and Montgomery, 2015). Ultimately, however, the morphology changes and/or filament fragmentation may be associated with improved fitness of strains in some cases. Early experiments with cyanobacteria demonstrated that targeted cell lysis at specialized cells along a filament can result in fragmentation or breakage of filaments into shorter filaments (Lamont, 1969; Bennett and Bogorad, 1973). Such filament fragmentation due to targeted cell lysis during the transition of *F. diplosiphon* cultures to RL was proposed to facilitate a reduction in the content of phycobiliproteins in the culture that are not optimal for growth in RL (Bennett and Bogorad, 1973). This morphological adaptation, thus, is associated with temporal tuning of the pigment content of the organism to the external light environment, which has been associated with optimizing photosynthetic efficiency (Campbell, 1996). Additionally, filament length regulation has been associated with fitness implications in a heterocyst-forming cyanobacterium, whereby the ability to restrict filament length under nitrogen-fixing conditions is associated with improved survival of *Anabaena* sp. strain PCC 7120 (Merino-Puerto et al., 2013).

Second Messengers are Correlated with the Regulation of Cyanobacterial Cell Shape

Second messengers serve as key signaling molecules used in cellular responses to external (or first messenger) cues. These molecules can be rapidly turned over in an energetically conservative manner to facilitate rapid changes in response to environmental fluctuations. Cyanobacteria contain a wide range of second messengers (Agostoni and Montgomery, 2014). Calcium is a ubiquitous second messenger utilized in response to a number of environmental cues that has been implicated in several cyanobacterial responses, including heterocyst development and homorgonia differentiation (reviewed by Agostoni and Montgomery, 2014). The second messengers guanosine pentaphosphate or tetraphosphate [(p)ppGpp] and cyclic AMP (cAMP) have been implicated specifically as mediators of cellular differentiation during cellular responses to nutrient availability (reviewed by Agostoni and Montgomery, 2014). Cyclic dimeric guanosine 3′,5′-monophosphate, i.e., c-di-GMP, presumably contributes to cell shape maintenance and cellular development in the filamentous cyanobacterium *Anaebaena* sp. PCC 7120 as deletion of putative c-di-GMP synthesis enzyme-encoding gene *all2874* results in reduced heterocyst development and smaller vegetative cells under high light intensity (Neunuebel and Golden, 2008). Notably, cAMP (Ohmori et al., 1988, 2002; Ohmori and Okamoto, 2004; Okamoto et al., 2004; Terauchi and Ohmori, 2004) and c-di-GMP (Agostoni et al., 2013) levels are controlled by light in cyanobacteria. As many of these molecules have only recently begun to be the focus of studies in cyanobacteria (Agostoni and Montgomery, 2014), many additional second messenger-regulated responses are expected to emerge.

Coordination of Cell Cycle, Cell Division, and Central Carbon Metabolism with Cell Dimensions

Cellular dimensions are largely heritable across a range of organisms, including prokaryotes and eukaryotes (Zaritsky, 1975;

Fantes, 1977; Koch, 1996; Chien et al., 2012; Marshall et al., 2012; Turner et al., 2012; Lloyd, 2013; Campos et al., 2014). Cell size control is a core feature of the bacterial cell cycle under defined growth conditions (Campos et al., 2014). Bacterial cell size homeostasis appears to be regulated by reproducible and constant cell size extension (Campos et al., 2014; Soifer et al., 2014; Taheri-Araghi et al., 2015). Coordination of cell growth or elongation and division during the cell cycle maintains and/or regulates cell size in bacteria, and can be impacted by nutrient availability or central metabolism (Weart et al., 2007; Siegal-Gaskins and Crosson, 2008; Chien et al., 2012; Yao et al., 2012; Hill et al., 2013; Robert et al., 2014).

Cell division and elongation in cyanobacteria have also been correlated (Koksharova and Wolk, 2002; Mazouni et al., 2004; Miyagishima et al., 2005; Marbouty et al., 2009; Gorelova et al., 2013). Although disruptions in cell division can result in elongated or filamentous cells, in some cases, the reversion of the impairment in cell division results in the restoration of the original cellular dimensions, providing evidence for the recognized strong regulation over cell size (Goclaw-Binder et al., 2012). Environmental cues such as nutrient availability can impact cell division and thereby impact cellular morphology in cyanobacteria (Goclaw-Binder et al., 2012), as described above for other bacterial systems.

An impact of central carbon metabolism on the cell cycle can affect cell size (reviewed by Vadia and Levin, 2015). While definitive experiments linking central carbon metabolism with cell size in cyanobacteria lag behind assessment in other systems, interesting phenomena have been reported. Cell division regulation that is associated with alterations in cell size has been shown to be impacted by sugar metabolism in *Synechocystis* (Osanai et al., 2013). This observed phenotype is driven by overexpression of a sigma factor gene, i.e., *sigE*, which impacts gene expression through interacting with the RNA polymerase (Osanai et al., 2013).

Carbon-Concentrating Mechanism and Correlations with Cell Shape

In cyanobacteria, carboxysomes are subcellular microcompartments centrally located in the cytoplasm of cells (**Figure 1**) and which are associated with the carbon-concentrating mechanism (CCM) and carbon fixation in cyanobacteria (Rae et al., 2013). In some studies, changes in carboxysome quantity and carboxysomal structural defects are apparent in elongated cell division mutants lacking *ftn2* and *ftn6* (Gorelova et al., 2013). Additionally, a disruption of apposite spatial distribution of carboxysomes in a mutant with cytoskeleton-associated defects in cellular morphology has been observed (Savage et al., 2010). Such observations suggest interactions between cellular shape and/or division with the spatial organization of carboxysomes central to carbon metabolism.

The CCM can also include alterations to inorganic carbon (Ci) uptake or active transport. Low Ci levels result in a reduced trichome size in *Arthrospira platensis* (Ma and Gao, 2014). This reduction of trichome size is hypothesized to increase the surface area to cellular volume ratio. Such a response may serve in part to sustain photochemical efficiency and reduce potential damage under high light conditions, which are associated with an increased CCM (Beardall, 1991; Qiu and Liu, 2004).

Conclusion

Cyanobacteria and other cells can alter cellular dimensions and/or shape in response to environmental cues. In some cases this morphological adaptation is linked to the production of cells tuned to the external environment to increase fitness. In other cases, the fitness implications of the energy-requiring morphological adaptations are less clear or only beginning to emerge. For example, *F. diplosiphon* can alter cellular morphology in response to wavelength- or intensity-dependent cues that appear to be correlated with regulating cellular capacity for photosynthetic protein accumulation, which is linked to tuning the capacity of photosynthetic efficiency and associated carbon metabolism to external cues. Other cyanobacteria can alter trichome spiral length and/or filament morphology to reduce surface area exposed to potential damaging light, such as UV or high intensity light. Cells such as *A. platensis* and *C. raciborskii* can alter trichome size in response to cues such as light intensity and Ci availability, presumably to limit damage and alter surface to volume ratios that may alter resource acquisition to sustain photosynthesis. Additional analyses of the mechanisms by which morphological changes are regulated and the potential fitness costs and/or benefits may result in targets for engineering cyanobacterial strains for use in biotechnological applications, in addition to providing greater insight into the roles of these adaptations in natural contexts.

Acknowledgments

Work in the author's laboratory on the impacts of light on growth and development of cyanobacteria is supported by the National Science Foundation (grant no. MCB-1243983 to BM), whereas work on adapting cyanobacterial cultures for biotechnological applications is supported by the U.S. Department of Energy, Chemical Sciences, Geosciences and Biosciences Division, Office of Basic Energy Sciences, Office of Science (grant no. DE-FG02-91ER20021 to BL). I would like to thank Dr. Bagmi Pattanaik for obtaining the transmission electron micrograph of *Fremyella diplosiphon*.

References

Adams, D. G., and Duggan, P. S. (2008). Cyanobacteria–bryophyte symbioses. *J. Exp. Bot.* 59, 1047–1058. doi: 10.1093/jxb/ern005

Agostoni, M., Koestler, B. J., Waters, C. M., Williams, B. L., and Montgomery, B. L. (2013). Occurrence of cyclic di-GMP-modulating output domains in cyanobacteria: an illuminating perspective. *mBio* 4, e00451-00413. doi: 10.1128/mBio.00451-13

Agostoni, M., and Montgomery, B. (2014). Survival strategies in the aquatic and terrestrial world: the impact of second messengers on cyanobacterial processes. *Life* 4, 745–769. doi: 10.3390/life4040745

Aldea, M., Garrido, T., Hernandez-Chico, C., Vicente, M., and Kushner, S. (1989). Induction of a growth-phase-dependent promoter triggers transcription of bolA, an *Escherichia coli* morphogene. *EMBO J.* 8, 3923–3931.

Aldea, M., Hernández-Chico, C., De La Campa, A. G., Kushner, S. R., and Vicente, M. (1988). Identification, cloning, and expression of bolA, an ftsZ-dependent morphogene of *Escherichia coli*. *J. Bacteriol.* 170, 5169–5176.

Alvey, R. M., Karty, J. A., Roos, E., Reilly, J. P., and Kehoe, D. M. (2003). Lesions in phycoerythrin chromophore biosynthesis in *Fremyella diplosiphon* reveal coordinated light regulation of apoprotein and pigment biosynthetic enzyme gene expression. *Plant Cell* 15, 2448–2463. doi: 10.1105/tpc.015016

Beardall, J. (1991). Effects of photon flux density on the 'CO_2-concentrating mechanism' of the cyanobacterium *Anabaena variabilis*. *J. Plankton Res.* 13, 133–141.

Bennett, A., and Bogorad, L. (1973). Complementary chromatic adaptation in a filamentous blue-green alga. *J. Cell Biol.* 58, 419–435.

Bittencourt-Oliveira, M., Buch, B., Hereman, T., Arruda-Neto, J., Moura, A., and Zocchi, S. (2012). Effects of light intensity and temperature on *Cylindrospermopsis raciborskii* (cyanobacteria) with straight and coiled trichomes: growth rate and morphology. *Braz. J. Biol.* 72, 343–351. doi: 10.1590/S1519-69842012000200016

Bordowitz, J. R., and Montgomery, B. L. (2008). Photoregulation of cellular morphology during complementary chromatic adaptation requires sensor-kinase-class protein RcaE in *Fremyella diplosiphon*. *J. Bacteriol.* 190, 4069–4074. doi: 10.1128/jb.00018-08

Bordowitz, J. R., Whitaker, M. J., and Montgomery, B. L. (2010). Independence and interdependence of the photoregulation of pigmentation and development in *Fremyella diplosiphon*. *Commun. Integr. Biol.* 3, 151–153. doi: 10.4161/cib.3.2.10367

Busch, A. W. U., and Montgomery, B. L. (2015). Interdependence of tetrapyrrole metabolism, the generation of oxidative stress and the mitigative oxidative stress response. *Redox Biol.* 4, 260–271. doi: 10.1016/j.redox.2015.01.010

Cabeen, M. T., and Jacobs-Wagner, C. (2005). Bacterial cell shape. *Nat. Rev. Microbiol.* 3, 601–610. doi: 10.1038/nrmicro1205

Cabeen, M. T., and Jacobs-Wagner, C. (2007). Skin and bones: the bacterial cytoskeleton, cell wall, and cell morphogenesis. *J. Cell Biol.* 179, 381–387. doi: 10.1083/jcb.200708001

Campbell, D. (1996). Complementary chromatic adaptation alters photosynthetic strategies in the cyanobacterium Calothrix. *Microbiology* 142, 1255–1263.

Campos, M., Surovtsev, I. V., Kato, S., Paintdakhi, A., Beltran, B., Ebmeier, S. E., et al. (2014). A constant size extension drives bacterial cell size homeostasis. *Cell* 159, 1433–1446. doi: 10.1016/j.cell.2014.11.022

Carballido-López, R. (2006). Orchestrating bacterial cell morphogenesis. *Mol. Microbiol.* 60, 815–819. doi: 10.1111/j.1365-2958.2006.05161.x

Carol, R. J., and Dolan, L. (2006). The role of reactive oxygen species in cell growth: lessons from root hairs. *J. Exp. Bot.* 57, 1829–1834. doi: 10.1093/jxb/erj201

Chien, A.-C., Hill, N. S., and Levin, P. A. (2012). Cell size control in bacteria. *Curr. Biol.* 22, R340–R349. doi: 10.1016/j.cub.2012.02.032

Cobley, J. G., Clark, A. C., Weerasurya, S., Queseda, F. A., Xiao, J. Y., Bandrapali, N., et al. (2002). CpeR is an activator required for expression of the phycoerythrin operon (cpeBA) in the cyanobacterium *Fremyella diplosiphon* and is encoded in the phycoerythrin linker-polypeptide operon (cpeCDESTR). *Mol. Microbiol.* 44, 1517–1531. doi: 10.1046/j.1365-2958.2002.02966.x

Daniel, J. W., and Rusch, H. P. (1962). Method for inducing sporulation of pure cultures of the myxocmycete *Physarum polycephalum*. *J. Bacteriol.* 83, 234–240.

Desnos, T., Orbović, V., Bellini, C., Kronenberger, J., Caboche, M., Traas, J., et al. (1996). Procuste1 mutants identify two distinct genetic pathways controlling hypocotyl cell elongation, respectively in dark- and light-grown *Arabidopsis* seedlings. *Development* 122, 683–693.

Dyachok, J., Zhu, L., Liao, F., He, J., Huq, E., and Blancaflor, E. B. (2011). SCAR mediates light-induced root elongation in *Arabidopsis* through photoreceptors and proteasomes. *Plant Cell* 23, 3610–3626. doi: 10.1105/tpc.111.088823

Fantes, P. A. (1977). Control of cell size and cycle time in *Schizosaccharomyces pombe*. *J. Cell Sci.* 24, 51–67.

Flores, E., and Herrero, A. (2010). Compartmentalized function through cell differentiation in filamentous cyanobacteria. *Nat. Rev. Microbiol.* 8, 39–50. doi: 10.1038/nrmicro2242

Foreman, J., Demidchik, V., Bothwell, J. H. F., Mylona, P., Miedema, H., Torres, M. A., et al. (2003). Reactive oxygen species produced by NADPH oxidase regulate plant cell growth. *Nature* 422, 442–446. doi: 10.1038/nature01485

Freire, P., Moreira, R. N., and Arraiano, C. M. (2009). BolA inhibits cell elongation and regulates MreB expression levels. *J. Mol. Biol.* 385, 1345–1351. doi: 10.1016/j.jmb.2008.12.026

Gao, K., Yu, H., and Brown, M. T. (2007). Solar PAR and UV radiation affects the physiology and morphology of the cyanobacterium *Anabaena* sp. *PCC* 7120. *J. Photochem. Photobiol.* B 89, 117–124. doi: 10.1016/j.jphotobiol.2007.09.006

Gendreau, E., Traas, J., Desnos, T., Grandjean, O., Caboche, M., and Höfte, H. (1997). Cellular basis of hypocotyl growth in *Arabidopsis thaliana*. *Plant Physiol.* 114, 295–305.

Goclaw-Binder, H., Sendersky, E., Shimoni, E., Kiss, V., Reich, Z., Perelman, A., et al. (2012). Nutrient-associated elongation and asymmetric division of the cyanobacterium *Synechococcus* PCC 7942. *Environ. Microbiol.* 14, 680–690. doi: 10.1111/j.1462-2920.2011.02620.x

González, A., Teresa Bes, M., Barja, F., Peleato, M. L., and Fillat, M. F. (2010). Overexpression of FurA in *Anabaena* sp. *PCC* 7120 reveals new targets for this regulator involved in photosynthesis, iron uptake and cellular morphology. *Plant Cell Physiol.* 51, 1900–1914. doi: 10.1093/pcp/pcq148

Gorelova, O. A., Baulina, O. I., Rasmussen, U., and Koksharova, O. A. (2013). The pleiotropic effects of ftn2 and ftn6 mutations in cyanobacterium *Synechococcus* sp. PCC 7942. *Protoplasma* 250, 931–942. doi: 10.1007/s00709-012-0479-2

Gutu, A., and Kehoe, D. M. (2012). Emerging perspectives on the mechanisms, regulation, and distribution of light color acclimation in cyanobacteria. *Mol. Plant* 5, 1–13. doi: 10.1093/mp/ssr054

He, Y. Y., and Häder, D. P. (2002). Involvement of reactive oxygen species in the UV-B damage to the cyanobacterium *Anabaena* sp. *J. Photochem. Photobiol. B* 66, 73–80. doi: 10.1016/S1011-1344(01)00278-0

Hill, N. S., Buske, P. J., Shi, Y., and Levin, P. A. (2013). A moonlighting enzyme links *Escherichia coli* cell size with central metabolism. *PLoS Genet.* 9:e1003663. doi: 10.1371/journal.pgen.1003663

Hu, B., Yang, G., Zhao, W., Zhang, Y., and Zhao, J. (2007). MreB is important for cell shape but not for chromosome segregation of the filamentous cyanobacterium *Anabaena* sp. *PCC* 7120. *Mol. Microbiol.* 63, 1640–1652. doi: 10.1111/j.1365-2958.2007.05618.x

Kadota, A., and Wada, M. (1992). Reorganization of the cortical cytoskeleton in tip-growing fern protonemal cells during phytochrome-mediated phototropism and blue light-induced apical swelling. *Protoplasma* 166, 35–41. doi: 10.1007/bf01320140

Kaplan-Levy, R., Hadas, O., Summers, M., Rücker, J., and Sukenik, A. (2010). "Akinetes: dormant cells of cyanobacteria," in *Dormancy and Resistance in Harsh Environments*, eds E. Lubzens, J. Cerda, and M. Clark (Berlin: Springer), 5–27.

Kehoe, D. M., and Grossman, A. R. (1996). Similarity of a chromatic adaptation sensor to phytochrome and ethylene receptors. *Science* 273, 1409–1412.

Koch, A. L. (1996). What size should a bacterium be? A question of scale. *Annu. Rev. Microbiol.* 50, 317–348. doi: 10.1146/annurev.micro.50.1.317

Koksharova, O. A., Klint, J., and Rasmussen, U. (2007). Comparative proteomics of cell division mutants and wild-type of *Synechococcus* sp. strain *PCC* 7942. *Microbiology* 153, 2505–2517. doi: 10.1099/mic.0.2007/007039-0

Koksharova, O. A., and Wolk, C. P. (2002). A novel gene that bears a DnaJ motif influences cyanobacterial cell division. *J. Bacteriol.* 184, 5524–5528. doi: 10.1128/JB.184.19.5524-5528.2002

Kumar, K., Mella-Herrera, R. A., and Golden, J. W. (2010). Cyanobacterial heterocysts. *Cold Spring Harb. Perspect. Biol.* 2, a000315. doi: 10.1101/cshperspect.a000315

Lamont, H. (1969). Sacrificial cell death and trichome breakage in an oscillatoriacean blue-green alga: the role of murein. *Arch. Mikrobiol.* 69, 237–259. doi: 10.1007/bf00408976

Li, H. B., Qin, Y. M., Pang, Y., Song, W. Q., Mei, W. Q., and Zhu, Y. X. (2007). A cotton ascorbate peroxidase is involved in hydrogen peroxide homeostasis during fibre cell development. *New Phytol.* 175, 462–471. doi: 10.1111/j.1469-8137.2007.02120.x

Lindeboom, J. J., Nakamura, M., Hibbel, A., Shundyak, K., Gutierrez, R., Ketelaar, T., et al. (2013). A mechanism for reorientation of cortical microtubule arrays driven by microtubule severing. *Science* 342, 1245533. doi: 10.1126/science.1245533

Lloyd, A. C. (2013). The regulation of cell size. *Cell* 154, 1194–1205. doi: 10.1016/j.cell.2013.08.053

Ma, Z., and Gao, K. (2009). Photoregulation of morphological structure and its physiological relevance in the cyanobacterium *Arthrospira* (Spirulina) *platensis*. *Planta* 230, 329–337. doi: 10.1007/s00425-009-0947-x

Ma, Z., and Gao, K. (2010). Spiral breakage and photoinhibition of *Arthrospira platensis* (Cyanophyta) caused by accumulation of reactive oxygen species under solar radiation. *Environ. Exp. Bot.* 68, 208–213. doi: 10.1016/j.envexpbot.2009.11.010

Ma, Z., and Gao, K. (2014). Carbon limitation enhances CO_2 concentrating mechanism but reduces trichome size in *Arthrospira platensis* (cyanobacterium). *J. Appl. Phycol.* 26, 1465–1472. doi: 10.1007/s10811-013-0181-6

Marbouty, M., Saguez, C., Cassier-Chauvat, C., and Chauvat, F. (2009). ZipN, an FtsA-like orchestrator of divisome assembly in the model cyanobacterium *Synechocystis* PCC 6803. *Mol. Microbiol.* 74, 409–420. doi: 10.1111/j.1365-2958.2009.06873.x

Margolin, W. (2009). Sculpting the bacterial cell. *Curr. Biol.* 19, R812–R822. doi: 10.1016/j.cub.2009.06.033

Marshall, W., Young, K., Swaffer, M., Wood, E., Nurse, P., Kimura, A., et al. (2012). What determines cell size? *BMC Biol.* 10:101. doi: 10.1186/1741-7007-10-101

Marwan, W. (2001). "Photomovement and photomorphogenesis in *Physarum polycephalum*: targeting of cytoskeleton and gene expression by light," in *Comprehensive Series in Photosciences*, eds D. P. Häder and A. M. Breure (Amsterdam: Elsevier), 561–587.

Mathur, J. (2004). Cell shape development in plants. *Trends Plant Sci.* 9, 583–590. doi: 10.1016/j.tplants.2004.10.006

Mazouni, K., Domain, F., Cassier-Chauvat, C., and Chauvat, F. (2004). Molecular analysis of the key cytokinetic components of cyanobacteria: FtsZ, ZipN and MinCDE. *Mol. Microbiol.* 52, 1145–1158. doi: 10.1111/j.1365-2958.2004.04042.x

Meeks, J. C., and Elhai, J. (2002). Regulation of cellular differentiation in filamentous cyanobacteria in free-living and plant-associated symbiotic growth states. *Mol. Microbiol. Biol. Rev.* 66, 94–121. doi: 10.1128/mmbr.66.1.94-121.2002

Merino-Puerto, V., Herrero, A., and Flores, E. (2013). Cluster of genes that encode positive and negative elements influencing filament length in a heterocyst-forming cyanobacterium. *J. Bacteriol.* 195, 3957–3966. doi: 10.1128/JB.00181-13

Miyagishima, S. Y., Wolk, C. P., and Osteryoung, K. W. (2005). Identification of cyanobacterial cell division genes by comparative and mutational analyses. *Mol. Microbiol.* 56, 126–143. doi: 10.1111/j.1365-2958.2005.04548.x

Montgomery, B. L. (2008). Shedding new light on the regulation of complementary chromatic adaptation. *Cent. Eur. J. Biol.* 3, 351–358. doi: 10.2478/s11535-008-0039-0

Müller, K., Linkies, A., Vreeburg, R. A., Fry, S. C., Krieger-Liszkay, A., and Leubner-Metzger, G. (2009). In vivo cell wall loosening by hydroxyl radicals during cress seed germination and elongation growth. *Plant Physiol.* 150, 1855–1865. doi: 10.1104/pp.109.139204

Neunuebel, M. R., and Golden, J. W. (2008). The *Anabaena* sp. strain *PCC* 7120 gene all2874 encodes a diguanylate cyclase and is required for normal heterocyst development under high-light growth conditions. *J. Bacteriol.* 190, 6829–6836. doi: 10.1128/jb.00701-08

Nishihama, R., Ishizaki, K., Hosaka, M., Matsuda, Y., Kubota, A., and Kohchi, T. (2015). Phytochrome-mediated regulation of cell division and growth during regeneration and sporeling development in the liverwort *Marchantia polymorpha*. *J. Plant Res.* 1–15. doi: 10.1007/s10265-015-0724-9

Ohmori, M., Ohmori, K., and Hasunuma, K. (1988). Rapid change in cyclic 3′, 5′-AMP concentration triggered by a light-off or light-on signal in *Anabaena cylindrica*. *Arch. Microbiol.* 150, 203–204.

Ohmori, M., and Okamoto, S. (2004). Photoresponsive cAMP signal transduction in cyanobacteria. *Photochem. Photobiol. Sci.* 3, 503–511. doi: 10.1039/b401623h

Ohmori, M., Terauchi, K., Okamoto, S., and Watanabe, M. (2002). Regulation of cAMP-mediated photosignaling by a phytochrome in the cyanobacterium *Anabaena cylindrica*. *Photochem. Photobiol.* 75, 675–679. doi: 10.1562/0031-8655(2002)075<0675:ROCMPB>2.0.CO;2

Okamoto, S., Kasahara, M., Kamiya, A., Nakahira, Y., and Ohmori, M. (2004). A phytochrome-like protein AphC triggers the cAMP signaling induced by far-red light in the cyanobacterium *Anabaena* sp. strain *PCC*7120. *Photochem. Photobiol.* 80, 429–433. doi: 10.1562/0031-8655(2004)080<0429:APPATT>2.0.CO;2

Osanai, T., Kuwahara, A., Iijima, H., Toyooka, K., Sato, M., Tanaka, K., et al. (2013). Pleiotropic effect of sigE over-expression on cell morphology, photosynthesis and hydrogen production in *Synechocystis* sp. *PCC* 6803. *Plant J.* 76, 456–465. doi: 10.1111/tpj.12310

Osborn, M. J., and Rothfield, L. (2007). Cell shape determination in *Escherichia coli*. *Curr. Opin. Microbiol.* 10, 606–610. doi: 10.1016/j.mib.2007.09.004

Pattanaik, B., and Montgomery, B. L. (2010). A novel role for a TonB-family protein and photoregulation of iron acclimation in *Fremyella diplosiphon*. *Plant Signal. Behav.* 5, 851–853. doi: 10.4161/psb.5.7.11827

Pattanaik, B., Whitaker, M. J., and Montgomery, B. L. (2011a). Convergence and divergence of the photoregulation of pigmentation and cellular morphology in *Fremyella diplosiphon*. *Plant Signal. Behav.* 6, 2038–2041. doi: 10.4161/psb.6.12.18239

Pattanaik, B., Whitaker, M. J., and Montgomery, B. L. (2011b). Regulation of phycoerythrin synthesis and cellular morphology in *Fremyella diplosiphon* green mutants. *Biochem. Biophys. Res. Commun.* 413, 182–188. doi: 10.1016/j.bbrc.2011.08.051

Pattanaik, B., Whitaker, M. J., and Montgomery, B. L. (2012). Light quantity affects the regulation of cell shape in *Fremyella diplosiphon*. *Front. Microbiol.* 3:170. doi: 10.3389/fmicb.2012.00170

Poetsch, B., Schreckenbach, T., and Werenskiold, A. K. (1989). Photomorphogenesis in *Physarum polycephalum*: temporal expression pattern of actin, α- and β-tubulin. *Eur. J. Biochem.* 179, 141–146. doi: 10.1111/j.1432-1033.1989.tb14531.x

Postius, C., Neuschaefer-Rube, O., Haid, V., and Böger, P. (2001). N2-fixation and complementary chromatic adaptation in non-heterocystous cyanobacteria from Lake Constance. *FEMS Microbiol. Ecol.* 37, 117–125. doi: 10.1111/j.1574-6941.2001.tb00859.x

Potocký, M., Jones, M. A., Bezvoda, R., Smirnoff, N., and Žárský, V. (2007). Reactive oxygen species produced by NADPH oxidase are involved in pollen tube growth. *New Phytol.* 174, 742–751. doi: 10.1111/j.1469-8137.2007.02042.x

Putzer, H., Verfuerth, C., Claviez, M., and Schreckenbach, T. (1984). Photomorphogenesis in *Physarum*: induction of tubulins and sporulation-specific proteins and of their mRNAs. *Proc. Natl. Acad. Sci. U.S.A.* 81, 7117–7121.

Qiu, B., and Liu, J. (2004). Utilization of inorganic carbon in the edible cyanobacterium Ge-Xian-Mi (Nostoc) and its role in alleviating photo-inhibition. *Plant Cell Environ.* 27, 1447–1458. doi: 10.1111/j.1365-3040.2004.01248.x

Rae, B. D., Long, B. M., Whitehead, L. F., Forster, B., Badger, M. R., and Price, G. D. (2013). Cyanobacterial carboxysomes: microcompartments that facilitate CO_2 fixation. *J. Mol. Microbiol. Biotechnol.* 23, 300–307. doi: 10.1159/000351342

Rastogi, R. P., Incharoensakdi, A., and Madamwar, D. (2014). Responses of a rice-field cyanobacterium *Anabaena siamensis* TISTR-8012 upon exposure to PAR and UV radiation. *J. Plant Physiol.* 171, 1545–1553. doi: 10.1016/j.jplph.2014.07.011

Rastogi, R. P., Singh, S. P., Häder, D.-P., and Sinha, R. P. (2010). Detection of reactive oxygen species (ROS) by the oxidant-sensing probe 2′,7′-dichlorodihydrofluorescein diacetate in the cyanobacterium *Anabaena variabilis* PCC 7937. *Biochem. Biophys. Res. Commun.* 397, 603–607. doi: 10.1016/j.bbrc.2010.06.006

Rausenberger, J., Hussong, A., Kircher, S., Kirchenbauer, D., Timmer, J., Nagy, F., et al. (2010). An integrative model for phytochrome B mediated photomorphogenesis: from protein dynamics to physiology. *PLoS ONE* 5:e10721. doi: 10.1371/journal.pone.0010721

Reed, J. W., Nagatani, A., Elich, T. D., Fagan, M., and Chory, J. (1994). Phytochrome A and phytochrome B have overlapping but distinct functions in *Arabidopsis* development. *Plant Physiol.* 104, 1139–1149.

Rippka, R., Deruelles, J., Waterbury, J. B., Herdman, M., and Stanier, R. Y. (1979). Generic assignments, strain histories and properties of pure cultures of cyanobacteria. *Microbiology* 111, 1–61. doi: 10.1099/00221287-111-1-1

Robert, L., Hoffmann, M., Krell, N., Aymerich, S., Robert, J., and Doumic, M. (2014). Division in *Escherichia coli* is triggered by a size-sensing rather than a timing mechanism. *BMC Biol.* 12:17. doi: 10.1186/1741-7007-12-17

Sachetto-Martins, G., Franco, L. O., and De Oliveira, D. E. (2000). Plant glycine-rich proteins: a family or just proteins with a common motif? *Biochim. Biophys. Acta* 1492, 1–14. doi: 10.1016/S0167-4781(00)00064-6

Santos, J. M., Freire, P., Vicente, M., and Arraiano, C. M. (1999). The stationary-phase morphogene bolA from *Escherichia coli* is induced by stress during early stages of growth. *Mol. Microbiol.* 32, 789–798.

Savage, D. F., Afonso, B., Chen, A. H., and Silver, P. A. (2010). Spatially ordered dynamics of the bacterial carbon fixation machinery. *Science* 327, 1258–1261. doi: 10.1126/science.1186090

Seib, L. O., and Kehoe, D. M. (2002). A turquoise mutant genetically separates expression of genes encoding phycoerythrin and its associated linker peptides. *J. Bacteriol.* 184, 962–970. doi: 10.1128/jb.184.4.962-970.2002

Semighini, C. P., and Harris, S. D. (2008). Regulation of apical dominance in *Aspergillus nidulans* hyphae by reactive oxygen species. *Genetics* 179, 1919–1932. doi: 10.1534/genetics.108.089318

Siegal-Gaskins, D., and Crosson, S. (2008). Tightly regulated and heritable division control in single bacterial cells. *Biophys. J.* 95, 2063–2072. doi: 10.1529/biophysj.108.128785

Singh, S. P., Miller, H. L., and Montgomery, B. L. (2013). Temporal dynamics of changes in reactive oxygen species (ROS) levels and cellular morphology are coordinated during complementary chromatic acclimation in *Fremyella diplosiphon. Photosyn. Res.* 1–10. doi: 10.1007/s11120-013-9938-7 [Epub ahead of print].

Singh, S. P., and Montgomery, B. L. (2011). Determining cell shape: adaptive regulation of cyanobacterial cellular differentiation and morphology. *Trends Microbiol.* 19, 278–285. doi: 10.1016/j.tim.2011.03.001

Singh, S. P., and Montgomery, B. L. (2012). Reactive oxygen species are involved in the morphology-determining mechanism of *Fremyella diplosiphon* cells during complementary chromatic adaptation. *Microbiology* 158, 2235–2245. doi: 10.1099/mic.0.060475-0

Singh, S. P., and Montgomery, B. L. (2014). Morphogenes bolA and mreB mediate the photoregulation of cellular morphology during complementary chromatic acclimation in *Fremyella diplosiphon. Mol. Microbiol.* 93, 167–182. doi: 10.1111/mmi.12649

Smith, L. G. (2003). Cytoskeletal control of plant cell shape: getting the fine points. *Curr. Opin. Plant Biol.* 6, 63–73. doi: 10.1016/S1369-5266(02)00012-2

Soifer, I., Robert, L., Barkai, N., and Amir, A. (2014). Single-cell analysis of growth in budding yeast and bacteria reveals a common size regulation strategy. arXiv preprint. arXiv:1410.4771

Speranza, A., Crinelli, R., Scoccianti, V., and Geitmann, A. (2012). Reactive oxygen species are involved in pollen tube initiation in kiwifruit. *Plant Biol. (Stuttg.)* 14, 64–76. doi: 10.1111/j.1438-8677.2011.00479.x

Stewart, G. C. (2005). Taking shape: control of bacterial cell wall biosynthesis. *Mol. Microbiol.* 57, 1177–1181. doi: 10.1111/j.1365-2958.2005.04760.x

Taheri-Araghi, S., Bradde, S., Sauls, J. T., Hill, N. S., Levin, P. A., Paulsson, J., et al. (2015). Cell-size control and homeostasis in bacteria. *Curr. Biol.* 25, 385–391. doi: 10.1016/j.cub.2014.12.009

Tandeau de Marsac, N. (1991). "Chromatic adaptation by cyanobacteria," in *Cell Culture and Somatic Cell Genetics of Plants*, eds L. Bogorad and I. K. Vasil (New York: Academic Press), 417–446.

Terauchi, K., and Ohmori, M. (2004). Blue light stimulates cyanobacterial motility via a cAMP signal transduction system. *Mol. Microbiol.* 52, 303–309. doi: 10.1111/j.1365-2958.2003.03980.x

Terenna, C. R., Makushok, T., Velve-Casquillas, G., Baigl, D., Chen, Y., Bornens, M., et al. (2008). Physical mechanisms redirecting cell polarity and cell shape in fission yeast. *Curr. Biol.* 18, 1748–1753. doi: 10.1016/j.cub.2008.09.047

Turner, J. J., Ewald, J. C., and Skotheim, J. M. (2012). Cell size control in yeast. *Curr. Biol.* 22, R350–R359. doi: 10.1016/j.cub.2012.02.041

Ueki, N., and Nishii, I. (2009). Controlled enlargement of the glycoprotein vesicle surrounding a *Volvox* embryo requires the InvB nucleotide-sugar transporter and is required for normal morphogenesis. *Plant Cell* 21, 1166–1181. doi: 10.1105/tpc.109.066159

Vadia, S., and Levin, P. A. (2015). Growth rate and cell size: a re-examination of the growth law. *Curr. Opin. Microbiol.* 24, 96–103. doi: 10.1016/j.mib.2015.01.011

Walters, K. J., Whitaker, M. J., Singh, S. P., and Montgomery, B. L. (2013). Light intensity and reactive oxygen species are centrally involved in photoregulatory responses during complementary chromatic adaptation in *Fremyella diplosiphon. Commun. Integr. Biol.* 6, e25005. doi: 10.4161/cib.25005

Weart, R. B., Lee, A. H., Chien, A. C., Haeusser, D. P., Hill, N. S., and Levin, P. A. (2007). A metabolic sensor governing cell size in bacteria. *Cell* 130, 335–347. doi: 10.1016/j.cell.2007.05.043

Whitaker, M. J., Bordowitz, J. R., and Montgomery, B. L. (2009). CpcF-dependent regulation of pigmentation and development in *Fremyella diplosiphon. Biochem. Biophys. Res. Commun.* 389, 602–606. doi: 10.1016/j.bbrc.2009.09.030

Whitaker, M. J., Pattanaik, B., and Montgomery, B. L. (2011). Characterization of green mutants in *Fremyella diplosiphon* provides insight into the impact of phycoerythrin deficiency and linker function on complementary chromatic adaptation. *Biochem. Biophys. Res. Commun.* 404, 52–56. doi: 10.1016/j.bbrc.2010.11.056

Wong, F. C. Y., and Meeks, J. C. (2002). Establishment of a functional symbiosis between the cyanobacterium *Nostoc punctiforme* and the bryophyte *Anthoceros punctatus* requires genes involved in nitrogen control and initiation of heterocyst differentiation. *Microbiology* 148, 315–323.

Wu, H., Gao, K., Villafane, V. E., Watanabe, T., and Helbling, E. W. (2005). Effects of solar UV radiation on morphology and photosynthesis of filamentous cyanobacterium *Arthrospira platensis. Appl. Environ. Microbiol.* 71, 5004–5013. doi: 10.1128/aem.71.9.5004-5013.2005

Yao, Z., Davis, R. M., Kishony, R., Kahne, D., and Ruiz, N. (2012). Regulation of cell size in response to nutrient availability by fatty acid biosynthesis in *Escherichia coli. Proc. Natl. Acad. Sci. U.S.A.* 109, E2561–E2568. doi: 10.1073/pnas.1209742109

Young, K. D. (2006). The selective value of bacterial shape. *Microbiol. Mol. Biol. Rev.* 70, 660–703. doi: 10.1128/MMBR.00001-06

Young, K. D. (2007). Bacterial morphology: why have different shapes? *Curr. Opin. Microbiol.* 10, 596–600. doi: 10.1016/j.mib.2007.09.009

Zaritsky, A. (1975). On dimensional determination of rod-shaped bacteria. *J. Theor. Biol.* 54, 243–248.

8

The DnaA tale

Flemming G. Hansen[1] and Tove Atlung[2]*

[1] Department of Bioengineering, Technical University of Denmark, Lyngby, Denmark, [2] Department of Science and Environment, Roskilde University, Roskilde, Denmark

***Correspondence:**
Flemming G. Hansen
fgha@dtu.dk*

More than 50 years have passed since the presentation of the Replicon Model which states that a positively acting initiator interacts with a specific site on a circular chromosome molecule to initiate DNA replication. Since then, the origin of chromosome replication, *oriC*, has been determined as a specific region that carries sequences required for binding of positively acting initiator proteins, DnaA-boxes and DnaA proteins, respectively. In this review we will give a historical overview of significant findings which have led to the very detailed knowledge we now possess about the initiation process in bacteria using *Escherichia coli* as the model organism, but emphasizing that virtually all bacteria have DnaA proteins that interacts with DnaA boxes to initiate chromosome replication. We will discuss the *dnaA* gene regulation, the special features of the *dnaA* gene expression, promoter strength, and translation efficiency, as well as, the DnaA protein, its concentration, its binding to DnaA-boxes, and its binding of ATP or ADP. Furthermore, we will discuss the different models for regulation of initiation which have been proposed over the years, with particular emphasis on the Initiator Titration Model.

Keywords: DnaA protein, DnaA box, *dnaA* gene, DnaA ADP/ATP, *oriC*, initiation control models, initiator titration, cell cycle

INTRODUCTION

The Replicon Model hypothesizes that to initiate DNA replication a positively acting initiator molecule interacts with a specific site, the replicator, on a circular chromosome (**Figure 1**; Jacob et al., 1963). The model was probably designed based on two main findings: (1) that the bacterial chromosome was circular (Cairns, 1963); and (2) that arrest of protein and/or RNA synthesis led to inhibition of initiation but allowed termination of ongoing rounds of replication (Maaløe and Hanawalt, 1961).

Since then, several models, to be addressed later, have been designed to explain some of the characteristics of the coupling between the cell cycle and the growth of *Escherichia coli* and possibly also several other bacteria. *Escherichia coli* can grow at very different growth rates depending on the carbon source and other nutrients present in the growth medium. The generation time in rich media, such as LB medium, is ~20 min at 37°C whereas generation times in minimal media with poor carbon sources can be several hours. At all these generation times it is characteristic for *E. coli* (and most other bacteria) that fast growing cells are big and slowly growing cells are small, but the size of a baby cell will always be very close to half the size of the mother cell independent of the growth rate (Schaechter et al., 1958; Maaløe and Kjeldgaard, 1966).

It should be emphasized that the present review is the story about DnaA and *oriC*. For more detailed reviews see e.g., Messer (2002), Skarstad and Katayama (2013), or Leonard and Grimwade (2015).

FIGURE 1 | The replicon model. A drawing in the sand (provided by M. Mechali to an EMBO meeting report by Skarstad et al. (2003, Copyright, John Wiley and Sons) to celebrate the 40th anniversary of the model.

THE CELL CYCLE

A very important early input to characterization of the bacterial cell cycle was obtained in the work with the baby cell machine; a membrane elution technique (Helmstetter and Cummings, 1963, 1964). In this method the cells of an exponentially growing culture are attached to a membrane filter, the filter holder is inverted and growth medium is passed through the filter. The bacteria grow on the filter and one daughter cell detaches upon cell division and is collected. The main findings were that if cells grew with generation times shorter than 60 min the time it took to replicate the chromosome was close to 40 min, this was termed the C-period, and that the time it took from termination of chromosome replication to cell division was close to 20 min, termed the D-period (Helmstetter, 1967; Helmstetter and Cooper, 1968; Helmstetter et al., 1968). These primary studies were carried out using the *Escherichia coli* B/r strain that behaved very well in the baby cell machine, i.e., the mother cells were firmly attached to the filter (Helmstetter and Cummings, 1963, 1964). Two different approaches have been used in baby cell experiments. One where DNA was labeled with radioactivity before the cells were attached to the filter and radioactivity was determined in the effluent at different times, the other (and conceptually simpler) where newborn cells were collected on ice to stop growth and an experiment was started with bacterial babies transferred to the desired growth temperature getting synchronized cell division (Clark and Maaløe, 1967). This has been the preferred method for later experiments addressing cell cycle parameters. Cells with generation times longer than 60 min very often have a period with only one non-replicating chromosome, this period was termed the B-period (Helmstetter and Pierucci, 1976). Cells growing with generation times longer than 60 min have longer C- and D-periods than given above, and the B-, C-, and D-period show some variation in different strains of *E. coli* (Helmstetter, 1996; Michelsen et al., 2003).

THE INITIATION MASS

The determination of the cell cycle parameters (see above) and the very careful measurements of DNA and cell mass in a study of macromolecular concentrations at different growth rates of *Salmonella typhimurium* (Schaechter et al., 1958) led to the suggestion that bacteria like *E. coli* and to *S. typhimurium*, initiate replication when a certain mass (or volume) is present per origin of replication, the initiation mass (Donachie, 1968) or the critical volume (Pritchard et al., 1969).There has been some dispute whether the initiation mass is the same at all growth rates (Wold et al., 1994). However, it is generally accepted that the initiation mass does not vary very much in *E. coli* K-12 growing at different growth rates (Herrick et al., 1996).

PHENOMENOLOGY OF CONTROL OF CHROMOSOME REPLICATION

The results obtained with the baby cell machine can only be explained if all chromosomal origins in a cell are initiated once and only once per cell cycle. The length of the C and D periods for fast growing bacteria implies that these cells contain several origins and have several ongoing replication forks. That this is the case was demonstrated clearly by flow cytometry of cultures treated with rifampicin that inhibits initiation of chromosome replication but allows ongoing replication forks to terminate (Skarstad et al., 1986). To determine origins per cell at the time of drug addition it is necessary also to stop cell division (Løbner-Olesen et al., 1989) by simultaneous addition of cephalexin (Begg and Donachie, 1985). In the fast growing cells which carried more than one origin at the time of initiation (and rifampicin addition) flow cytometry showed in addition that all origins were initiated synchronously, i.e., cells had either 2^n or $2^{(n+1)}$ replicated chromosomes (Skarstad et al., 1986) which (with simultaneous cephalexin addition) will be equal to the number of origins per cell (Skarstad et al., 1986; Løbner-Olesen et al., 1989). Thus, a model for control of chromosome replication should satisfy what we know about the cell cycle and the behavior of chromosomes and origins. However, before going into the modeling mode we will give a relatively detailed description of the *dnaA* gene and its gene product, which is a positively acting initiator, and the origin of replication, which is the replicator. In Francois Jacob's laboratory, Masamichi Kohiyama isolated temperature sensitive DNA replication mutants some of which continued to replicate at a non-permissive growth temperature until they terminated replication, presumably at the terminus of replication (Jacob et al., 1963; Kohiyama et al., 1963). Some of these mutants were *dnaA* mutants. This historical tale is about the initiator, the DnaA protein and its interaction with the replicator, the origin of replication - *oriC*, and will focus on details about the *dnaA* gene encoding the DnaA protein as well as the requirements for the interaction of the DnaA protein with *oriC*.

THE *dnaA* MUTANTS

The temperature sensitive K-12 strains CRT46 and CRT83, were the first strains carrying mutations in the *dnaA* gene, *dnaA46*, and *dnaA83*, respectively. These mutations were genetically mapped and found to be located in the vicinity of the *ilv* locus (Hirota et al., 1968). In the following years several strains carrying mutations in the *dnaA* gene (see **Figure 5** below) were isolated in different laboratories (Kuempel, 1969; Carl, 1970; Wechsler and Gross, 1971; Beyersmann et al., 1974; Sevastopoulos et al., 1977). The PC2 strain (Carl, 1970) exhibits a slow stop phenotype similar to that of *dnaA* mutants, but carries a mutation in the *dnaC* gene that encodes the loading factor for DnaB, the DNA helicase.

FINE MAPPING OF THE *dnaA GENE*

The genetic structure of the *dnaA* region of the chromosome was characterized using specialized transducing λ bacteriophages, λ*tna* (Hansen and von Meyenburg, 1979; Ream et al., 1980). The chromosomal genes carried on the phages were identified by recombinational rescue of mutations known to be located close to the *ilv* locus, and mapped relative to an *Eco*RI restriction enzyme map constructed from cleavage of different transducing phages. In this study the size of gene products, the proteins, were also characterized by introducing the λ*tna* phages into UV-killed bacteria which had degraded all chromosomal DNA (Murialdo and Siminovitch, 1972). The *dnaA* and *dnaN* genes were also characterized by a Japanese group (Sakakibara and Mizukami, 1980). Several recombinant plasmids were constructed using restriction enzymes and DNA from different specialized transducing λ*tna* phages. These plasmids were used to determine the precise position of more genes by complementation of mutants and characterization of gene products in the maxicell system (Sancar et al., 1979; Hansen F. G. et al., 1982; Hansen et al., 1985). The *dnaA* region as characterized by λ*tna* phages and plasmids derived from the phages is shown in **Figure 2**.

THE ORIGIN OF REPLICATION—*oriC*

The origin of replication was found to be located close to the *ilv* genes in several studies during the 1970s (e.g., Bird et al., 1972; Marsh and Worcel, 1977). The *oriC* symbol was introduced for the origin of replication (Hiraga, 1976). A minichromosome, pSY211, carrying *oriC* was constructed by combining an ~9.5 kb *Eco*RI fragment with a non-replicating fragment conferring ampicillin resistance (Yasuda and Hirota, 1977). It was also obtained on specialized transducing λ*asn* phages (von Meyenburg et al., 1979) which could be established as minichromosomes when transduced into a recombination deficient *recA* strain lysogenic for λ and requiring asparagine. Such strains are heat-sensitive due to the presence of the λcI_{857} gene on the λ-minichromosome. Small minichromosomes, e.g., pCM959 (von Meyenburg et al., 1979; Buhk and Messer, 1983) carrying exclusively chromosomal DNA, were isolated by selecting heat-resistant Asn^+ clones (von Meyenburg et al., 1979). Further characterization of *oriC* was carried out by deletion analysis of a plasmid carrying a fragment with a selectable marker and the 9.5 kb *Eco*RI fragment from λ*asn132* (Messer et al., 1978, 1979). In this way *oriC* could be located between a *Bam*HI and a *Xho*I site and this 422 bp region was sequenced (Meijer et al., 1979; Messer et al., 1979; Sugimoto et al., 1979)[1]. Further deletion analysis limited the *oriC* region to 245 bp (Tabata et al., 1983).

THE DnaA-BOXES

The sequence of the *E. coli oriC* was the first origin of replication to be reported for a cellular organism. The next origin to be sequenced was that of *Salmonella typhimurium* (Zyskind and Smith, 1980). Similarly to the *E. coli oriC* the *Salmonella* sequence carried an unexpected high density of GATC sites (**Figure 3**), where the position of most were conserved when compared to *E. coli*.

Sequencing other origins of the *Enterobacteriacea* family (Zyskind et al., 1981; Cleary et al., 1982) revealed a **r**epeated sequence (R-box) which was later renamed the DnaA-box. Since these early days hundreds of bacteria have been sequenced. Virtually all bacteria carry a *dnaA* gene. The approximate position of the origin of replication can in many cases be determined by the Base-Skew-Method (McLean et al., 1998) and in most cases the precise location of the origin can be identified by the presence of DnaA-boxes which are also present in virtually all bacteria.

DnaA SETS THE INITIATION MASS—IS THE MASTER REGULATOR OF INITIATION

The first evidence that DnaA was not just essential for the initiation process but could be the initiator regulating initiation came from the studies of *dnaA*(Ts) mutants at permissive temperature which showed that initiation mass increased with increasing growth temperature (Hansen and Rasmussen, 1977; Tippe-Schindler et al., 1979). Further support for DnaA as the initiator was provided by the *dnaA*cos mutant (a pseudorevertant of *dnaA46*) which has a decreased initiation mass at low temperature (Kellenberger-Gujer et al., 1978). The hypothesis that DnaA was the initiator was disputed after the first studies of strains carrying DnaA overproducing plasmids since they did not show an increase in DNA per mass as expected

[1]At the Cold Spring Harbor meeting in 1978 where the 422 bp sequence carrying *oriC* was presented (Messer et al., 1979), Hirota's group presented the complete sequence of a 2.1 kb BamHI fragment. After the meeting K. von Meyenburg and F. G. Hansen went to Arthur Kornberg's laboratory at Stanford University where we after comparing the sequences could see that the small 91 bp BamHI fragment was missing from the original presentation of the Japanese group. Therefore minichromosome pSY211, which was available at Stanford University, was restricted with *Bam*HI and analyzed on non-denaturing poly-acrylamide gels that allowed the separation of very small fragments. F.G.Hansen found that minichromosome pSY211 also contained 2 two small BamHI fragment. The sequence of the 91 bp BamHI fragments was added to the final report from the Japanese group (Hirota et al., 1979) and referred to as "Hausen and Schaller, personal communication."

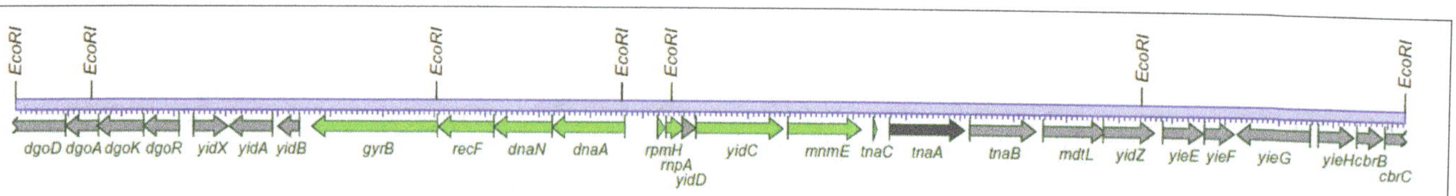

FIGURE 2 | The *dnaA* region of the *E. coli* chromosome. 26,332 bp between the outmost *Eco*RI sites is the chromosomal DNA present on two λ*tna* phages (λ*tna* 552 and λ *tna* 530). The drawing is based on the published complete *E. coli* sequence version (acc.no. U00096.2). The genes (proteins) stained green are those that could be identified on SDS-polyacrylamide gels. The light gray proteins and the dark gray protein encoded by the *tnaA* gene were not visible (Hansen and von Meyenburg, 1979; Ream et al., 1980).

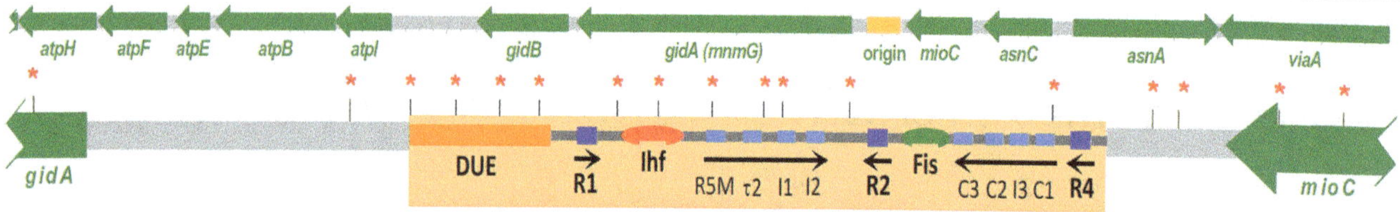

FIGURE 3 | The region of the *E. coli* chromosome carrying *oriC*, the origin of replication, and neighboring genes. A recent review presents a detailed walkthrough of the various binding sites in *oriC* and the interactions with the DnaA protein and other factors in forming the orisome that is required to initiate replication (Leonard and Grimwade, 2015). DnaA boxes with high (dark blue boxes) and low (light blue boxes) affinity, respectively; positions of GATC sites (*). However, in an exponentially growing culture containing cells of all ages only the R1, R2, and R4 are bound by DnaA protein (Samitt et al., 1989).

from a decreased initiation mass (Churchward et al., 1983). An analysis measuring also the actual concentration of *oriC* DNA per mass (Atlung et al., 1987) revealed that initiation was stimulated in cells overproducing DnaA protein, but the replication speed was decreased so much that there was virtually no increase in DNA per mass. A later analysis where we used *plac* controlled expression to increase the DnaA protein concentration from 1.2- to 4-fold the normal (Atlung and Hansen, 1993) showed that within a limited range the origins per mass responded to the DnaA protein concentration (see **Figure 4**) showing that DnaA protein is indeed the limiting factor for initiation. Also in these experiments replication speed is decreased inversely with the higher DnaA protein concentration. Supporting this finding are the following experiments: (1) where wild type DnaA protein was induced to different subnormal levels in a *dnaA46* mutant at 42°C which showed low DNA per mass (Løbner-Olesen et al., 1989); (2) lowering the expression in a *dnaA*$^+$ strain by interfering with the expression of the messenger of the chromosomal *dnaA* gene using the tCRISPRi system (Si et al., 2017); and 3. using a slowly growing *dnaA204* mutant where the DnaA protein concentration was significantly reduced, due to instability of the protein, causing an increase in the initiation mass and thus a decrease in origin to mass ratio (Torheim et al., 2000).

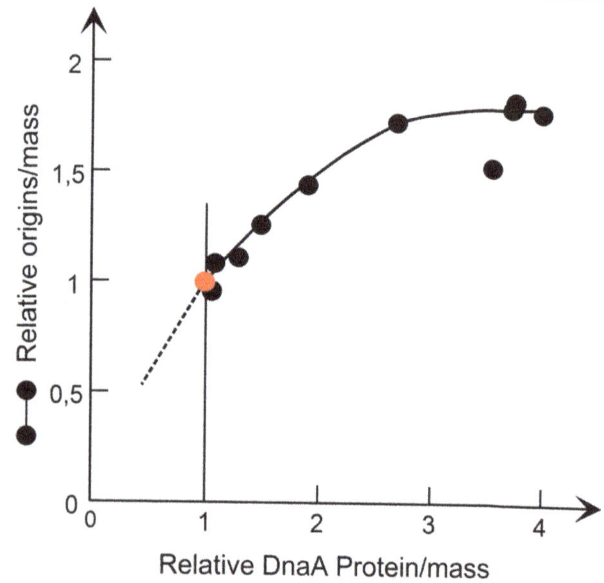

FIGURE 4 | The origin/mass responds to an increase in DnaA protein. The DnaA protein was synthesized from a *lacP* controlled *dnaA* gene and origins/mass was determined by flow cytometry. The data are redrawn from Figure 4 in Atlung and Hansen (1993, Copyright, American Society for Microbiology).

THE DnaA PROTEIN STRUCTURE

The DnaA protein has four domains (Messer et al., 1999). The domain structure was first suggested by the comparison of the *E. coli* and *B. subtilis* proteins (Ogasawara et al., 1985) which showed that there is a moderately conserved N-terminal region and a more highly conserved large C-terminal region separated by a region of highly variable sequence and length. More closely related proteins show a high conservation of the N- and C-terminal domains—e.g., the Enterobacterial proteins have virtually identical sequences in these domains—but many amino acid substitutions, deletions and insertions in the variable region (Skovgaard and Hansen, 1987; Skovgaard, 1990). The C-terminal region contains domains III and IV, with an AAA+ family ATPase motif and the DNA binding domain respectively (Messer et al., 1999). It took a long time after the initial purification of the DnaA protein (Chakraborty et al., 1982; Fuller and Kornberg, 1983) before the first physical structures of DnaA were obtained and as yet no structure of the full length protein has been reported. The structure of domain IV from *E. coli* was determined by NMR and (Obita et al., 2002) and a year later the crystal structure of domain IV in complex with a DnaA-box was obtained (Fujikawa et al., 2003). This confirmed that the DNA binding is mediated by a combination of a basic loop and a helix-turn-helix motif. The structure of domain III has only been obtained from *Aquifex aeolicus* (Erzberger et al., 2002). When bound to ATP, but not to ADP, DnaA forms a super-helical structure with four monomers per turn (Erzberger et al., 2006; see Mott and Berger, 2007; for review and for the best figure). The structure of domain I has been determined for *E. coli* by NMR (Abe et al., 2007) and from three other bacterial species (see Zawilak-Pawlik et al., 2017 for review).

FUNCTION OF THE DIFFERENT DOMAINS

The first indication that each of the domains is essential for the activity of the DnaA protein came from the mapping and sequencing of the classical *dnaA*Ts mutants (Hansen et al., 1984, 1992), see **Figure 5** for position of the mutations and their amino acid changes. Since then many other *dnaA* mutants have been isolated both by different selections and by screening of *in vitro* generated mutants (e.g., reviewed in Erzberger et al., 2002; and additional mutants described in the following citations: Felczak and Kaguni, 2004; Asklund and Atlung, 2005; Felczak et al., 2005; Chodavarapu et al., 2013).

The C-terminal domain IV is alone responsible for sequence specific binding to double stranded DNA and the classical *dnaA*Ts mutations mapping here reduce the binding (Roth and Messer, 1995).

Domain III contains the AAA+ ATPase motif and is responsible for nucleotide binding (Sekimizu et al., 1987). DnaA binds both ATP and ADP with similar affinities, K_D 10–30 nM and 60–100 nM respectively (Sekimizu et al., 1987; Kawakami et al., 2006; Keyamura and Katayama, 2011), and the ATP form is required for the normal control of initiation of replication (Sekimizu et al., 1987). The ATP form facilitates cooperative binding to weak DnaA boxes (Margulies and Kaguni, 1996; Speck et al., 1999; McGarry et al., 2004), is necessary for unwinding of the DUE (Duplex Unwinding Element) in *oriC* (Sekimizu et al., 1987), and for binding to single stranded (ss) DNA in the DUE (Speck and Messer, 2001; Ozaki and Katayama, 2012). Domain III binds the ss DNA directly through specific amino acids, e.g., V211 and R245 (Ozaki et al., 2008; Duderstadt et al., 2011). The terminal part of Domain III also contains a sequence patch for interaction with the membrane (Garner and Crooke, 1996) which stimulates release of the nucleotide from DnaA (Sekimizu and Kornberg, 1988). Except for *dnaA167*, and possibly also *dnaA517*, all the other classical Ts *dnaA* mutants that map in domain III carry the same mutation, i.e., the A184V change close to the P-loop of the Walker A ATP binding motif and four different secondary mutations (**Figure 3**), and the DnaA46 protein has been shown to be defective in nucleotide binding (Hwang and Kaguni, 1988; Carr and Kaguni, 1996).

Domain II is thought to be a flexible linker connecting domains I and III. All parts of domain II (between residues 78 and 136) can be deleted without major effects on function as long as the linker is 21–27 amino acids long (Nozaki and Ogawa, 2008) and the entire YFP sequence can be inserted without any effect on function (Nozaki et al., 2009a).

Domain I has several functions in relation to protein-protein interactions (see Zawilak-Pawlik et al., 2017 for review). Domain I is responsible for dimerization of DnaA protein (Weigel et al.,

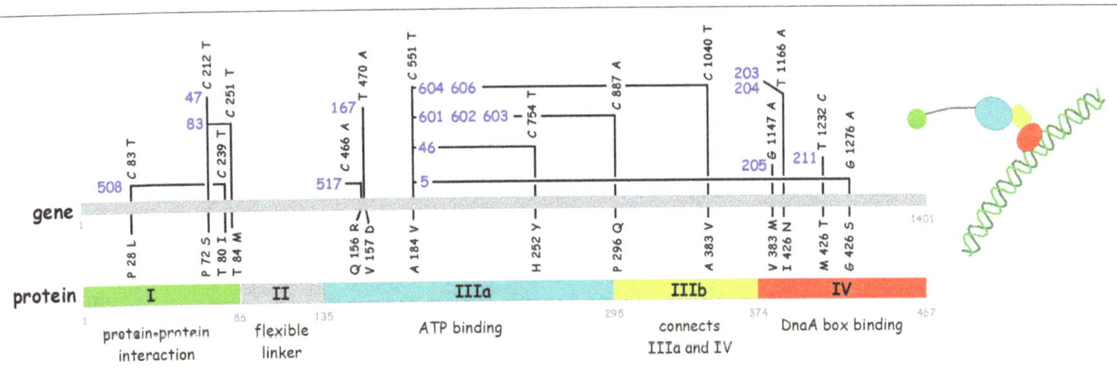

FIGURE 5 | Nucleotide and amino acid alterations in the *dnaA* gene of sixteen temperature sensitive mutants. The mutant numbers are in blue. References for the isolation of the different mutants are: *dnaA508*, *dnaA517* (Wechsler and Gross, 1971); *dnaA47* (Kuempel, 1969); *dnaA46*, *dnaA83* (Kohiyama et al., 1966); *dnaA167* (Abe and Tomizawa, 1971); *dnaA601*, *dnaA602*, *dnaA603*, *dnaA604*, *dnaA606* (Sevastopoulos et al., 1977); *dnaA203*, *dnaA204*, *dnaA205*, *dnaA211* (Beyersmann et al., 1974). The sequences of 12 of these mutants are described by Hansen et al. (1992). The four mutations *dnaA47*, *dnaA83*, *dnaA517*, and *dnaA603* were sequenced using PCR fragments obtained from the original mutant strains (this work). The domains in the DnaA protein and the borders between them are shown in the lower part of the figure.

1999). The dimerization is essential for the function of the protein in initiation of replication as deletion of the first 24 amino acids renders the protein inactive in complementation of the *dnaA46* mutant (Weigel et al., 1999). Residues 24–86 are essential for interaction with DnaB, but recruitment of DnaB at *oriC* also requires the first 23 amino acids (Seitz et al., 2000). The DiaA (*D*naA *i*nitiator *a*ssociating protein) also interacts with domain I. The *diaA* gene was identified in a selection for suppressors of the *dnaAcos* mutant that over initiates (Ishida et al., 2004). The DiaA protein is not essential, but stimulates initiation of replication *in vitro*, and a *diaA* null mutant is slightly impaired in initiation and shows moderate asynchrony (Ishida et al., 2004). The DiaA protein is a tetramer (Keyamura et al., 2007, 2009). The amino acids in DnaA involved in dimerization and in DiaA/DnaB interaction are located on opposite faces of the 3 D structure of domain I, therefore the DiaA tetramer protein can act as a connector for more distantly located DnaA dimers and monomers.

AUTOREGULATION OF DnaA PROTEIN SYNTHESIS

The first indication that the *dnaA* gene is autoregulated came from the studies of *dnaA*(Ts) mutants. DnaA protein continues to be synthesized at non-permissive temperatures where it is inactive, but in some mutants it can be reactivated when brought back to a permissive growth temperature (Hanna and Carl, 1975; Messer et al., 1975). This reversible initiation capacity, which could be observed in the absence of protein synthesis at the permissive temperature, was higher than expected from the mass increase of the culture and suggested that the DnaA protein synthesis was derepressed at the high growth temperature. These findings led to a detailed study of the *dnaA46* mutant at permissive, intermediate, and non-permissive growth temperatures (Hansen and Rasmussen, 1977). The results obtained in this study suggested that the DnaA46 protein became less and less active—the higher the growth temperature—resulting in gradually lower DNA and origin concentrations at the intermediate temperatures. In parallel with the decrease in DnaA protein activity an increase in initiation capacity was observed. Based on these findings Hansen and Rasmussen (1977) proposed that the DnaA protein had a positive role in initiation of replication and a negative role in its own synthesis.

The *dnaA* promoter region and the *dnaA* gene was sequenced (Hansen E. B. et al., 1982; Hansen F. G. et al., 1982) and two promoters giving transcripts entering the *dnaA* gene were identified. The DnaA promoter region carries 9 GATC sites within 225 bp and between the two promoters a sequence homologous to repeats (DnaA-boxes) in the *oriC* region was identified (box 1 in **Figure 6**). Both promoters are negatively regulated by DnaA protein (Atlung et al., 1984, 1985; Braun et al., 1985; Kücherer et al., 1986). In these studies the *dnaA* transcription was found to be increased up to 5-fold in *dnaA*Ts mutants at non-permissive temperature and to be repressed 4- to 5-fold by overproduction of DnaA protein. The DnaA-box is essential for the autoregulation of the *dnaA* gene (Atlung et al., 1984, 1985; Braun et al., 1985). A closer look at the sequence of the *dnaA2p* promoter region reveals interesting features. Two GATC sites are present in this promoter, one in the−10 sequence and the other in the −35 sequence, and methylation enhances transcription 2-fold from this promoter both *in vivo* and *in vitro* (Braun and Wright, 1986). Furthermore, DnaA protein binds with high affinity to sequences upstream of the *dnaA2p* promoter (Speck et al., 1999). This *in vitro* study showed that DnaA-ATP as well as DnaA-ADP binds to what they call DnaA box 1 and DnaA box 2, whereas a box they call "DnaA box a" that has two misfits to the consensus sequence (see below) only binds DnaA-ATP. From a comparison of the *dnaA2p* promoter region with the *mioC* promoter region (see below) as well as from a closer look at the *in vitro* binding data (Speck et al., 1999) we suggest that the region act as having four closely spaced DnaA boxes upstream of the *dnaA2p* promoter region one of which requires DnaA-ATP for binding (**Figure 6**) and two sites, box b and box c with no homology to DnaA boxes, which also requires DnaA-ATP. The DnaA-ATP form is required for repression of both promoters *in vitro* (Speck et al., 1999) in accordance with the location of the different kinds of DnaA-boxes.

OTHER FEATURES OF *dnaA* GENE EXPRESSION

Transcription

The *dnaA* promoter region is fairly strong, giving transcription corresponding to 50% of the *tet* promoter in pBR322 (Atlung et al., 1984). The *dnaA* gene proximal promoter p2 is approximately 3-fold stronger than the p1 promoter (Atlung

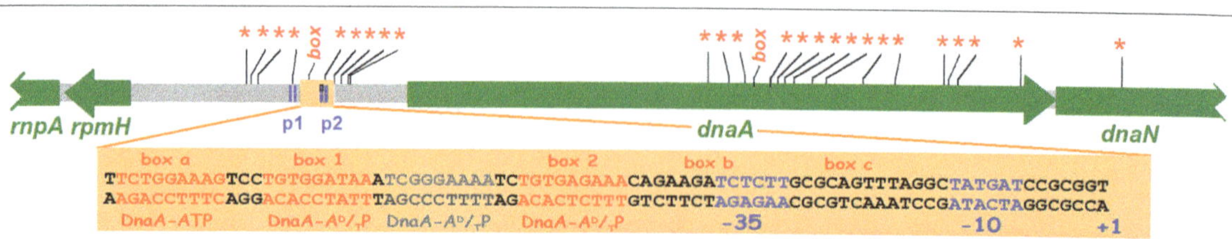

FIGURE 6 | The *dnaA* region with GATC sites (*) and consensus DnaA boxes (box). The position of *dnaA* promoters p1 and p2 are indicated (Hansen F. G. et al., 1982). The light orange rectangle shows the p2 promoter and potential DnaA boxes upstream of and overlapping the promoter. box1, box2 and the region in between are protected both by DnaA-ADP and DnaA-ATP, the boxes a, b, and c require DnaA-ATP (Speck et al., 1999). The CCGCGG sequence upstream of the +1 position is a Travers box (Travers, 1980).

et al., 1985; Kücherer et al., 1986; Chiaramello and Zyskind, 1990). *dnaA* transcription has been reported to be growth rate regulated decreasing with decreasing growth rate (Chiaramello and Zyskind, 1990; Polaczek and Wright, 1990). The p2 promoter carries a discriminator "Travers box" (Travers, 1980) downstream of the −10 sequence and transcription is, like that of the growth rate regulated *rrnP1* promoter, almost completely inhibited by high ppGpp levels.

Translation and mRNA Stability

The coding sequence of *dnaA* has a GUG start codon and a rather poor ribosome binding site (GGAG). The efficiency of translation initiation in mRNA with a SD sequence is 2- to 3-fold lower with a GUG start codon relative to an AUG start codon (Reddy et al., 1985; Donnell and Janssen, 2001). That this is similar for the *dnaA* mRNA is supported by the finding that the *dnaA508cos* mutant carrying a GUG to AUG mutation has an increased DnaA protein level sufficient to give auto-suppression of the temperature sensitivity (Eberle et al., 1989). This is similar to suppression of the temperature sensitivity obtained by having a *dnaA*(Ts) gene on pBR322 derived plasmids and on the chromosome (Hansen et al., 1992; Nyborg et al., 2000).

In addition to the low translation frequency the production of DnaA protein from the mRNA is limited by a short mRNA half-life. The chemical half-life is in the low range, in minimal medium it was amongst the shortest 1% of all mRNAs (1/3 of the mean half-life, Bernstein et al., 2002).

The wild type DnaA protein is stable (Atlung and Hansen, 1999) and is present in K-12 at ~200 monomeric molecules per *oriC* (Hansen et al., 1991b). Therefore, the cell has to produce 100 new molecules per *dnaA* gene per generation. The above characteristics of DnaA expression, moderately high transcription rate, low translation initiation frequency and highly unstable mRNA we estimate to give on the average one protein molecule per transcript. These are the features needed to ensure a well-regulated production of protein over the cell cycle that responds smoothly to regulatory changes in transcription initiation (see **Figure 8**).

DnaA PROTEIN AT DIFFERENT GROWTH RATES

DnaA protein concentration in several different *E. coli* K-12 strains and in *Salmonella typhimurium* was determined by immunoblot analysis and found to be very similar in all these strains (Hansen et al., 1991b). This study furthermore showed that the concentration of DnaA protein in all the strains was the same at all growth rates (doubling times between 150 and 22 min). Similar studies using four different primary DnaA antibodies gave the same results (Herrick et al., 1996). The constant DnaA protein concentration might intuitively be in conflict with increasing levels of mRNA with increasing growth rate. But you have to consider two factors: (1) the time available for protein synthesis, i.e., the generation time, (2) the translation efficiency, i.e., the number of proteins produced per mRNA at the different growth rates (see **Table 1**). If the translation efficiency is the same at 30 and 60 min doubling times, you will need twice as many mRNAs/mass to produce the same number of protein molecules/mass in the faster growing cells. It has been calculated that the average mRNAs are translated more efficiently in faster growing cultures, close to proportionality between number of translations and growth rate (Bremer and Dennis, 1996). The *dnaA* mRNA is, however, very different from the average mRNA in translation efficiency, we estimated it to be one translation per mRNA (see above) compared to the 30–90 translations for the average mRNA. It is therefore likely that *dnaA* mRNA efficiency might be more similar at different growth rates.

TABLE 1 | DnaA protein production at different growth rates.

		Generation time (min)			
Row	**Entity**	**100**	**60**	**30**	**Reference/derivation**
1	Relative total RNA/mass	0,6	0,74	1	Bremer and Dennis, 1996
2	% mRNA of total RNA	2	2	2	Bremer and Dennis, 1996
3	Relative total mRNA/mass	0,6	0,75	1	From rows 1 and 2
4	Average no of translations per mRNA	27	33	71	Bremer and Dennis, 1996
5	Relative *dnaA* mRNA/total RNA	0,27	0,37	1	Chiaramello and Zyskind, 1990
6	Relative *dnaA* mRNA/mass	0,45	0,5	1	from rows 1 and 5
7	Relative DnaA per mass	1	1	1	Hansen et al., 1991b

INITIATOR TITRATION

The first studies on titration of DnaA protein by DnaA boxes *in vivo* were conducted looking at derepression of the *dnaA* gene by DnaA boxes from *oriC—mioC* and the *dnaA* promoter region using a single copy *dnaA-lacZ* fusion (Hansen et al., 1987) and lead to the formulation of the initiator-titration model for control of initiation of replication (see below, Hansen et al., 1991a).

This was followed by two studies aimed at identifying other high affinity DnaA binding sites on the chromosome. In the first only one fragment was obtained and analyzed (Kitagawa et al., 1996). It carried what was later named the *datA* (DnaA titration) locus. The *datA* locus contains two consensus DnaA boxes and four 1-misfit boxes (see **Figure 14**) but was found to titrate much more DnaA protein than the *oriC mioC* region (Kitagawa et al., 1996). The other study, using a fishing method to isolate high affinity DNA fragments, identified five loci (in addition to *oriC*) including *datA* and the *mutH* promoter (Roth and Messer, 1998) that was later found in the search for DARSs (DnaA Reactivating Sequences, see below; Fujimitsu and Katayama, 2004). The *in vivo* titration ability—effect on *dnaA* transcription—of these other four loci has not been studied.

The consensus DnaA box TTWTNCACA was suggested by Schaper and Messer (1995) based on binding of purified DnaA protein to 21 bp oligonucleotides. There are 307 of these consensus DnaA boxes on the *E. coli* chromosome. The DnaA box quality depends on the base in the fifth position and on the neighboring base sequence. Oligonucleotides which carried a DnaA box of the R1-R4 type (TTATCCACA) and had the neighboring bases as present in *oriC* were very efficient in binding the protein (K_D 10^{-9} M). Similar experiments were carried out using 17 bp double stranded oligonucleotides with 6 bp upstream and 2 bp downstream of the four consensus boxes starting with TTAT and the *mioC* (R5) box TTTTCCACA as well as with 104 bp PCR fragments with identical sequences except for the 8 different consensus boxes in the middle of the PCR fragment (Hansen et al., 2006). In the boxes starting with TTAT a C in the fifth position (R1-R4 type) is better than A (R2) or G which is better than T. In the boxes of the *mioC* type starting with TTTT, C in the fifth position is much better than the other possibilities, which show very weak or unspecific binding. Plasmids carrying the *mioC* promoter with the 8 different consensus boxes gave varying degrees of titration of DnaA protein that was in good accordance with the *in vitro* determined relative binding affinities, and the degree of repression of transcription from the *mioC* promoter also varied as expected from the binding affinities.

As noted above the *mioC* promoter region carries one consensus DnaA box and the *oriC* region carries three boxes, which in the gel shift assays were found to be very efficient binders. Nevertheless, the plasmid carrying the *mioC* promoter region is as efficient in titrating DnaA protein as a plasmid carrying an inactivated *oriC* region (Hansen et al., 2006). This study also demonstrated that sequences ~100 bp upstream of the promoter contributed to efficient titration and thus derepression of the *dnaA-lacZ* promoter. An even more detailed study (Hansen et al., 2007) led to definition of DnaA boxes in the *mioC* region with several misfits (**Figure 7**). The number of DnaA proteins per origin has been determined to be ~200 monomers in several different *E. coli* K-12 derivatives growing at different growth rates (Hansen et al., 1991b). The plasmid with the *mioC* promoter region increases *dnaA* gene expression 1.4-fold corresponding to 80 DnaA monomers and as the pBR322 derived plasmid carrying the *mioC* promoter region is present in ~10 copies/*oriC* (Atlung et al., 1999) it follows that each *mioC* promoter region titrates approximately eight DnaA monomers. Similar calculations for the titration caused by the *dnaA* promoter region (Hansen et al., 1987) suggest that the *dnaA* promoter region titrates 3–4 DnaA proteins.

The presence of plasmids carrying DnaA boxes affects initiation of replication leading to increased initiation mass in parallel with the increased *dnaA* gene expression (Christensen et al., 1999; Morigen et al., 2003).

As mentioned above the *datA* region that carries two consensus DnaA boxes surrounded by closely spaced boxes with up to several misfits (see **Figure 14**) titrates large amounts of DnaA protein and plasmids carrying these boxes are difficult to handle in wild type cells.

GATC SITES, SeqA PROTEIN, AND THE ECLIPSE PERIOD

The average density of the GATC sequence in a DNA sequence containing equal amounts of the four bases is one in 256 bp. The density of GATC sites in the *oriC* sequence and the *dnaA* promoter sequence is approx. ten times higher (**Figures 3**, **6**). Unmethylated plasmids carrying the *oriC* or the pBR322 origin cannot transform Dam^+ strains, and fully methylated plasmids cannot transform a Dam^- strain. In both of these cases the plasmids will become hemimethylated upon replication in the recipient cell, therefore it was concluded that full methylation was essential for the establishment of the plasmids (Russell and Zinder, 1987). The *oriC* and the *dnaA* regions become hemimethylated after initiation of replication and stays hemimethylated (are sequestered) for a considerable time period (1/3 and 1/5 of a doubling time, respectively) while regions with few GATC-sites become methylated within 1–2 min (Lark, 1968; Campbell and Kleckner, 1990).

The SeqA protein binds to closely spaced hemi-methylated GATC sites (Brendler et al., 2000) and the *oriC* and *dnaA* promoter regions contains several closely spaced GATC sites (see **Figures 3**, **6**). These regions were "sequestered" in much shorter periods when the *seqA* gene was mutated or absent (Lu et al., 1994; von Freiesleben et al., 1994). A flow cytometric study of the re-initiation kinetics in several *dnaA*(Ts) mutants

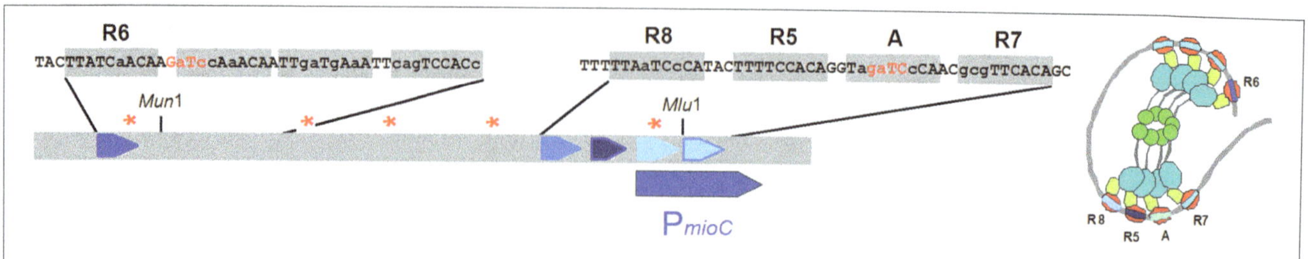

FIGURE 7 | The five experimentally verified DnaA boxes in the *mioC* region are named above the sequence boxes (Hansen et al., 2007) but also potential box sequences to the right of the R6 box are indicated. GATC sites are indicated with red stars (or red letters). To the right is a speculative view of DnaA protein bound to the DnaA-boxes in the *mioC* region (see **Figure 5** for domain colors).

which had been initiation aligned at high temperatures showed that after the immediate initiation of one round of replication at the permissive temperature there was a constant number of origins until a second initiation took place (Hansen, 1995). This period, the eclipse, was studied in more detail in strains producing normal, less than normal, and higher than normal Dam-methylase (von Freiesleben et al., 2000a). These experiments showed clearly that the length of the eclipse (sequestration) period was inversely related to the cellular level of Dam-methylase.

THE INITIATION CASCADE

Flow cytometry studies of wild type and *dam* mutant cells using synchronous cultures obtained by the baby cell machine (Løbner-Olesen et al., 1994) gave the experimental evidence for the initiation cascade as a model for how *E. coli* containing multiple origins initiates all of these once and only once per cell cycle (Hansen et al., 1991a). The cascade states that in wild type cells initiation of one origin will lead to release of DnaA protein bound to that origin, these DnaA molecules cannot bind to the two new *oriC* sequences due to the sequestration of the hemimethylated DNA by the SeqA protein. Therefore, the level of unbound DnaA protein is raised and will promote initiation at another "old" fully methylated origin. Thus, initiation will continue until all origins have initiated. To prevent reinitiation in the same cell cycle the availability of unbound initiation competent DnaA protein must be lowered before the end of the eclipse period.

The initiation cascade also explains the lack of incompatibility between minichromosomes and the chromosome despite the fairly high copy number of the minichromosomes (average 8–10 copies per chromosomal *oriC*; Løbner-Olesen et al., 1987). Before the isolation and characterization of the first minichromosomes it was expected, based on the observations with plasmids, that there would be incompatibility between the chromosomal and minichromosomal *oriC*s, i.e., that the presence of minichromosomes would inhibit initiation of chromosomal replication. Baby cell experiments showed that the minichromosomes initiate at the same time in the cell cycle as the chromosome (Leonard and Helmstetter, 1986), and flow cytometric analysis showed that the presence of a minichromosome only has a very small effect on the synchronous initiation of 4 or 8 chromosomal origins in fast growing cells (Løbner-Olesen and von Freiesleben, 1996; Skarstad and Løbner-Olesen, 2003). In contrast DnaA boxes carried on plasmids like pBR322 that replicate randomly during the cell cycle disturbs the initiation cascade leading to asynchrony (Christensen et al., 1999; Morigen et al., 2003).

SYNTHESIS OF DnaA PROTEIN DURING THE CELL CYCLE

Transcript analysis using an initiation aligned *dnaC2* mutant showed that after 30 min at 40°C transcription of the *mioC* gene was completely shut down, *dnaA* transcription was halved and *gidA* transcription was virtually unaffected (Theisen et al., 1993). This suggests that DnaA protein synthesized during the incubation of the culture at 40°C will be filling the DnaA boxes in the *mioC* promoter region effectively repressing all transcription, while the *dnaA* promoter will be less affected since it can only be partially repressed. The *gidA* promoter, that has no DnaA boxes, is not affected at the high temperature. Soon after initiation at the permissive temperature, there was virtually no transcription from the *dnaA* promoter and only a little from the *gidA* promoter whereas *mioC* transcription was restored after a few minutes. One minute after initiation the *dnaA* gene is replicated and the *dnaA* promoter region will become hemimethylated and sequestered by SeqA for a considerable period (~10 min; Campbell and Kleckner, 1990; Lu et al., 1994) inactivating the two promoters. The *gidA* promoter is located close to the sequestered *oriC* and is probably partially co-sequestered with the *oriC* region for some time. The *mioC* promoter is not sequestered and repression will be reduced by decreased DnaA activity due to titration and RIDA as replication proceeds. The findings obtained with the *dnaC* aligned system was confirmed using synchronous cultures obtained with the baby cell machine (Theisen et al., 1993). A considerable number of non-sequestered genes was also analyzed using these two synchronization techniques (Zhou et al., 1997) but no dramatic changes during the cell cycle were observed.

The *dnaA* transcription data presented above (Theisen et al., 1993) allow us to speculate about the relative numbers of DnaA protein and DnaA boxes during the cell cycle in the simple situation of a slowly growing culture. Assuming that DnaA protein is synthesized in proportion to the DnaA messenger we can predict how the number of DnaA proteins increases during a cell cycle (**Figure 8**). There is a period before initiation where *dnaA* messenger and thus protein is accumulated slowly due to autorepression. Soon after initiation the *dnaA* promoter region is sequestered and both *dnaA* promoters are inactivated. When the *dnaA* promoters are released from sequestration following methylation, the DnaA protein synthesis rate will be high, due to

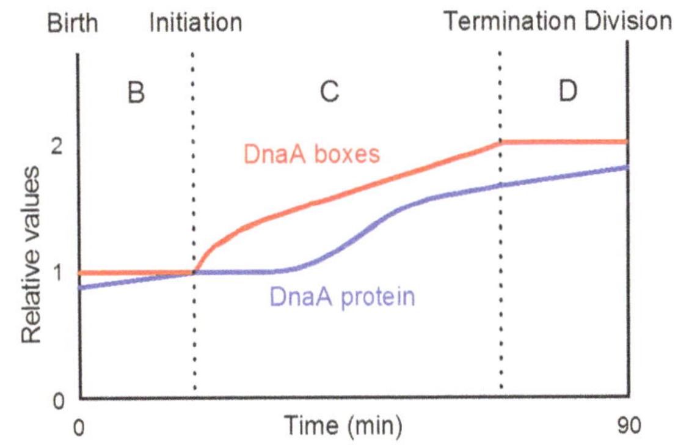

FIGURE 8 | A simplistic view of when initiation takes place in the cell cycle of slowly growing bacteria. The DnaA protein curve is drawn on the basis of the data presented by Theisen et al. (1993). DnaA-boxes with any distribution will be doubled during the C-period; the actual distribution of all consensus DnaA-boxes and selected groups of consensus boxes are presented in **Figure 10**; the figure is redrawn from Figure 2A in Herrick et al. (1996, Copyright, John Wiley and Sons).

a low concentration of DnaA protein relative to titrating DnaA boxes. The level of DnaA-ATP, that is the active form repressing the *dnaA* promoter, will also be low, due to RIDA and DDAH (see later).

MODELS FOR CONTROL OF INITIATION OF CHROMOSOME REPLICATION

The Replicon Model (Jacob et al., 1963) states that an initiator structure has to be build to interact with the origin to initiate replication. At initiation new origins are synthesized and the initiator structure is destroyed or at least prevented from reinitiating at newly synthesized origins. New initiator structures have to be constructed/reformed before initiation can take place in the following cell cycle. Basically such a model is valid but at the time it was presented the molecular basis for the initiator structure could not be addressed.

A model which attracted a lot of attention at these early times was the Inhibitor Dilution Model (Pritchard et al., 1969). It postulates that a fixed amount of messengers coding for inhibitor molecules are synthesized in a burst soon after initiation. These inhibitors would be diluted during growth of the cell and when the concentration reaches a threshold allow initiation to take place followed by the burst of new inhibitors. The molecular mechanism for how to synthesize a "fixed" amount of inhibitor after initiation was not discussed. This model inspired formulation of new models for how initiation in *E. coli* could be controlled. The inhibitor dilution model has later been formally rejected by Margalit and Grover (1987).

The autorepressor model (Sompayrac and Maaløe, 1973) hypothesizes the existence of an operon with two genes, one coding for a repressor protein, the other for an initiator protein (**Figure 9**). The repressor interacts with an operator in the promoter and represses its own synthesis to keep the repressor concentration constant and thus the production of initiator molecules constant. The model proposes that the initiator molecules build an initiator structure which will be ready to initiate when the cell reaches its initiation mass and will be destroyed at initiation. After initiation the cell can start building a new initiator structure which can be used to initiate the following replication cycle. An analysis of the model using a stochastic approach suggested that the synthesis of 250 initiator molecules shows the best fit to the experimental observations in a culture synchronized with the baby cell machine (Clark and Maaløe, 1967). This number is surprisingly close to the number of DnaA proteins per *oriC* found in several *E. coli* strains (Hansen et al., 1991b). The autorepressor model has been tested in a very detailed theoretical study by Margalit et al. (1984) in which it was concluded that the model could be valid provided that the location of the operon is close to *oriC*, that the promoter is moderate to strong, translation efficiency is low, and that the repressor binds with a moderate affinity ($K_D \sim 10^{-9}$ M); the *dnaA* gene and DnaA protein fulfills all of these requirements.

The model published by Mahaffy and Zyskind (1989) was based on data stating that the DnaA protein concentration increases at increasing growth rates (Chiaramello and Zyskind, 1989). This is at variance with the now generally accepted invariance of DnaA protein concentration in K-12. It should be noted that Mahaffy and Zyskind (1989) had to use different nucleotide bound forms of DnaA to get the model to fit with the different DnaA concentrations at different growth rates.

The Initiator Titration Model, which we will cover in some detail, was first presented at the EMBO workshop in Segovia, Spain in 1987 and it did not change before it was published (Hansen et al., 1991a).

The model hypothesizes that:

1. *The initiator—DnaA protein—has a high affinity for specific sites—DnaA boxes on the chromosome.*
2. *The initiator—DnaA protein—has a lower affinity for sites involved in the formation of the initiation complex.*
3. *DnaA protein bound to DnaA boxes is released (pushed off) by replication.*
4. *A newly replicated origin is refractory to initiation for the period it takes to regain a normal topology.*

The model was analyzed by computer simulations using a stochastic approach. A main goal was to find ONE set of parameters to put into *in silico* experiments which could be compared to different *in vivo* experimental data. To this end the comprehensive theoretical study which concluded that the autorepressor model might be a possible model (Margalit et al., 1984) was very helpful. Some of the parameters we have used are presented in **Table 2**. When we designed parameters for the model we hypothesized that the active form of DnaA protein was a tetramer and most often we used 75 DnaA-boxes distributed on the chromosome, usually with a higher occurrence of DnaA-boxes close to the origin. The actual consensus DnaA box distribution on the chromosome is presented in **Figure 10**.

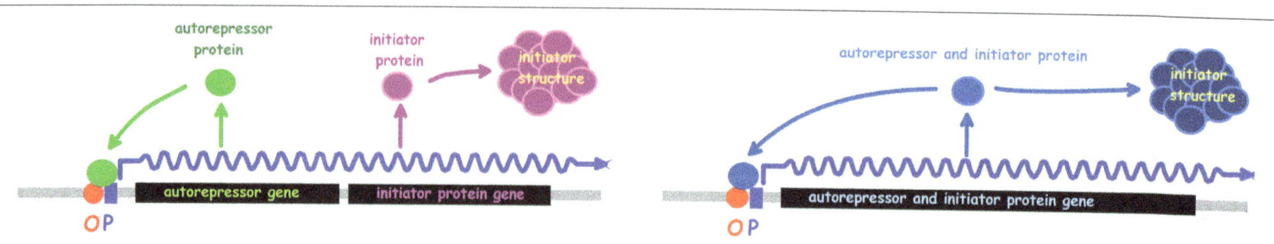

FIGURE 9 | The autorepressor model. The left drawing presents the model in its original formulation (Sompayrac and Maaløe, 1973). The right drawing shows an alternative where the autorepressor and initiator is the same protein.

TABLE 2 | Selected parameters used in the Initiator Titration Model.

Input variables		Calculated numbers	
τ	Doubling time	V	Cell volume
C	Replication time	T	Age of cells
D	Time from termination to cell division	T_n's	Replication fork positions
A_i	Number of DnaA molecules required for initiation	n	DnaA boxes per growing cell
K_D	Dissociation constant for DnaA-DnaA box interaction	G	Genome equivalents per cell
k_{dnaA1p}	Promoter constant *dnaA1p* (3)[a]	g	*dnaA* genes per cell
k_{dnaA2p}	Promoter constant *dnaA2p* (15)[a]	A	DnaA proteins per cell
k_{dnaAcp}	Promoter constant *dnaA* constitutive component (3)[a]	m	*dam* genes per cell
k_{dam}	Promoter constant *damP* (0.5)[a]	δ	Fraction of time the *dnaA* promoters are open (calculated using K_D, A, n, V)

[a] *The numbers in parentheses are relative promoter activities.*

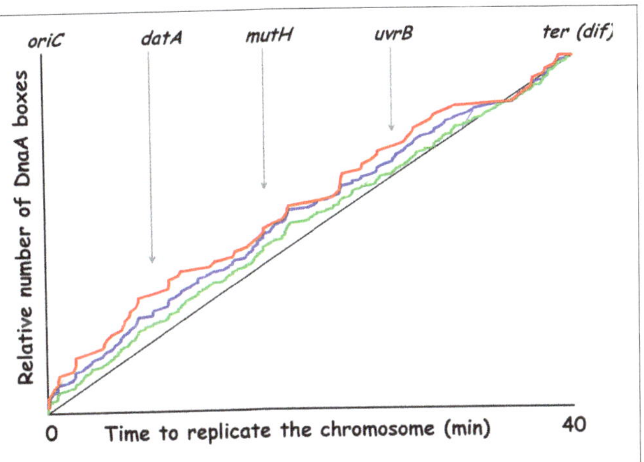

FIGURE 10 | Replication time for DnaA boxes on the K-12 chromosome. Green curve: 307 consensus DnaA boxes (TTWTNCACA); blue curve: 164 DnaA boxes (TTATNCACA + TTTTCCACA); red curve: 78 DnaA boxes (TTATNCACA where N is either C, A, or G). Genes in or lying close to chromosomal regions involved in nucleotide conversions are indicated.

In the computer experiments we started with one predefined unit baby cell that was allowed to run through 50 generations where one randomly chosen daughter was discarded. After 50 generations randomly selected daughters (baby cells) were saved in an array until arriving at a chosen cell number (e.g., 1,000 cells). These computer cells can be used directly as a perfectly synchronized culture, or they can be randomized to obtain an exponential culture by starting the baby cells one by one at different times. The computer culture grows and accumulates cells, until the cell number has doubled, then the computer culture is diluted 2-fold.

The *in silico* experiment shown in **Figure 11B** was first carried out with the assumption that the mother cell divided into two equally sized baby cells. To our surprise these simulations resulted in a computer culture behaving incredibly synchronous and that kept the synchrony for a much longer time than should be expected (data not shown). This *in silico* result was the reason why we changed the cell division mode to have a stochastic positioning of the division septum. To illustrate our stochastic approach we present an example where we use experimental data from the literature to address cell size variation after cell division. The cell length of mothers and daughters of slowly growing *E. coli* B/r cells were analyzed (Schaechter et al., 1962; Koppes et al., 1978) and found to exhibit some variation (coefficient of variation, CV, of 8–10%) in the positioning of the septum. **Figure 12** demonstrates how we handle every single cell at division.

Consequences of Asymmetric Divisions

The synthesis of DnaA protein can be described by the formulas:

$$\frac{dDnaA}{dt} = C_{cell} * C_{promoter} \quad (1)$$

C_{cell} is a component describing the general status of the cell at a given time point:

$$C_{cell} = \frac{PSS * g}{G} \quad (2)$$

where PSS is the amount of ribosomes, RNA polymerase, and other proteins, as well as nucleotides, amino acids and other small molecules needed to synthesize proteins, RNA, and DNA. The PSS will be proportional to the cell size. The g is the actual number of *dnaA* genes in the cell at any given time in the cell cycle, and G is the actual number of genome equivalents in the cell which will be proportional to all the other genes. And the promoter component:

$$C_{promoter} = k_{\mu} \times \alpha \left(\delta(k_{dnaA1p} + k_{dnaA2p}) + k_{dnaAc}\right) \quad (3)$$

that is growth rate dependent (k_{μ}), α is 0 if the promoter region is sequestered or 1 if the region is accessible, the k_{dnaA1} and k_{dnaA2} are promoter constants, k_{dnaAc} describes the DnaA protein synthesis that cannot be repressed. For each cell in the computer at each time step as the cell grows the degree of promoter openness (δ) will be calculated from K_D, the number of DnaA proteins, and the number of DnaA boxes.

Thus if a cell divides asymmetrically one of the babies will become bigger and will therefore contain more PSS, but each baby will be born with the same number of number of *dnaA* genes (g) and the same amount of genome equivalents (G). The net result of the asymmetric division will be that the big daughter synthesizes DnaA protein a little faster than the smaller and therefore will be ready for the next initiation a little faster than the other. The overall result of this is that the two cells have about the same size (but not the same age) at the next initiation.

Today we know much more than we did in 1987, and we have to include the different nucleotide bound forms of DnaA protein (see later). However, the model with its original parameter setting can simulate several experiments as published earlier (Hansen et al., 1991a) and the one shown in **Figure 11**.

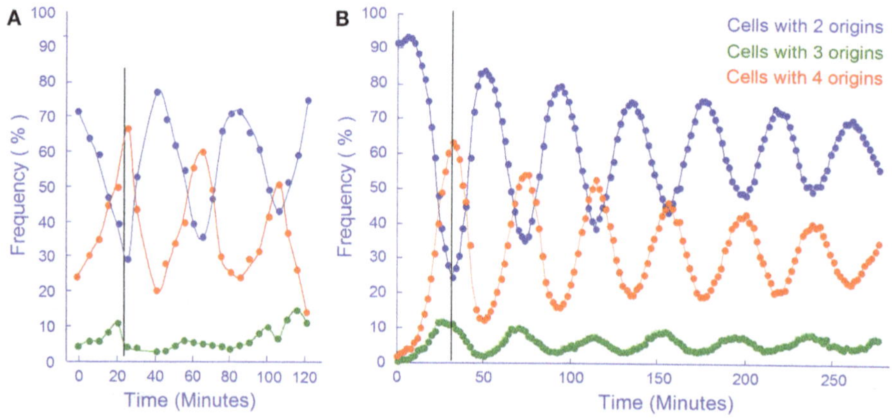

FIGURE 11 | *In vivo* and *in silico* synchronized cultures. **(A)** Samples from a synchronous culture obtained using the baby cell machine were incubated with rifampicin and the frequency of cells containing different numbers of origins were determined by flow cytometry (redrawn from Figure 4A in Løbner-Olesen et al., 1994, Copyright, John Wiley and Sons). **(B)** Same as A but using computer simulation of the Initiator Titration Model to obtain baby cells. The cell size variation in this simulation had a CV of 6.3.

```
c->volum_start=c->volum_start*pow(2.0,(double)c->tim[0]/(double)congen_time);

for(i=0;i<No_of_random_no;++i){
   rnd=(double)rand();
   rnd/=(double)RAND_MAX;
   rnd1+=rnd;
}
rnd1/=No_of_random_no;
rnd1*=0.4;
z->volum_start=c->volum_start*(0.3+rnd1);
c->volum_start-=z->volum_start;
```

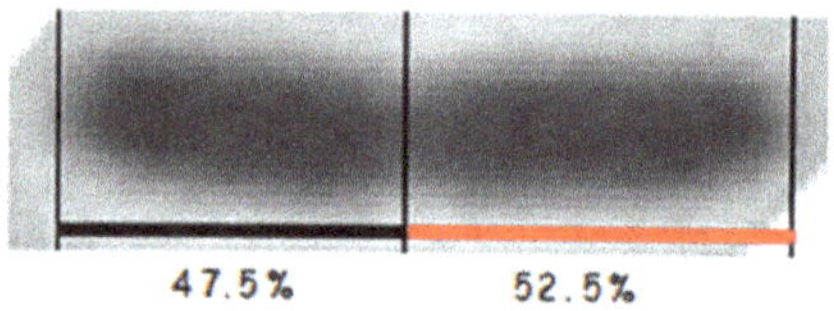

FIGURE 12 | The picture shows a constricted cell of *E. coli* which will divide asymmetrically. The cell length and the position of the constriction was found using a macro written for the Image-Pro Plus software (Nielsen and Hansen, 2010). To the left is C-code showing how we handle cell division in our simulation. c is the starting cell that increases exponentially in size with a constant generation time. At division a sum of random numbers are generated and divided with the number of random numbers. For this calculation we assume that the division take place in the central 40% of the cell (~the multiplication by 0.4), thus one of new cells will inherit 30% of the size of the old cell plus what came out of the random number considerations. If the No_of_random_no is 8 or 13 it will give CVs of 8.3 or 6.3, respectively.

RECENT ADDITIONAL FEATURES OF REGULATION OF INITIATION BY DnaA

The initiator titration model was formulated based on the concept that the concentration of DnaA protein determines the initiation mass and the findings of titration of DnaA protein by plasmids carrying DnaA boxes. The model incorporated detailed knowledge about regulation of *dnaA* gene expression, the eclipse period and the role of Dam, and the initiation synchrony of multiple origins in fast growing cells. In these last sections we will review the newer findings about the role of the *datA* locus as well as the DARS1 and DARS2 loci and the interconversion of the two nucleotide forms of DnaA (**Figure 13**) and evaluate their respective contributions to initiation control before concluding remarks about revision of the model for control of initiation to include these aspects.

DnaA-ATP HYDROLYSIS—Hda AND *datA*

Replicative inactivation of DnaA (RIDA) through hydrolysis of DnaA-ATP to DnaA-ADP was first discovered *in vitro* and shown to be dependent on replication and to require the "IdaB" (now Hda) protein and the β-clamp (DnaN; Katayama et al., 1998). A large increase in the DnaA-ATP form is found at high temperature in the *dnaN59*(Ts) and several other temperature sensitive replication mutants (Katayama et al., 1998; Kurokawa et al., 1999), showing that RIDA is also working *in vivo*. The *hda* gene was found in a search for clones that complemented (new) mutants defective in F plasmid maintenance at permissive temperature and were Ts for growth (Kato and Katayama, 2001). The gene was named *hda* (homologous to DnaA) and the Hda protein has extensive homology to domain III of DnaA. The Hda protein was shown to be identical to the IdaB protein through *in vitro* complementation, and an *hda* null mutant was shown to have a high percentage of DnaA-ATP (Kato and Katayama, 2001). In this study the *hda*(Ts) strain was also characterized *in vivo* and found to have increased *oriC* DNA per mass and increased *oriC*/*terC* indicating "abortive replication."

Until recently, there has been some controversy regarding the essentiality of the *hda* gene. In the initial identification of *hda* (Kato and Katayama, 2001) it was found that the gene could not

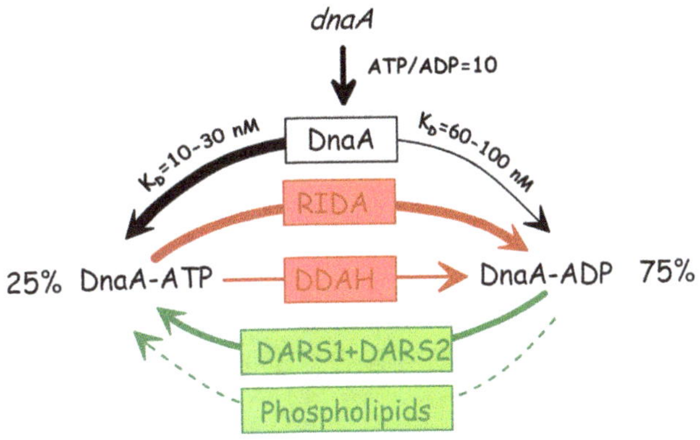

FIGURE 13 | The three states of the DnaA protein. The thickness of the arrows indicate the approximate relative contributions of the different elements to the interconversion of the different forms. Stipled arrow: the relative contribution of phospholipid mediated nucleotide exchange (Sekimizu and Kornberg, 1988; Aranovich et al., 2006, 2015) is unknown. RIDA: Replicative inactivation of DnaA (Katayama et al., 1998; Kato and Katayama, 2001); DDAH: *datA* dependent ATP hydrolysis (Kasho and Katayama, 2013); DARS: DnaA Reactivating sequence (Fujimitsu et al., 2009).

be deleted and only conditional mutants could be obtained. A little later isolation of an *hda* null mutant was reported (Camara et al., 2003). The *hda* null mutants are, however, slowly growing under standard conditions (aerobically on rich medium at 37°C) and genetically very unstable and quickly acquire suppressor mutations of many different types (Riber et al., 2006; Charbon et al., 2011). The growth defects are due to replication fork collapse caused by oxidative DNA damage so *hda* mutants have virtually no growth defects when grown anaerobically (Charbon et al., 2014) and also grow well in minimal media (Charbon et al., 2017). When grown under either of these permissive conditions the *hda* mutants still exhibit a decreased initiation mass (0.70–0.75) and moderate asynchrony.

The *datA* locus was originally believed to exert its effect on regulation of initiation through titration of DnaA protein, similar to, but much stronger than, the titration by plasmids with the *mioC* or *dnaA* promoters (Kitagawa et al., 1998; Morigen et al., 2005). Later it was found that *datA* mediates DnaA-ATP hydrolysis of DNA bound protein, a process named DDAH (Kasho and Katayama, 2013). Only two of the five original DnaA-boxes (box 2 and 3) were absolutely required together with the IHF binding site (IBS) located between them. Further studies showed that a 2-misfit box next to box 2 (box 7, see **Figure 14**) was also essential for the DDAH activity (Kasho et al., 2017). The contribution of DDAH to DnaA-ATP hydrolysis is fairly small, in an otherwise wt strain deletion of *datA* increased DnaA-ATP a few % (Katayama et al., 2001) and when the *datA* deletion was introduced into the *rnhA* Δ*oriC*- Δ*hda* strain the percentage of DnaA-ATP was increased from 72 to ~90% (Kasho and Katayama, 2013).

Deletion of the *datA* locus has no effect on the growth rate of the bacteria but does affect the initiation control, leading to a decrease in the initiation mass (Kitagawa et al., 1998; Morigen et al., 2005; Frimodt-Møller et al., 2016). There is some controversy as to the degree of asynchrony conferred by deletion of *datA*. The first study which found a rather pronounced asynchrony in Δ*datA* cells (Kitagawa et al., 1998) also showed that such cells have close to the same DNA content and the same cell size distribution as wild type cells. Morigen et al. (2005) showed that much of the apparent asynchrony was due to rifampicin resistant initiations. It should be noted (emphasized) that rifampicin resistant initiations lead to overestimation of the effect on initiation mass. Rifampicin resistant initiation of chromosome replication has also been observed in *ihf* mutants, however, such initiations did not take place in strains carrying *ihf* and *dnaA46* mutations (von Freiesleben et al., 2000b) suggesting that the rifampicin resistant initiation phenotype of *ihf* mutants might be due to reduced function of *datA* in DDAH.

The phenotype of the Δ*datA* strain must primarily be due to the lack of DDAH activity since different point mutations inactivating the DDHA activity—mutation of the IBS, or inactivation of box 2 or box 7 have the same phenotype (Ogawa et al., 2002; Nozaki et al., 2009b; Kasho et al., 2017). The apparent very high titration activity of plasmids with the intact *datA* locus can largely be explained by the DDAH activity which lowers the DnaA-ATP level and gives increased expression from promoters which require this form of DnaA for repression. It is, however, clear that the *datA* locus also directly titrates DnaA protein by binding to the boxes, both from the initial deletion analysis of the *datA* plasmids where deletions outside the box 1–4 region gave reduced titration (Kitagawa et al., 1996) and from the analysis of plasmids with a box 2 or an IBS point mutation that only reduced titration (Nozaki et al., 2009b).

DnaA-ADP REJUVENATION

The first DARS DnaA-reactivating sequence, which is present in the ColE1 plasmid origin, was found somewhat fortuitously in a search for factors mediating *in vitro* DnaA-ADP to ATP exchange (Fujimitsu and Katayama, 2004). The full ColE1 DARS was delimited to a 70 bp sequence with three DnaA boxes, where two of the boxes in correct orientation were essential for activity. Subsequently two chromosomal DARSs were identified in the genome sequence amongst 52 pairs of similarly arranged DnaA boxes with 0 or 1 misfit to the sequence TTATNCACA (Fujimitsu et al., 2009). The DARS1 sequence is located in the *uvrB* promoter region and DARS2 in the *mutH* promoter (**Figures 10, 15**), both previously identified as sequences that interact with DnaA (van den Berg et al., 1985; Roth and Messer, 1998). The presence of extra DARS1 or DARS2 sequences on low copy number plasmids leads to decreased initiation mass (Fujimitsu et al., 2009; Charbon et al., 2011) and introduction of higher copy number plasmids with DARS1 or DARS2 is deleterious to the cells (van den Berg et al., 1985; Fujimitsu et al., 2009). Deletion of either DARS1 or DARS2 from the chromosome increases initiation mass by ~20% (Fujimitsu et al., 2009; Frimodt-Møller et al., 2016) and the two DARS sequences act synergistically since deletion of both increase initiation mass to close to 150% of the normal (Fujimitsu et al., 2009). Deletion of either DARS sequence reduce the DnaA-ATP percentage in the *rnhA* Δ*oriC*-Δ*hda* strain, and deletion of both DARS's almost returns the DnaA-ATP level to that of the *hda*$^{+}$ strain (Fujimitsu et al., 2009). The effect of the DARS deletions, especially DARS2, is sufficient to give suppression of

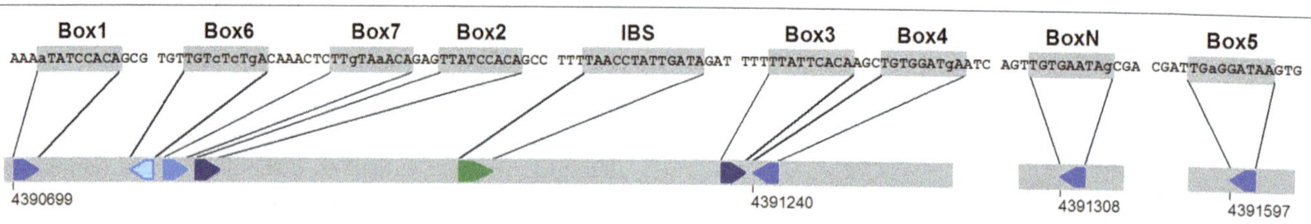

FIGURE 14 | The *datA* region responsible for DDAH (datA dependent ATP hydrolysis, Kitagawa et al., 1998; Morigen et al., 2005). The DnaA boxes 2, 7, 3, and IBS are absolutely required for DDAH activity (Kasho et al., 2017). The coordinates are from *E. coli* K12 (acc.no. U00096.2). IBS: IHF binding site. The color code of the boxes is the same as in **Figure 7**. An additional one-misfit box (BoxN) is also indicated.

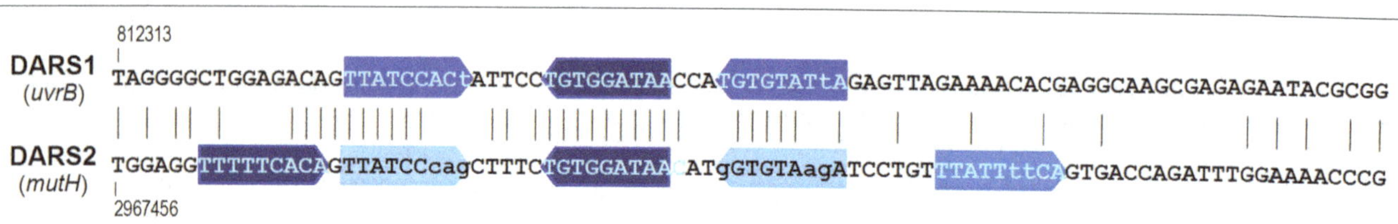

FIGURE 15 | The DARSs (DnaA reactivating sequences). The coordinates given above and below the DARS sequences (Fujimitsu et al., 2009) are from *E. coli* K12 (acc.no. U00096.2). Vertical lines between the sequences indicate identical bases. See **Figure 7** for color code for the boxes.

the lethality associated with lack of Hda mediated ATP hydrolysis (Fujimitsu et al., 2009; Charbon et al., 2011).

Intriguingly, it is only plasmids with DARS2, not with DARS1, that give asynchrony (Fujimitsu et al., 2009; Charbon et al., 2011), and only the DARS2 deletion that leads to asynchrony (Fujimitsu et al., 2009; Frimodt-Møller et al., 2016). This might be due to a higher efficiency of DARS2 in DnaA rejuvenation giving a stronger effect on DnaA-ATP % (Fujimitsu et al., 2009). It should be noted that all the data concerning initiation mass and asynchrony referred to above are from cells grown in minimal glucose casamino acids supplemented medium at 37°C. No systematic studies of the effect of DARS have been published for other growth media—i.e., slow growing cells.

We have summarized the effects of the different elements, de novo synthesis, RIDA, DDAH, and DARS on the interconversion of the two nucleotide bound forms of DnaA in **Figure 13**.

INITIATOR TITRATION REVISITED

A fundamental feature of the initiator titration model is that at the end of the eclipse (*oriC* sequestration) period the replication process has increased the number of DnaA boxes sufficiently to titrate so many DnaA proteins that there is not enough available for the low affinity binding at *oriC*. The additional feature with the two nucleotide forms of the DnaA protein and their interconversion does not essentially change this concept, in this case the amount of DnaA-ATP protein has to be reduced below a threshold during the eclipse period. Let us first consider a relatively slowly growing cell, e.g., 60 min doubling with a C time of 40 min. As soon as replication starts RIDA begins to work due to the presence of β-clamps remaining on the DNA after passage of the replisome, and when the *datA* locus is duplicated (8 min after initiation) the DDAH process will increase the hydrolysis rate a little. During these first approx. 10 min there is no DnaA protein synthesis and only one copy of the DARS sequences, and therefore the DnaA-ATP level will fall quickly. When the *dnaA* gene is no longer sequestered the DnaA-ATP dependent autorepression, will be low, and the DnaA-ATP level will start to increase due to synthesis of new proteins. At 16 min DARS2 is duplicated. In this relatively slow growing cell DARS2 might, however, be inefficient in nucleotide release since it requires Fis for this activity (Kasho et al., 2014) and Fis concentration is low in slow growing cells (Ball et al., 1992). At 26 min also DARS1 is duplicated and this will increase the building up of DnaA-ATP levels. At 40 min replication terminates and soon after RIDA will cease due to unloading of the β-clamps from the DNA which only takes a few minutes (Leu et al., 2000)[2]. In faster growing cells with overlapping replication cycles RIDA will be active throughout the cell cycle, but will vary with the number of replication forks, e.g., in a cell growing with 25 min doubling time the old forks will terminate 15 min after initiation, reducing the RIDA activity to two thirds, thus accelerating the buildup of DnaA-ATP. Unfortunately the variation in DnaA-ATP levels over the cell cycle has not been investigated using baby cells, it has only been measured in *dnaC* initiation synchronized cells (Kurokawa et al., 1999), which is an artificial situation probably with an unnaturally high DnaA-ATP starting level.

It will be interesting to incorporate the DnaA-ATP/ADP conversion systems into the initiator titration model and see how that behaves compared to the original initiator titration model,

[2]The calculated average residence time for the β-clamps (3.5–7 min) is based on a very careful determination of the number of molecules of different polymerase subunits per cell, giving a number of 350 β-clamp dimers/cell in an exponentially growing culture of C600 in LB medium. However, the final calculation uses only 2–4,000 Okazaki fragments produced per cell in 40 min, corresponding to only one replicating chromosome per cell, but C600 growing in LB medium has mostly 8 origins per cell, and thus 4 pairs of replisomes, so the average residence time therefore has to be 3- to 4-fold shorter.

and to a version of the model where we insert the actual high affinity DnaA-box distribution.

It is clear that the original initiator titration model can to a large extent simulate the behavior of *E. coli* during exponential growth. And the *dnaA46* mutant with a protein that does not bind ATP (*in vitro*) is quite fit at low temperature, growing with an almost normal growth rate and having a reasonably uniform cell size distribution despite the asynchrony of initiation (and increased initiation mass). This raises the question why has *E. coli* (and other bacteria) evolved an additional level of regulation with the DnaA-ATP/DnaA-ADP interconversion systems? It will be interesting to see if these systems are necessary and under which growth conditions—how does an *E. coli* strain that is deleted for *hda*, *datA*, and the two DARSs behave? It will probably be necessary to simultaneously reduce the *dnaA* gene expression to get a cell with a normal initiation mass avoiding the side effects of over initiation.

AUTHOR CONTRIBUTIONS

All authors listed have made a substantial, direct and intellectual contribution to the work, and approved it for publication.

ACKNOWLEDGMENTS

The authors are particularly grateful to Knud V. Rasmussen who as a Master project supervisor started our scientific entry into the field of initiation of chromosome replication and to Kaspar von Meyenburg who took part in much of the early work on the subject. We are also grateful to deceased Walter Messer who has participated in a number of open discussions of what his group did at the Max Planck Institute in Berlin and what we did in Copenhagen. At different periods we have also had close collaboration with Anders Løbner-Olesen, Ole Skovgaard, and Bjarke Bak Christensen.

REFERENCES

Abe, M., and Tomizawa, J. (1971). Chromosome replication in *E. coli* K12 mutant affected in DNA initiation. *Genetics* 69, 1–15.

Abe, Y., Jo, T., Matsuda, Y., Matsunaga, C., Katayama, T., and Ueda, T. (2007). Structure and function of DnaA N-terminal domains: specific sites and mechanisms in inter-DnaA interaction and in DnaB helicase loading on oriC. *J. Biol. Chem.* 282, 17816–17827. doi: 10.1074/jbc.M701841200

Aranovich, A., Braier-Marcovitz, S., Ansbacher, E., Granek, R., Parola, A. H., and Fishov, I. (2015). N-terminal-mediated oligomerization of DnaA drives the occupancy-dependent rejuvenation of the protein on the membrane. *Biosci. Rep.* 35:e00250. doi: 10.1042/BSR20150175

Aranovich, A., Gdalevsky, G. Y., Cohen-Luria, R., Fishov, I., and Parola, A. H. (2006). Membrane-catalyzed nucleotide exchange on DnaA. Effect of surface molecular crowding. *J. Biol. Chem.* 281, 12526–12534. doi: 10.1074/jbc.M510266200

Asklund, M., and Atlung, T. (2005). New non-detrimental DNA-binding mutants of the *Escherichia coli* initiator protein DnaA. *J. Mol. Biol.* 345, 717–730. doi: 10.1016/j.jmb.2004.10.092

Atlung, T., Christensen, B. B., and Hansen, F. G. (1999). Role of the Rom protein in copy number control of plasmid pBR322 at different growth rates in *Escherichia coli* K-12. *Plasmid* 41, 110–119. doi: 10.1006/plas.1998.1386

Atlung, T., Clausen, E., and Hansen, F. G. (1984). "Autorepression of the *dnaA* gene of *Escherichia coli*," in *Proteins Involved in DNA Replication*, eds U. Huebscher and S. Spadari (New York, NY; London: Plenum Press), 199–207.

Atlung, T., Clausen, E. S., and Hansen, F. G. (1985). Autoregulation of the *dnaA* gene of *Escherichia coli*. *Mol. Gen. Genet.* 200, 442–450. doi: 10.1007/BF00425729

Atlung, T., and Hansen, F. G. (1993). Three distinct chromosome replication states are induced by increasing concentrations of DnaA protein in *Escherichia coli*. *J. Bacteriol.* 175, 6537–6545. doi: 10.1128/jb.175.20.6537-6545.1993

Atlung, T., and Hansen, F. G. (1999). Low-temperature-induced dnaa protein synthesis does not change initiation mass in *Escherichia coli* K-12 low-temperature-induced dnaa protein synthesis does not change initiation mass in *Escherichia coli* K-12. *J. Bacteriol.* 181, 5557–5562.

Atlung, T., Løbner-Olesen, A., and Hansen, F. G. (1987). Overproduction of DnaA protein stimulates initiation of chromosome and minichromosome replication in *E. coli*. *Mol. Gen. Genet.* 206, 51–59. doi: 10.1007/BF00326535

Ball, C. A., Osuna, R., Ferguson, K. C., and Johnson, R. C. (1992). Dramatic change of fis level upon nutrient upshift in *Escherichia coli*. *J. Bacteriol.* 174, 8043–8056. doi: 10.1128/jb.174.24.8043-8056.1992

Begg, K. J., and Donachie, W. D. (1985). Cell shape and division in *Escherichia coli*: experiments with shape and division mutants. *J. Bacteriol.* 163, 615–622.

Bernstein, J. A., Khodursky, A. B., Lin, P. H., Lin-Chao, S., and Cohen, S. N. (2002). Global analysis of mRNA decay and abundance in *Escherichia coli* at single-gene resolution using two-color fluorescent DNA microarrays. *Proc. Natl. Acad. Sci. U.S.A.* 99, 9697–9702. doi: 10.1073/pnas.112318199

Beyersmann, D., Messer, W., and Schlicht, M. (1974). Mutants of *Escherichia coli* B/r defective in DNA initiation: dnaI, a new gene for replication. *J. Bacteriol.* 118, 783–789.

Bird, R. E., Louarn, J., Maruscelli, J., and Caro, L. (1972). Origin and sequence of chromosome replication in *Escherichia coli*. *J.Mol.Biol.* 70, 549–566. doi: 10.1016/0022-2836(72)90559-1

Braun, R. E., O'Day, K., and Wright, A. (1985). Autoregulation of the DNA replication gene *dnaA* in *E. coli*. *Cell* 40, 159–169.

Braun, R. E., and Wright, A. (1986). DNA methylation differentially enhances the expression of one of the two *E. coli dnaA* promoters *in vivo* and *in vitro*. *Mol. Gen. Genet.* 202, 246–250. doi: 10.1007/BF00331644

Bremer, H., and Dennis, P. P. (1996). "Modulation of chemical composition and other parameters of the cell by growth rate," in *Escherichia coli and Salmonella: Cellular and Molecular Biology*, eds F. C. Neidhardt, R. I. Curtiss, J. L. Ingraham, E. C. C. Lin, K. Brooks Low, B. Magasanik et al. (Washington, DC: American Society for Microbiology), 1553–1569.

Brendler, T., Sawitzke, J., Sergueev, K., and Austin, S. (2000). A case for sliding SeqA tracts at anchored replication forks during *Escherichia coli* chromosome replication and segregation. *EMBO J.* 19, 6249–6258. doi: 10.1093/emboj/19.22.6249

Buhk, H.-J., and Messer, W. (1983). The replication origin region of *Escherichia coli*: nucleotide sequence and functional units. *Gene* 24, 265–279. doi: 10.1016/0378-1119(83)90087-2

Cairns, J. (1963). The chromosome of *Escherichia coli*. *Cold Spring Harb. Symp. Quant. Biol.* 28, 43–46. doi: 10.1101/SQB.1963.028.01.011

Camara, J. E., Skarstad, K., and Crooke, E. (2003). Controlled initiation of chromosomal replication in *Escherichia coli* requires functional Hda protein. *J. Bacteriol.* 185, 3244–3248. doi: 10.1128/JB.185.10.3244-3248.2003

Campbell, J. L., and Kleckner, N. (1990). *E. coli oriC* and the *dnaA* gene promoter are sequestered from dam methyltransferase following the passage of the chromosomal replication fork. *Cell* 62, 967–979. doi: 10.1016/0092-8674(90)90271-F

Carl, P. L. (1970). *E. coli* mutants with temperature sensitive synthesis of DNA. *Mol. Gen. Genet.* 109, 107–122. doi: 10.1007/BF00269647

Carr, K. M., and Kaguni, J. M. (1996). The A184V missense mutation of the *dnaA5* and *dnaA46* alleles confers a defect in ATP binding and thermolability in initiation of *Escherichia coli* DNA replication. *Mol. Microbiol.* 20, 1307–1318. doi: 10.1111/j.1365-2958.1996.tb02649.x

Chakraborty, T., Yoshinaga, K., Lother, H., and Messer, W. (1982). Purification of the *E. coli dnaA* gene product. *EMBO J.* 1, 1545–1549.

Charbon, G., Bjørn, L., Mendoza-Chamizo, B., Frimodt-Møller, J., and Løbner-Olesen, A. (2014). Oxidative DNA damage is instrumental in hyperreplication stress-induced inviability of *Escherichia coli*. *Nucleic Acids Res.* 42, 13228–13241. doi: 10.1093/nar/gku1149

Charbon, G., Campion, C., Chan, S. H. J., Bjørn, L., Weimann, A., da Silva, L., et al. (2017). Re-wiring of energy metabolism promotes viability during hyperreplication stress in *E. coli*. *PLoS Genet.* 13:e1006590. doi: 10.1371/journal.pgen.1006590

Charbon, G., Riber, L., Cohen, M., Skovgaard, O., Fujimitsu, K., Katayama, T., et al. (2011). Suppressors of DnaA(ATP) imposed overinitiation in *Escherichia coli*. *Mol. Microbiol.* 79, 914–928. doi: 10.1111/j.1365-2958.2010.07493.x

Chiaramello, A. E., and Zyskind, J. W. (1989). Expression of *Escherichia coli dnaA* and *mioC* genes as function of growth rate. *J. Bacteriol.* 171, 4272–4280. doi: 10.1128/jb.171.8.4272-4280.1989

Chiaramello, A. E., and Zyskind, J. W. (1990). Coupling of DNA replication to growth rate in *Escherichia coli*: a possible role for guanosine tetraphosphate. *J. Bacteriol.* 172, 2013–2019. doi: 10.1128/jb.172.4.2013-2019.1990

Chodavarapu, S., Felczak, M. M., Simmons, L. A., Murillo, A., and Kaguni, J. M. (2013). Mutant DnaAs of *Escherichia coli* that are refractory to negative control. *Nucleic Acids Res.* 41, 10254–10267. doi: 10.1093/nar/gkt774

Christensen, B. B., Atlung, T., and Hansen, F. G. (1999). DnaA boxes are important elements in setting the initiation mass of *Escherichia coli*. *J. Bacteriol.* 181, 2683–2688.

Churchward, G., Holmans, P., and Bremer, H. (1983). Increased expression of the *dnaA* gene has no effect on DNA replication in a *dnaA*+ strain of *Escherichia coli*. *Mol. Gen. Genet.* 192, 506–508. doi: 10.1007/BF00392197

Clark, D. J., and Maaløe, O. (1967). DNA replication and the division cycle in *Escherichia coli*. *J. Mol. Biol.* 23, 99–112. doi: 10.1016/S0022-2836(67)80070-6

Cleary, J. M., Smith, D. W., Harding, N. E., and Zyskind, J. W. (1982). Primary structure of the chromosomal origins (*oriC*) of *Enterobacter aerogenes* and *Klebsiella pneumoniae*: comparisons and evolutionary relationships. *J. Bacteriol.* 150, 1467–1471.

Donachie, W. D. (1968). Relationship between cell size and time of initiation of DNA replication. *Nature* 219, 1077–1079. doi: 10.1038/2191077a0

Donnell, S. M. O., and Janssen, G. R. (2001). The initiation codon affects ribosome binding and translational efficiency in *Escherichia coli* of cI mRNA with or without the 5′ untranslated leader. *J. Bacteriol.* 183, 1277–1283. doi: 10.1128/JB.183.4.1277-1283.2001

Duderstadt, K. E., Chuang, K., and Berger, J. M. (2011). DNA stretching by bacterial initiators promotes replication origin opening. *Nature* 478, 209–213. doi: 10.1038/nature10455

Eberle, H., van de Merwe, W., Madden, K., Kampo, G., Wright, L., and Donlon, K. (1989). The nature of an intragenic suppressor of the *Escherichia coli dnaA508* temperature-sensitive mutation. *Gene* 84, 237–245. doi: 10.1016/0378-1119(89)90497-6

Erzberger, J. P., Mott, M. L., and Berger, J. M. (2006). Structural basis for ATP-dependent DnaA assembly and replication-origin remodeling. *Nat. Struct. Mol. Biol.* 13, 676–683. doi: 10.1038/nsmb1115

Erzberger, J. P., Pirruccello, M. M., and Berger, J. M. (2002). The structure of bacterial DnaA: implications for general mechanisms underlying DNA replication initiation. *EMBO J.* 21, 4763–4773. doi: 10.1093/emboj/cdf496

Felczak, M. M., and Kaguni, J. M. (2004). The box VII motif of *Escherichia coli* DnaA protein is required for DnaA oligomerization at the *E. coli* replication origin. *J. Biol. Chem.* 279, 51156–51162. doi: 10.1074/jbc.M409695200

Felczak, M. M., Simmons, L. A., and Kaguni, J. M. (2005). An essential tryptophan of *Escherichia coli* DnaA protein functions in oligomerization at the *E. coli* replication origin. *J. Biol. Chem.* 280, 24627–24633. doi: 10.1074/jbc.M503684200

Frimodt-Møller, J., Charbon, G., Krogfelt, K. A., and Løbner-Olesen, A. (2016). DNA replication control is linked to genomic positioning of control regions in *Escherichia coli*. *PLoS Genet.* 12:e1006286. doi: 10.1371/journal.pgen.1006286

Fujikawa, N., Kurumizaka, H., Nureki, O., Terada, T., Shirouzu, M., Katayama, T., et al. (2003). Structural basis of replication origin recognition by the DnaA protein. *Nucleic Acids Res.* 31, 2077–2086. doi: 10.1093/nar/gkg309

Fujimitsu, K., and Katayama, T. (2004). Reactivation of DnaA by DNA sequence-specific nucleotide exchange *in vitro*. *Biochem. Biophys. Res. Commun.* 322, 411–419. doi: 10.1016/j.bbrc.2004.07.141

Fujimitsu, K., Senriuchi, T., and Katayama, T. (2009). Specific genomic sequences of *E. coli* promote replicational initiation by directly reactivating ADP-DnaA. *Genes Dev.* 23, 1221–1233. doi: 10.1101/gad.1775809

Fuller, R. S., and Kornberg, A. (1983). Purified *dnaA* protein in initiation of replication at the *Escherichia coli* chromosomal origin of replication. *Proc. Natl. Acad. Sci. U.S.A.* 80, 5817–5821. doi: 10.1073/pnas.80.19.5817

Garner, J., and Crooke, E. (1996). Membrane regulation of the chromosomal replication activity of *E. coli* DnaA requires a discrete site on the protein. *EMBO J.* 15, 3477–3485.

Hanna, M. H., and Carl, P. L. (1975). Reinitiation of deoxyribonucleic acid synthesis by deoxyribonucleic acid initiation mutants of *E. coli*: role of ribonucleic acid synthesis, protein synthesis, and cell division. *J. Bacteriol.* 121, 219–226.

Hansen, E. B., Atlung, T., Hansen, F. G., Skovgaard, O., and von Meyenburg, K. (1984). Fine structure genetic map and complementation analysis of mutations in the *dnaA* gene of *Escherichia coli*. *Mol. Gen. Genet.* 196, 387–396. doi: 10.1007/BF00436184

Hansen, E. B., Hansen, F. G., and von Meyenburg, K. (1982). The nucleotide sequence of the *dnaA* gene and the first part of the *dnaN* gene of *Escherichia coli* K12. *Nucleic Acids Res.* 10, 7373–7385. doi: 10.1093/nar/10.22.7373

Hansen, F. G. (1995). Reinitiation kinetics in eight *dnaA*(Ts) mutants of *Escherichia coli*: rifampicin resistant initiation of chromosome replication. *Mol. Microbiol.* 15, 133–140. doi: 10.1111/j.1365-2958.1995.tb02227.x

Hansen, F. G., Atlung, T., Braun, R. E., Wright, A., Hughes, P., and Kohiyama, M. (1991b). Initiator (DnaA) protein concentration as a function of growth rate in *Escherichia coli* and *Salmonella typhimurium*. *J. Bacteriol.* 173, 5194–5199. doi: 10.1128/jb.173.16.5194-5199.1991

Hansen, F. G., Christensen, B. B., and Atlung, T. (1991a). The initiator titration model: computer simulation of chromosome and minichromosome control. *Res. Microbiol.* 142, 161–167. doi: 10.1016/0923-2508(91)90025-6

Hansen, F. G., Christensen, B. B., and Atlung, T. (2007). Sequence characteristics required for cooperative binding and efficient *in vivo* titration of the replication initiator protein DnaA in *E. coli*. *J. Mol. Biol.* 367, 942–952. doi: 10.1016/j.jmb.2007.01.056

Hansen, F. G., Christensen, B. B., Nielsen, C. B., and Atlung, T. (2006). Insights into the quality of DnaA boxes and their cooperativity. *J. Mol. Biol.* 355, 85–95. doi: 10.1016/j.jmb.2005.10.036

Hansen, F. G., Hansen, E. B., and Atlung, T. (1982). The nucleotide sequence of the *dnaA* gene promoter and of the adjacent *rpmH* gene, coding for the ribosomal protein L34, of *Escherichia coli*. *EMBO J.* 1, 1043–1048.

Hansen, F. G., Hansen, E. B., and Atlung, T. (1985). Physical mapping and DNA sequence of the gene *rnpA*, that encodes the protein component of ribonuclease P in *Escherichia coli*. *Gene* 38, 85–93. doi: 10.1016/0378-1119(85)90206-9

Hansen, F. G., Koefoed, S., and Atlung, T. (1992). Cloning and nucleotide sequence determination of twelve mutant *dnaA* genes of *Escherichia coli*. *Mol. Gen. Genet.* 234, 14–21.

Hansen, F. G., Koefoed, S., Sørensen, L., and Atlung, T. (1987). Titration of DnaA protein by *oriC* DnaA-boxes increases *dnaA* gene expression in *Escherichia coli*. *EMBO J.* 6, 255–258.

Hansen, F. G., and Rasmussen, K. V. (1977). Regulation of the *dnaA* product in *E. coli*. *Mol. Gen. Genet.* 155, 219–225. doi: 10.1007/BF00393163

Hansen, F. G., and von Meyenburg, K. (1979). Characterization of the *dnaA*, *gyrB* and other genes in the *dnaA* region of the *Escherichia coli* chromosome on specialized transducing phages lambda-*tna*. *Mol. Gen. Genet.* 175, 135–144. doi: 10.1007/BF00425529

Helmstetter, C. E. (1967). Rate of DNA synthesis during the division cycle of *E. coli* B/r. *J. Mol. Biol.* 24, 417–427. doi: 10.1016/0022-2836(67)90228-8

Helmstetter, C. E. (1996). "Timing of synthetic activities in the cell cycle," in *Escherichia coli and Salmonella Cellular and Molecular Biology*, eds F. C. Neidhardt, J. L. Ingraham, E. C. C. Lin, K. B. Low, B. Magasanik, W. S. Reznikoff et al. (Washington, DC: ASM Press), 1627–1639.

Helmstetter, C. E., and Cooper, S. (1968). DNA synthesis during the division cycle of rapidly growing *Escherichia coli* B/r. *J. Mol. Biol.* 31, 507–518. doi: 10.1016/0022-2836(68)90424-5

Helmstetter, C. E., Cooper, S., Pierucci, D., and Revelas, E. (1968). The bacterial life cycle. *Cold Spring Harb. Symp. Quant. Biol.* 33, 809–822. doi: 10.1101/SQB.1968.033.01.093

Helmstetter, C. E., and Cummings, D. J. (1963). Bacterial synchronization by selection of cells at division. *Proc. Natl. Acad. Sci. U.S.A.* 50, 767–774. doi: 10.1073/pnas.50.4.767

Helmstetter, C. E., and Cummings, D. J. (1964). An improved method for the selection of bacterial cells at division. *Biochim. Biophys. Acta* 82, 608–610. doi: 10.1016/0304-4165(64)90453-2

Helmstetter, C. E., and Pierucci, O. (1976). DNA synthesis during the division cycle of three substrains of *Escherichia coli* B/r. *J. Mol. Biol.* 102, 477–486. doi: 10.1016/0022-2836(76)90329-6

Herrick, J., Kohiyama, M., Atlung, T., and Hansen, F. G. (1996). The initiation mess? *Mol. Microbiol.* 19, 659–666. doi: 10.1046/j.1365-2958.1996.432956.x

Hiraga, S. (1976). Novel F prime factors able to replicate in *E. coli* Hfr strains. *Proc. Natl. Acad. Sci. U.S.A.* 73, 198–202. doi: 10.1073/pnas.73.1.198

Hirota, Y., Ryter, A., and Jacob, F. (1968). Thermosensitive mutants of *E. coli* affected in the process of DNA synthesis and cellular division. *Cold Spring Harb. Symp. Quant. Biol.* 33, 677–693. doi: 10.1101/SQB.1968.033.01.077

Hirota, Y., Yasuda, S., Yamada, M., Nishimura, A., Sugimoto, K., Sugisaki, H., et al. (1979). Structural and functional properties of the *Escherichia coli* origin of DNA replication. *Cold Spring Harb. Symp. Quant. Biol.* 43, 129–138. doi: 10.1101/SQB.1979.043.01.019

Hwang, D. S., and Kaguni, J. M. (1988). Interaction of dnaA46 protein with a stimulatory protein in replication from the *Escherichia coli* chromosomal origin. *J. Biol. Chem.* 263, 10633–10640.

Ishida, T., Akimitsu, N., Kashioka, T., Hatano, M., Kubota, T., Ogata , Y., et al. (2004). DiaA, a novel DnaA-binding protein, ensures the timely initiation of *Escherichia coli* chromosome replication. *J. Biol. Chem.* 279, 45546–45555. doi: 10.1074/jbc.M402762200

Jacob, F., Brenner, S., and Cuzin, F. (1963). On the regulation of DNA replication in bacteria. *Cold Spring Harb. Symp. Quant. Biol.* 28, 329–348. doi: 10.1101/SQB.1963.028.01.048

Kasho, K., Fujimitsu, K., Matoba, T., Oshima, T., and Katayama, T. (2014). Timely binding of IHF and Fis to DARS2 regulates ATP-DnaA production and replication initiation. *Nucleic Acids Res.* 42, 13134–13149. doi: 10.1093/nar/gku1051

Kasho, K., and Katayama, T. (2013). DnaA binding locus datA promotes DnaA-ATP hydrolysis to enable cell cycle-coordinated replication initiation. *Proc. Natl. Acad. Sci. U.S.A.* 110, 936–941. doi: 10.1073/pnas.1212070110

Kasho, K., Tanaka, H., Sakai, R., and Katayama, T. (2017). Cooperative DnaA binding to the negatively supercoiled datA locus stimulates DnaA-ATP hydrolysis. *J. Biol. Chem.* 292, 1251–1266. doi: 10.1074/jbc.M116.762815

Katayama, T., Fujimitsu, K., and Ogawa, T. (2001). Multiple pathways regulating DnaA function in *Escherichia coli*: distinct roles for DnaA titration by the datA locus and the regulatory inactivation of DnaA. *Biochimie* 83, 13–17. doi: 10.1016/S0300-9084(00)01206-2

Katayama, T., Kubota, T., Kurokawa, K., Crooke, E., and Sekimizu, K. (1998). The initiator function of DnaA protein is negatively regulated by the sliding clamp of the *E. coli* Chromosomal replicase. *Cell* 94, 61–71. doi: 10.1016/S0092-8674(00)81222-2

Kato, J., and Katayama, T. (2001). Hda, a novel DnaA-related protein, regulates the replication cycle in *Escherichia coli*. *EMBO J.* 20, 4253–4262. doi: 10.1093/emboj/2

Kawakami, H., Ozaki S., Suzuki,S., Nakamura, K., Senriuchi, T., Su'etsugu, M., et al. (2006). The exceptionally tight affinity of DnaA for ATP/ADP requires a unique aspartic acid residue in the AAA+ sensor 1 motif. *Mol. Microbiol.* 62, 1310–1324. doi: 10.1111/j.1365-2958.2006.05450.x

Kellenberger-Gujer, G., Podhajska, A. J., and Caro, L. (1978). A cold sensitive *dnaA* mutant of *E. coli* which overinitiates chromosome replication at low temperature. *Mol. Gen. Genet.* 162, 9–16. doi: 10.1007/BF00333845

Keyamura, K., Abe, Y., Higashi, M., Ueda, T., and Katayama, T. (2009). DiaA dynamics are coupled with changes in initial origin complexes leading to helicase loading. *J. Biol. Chem.* 284, 25038–25050. doi: 10.1074/jbc.M109.002717

Keyamura, K., Fujikawa, N., Ishida, T., Ozaki, S., Su'etsugu, M., Fujimitsu K., et al. (2007). The interaction of DiaA and DnaA regulates the replication cycle in *E. coli* by directly promoting ATP-DnaA-specific initiation complexes. *Genes Dev.* 21, 2083–2099. doi: 10.1101/gad.1561207

Keyamura, K., and Katayama, T. (2011). DnaA protein DNA-binding domain binds to Hda protein to promote inter-AAA+ domain interaction involved in regulatory inactivation of DnaA. *J. Biol. Chem.* 286, 29336–29346. doi: 10.1074/jbc.M111.233403

Kitagawa, R., Mitsuki, H., Okazaki, T., and Ogawa, T. (1996). A novel DnaA protein-binding site at 94.7 min on the *Escherichia coli* chromosome. *Mol. Microbiol.* 19, 1137–1147. doi: 10.1046/j.1365-2958.1996.453983.x

Kitagawa, R., Ozaki, T., Moriya, S., and Ogawa, T. (1998). Negative control of replication initiation by a novel chromosomal locus exhibiting exceptional affinity for *Escherichia coli* DnaA protein. *Genes Dev.* 12, 3032–3043. doi: 10.1101/gad.12.19.3032

Kohiyama, M., Cousin, D., Ryter, A., and Jacob, F. (1966). Mutants thermosensibles d'E. coli K-12. I. Isolement et characterisation rapide. *Ann. Inst. Pasteur* 110, 465–486.

Kohiyama, M., Lanfrom, H., Brenner, S., and Jacob, F. (1963). Modifications de fonctions indispensables chez des mutants thermosensibles *D'escherichia Coli.* sur une mutation emp echant la r'eplication du chromosome bact'erien. *C. R. Hebd. Seances Acad. Sci.* 257, 1979–1981.

Koppes, L. J. H., Woldringh, C. L., and Nanninga, N. (1978). Size variation and correlation of different cell cycle events in slow-growing *Escherichia coli*. *J. Bacteriol.* 134, 423–433.

Kücherer, C., Lother, H., Kölling, R., Schauzu, M. A., and Messer, W. (1986). Regulation of transcription of the chromosomal *dnaA* gene of *Escherichia coli*. *Mol. Gen. Genet.* 205, 115–121. doi: 10.1007/BF02428040

Kuempel, P. L. (1969). Temperature-sensitive initiation of chromosome replication in a mutant of *Escherichia coli*. *J. Bacteriol.* 100, 1302–1310.

Kurokawa, K., Nishida, S., Emoto, A., Sekimizu, K., and Katayama, T. (1999). Replication cycle-coordinated change of the adenine nucleotide-bound forms of DnaA protein in *Escherichia coli*. *EMBO J.* 18, 6642–6652. doi: 10.1093/emboj/18.23.6642

Lark, C. (1968). Studies on the *in vivo* methylation of DNA in *Escherichia coli* 15T-. *J. Mol. Biol.* 31, 389–399. doi: 10.1016/0022-2836(68)90416-6

Leonard, A. C., and Grimwade, J. E. (2015). The orisome: structure and function. *Front. Microbiol.* 6:545. doi: 10.3389/fmicb.2015.00545

Leonard, A. C., and Helmstetter, C. E. (1986). Cell cycle-specific replication of *E. coli* minichromosomes. *Proc. Natl Acad. Sci. U.S.A.* 83, 5101–5105. doi: 10.1073/pnas.83.14.5101

Leu, F. P., Hingorani, M. M., Turner, J., and O'donnell, M. (2000). The delta subunit of DNA polymerase III holoenzyme serves as a sliding clamp unloader in *Escherichia coli*. *J. Biol. Chem.* 275, 34609–34618. doi: 10.1074/jbc.M005495200

Løbner-Olesen, A., Atlung, T., and Rasmussen, K. V. (1987). Stability and replication control of *E. coli* minichromosomes. *J. Bacteriol.* 169, 2835–2842. doi: 10.1128/jb.169.6.2835-2842.1987

Løbner-Olesen, A., Hansen, F. G., Rasmussen, K. V., Martin, B., and Kuempel, P. L. (1994). The initiation cascade for chromosome replication in wild-type and Dam methyltransferase deficient *Escherichia coli* cells. *EMBO J.* 13, 1856 1862.

Løbner-Olesen, A., Skarstad, K., Hansen, F. G., von Meyenburg, K., and Boye, E. (1989). The DnaA protein determines the initiation mass of *Escherichia coli* K-12. *Cell* 57, 881–889. doi: 10.1016/0092-8674(89)90802-7

Løbner-Olesen, A., and von Freiesleben, U. (1996). Chromosomal replication incompatibility in Dam methyltransferase deficient *Escherichia coli* cells. *EMBO J.* 15, 5999–6008.

Lu, M., Campbell, J. L., Boye, E., and Kleckner, N. (1994). SeqA: a negative modulator of replication initiation in *E. coli*. *Cell* 77, 413–426. doi: 10.1016/0092-8674(94)90156-2

Maaløe, O., and Hanawalt, P. C. (1961). Thymine deficiency and the normal DNA replication cycle I. *J. Mol. Biol.* 3, 144–155. doi: 10.1016/S0022-2836(61)80041-7

Maaløe, O., and Kjeldgaard, N. O. (1966). *Control of Macromolecular Synthesis.* New York, NY: Amsterdam;Benjamin.

Mahaffy, J. M., and Zyskind, J. W. (1989). A model for the initiation of replication in *Escherichia coli*. *J. Theor. Biol.* 140, 453–477. doi: 10.1016/S0022-5193(89)80109-2

Margalit, H., and Grover, N. B. (1987). Initiation of chromosome replication in bacteria: analysis of an inhibitor control model. *J. Bacteriol.* 169, 5231–5240. doi: 10.1128/jb.169.11.5231-5240.1987

Margalit, H., Rosenberger, R. F., and Grover, N. B. (1984). Initiation of DNA replication in bacteria: analysis of an autorepressor model. *J. Theor. Biol.* 111, 183–199. doi: 10.1016/S0022-5193(84)80204-0

Margulies, C., and Kaguni, J. M. (1996). Ordered and sequential binding of DnaA protein to *oriC*, the chromosomal origin of *Escherichia coli*. *J. Biol. Chem.* 271, 17035–17040. doi: 10.1074/jbc.271.29.17035

Marsh, R. C., and Worcel, A. (1977). A DNA fragment containing the origin of replication of the *E. coli* chromosome. *Proc. Natl. Acad. Sci. U.S.A.* 74, 2720–2724. doi: 10.1073/pnas.74.7.2720

McGarry, K. C., Ryan, V. T., Grimwade, J. E., and Leonard, A. C. (2004). Two discriminatory binding sites in the *Escherichia coli* replication origin are required for DNA strand opening by initiator DnaA-ATP. *Proc. Natl. Acad. Sci. U.S.A.* 101, 2811–2816. doi: 10.1073/pnas.0400340101

McLean, M. J., Wolfe, K. H., and Devine, K. M. (1998). Base composition skews, replication orientation, and gene orientation in 12 prokaryote genomes. *J. Mol. Evol.* 47, 691–696. doi: 10.1007/PL00006428

Meijer, M., Beck, E., Hansen, F. G., Bergmans, H. E., Messer, W., von Meyenburg, K., et al. (1979). Nucleotide sequence of the origin of replication of the *E. coli* K-12 chromosome. *Proc. Natl. Acad. Sci. U.S.A.* 76, 580–584. doi: 10.1073/pnas.76.2.580

Messer, W. (2002). The bacterial replication initiator DnaA. DnaA and oriC, the bacterial mode to initiate DNA replication. *FEMS Microbiol. Rev.* 26, 355–374. doi: 10.1016/S0168-6445(02)00127-4

Messer, W., Bergmans, H. E., Meijer, M., Womack, J. E., Hansen, F. G., and von Meyenburg, K. (1978). Minichromosomes: plasmids which carry the *E. coli* replication origin. *Mol. Gen. Genet.* 162, 269–275. doi: 10.1007/BF002 68852

Messer, W., Blaesing, F., Majka, J., Nardmann, J., Schaper, S., Schmidt, A., et al. (1999). Functional domains of DnaA proteins. *Biochimie* 81, 819–825. doi: 10.1016/S0300-9084(99)00215-1

Messer, W., Dankwarth, L., Tippe-Schindler, R., Womack, J. E., and Zahn, G. (1975). Regulation of the initiation of DNA replication in *E. coli*. Isolation of iRNA and the control of iRNA synthesis. *ICN-UCLA Symp. Mol. Cell. Biol.* 3, 602–617.

Messer, W., Meijer, M., Bergmans, H. E., Hansen, F. G., von Meyenburg, K., Beck, E., et al. (1979). Origin of replication, *oriC*, of the *Escherichia coli* K- 12 chromosome: nucleotide sequence. *Cold Spring Harb. Symp. Quant. Biol.* 43, 139–145. doi: 10.1101/SQB.1979.043.01.020

Michelsen, O., Teixeira de Mattos, M. J., Jensen, P. R., and Hansen, F. G. (2003). Precise determinations of C and D periods by flow cytometry in *Escherichia coli* K-12 and B/r. *Microbiology* 149, 1001–1010. doi: 10.1099/mic.0.26058-0

Morigen, Løbner-Olesen, A., and Skarstad, K. (2003). Titration of the *Escherichia coli* DnaA protein to excess *datA* sites causes destabilization of replication forks, delayed replication initiation and delayed cell division. *Mol. Microbiol.* 50, 349–362. doi: 10.1046/j.1365-2958.2003.03695.x

Morigen, Molina, F., and Skarstad, K. (2005). Deletion of the *datA* site does not affect once-per-cell-cycle timing but induces rifampin-resistant replication. *J. Bacteriol.* 187, 3913–3920. doi: 10.1128/JB.187.12.3913-3920.2005

Mott, M. L., and Berger, J. M. (2007). DNA replication initiation: mechanisms and regulation in bacteria. *Nat. Rev. Microbiol.* 5, 343–354. doi: 10.1038/nrmicro1640

Murialdo, H., and Siminovitch, L. (1972). The morphogenesis of bacteriophage lambda. IV. Identification of gene products and control of the expression of the morphogenetic information. *Virology* 48, 785–823. doi: 10.1016/0042-6822(72)90162-6

Nielsen, H. J., and Hansen, F. G. (2010). An automated and highly efficient method for counting and measuring fluorescent foci in rod-shaped bacteria. *J. Microsc.* 239, 194–199. doi: 10.1111/j.1365-2818.2010.03374.x

Nozaki, S., Niki, H., and Ogawa, T. (2009a). Replication initiator DnaA of *Escherichia coli* changes its assembly form on the replication origin during the cell cycle. *J. Bacteriol.* 191, 4807–4814. doi: 10.1128/JB.00435-09

Nozaki, S., and Ogawa, T. (2008). Determination of the minimum domain II size of *Escherichia coli* DnaA protein essential for cell viability. *Microbiology* 154, 3379–3384. doi: 10.1099/mic.0.2008/019745-0

Nozaki, S., Yamada, Y., and Ogawa, T. (2009b). Initiator titration complex formed at *datA* with the aid of IHF regulates replication timing in *Escherichia coli*. *Genes Cells* 14, 329–341. doi: 10.1111/j.1365-2443.2008.01269.x

Nyborg, M., Atlung, T., Skovgaard, O., and Hansen, F. G. (2000). Two types of cold sensitivity associated with the A184V change in the DnaA protein. *Mol. Microbiol.* 35, 1202–1210. doi: 10.1046/j.1365-2958.2000.01790.x

Obita, T., Iwura, T., Su'etsugu, M., Yoshida, Y., Tanaka, Y., Katayama, T., et al. (2002). Determination of the secondary structure in solution of the *Escherichia coli* DnaA DNA-binding domain. *Biochem. Biophys. Res. Commun.* 299, 42–48. doi: 10.1016/S0006-291X(02)02590-1

Ogasawara, N., Moriya, S., von Meyenburg, K., Hansen, F. G., and Yoshikawa, H. (1985). Conservation of genes and their organization in the chromosomal replication origin region of *Bacillus subtilis* and *Escherichia coli*. *EMBO J.* 4, 3345–3350.

Ogawa, T., Yamada, Y., Kuroda, T., Kishi, T., and Moriya, S. (2002). The *datA* locus predominantly contributes to the initiator titration mechanism in the control of replication initiation in *Escherichia coli*. *Mol. Microbiol.* 44, 1367–1375. doi: 10.1046/j.1365-2958.2002.02969.x

Ozaki, S., and Katayama, T. (2012). Highly organized DnaA-oriC complexes recruit the single-stranded DNA for replication initiation. *Nucleic Acids Res.* 40, 1648–1665. doi: 10.1093/nar/gkr832

Ozaki, S., Kawakami, H., Nakamura, K., Fujikawa, N., Kagawa, W., Park, S. Y., et al. (2008). A common mechanism for the ATP-DnaA-dependent formation of open complexes at the replication origin. *J. Biol. Chem.* 283, 8351–8362. doi: 10.1074/jbc.M708684200

Polaczek, P., and Wright, A. (1990). Regulation of expression of the *dnaA* gene in *Escherichia coli*: role of the two promoters and the DnaA box. *New Biol.* 2, 564–582.

Pritchard, R. H., Barth, P. T., and Collins, J. (1969). Control of DNA synthesis in bacteria. *Symp. Soc. Gen. Microbiol.* 19, 263–297.

Ream, L. W., Margossian, L., Clark, A. J., Hansen, F. G., and von Meyenburg, K. (1980). Genetic and physical mapping of *recF* in *Escherichia coli* K- 12. *Mol. Gen. Genet.* 180, 115–121. doi: 10.1007/BF00267359

Reddy, P., Peterkofsky, A., and McKenney, K. (1985). Translational efficiency of the *Escherichia coli* adenylate cyclase gene: mutating the UUG initiation codon to GUG or AUG results in increased gene expression. *Proc. Natl. Acad. Sci. U.S.A.* 82, 5656–5660. doi: 10.1073/pnas.82.17.5656

Riber, L., Olsson, J. A., Jensen, R. B., Skovgaard, O., Dasgupta, S., and Marinus, M. G. (2006). Hda-mediated inactivation of the DnaA protein and *dnaA* gene autoregulation act in concert to ensure homeostatic maintenance of the *Escherichia coli* chromosome. *Genes Dev.* 20, 2121–2134. doi: 10.1101/gad.379506

Roth, A., and Messer, W. (1995). The DNA binding domain of the initiator protein DnaA. *EMBO J.* 14, 2106–2111.

Roth, A., and Messer, W. (1998). High-affinity binding sites for the initiator protein DnaA on the chromosome of *Escherichia coli*. *Mol. Microbiol.* 28, 395–401. doi: 10.1046/j.1365-2958.1998.00813.x

Russell, D. W., and Zinder, N. D. (1987). Hemimethylation prevents DNA replication in *E. coli*. *Cell* 50, 1071–1079. doi: 10.1016/0092-8674(87)90173-5

Sakakibara, Y., and Mizukami, T. (1980). A temperature sensitive *Escherichia coli* mutant defective in DNA replication: *dnaN*, a new gene adjacent to the *dnaA* gene. *Mol. Gen. Genet.* 178, 541–553. doi: 10.1007/BF00337859

Samitt, C. E., Hansen, F. G., Miller, J. F., and Schaechter, M. (1989). *In vivo* studies of DnaA binding to the origin of replication of *Escherichia coli*. *EMBO J.* 8, 989–993.

Sancar, A., Hack, A. M., and Rupp, W. D. (1979). Simple method for identification of plasmid-coded proteins. *J. Bacteriol.* 137, 692–693.

Schaechter, M., Maaløe, O., and Kjeldgaard, N. O. (1958). Dependency on medium and temperature of cell size chemical composition during balanced growth of *Salmonella typhimurium*. *J. Gen. Microbiol.* 19, 592–606. doi: 10.1099/00221287-19-3-592

Schaechter, M., Williamson, J. P., Hood, J. R., and Koch, A. L. (1962). Growth, cell and nuclear divisions in some bacteria. *J. Gen. Microbiol.* 29, 421–434. doi: 10.1099/00221287-29-3-421

Schaper, S., and Messer, W. (1995). Interaction of the initiator protein DnaA of *Escherichia coli* with its DNA target. *J. Biol. Chem.* 270, 17622–17626. doi: 10.1074/jbc.270.29.17622

Seitz, H., Weigel, C., and Messer, W. (2000). The interaction domains of the DnaA and DnaB replication proteins of *Escherichia coli*. *Mol. Microbiol.* 37, 1270–1279. doi: 10.1046/j.1365-2958.2000.02096.x

Sekimizu, K., Bramhill, D., and Kornberg, A. (1987). ATP activates dnaA protein in initiating replication of plasmids bearing the origin of the *E. coli* chromosome. *Cell* 50, 259–265. doi: 10.1016/0092-8674(87)90221-2

Sekimizu, K., and Kornberg, A. (1988). Cardiolipin activation of dnaA protein, the initiation protein of replication in *Escherichia coli*. *J. Biol. Chem.* 263, 7131–7135.

Sevastopoulos, C. G., Wehr, C. T., and Glaser, D. A. (1977). Largescale automated isolation of *Escherichia coli* mutants with thermosensitive DNA replication. *Proc. Natl. Acad. Sci. U.S.A.* 74, 3485–3489. doi: 10.1073/pnas.74.8.3485

Si, F., Li, D., Cox, S. E., Sauls, J. T., Azizi, O., Sou, C., et al. (2017). Invariance of initiation mass and predictability of cell size in *Escherichia coli*. *Curr. Biol.* 27, 1278–1287. doi: 10.1016/j.cub.2017.03.022

Skarstad, K., Boye, E., and Fanning, E. (2003). Circles in the sand. *EMBO Rep.* 4, 661–665. doi: 10.1038/sj.embor.embor888

Skarstad, K., Boye, E., and Steen, H. B. (1986). Timing of initiation of chromosome replication in individual *E. coli* cells. *EMBO J.* 5, 1711–1717.

Skarstad, K., and Katayama, T. (2013). Regulating DNA replication in bacteria. *Cold Spring Harb. Perspect. Biol.* 5, 1–17. doi: 10.1101/cshperspect.a012922

Skarstad, K., and Løbner-Olesen, A. (2003). Stable co-existence of separate replicons in *Escherichia coli* is dependent on once-per-cell-cycle initiation. *EMBO J.* 22, 140–150. doi: 10.1093/emboj/cdg003

Skovgaard, O. (1990). Nucleotide sequence of a *Proteus mirabilis* fragment homologous to the *60K-rnpA-rpmH-dnaA-dnaN-recF-gyrB* region of *Escherichia coli*. *Gene* 93, 27–34. doi: 10.1016/0378-1119(90)90131-A

Skovgaard, O., and Hansen, F. G. (1987). Comparison of *dnaA* nucleotide sequences of *Escherichia coli*, *Salmonella typhimurium*, and *Serratia Marcescens*. *J. Bacteriol.* 169, 3976–3981. doi: 10.1128/jb.169.9.3976-3981.1987

Sompayrac, L., and Maaloe, O. (1973). Autorepressor model for control of DNA replication. *Nat. New Biol.* 241, 133–135. doi: 10.1038/newbio241133a0

Speck, C., and Messer, W. (2001). Mechanism of origin unwinding: sequential binding of DnaA to double- and single-stranded DNA. *EMBO J.* 20, 1469–1476. doi: 10.1093/emboj/20.6.1469

Speck, C., Weigel, C., and Messer, W. (1999). ATP- and ADP-DnaA protein, a molecular switch in gene regulation. *EMBO J.* 18, 6169–6176. doi: 10.1093/emboj/18.21.6169

Sugimoto, K., Oka, A., Sugisaki, H., Takanami, M., Nishimura, A., Yasuda, Y., et al. (1979). Nucleotide sequence of *Escherichia coli* replication origin. *Proc. Natl. Acad. Sci. U.S.A.* 76, 575–579. doi: 10.1073/pnas.76.2.575

Tabata, S., Oka, A., Sugimoto, K., Takanami, M., Yasuda, S., and Hirota, Y. (1983). The 245 bp *oriC* sequence of the *E. coli* chromosome directs bidirectional replication at an adjacent region. *Nucleic Acids Res.* 11, 2617–2626. doi: 10.1093/nar/11.9.2617

Theisen, P. W., Grimwade, J. E., Leonard, A. C., Bogan, J. A., and Helmstetter, C. E. (1993). Correlation of gene transcription with the time of initiation of chromosome replication in *Escherichia coli*. *Mol. Microbiol.* 10, 575–584. doi: 10.1111/j.1365-2958.1993.tb00929.x

Tippe-Schindler, R., Zahn, G., and Messer, W. (1979). Control of the initiation of DNA replication in *E. coli*. I. Negative control of initiation. *Mol. Gen. Genet.* 168, 185–195. doi: 10.1007/BF00431444

Torheim, N. K., Boye, E., Løbner-Olesen, A., Stokke, T., and Skarstad, K. (2000). The *Escherichia coli* SeqA protein destabilizes mutant DnaA204 protein. *Mol. Microbiol.* 37, 629–638. doi: 10.1046/j.1365-2958.2000.02031.x

Travers, A. A. (1980). Promoter sequence for stringent control of bacterial ribonucleic acid synthesis. *J. Bacteriol.* 141, 973–976.

van den Berg, E. A., Geerse, R. H., Memelink, J., Bovenberg, R. A. L., Magnée, F. A., and van de Putte, P. (1985). Analysis of regulatory sequences upstream of the *E. coli* uvrB gene; involvement of the DnaA protein. *Nucleic Acids Res.* 13, 1829–1840. doi: 10.1093/nar/13.6.1829

von Freiesleben, U., Krekling, M. A., Hansen, F. G., and Løbner-Olesen, A. (2000a). The eclipse period of *Escherichia coli*. *EMBO J.* 19, 6240–6248. doi: 10.1093/emboj/19.22.6240

von Freiesleben, U., Rasmussen, K. V., Atlung, T., and Hansen, F. G. (2000b). Rifampicin resistant initiation of chromosome replication from *oriC* in *ihf* mutants. *Mol. Microbiol.* 37, 1087–1093. doi: 10.1046/j.1365-2958.2000.02060.x

von Freiesleben, U., Rasmussen, K. V., and Schaechter, M. (1994). SeqA limits DnaA activity in replication from *oriC* in *Escherichia coli*. *Mol. Microbiol.* 14, 763–772. doi: 10.1111/j.1365-2958.1994.tb01313.x

von Meyenburg, K., Hansen, F. G., Riise, E., Bergmans, H. E., Meijer, M., and Messer, W. (1979). Origin of replication, *oriC*, of the *E. coli* K-12 chromosome: genetic mapping and minichromosome replication. *Cold Spring Harb. Symp. Quant. Biol.* 43, 121–128. doi: 10.1101/SQB.1979.043.01.018

Wechsler, J. A., and Gross, J. D. (1971). *Escherichia coli* mutants temperature sensitive for DNA synthesis. *Mol. Gen. Genet.* 113, 273–284. doi: 10.1007/BF00339547

Weigel, C., Schmidt, A., Seitz, H., Tüngler, D., Welzeck, M., and Messer, W. (1999). The N-terminus promotes oligomerization of the *Escherichia coli* initiator protein DnaA. *Mol. Microbiol.* 34, 53–66. doi: 10.1046/j.1365-2958.1999.01568.x

Wold, S., Skarstad, K., Steen, H. B., Stokke, T., and Boye, E. (1994). The initiation mass for DNA replication in *Escherichia coli* K-12 is dependent on growth rate. *EMBO J.* 13, 2097–2102.

Yasuda, S., and Hirota, Y. (1977). Cloning and mapping of the replication origin of *Escherichia coli*. *Proc. Natl. Acad. Sci. U.S.A.* 74, 5458–5462. doi: 10.1073/pnas.74.12.5458

Zawilak-Pawlik, A., Nowaczyk, M., and Zakrzewska-Czerwinska, J. (2017). The role of the N-terminal domains of bacterial initiator DnaA in the assembly and regulation of the bacterial replication initiation complex. *Genes* 8, 1–18. doi: 10.3390/genes8050136

Zhou, P., Bogan, J. A., Welch, K., Pickett, S. R., Wang, H. J., Zaritsky, A., et al. (1997). Gene transcription and chromosome replication in *Escherichia coli*. *J. Bacteriol.* 179, 163–169. doi: 10.1128/jb.179.1.163-169.1997

Zyskind, J. W., Harding, N. E., Takeda, Y., Cleary, J. M., and Smith, D. W. (1981). The DNA replication origin region of the *Enterobacteriaceae*. *ICN-UCLA Symp. Mol. Cell. Biol.* 22, 13–25. doi: 10.1016/B978-0-12-583580-0.50007-8

Zyskind, J. W., and Smith, D. W. (1980). Nucleotide sequence of the *Salmonella typhimurium_* origin of DNA replication. *Proc. Natl. Acad. Sci. U.S.A.* 77, 2460–2464. doi: 10.1073/pnas.77.5.2460

9

Thymineless death, at the origin

Elena C. Guzmán* and Carmen M. Martín

Departamento de Bioquímica Biología Molecular y Genética, Facultad de Ciencias, Universidad de Extremadura, Badajoz, Spain

****Correspondence:***
Elena C. Guzmán,
Departamento de Bioquímica Biología Molecular y Genética, Facultad de Ciencias, Universidad de Extremadura, Badajoz 06071, Spain
eguzmac@gmail.com

Thymineless death (TLD) in bacteria has been a focus of research for decades. Nevertheless, the advances in the last 5 years, with *Escherichia coli* as the model organism, have been outstanding. Independent research groups have presented compelling results that establish that the initiation of chromosome replication under thymine starvation is a key element in the scenario of TLD. Here we review the experimental results linking the initiation of replication to the lethality under thymine starvation and the proposed mechanisms by which TLD occurs. The concept of this relationship was 'in the air,' but approaches were not sufficiently developed to demonstrate the crucial role of DNA initiation in TLD. Genome-wide marker frequency analysis and Two Dimensional agarose gel electrophoresis have been critical methods employed to reveal that initiation events and the degradation of the *oriC* region occur during thymine starvation. The relationships between these events and TLD have established them to be the main underlying causes of the lethality under thymine starvation. Furthermore, we summarize additional important findings from the study of different mutant strains, which support the idea that the initiation of chromosomal replication and TLD are connected.

Keywords: TLD, replication fork, initiation, *oriC*, rifampicin, DSBs, 2D gel DNA

Introduction

Thymineless death (TLD) is defined by the loss of viability that occurs in a culture of a *thyA* defective mutant strain when deprived of thymine (**Figure 1A**). It was first reported by Barner and Cohen 60 years ago (Barner and Cohen, 1954). In the ensuing 60 years a number of other laboratory groups have studied this phenomenon and have attempted to elucidate its mechanism. Throughout the years, TLD has been associated with DNA damage and DNA recombination structures, as well as their outcomes: SOS induction, filamentation, mutagenesis, loss of plasmids, or induction of suicide modules and prophages, among others (Ahmad et al., 1998). However, the relative contribution of these factors, individually or in combination, to TLD remains unknown. A novel and critical aspect has emerged in the last 5 years: the initiation of chromosomal replication. Abortive events in attempted initiation during thymine starvation may be associated with the observed degradation of the *oriC* DNA sequence that eventually leads to TLD.

The goal of this minireview is to detail the experiments and various approaches that establish the initiation of replication as a key element in the continuously evolving story of TLD.

The Unbalanced Growth Model

The first general model to explain TLD was the recognition of *unbalanced growth* generated under thymine starvation (Barner and Cohen, 1954). Certainly, DNA replication eventually stops

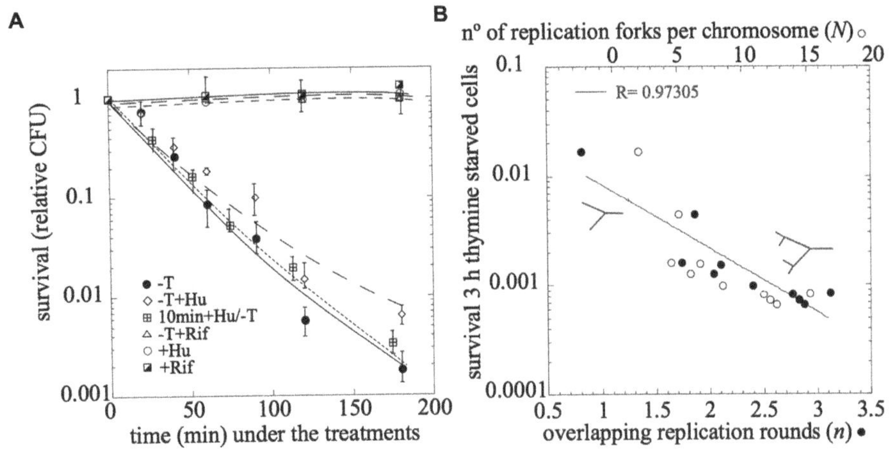

FIGURE 1 | (A) The relative cell survival after (•) thymine starvation, (◇) in the presence of hydroxyurea, (△) in the presence of rifampicin, (⊞) 10 min pretreatment with hydoxyurea and after addition of hydroxyurea 75 mM, (○) addition of hydroxyurea 75 mM, (◪) addition of rifampicin. **(B)** The relationship between the number of replication rounds per chromosome, *n*, or the number of replication forks per chromosome, *N*, and the relative viability after 3 h of thymine starvation. The red drawing corresponds to the chromosome configuration with one or two cycles of chromosome replication. (adapted from Martín and Guzmán, 2011; Martín, 2014).

in thymine-starved cells, while the processes of DNA recombination and repair may continue (Nakayama et al., 1994), and neither RNA nor protein synthesis are greatly affected (Hanawalt et al., 1961; Maaloe and Hanawalt, 1961; Nakayama and Hanawalt, 1975). Nevertheless, the notion of *unbalanced growth* was too vague, and it was shown that it cannot be the sole cause of TLD. Various experimental conditions suppressed TLD while maintaining an imbalance between DNA and protein synthesis. Examples of these conditions include the starvation of thymine in the presence of chloramphenicol (Okagaki et al., 1960) during which RNA synthesis remains active, or the heat inactivation of the proteins required for chromosomal initiation, DnaA or DnaC, (Bouvier and Sicard, 1975) during which RNA and protein synthesis are not affected. None of these conditions would predict the suppression of TLD according to the *unbalanced growth* model.

TLD is Related to DNA Replication

Thymine is exclusively incorporated into DNA during the replication process; therefore, TLD has been associated for many years with DNA replication. Early observations described conditions under which TLD was suppressed in cells that had completed replication rounds (Maaloe and Hanawalt, 1961). Additionally, TLD has been shown to be affected differently under inactivation of several replication proteins (Bouvier and Sicard, 1975; Nakayama et al., 1994). Thus, a direct involvement of the chromosomal replication process in TLD is apparent.

TLD Correlates with the Number of Replication Forks, but they are Not Required to be Fully Active

A relationship between the magnitude of the lethality under thymine starvation and the number of replication forks has been well established. Both parameters were determined in the strain MG1693 *thyA175* grown under various conditions to achieve different numbers of replication rounds per chromosome, *n*, (Sueoka and Yoshikawa, 1965). Survival after thymine starvation (**Figure 1A**) was found to inversely correlate with *n* and the number of forks per cell, *N*, before thymine starvation (**Figure 1B**; Martín and Guzmán, 2011; Martín, 2014). However, it is not clear whether the activity of the replication forks is required for the lethality. TLD is not suppressed when thymine starvation is accomplished under conditions of DNA inhibition, such as the addition of hydroxyurea (Morganroth and Hanawalt, 2006; Kuong and Kuzminov, 2009; Martín and Guzmán, 2011; **Figure 1A**) or the incubation of a *dnaBts* mutant at 42°C (Bouvier and Sicard, 1975). Controversy has arisen because the two approaches used to inactivate the replication forks do not show a clear-cut inhibition of DNA synthesis (Kuong and Kuzminov, 2009, 2010). Nevertheless, the complete absence of effect on TLD either after hydroxyurea addition or incubation of *dnaBts* at 42°C suggests that TLD is not dependent on the level of activity of the replication forks. Furthermore, the observation that the addition of hydroxyurea 10 min before thymine starvation does not change the kinetics of TLD supports this idea (**Figure 1A**; Martín and Guzmán, 2011; Martín, 2014).

DNA Fragmentation and Recombinant DNA Intermediates are Not Sufficient to Account for TLD

When considering the idea that the replication forks are targeted during thymine starvation, the primary assumption is that TLD results from DNA damage brought about by thymine starvation on its target. What effects could thymine starvation produce on the replication forks? Different models have suggested two primary sources of TLD that are not mutually exclusive: DNA breakage and DNA recombination intermediates, which have been associated either with RNA synthesis during thymine starvation (Nakayama and Hanawalt, 1975) or to different recombination pathways such as RecA/RecBCD or RecFOR (Fonville et al., 2010; Kuong and Kuzminov, 2010).

DNA breakage has been observed under thymine starvation; thus, the occurrence of single-strand breaks (SSBs), DNA single-strand gaps (DNA ss-gaps; Nakayama and Hanawalt, 1975), and double strand breaks (DSBs; Guarino et al., 2007; Kuong and Kuzminov, 2010; Martín and Guzmán, 2011) have been shown to occur during *thymineless* incubation of bacteria. Furthermore, Nakayama et al. (1994) demonstrated the presence of DNA recombination intermediates, known as "non-migrating DNA" (nmDNA), in thymine-starved cells. Nevertheless, Martín and Guzmán (2011) have shown that neither the number of DSBs nor the level of nmDNA correlated with TLD. DSBs appeared in the DNA of thymine-starved cells in the presence of rifampicin, a drug that prevents TLD. Thus, DSBs might be necessary but not sufficient to cause TLD. In addition, DNA recombination intermediates (nmDNA) were not observed under thymine starvation when hydroxyurea was present (Martín and Guzmán, 2011). Because death still occurs under thymine starvation in the presence of hydroxyurea (**Figure 1A**), DNA recombination intermediates may be associated with, but are not essential for, TLD. Thus, what is the critical condition for TLD to occur?

Initiation of Replication is a Key Element in TLD

Although TLD has been associated for many years with replicating cells, and recent results have demonstrated a correlation between TLD and the number of replication forks, two results have suggested that additional components of the replication process are involved in TLD. First, TLD is suppressed by inhibiting RNA or protein synthesis, or both, as is observed in experimental conditions including the presence of rifampicin (Hanawalt, 1963; Martín and Guzmán, 2011; Martín et al., 2014) or chloramphenicol or in amino acid-starved cells (Martín, 2014). Second, TLD is avoided by inactivating the DnaA protein (Bouvier and Sicard, 1975; Nakayama et al., 1994; Martín et al., 2014). The nexus between these three conditions (RNA and protein synthesis and the active form of DnaA protein) is their requirement at the initiation of replication. The inhibition of initiation by any of these treatments does not affect elongating replication forks, which can progress until the replication of the chromosome is completed. Thus, a correlation has been found between initiation of new replication rounds under thymine starvation and TLD. The causative link between DNA initiation and TLD has been established by several lines of evidence detailed below.

New Initiations Occur Under Thymine Starvation

The occurrence of new initiation events after restoring thymine to thymine-starved cells was first reported in the 1960s (Pritchard and Lark, 1964) and confirmed within the past 5 years (Martín and Guzmán, 2011; Martín et al., 2014). Furthermore, it has been shown by different approaches that initiation events do occur during thymine starvation.

(i) Replication runouts – This approach is based on the ΔG and the $\Delta G'$ values. Briefly, ΔG refers to the relative increase in the amount of DNA after the inhibition of new rounds of chromosome replication, a condition achieved by adding 150 $\mu g\ ml^{-1}$ of rifampicin to an exponentially growing culture (Sueoka and Yoshikawa, 1965; Zaritsky and Pritchard, 1971). $\Delta G'$ represents the relative increment of DNA in a culture subjected to 10 min of thymine starvation followed by restoration of thymine in the presence of 150 $\mu g\ ml^{-1}$ of rifampicin to inhibit new initiations. Thus, if new initiations occur in the thymine-starved cells, then the $\Delta G'$ value is expected to be higher than the ΔG value (Jiménez-Sánchez and Guzmán, 1988). The results in **Figure 2A** show that the value of $\Delta G'$ after the addition of thymine to 10-min thymine-starved cultures is higher than ΔG. These results show that new initiations occurred at a number of origins (*i*) under thymine starvation. Initiation events also occurred in the presence of hydroxyurea, but they were suppressed when the 10-min period of thymine starvation was performed in the presence of rifampicin or chloramphenicol. These results show a correlation between chromosomal initiation and lethality under thymine starvation (Martín and Guzmán, 2011; Martín et al., 2014).

Furthermore, it was shown that the number of initiations at *oriC* increased with the amount of treatment time (**Figure 2B**), correlating with a loss of colony-forming units on solid medium (**Figure 1A**). Flow cytometry profiles of the replication runouts after thymine addition to cultures previously exposed to increasing time periods of thymine starvation indicated that only the *thymineless*-initiations that had occurred during the first 30 min could be repaired to allow complete chromosome replications (Martín et al., 2014).

(ii) Visualization of the oriC replication intermediates under thymine starvation by 2D gels – The analysis of the replication fork progression at one specific position of the chromosome can be resolved by performing two-dimensional DNA gel electrophoresis (2D gels; Brewer and Fangman, 1987; Schwartzman et al., 2012). The results with 2D gels show the different DNA structures that occur in the *oriC* region during thymine starvation (**Figure 2C**). The bubble arcs detected after 40 min indicate initiation events at *oriC*, while the double-Y structures are most likely produced by two forks encountering each other at *oriC*. These structures indicate that replication rounds are initiated during thymine starvation in the *oriC* region, although the possibility of initiations at different locations cannot be excluded. A simple-Y arc, corresponding to accumulated Y-shaped replication intermediates, was clearly detected after 10 min of thymine starvation. This indicates the arrest of replication forks within the *oriC* region and also strongly suggests that some initiations occur outside of the restriction fragment but still in the vicinity of *oriC*.

The experiments using the mutant strains confirmed the occurrence of initiation during thymine starvation. Consistent with the suppression of TLD, none of the DNA intermediates observed in wild type strains under TLD

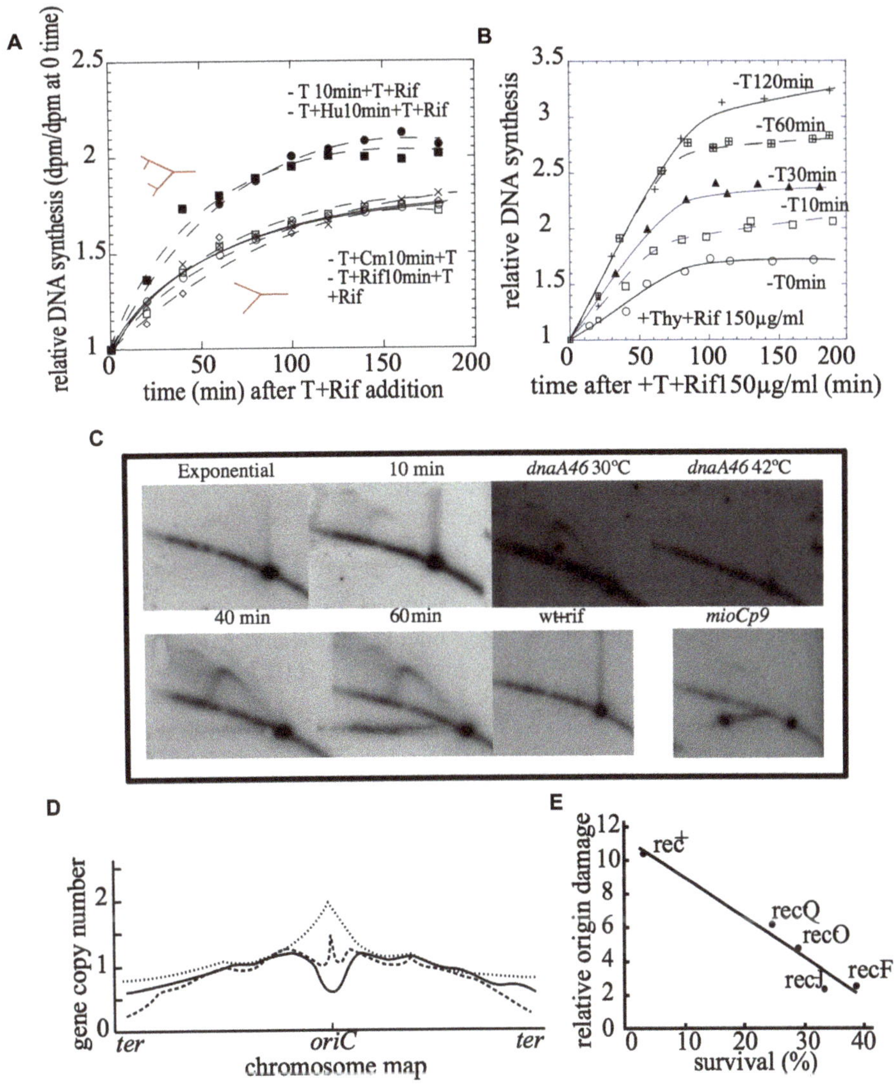

FIGURE 2 | (A) The relative DNA accumulation in the presence of rifampicin after thymine restoration to 10 min thymine-starved cells in the presence of rifampicin, chloramphenicol, hydroxyurea, or any drugs (Martín and Guzmán, 2011). **(B)** The relative DNA accumulation in the presence of rifampicin after thymine restoration to cells under thymine starvation for different extents of time (Martín et al., 2014). **(C)** The 2D gel DNA analysis of exponentially growing cells at 0, 10, 40, and 60 min of thymine starvation in the wild type strain; at 60 min and 30 or 42°C in the *dnaA46* thermosensitive mutant; in the presence of rifampicin and in the *mioCp9* mutant (adapted from Martín et al., 2014). **(D)** The Marker Frequency Analysis of the wild type strain (solid line), the *recA* (dotted line) mutant, or the *recBC* mutant (broken line) after 4 h of thymine starvation (adapted from Kuong and Kuzminov, 2012). **(E)** The correlation between the relative *origin* damage and percent of survival after 3 h of thymine starvation in the wild type strain and Rec defective mutants (*recQ, recO, recJ,* and *recF*) (adapted from Sangurdekar et al., 2010).

conditions were detected by 2D gels when initiations were inhibited by rifampicin or by DnaA inactivation (**Figure 2C**; Martín et al., 2014).

(iii) Copy number of oriC sequences increases during first 30 min under thymine starvation – The third approach providing evidence that new initiations occur during thymine starvation has been the quantification of the *ori/ter* ratio by performing either quantitative PCR (Sangurdekar et al., 2010) or Marker Frequency Analysis (dot-blot hybridization; Kuong and Kuzminov, 2012). The frequency of a gene along the chromosome, *f(a)*, follows a function that depends upon the number of replication rounds per chromosome, *n*, and the position of the gene relative to the origin of replication, *x*, being $f(a) = 2^{n(1-x)}$ (Sueoka and Yoshikawa, 1965).

Determination of the *ori/ter* ratio after 30 min of thymine starvation yielded a value higher than that obtained when the cells were growing exponentially (Kuong and Kuzminov, 2012). The overall results show that *thymineless*-initiation events do occur. With the passage of time, the number of origins decreases, revealing a progressive loss of DNA in the *oriC* sequence (Sangurdekar et al., 2010; Kuong and Kuzminov, 2012) as detailed below.

oriC is Degraded Under Thymine Starvation in a Rec-Dependent Manner

Marker Frequency Analysis on the scale of the whole chromosome by gene arrays is becoming the standard method of analyzing the replication pattern in bacteria. The results of the comparative genomic hybridization of chromosomal DNA after 3–4 h of thymine starvation revealed the loss of the *oriC* region (Sangurdekar et al., 2010; Kuong and Kuzminov, 2012; **Figure 2D**). Furthermore, this conclusion is supported by the loss of *oriC*-containing foci during TLD, which was revealed by fluorescence *in situ* hybridization (FISH; Fonville et al., 2010). An explanation for these observations could rely upon the degradation of the *oriC* region originated from DSBs nearby. An interesting question arises at this point: why and how is the *oriC* sequence preferentially degraded? One of the better-known triggers of chromosomal fragmentation is the incorporation of uracil into DNA. This occurs *dut* mutants, in such a way that the chromosome fragmentation exhibits a gradient that parallels the replication gradient (Kouzminova and Kuzminov, 2008). Thus, it may be possible that the distinctive degradation of the *oriC* region is explained by the AT-rich feature of the *oriC* sequence, which could allow for a very significant incorporation of uracil in the absence of thymine.

Several defective repair/recombination mutant strains have been assayed for DNA damage impacting the *oriC* region (Sangurdekar et al., 2010). These include those affecting enzymes that process and repair single-stranded gaps such a *recFOR*, *recJ*, or *recQ*. All of them are, to some extent, resistant to TLD and contain quantitatively lower origin DNA damage than in wild type strains (**Figure 2E**), thus connecting the extent of DNA damage at the *oriC* region to the lethality under thymine starvation. Interestingly, *recA* defective mutant starved for thymine for 4 h displayed an *ori/ter* ratio even higher than the initial value before treatment (Kuong and Kuzminov, 2012; **Figure 2D**). RecA is a protein that plays a central role in homologous DNA recombination and repair. Nevertheless, it has been reported that either it is not required for TLD (Nakayama et al., 1982) or it may partially palliate it (Fonville et al., 2010; Kuong and Kuzminov, 2010). Thus, the reported absence of *oriC* degradation in a *recA* mutant during thymine starvation would support the alleviated TLD observed in the *recA* mutant (Kuong and Kuzminov, 2012), consistent with the idea that TLD would depend on the origin degradation triggered by RecA-promoted recombinational misrepair.

The provocative exception is the feature exhibited by a *recBC* defective mutant, which has been described to be hypersensitive to TLD, although no *oriC* degradation is observed in this genetic background (Kuong and Kuzminov, 2010, 2012). The RecBCD complex has both helicase and exonuclease activities, and it initiates the repair of DSBs by homologous recombination in combination with RecA protein (Michel et al., 2007; Dillingham and Kowalczykowski, 2008). Because *recBC* mutant cells do not degrade double-strand DNA ends and cannot repair DSBs, the TLD in this defective mutant likely could be related to its inability to repair DSBs at the original replication forks, even though the *oriC* region is not degraded (**Figure 2D**).

The Transcription-Dependent Step of Initiation is the Target for Rifampicin Suppression of TLD

Thymineless death suppression by rifampicin was observed in early studies (Nakayama and Hanawalt, 1975), but its mechanism of action has not yet been elucidated. This problem has been analyzed by two different approaches.

(i) The effects of different concentrations of rifampicin – It has been shown that the activity of the RNA polymerase in thymine-starved cells modulates both the initiation of DNA replication under thymine starvation and TLD (Martín et al., 2014). Interestingly, suppression of TLD by rifampicin in a Δ*datA* mutant was achieved when treated with 1,000 μg ml^{-1} of the drug but not with 150 μg ml^{-1} (Martín and Guzmán, 2011). It has been suggested that in a Δ*datA* strain, less RNA transcription around *oriC* would be required for initiation to occur because more DnaA protein would be available to open the *oriC* sequence during initiation in this mutant (Morigen et al., 2005). Accordingly, the Δ*datA* strain initiates chromosome replication in the presence of 150 μg ml^{-1} rifampicin, explaining the observed TLD, while addition at 1,000 μg ml^{-1} completely inhibits the transcription requirement for initiation, hence suppressing TLD.

(ii) mioC and gid defective mutant strains – The importance of *mioC* and *gid* gene transcription for the initiation of chromosomal replication at *oriC* is widely accepted (Messer, 1972; Baker and Kornberg, 1988). Nevertheless, neither the deletion of *PmioC112* or *Pgid113* promoters, nor the constitutive transcription from the *mioCp9* promoter have a large effect on the cell cycle. However, reduction in initiation efficiency has been observed in rich medium (Bates et al., 1997; Molina et al., 1999; Su'etsugu et al., 2003). Under thymine starvation, altered *mioC* and *gid* transcription limited the initiation process and TLD was alleviated (Martín et al., 2014). The relevance of these results relies on the fact that TLD alleviation in these mutants must be related to the alteration of the normal transcription levels around *oriC*. This effect on TLD, combined with the broad reduction in chromosome initiation intermediates in these mutants (**Figure 2C**), strongly suggests that the transcription-dependent step of the initiation is the target for rifampicin suppression of TLD, hence being critical for TLD.

Conclusion

Overall, these experimental approaches pinpoint the initiations at the *oriC* region as the main targets for TLD in wild type strains. Thus, if DNA initiation is allowed under thymine starvation, death occurs likely due to the lethal consequences of the presence of DSBs, DNAss gaps, and DNA recombination intermediates at the *origin* that eventually result in *oriC* region degradation. If initiation is inhibited (*dnaA46* mutant, rifampicin, chloramphenicol) or impaired (*mioC, gid,* Δ*datA*, sub-inhibitory rifampicin concentrations), TLD is subsequently suppressed or alleviated, respectively. Thus, the observed correlation between TLD and the number of replication forks could reflect not only the importance of the forks as targets, but also the quantitative relationship between TLD, and the number of *origins* per chromosome, 2^n.

Regarding *thymineless*-initiation events in *thyA* mutants that are otherwise wild type cells, the observations could be divided into two stages. During the first 30–60 min following thymine starvation the instability of DNA ss-gaps and the resulting DNA degradation behind the replication forks (and/or different source; Kuong and Kuzminov, 2012), might provide the dNTPs necessary to support new initiations, residual DNA elongation and the repair of the replication forks. According to the results from different authors, initiation events (supposed to occur at the first stage) are required to yield TLD, as lethality has not been observed in thymine-starved cells under conditions inhibiting DNA initiation.

The second stage would proceed after 30–60 min when the new thymineless-initiation events would generate unrepaired DSBs or DNAss gaps together with unresolved DNA recombination intermediaries at the origin, somehow triggering the unique degradation of the *oriC* region that acts as the major lethal effect of thymine starvation. Supporting this explanation, it also has been shown that the extent of *origin* degradation (supposed to occur at the second stage) correlates with the magnitude of TLD (Sangurdekar et al., 2010). At this point, the role of the recombination/repair enzymes reflects the different sensitivities to thymine starvation described above. Accordingly, several independent pathways accounting for TLD has been proposed (Courcelle, 2005; Nakayama, 2005; Kuong and Kuzminov, 2010) and also supported by Rosenberg's lab (Fonville et al., 2010, 2011; Hamilton et al., 2013).

Several questions arise from this tentative model. The first one is whether new DNA initiation is a requisite for origin degradation. Second, what is the mechanism by which the *oriC* region is selectively degraded? Third, according to this proposal the DSBs located outside the origin region do not seem to be lethal or, alternatively, the inhibition of DNA initiation is counteracting their potentially lethal effect. Therefore, what is the mechanism of that phenomenon?

The advances in knowledge about TLD mechanisms have been impressive in the past 5 years. New technologies and approaches have evolved to provide novel insights, but TLD is still like a black hole - you know how you got into it but you never know where you will end up as time passes.

Acknowledgments

We wish to specially thank Phil Hanawalt, Arkady Khodursky, Alfonso Jiménez-Sánchez, and the reviewers for critically reading this manuscript, promoting constructive discussion and, especially to Andrei Kuzminov for providing such a wording suggestions. We thank the Servicio de Técnicas Aplicadas a la Biociencia (STAB) of the Universidad de Extremadura and Encarna Ferrera for technical assistance. This work was supported by grants GRU10058 from the Junta de Extremadura and BFU2007-63942 from the Ministerio de Educación y Ciencia to ECG.

References

Ahmad, S. I., Kirk, S. H., and Eisenstark, A. (1998). Thymine metabolism and thymineless death in prokaryotes and eukaryotes. *Annu. Rev. Microbiol.* 52, 591–625. doi: 10.1146/annurev.micro.52.1.591

Baker, T. A., and Kornberg, A. (1988). Transcriptional activation of initiation of replication from the E. *coli* chromosomal origin: an RNA-DNA hybrid near *oriC*. *Cell* 55, 113–123. doi: 10.1016/0092-8674(88)90014-1

Barner, H. D., and Cohen, S. S. (1954). The induction of thymine synthesis by T2 infection of a thymine requiring mutant of *Escherichia coli*. *J. Bacteriol.* 68, 80–88.

Bates, D., Boye, E., Asai, T., and Kogoma, T. (1997). The absence of effect of *gid* or *mioC* transcription on the initiation of chromosomal replication in *Escherichia coli*. *Proc. Natl. Acad. Sci. U.S.A.* 94, 12497–12502. doi: 10.1073/pnas.94.23.12497

Bouvier, F., and Sicard, N. (1975). Interference of *dna-ts* mutations of *Escherichia coli* with *thymineless death*. *J. Bacteriol.* 124, 1198–1204.

Brewer, B. J., and Fangman, W. L. (1987). The localization of replication origins on ARS plasmids in *S. cerevisiae*. *Cell* 51, 463–471. doi: 10.1016/0092-8674(87)90642-8

Courcelle, J. (2005). Recs preventing wrecks. *Mutant. Res.* 557, 217–227. doi: 10.1016/j.mrfmmm.2005.03.019

Dillingham, M. S., and Kowalczykowski, S. C. (2008). RecBCD enzyme and the repair of Double-Stranded DNA Breaks. *Microbiol. Mol. Biol. Rev.* 72, 642–671. doi: 10.1128/MMBR.00020-08

Fonville, N. C., Bates, D., Hastings, P. J., Hanawalt, P. C., and Rosenberg, S. M. (2010). Role of RecA and the SOS response in thymineless death in *Escherichia coli*. *PLoS Genet.* 6:e1000865. doi: 10.1371/journal.pgen.1000865

Fonville, N., Vaksman, Z., DeNapoli, J., Hastings, P. J., and Rosenberg, S. M. (2011). Pathways of resistance to thymineless death in *Escherichia coli* and the function of UvrD. *Genetics* 189, 23–36. doi: 10.1534/genetics.111.130161

Guarino, E., Salguero, I., Jiménez-Sánchez, A., and Guzmán, E. C. (2007). Double-strand break generation under deoxyribonucleotide starvation in *Escherichia coli*. *J. Bacteriol.* 189, 5782–5786. doi: 10.1128/JB.00411-07

Hamilton, H. M, Wilson, R., Blythe, M., Nehring, R. B., Fonville, N. C., Louis, E. J., et al. (2013). Thymineless death is inhibited by CsrA in Escherichia coli lacking the SOS response. *DNA Repair (Amst)* 12, 993–999. doi: 10.1016/j.dnarep.2013.08.011

Hanawalt, P. C. (1963). Involvement of synthesis of RNA in *thymineless death*. *Nature* 198, 286. doi: 10.1038/198286a0

Hanawalt, P. C., Maaløe, O., Cummings, D. J., and Schaechter, M. (1961). The normal DNA replication cycle. II. *J. Mol. Biol.* 3, 156–165. doi: 10.1016/S0022-2836(61)80042-9

Jiménez-Sánchez, A., and Guzmán, E. C. (1988). Direct procedure for the determination of the number of replication forks and the reinitiation fraction in bacteria. *Comput. Appl. Biosci.* 4, 431–433.

Kouzminova, E. A., and Kuzminov, A. (2008). Patterns of chromosomal fragmentation due to uracil-DNA incorporation reveal a novel mechanism of replication-dependent double-stranded breaks. *Mol. Microbiol.* 68, 202–215. doi: 10.1111/j.1365-2958.2008.06149.x

Kuong, K. J., and Kuzminov, A. (2009). Cyanide, peroxide and nitric oxide formation in solutions of hydroxyurea causes cellular toxicity and may contribute to its therapeutic potency. *J. Mol. Biol.* 390, 845–862. doi: 10.1016/j.jmb.2009.05.038

Kuong, K. J., and Kuzminov, A. (2010). Stalled replication fork repair and misrepair during thymineless death in *Escherichia coli*. *Genes Cells* 15, 619–634. doi: 10.1111/j.1365-2443.2010.01405.x

Kuong, K. J., and Kuzminov, A. (2012). Disintegration of nascent replication bubbles during thymine starvation triggers RecA- and RecBCD-dependent replication origin destruction. *J. Biol. Chem.* 287, 23958–23970. doi: 10.1074/jbc.M112.359687

Maaloe, O., and Hanawalt, P. C. (1961). Thymine deficiency and the normal DNA replication cycle. *I. J. Mol. Biol.* 3, 144–155. doi: 10.1016/S0022-2836(61)80041-7

Martín, C. M. (2014). *Role of DNA Replication Initiation on the Lethality Caused by Thymine Starvation*. Ph.D. thesis, Universidad de Extremadura, Badajoz.

Martín, C. M., and Guzmán, E. C. (2011). DNA replication initiation as a key element in thymineless death. *DNA Repair (Amst.)* 10, 94–101. doi: 10.1016/j.dnarep.2010.10.005

Martín, C. M., Viguera, E., and Guzmán, E. (2014). Rifampicin suppresses thymineless death by blocking the transcription-dependent step of chromosome initiation. *DNA Repair* 18, 10–71. doi: 10.1016/j.dnarep.2014.03.004

Messer, W. (1972). Initiation of deoxyribonucleic acid replication in *Escherichia coli* B-r: chronology of events and transcriptional control of initiation. *J. Bacteriol.* 112, 7–12.

Michel, B., Boubakri, H, Baharoglu, Z, LeMasson, M, and Lestini, R. (2007). Recombination proteins and rescue of arrested replication forks. *DNA Repair* 6, 967–980. doi: 10.1016/j.dnarep.2007.02.016

Molina, F., Jiménez-Sánchez, A., Zyskind, J. W., and Guzmán, E. C. (1999). Chromosomal insertions localized around *oriC* affect the cell cycle in *Escherichia coli*. *Biochimie* 81, 811–818. doi: 10.1016/S0300-9084(99)00216-3

Morganroth, P. A., and Hanawalt, P. C. (2006). Role of DNA replication and repair in thymineless death in *Escherichia coli*. *J. Bacteriol.* 188, 5286–5288. doi: 10.1128/JB.00543-06

Morigen, Molina, F., and Skarstad, K. (2005). Deletion of the *datA* site does not affect once-per-cell-cycle timing but induces rifampin-resistant replication. *J. Bacteriol.* 187, 3913–3920. doi: 10.1128/JB.187.12.3913-3920.2005

Nakayama, H. (2005). *Escherichia coli* RecQ helicase: a player in *thymineless death*. *Mutat. Res.* 577, 228–236. doi: 10.1016/j.mrfmmm.2005.02.015

Nakayama, H., and Hanawalt, P. C. (1975). Sedimentation analysis of deoxyribonucleic acid from thymine-starved *Escherichia coli*. *J. Bacteriol.* 121, 537–547.

Nakayama, H., Nakayama, K., Nakayama, R., and Nakayama, Y. (1982). Recombination-deficient mutations and thymineless death in *Escherichia coli*: reciprocal effects of *recBC* and *recF* and indifference of *recA* mutations. *Can. J. Microbiol.* 28, 425–430. doi: 10.1139/m82-064

Nakayama, K., Kusano, K., Irino, N., and Nakayama, H. (1994). Thymine starvation-induced structural changes in *Escherichia coli* DNA. Detection by pulsed field gel electrophoresis and evidence for involvement of homologous recombination. *J. Mol. Biol.* 243, 611–620. doi: 10.1016/0022-2836(94)90036-1

Okagaki, H., Tsubota, Y., and Sibataini, A. (1960). Unbalanced growth and bacterial death in thymine-deficient and ultraviolet irradiated *Escherichia coli*. *J. Bacteriol.* 80, 762–771.

Pritchard, R. H., and Lark, K. G. (1964). Induction of replication by thymine starvation at the chromosome origin in *Escherichia coli*. *J. Mol. Biol.* 9, 288–307. doi: 10.1016/S0022-2836(64)80208-4

Sangurdekar, D. P., Hamann, B. L., Smirnov, D., Srienc, F., Hanawalt, P. C., and Khodursky, A. B. (2010). *Thymineless death* is associated with loss of essential genetic information from the replication origin. *Mol. Microbiol.* 75, 1455–1467. doi: 10.1111/j.1365-2958.2010.07072.x

Schwartzman, J. B., Martínez-Robles, M. L., López, V., Hernández, P., and Krimer, D. B. (2012). 2D gel and their third-dimensional potential. *Methods* 57, 170–178. doi: 10.1016/j.ymeth.2012.03.013

Sueoka, N., and Yoshikawa, H. (1965). The chromosome of Bacillus subtilis. I. Theory of marker frequency analysis. *Genetics* 52, 747–757.

Su'etsugu, M., Emoto, A., Fujimitsu, K., Keyamura, K., and Katayama, T. (2003). Transcriptional control for initiation of chromosomal replication in *Escherichia coli*: fluctuation of the level of origin transcription ensures timely initiation. *Genes Cells* 8, 731–745. doi: 10.1046/j.1365-2443.2003.00671.x

Zaritsky, A., and Pritchard, R. (1971). Replication time of the chromosome in *thymineles* mutants of *Escherichia coli*. *J. Mol. Biol.* 60, 65–74. doi: 10.1016/0022-2836(71)90447-5

10

The empirical fluctuation pattern of *E. coli* division control

Jacopo Grilli[1], *Clotilde Cadart*[2,3], *Gabriele Micali*[4,5], *Matteo Osella*[6,7] *and Marco Cosentino Lagomarsino*[8,9,10*]

[1] Santa Fe Institute, Santa Fe, NM, United States, [2] Centre National de la Recherche Scientifique, Institut Curie, PSL Research University, UMR 144, Paris, France, [3] Institut Pierre-Gilles de Gennes, PSL Research University, Paris, France, [4] Department of Environmental Microbiology, Eawag, Dübendorf, Switzerland, [5] Department of Environmental Systems Science, ETH Zurich, Zurich, Switzerland, [6] Physics Department, University of Turin, Turin, Italy, [7] Istituto Nazionale di Fisica Nucleare Sezione di Torino, Turin, Italy, [8] Sorbonne Université, Paris, France, [9] Centre National de la Recherche Scientifique, UMR 7238, Paris, France, [10] IFOM, FIRC Institute of Molecular Oncology, Milan, Italy

***Correspondence:**
Marco Cosentino Lagomarsino
marco.cosentino-lagomarsino@ifom.eu

In physics, it is customary to represent the fluctuations of a stochastic system at steady state in terms of linear response to small random perturbations. Previous work has shown that the same framework describes effectively the trade-off between cell-to-cell variability and correction in the control of cell division of single *E. coli* cells. However, previous analyses were motivated by specific models and limited to a subset of the measured variables. For example, most analyses neglected the role of growth rate variability. Here, we take a comprehensive approach and consider several sets of available data from both microcolonies and microfluidic devices in different growth conditions. We evaluate all the coupling coefficients between the three main measured variables (interdivision times, cell sizes and individual-cell growth rates). The linear-response framework correctly predicts consistency relations between *a priori* independent experimental measurements, which confirms its validity. Additionally, the couplings between the cell-specific growth rate and the other variables are typically non zero. Finally, we use the framework to detect signatures of mechanisms in experimental data involving growth rate fluctuations, finding that (1) noise-generating coupling between size and growth rate is a consequence of inter-generation growth rate correlations and (2) the correlation patterns agree with a near-adder model where the added size has a dependence on the single-cell growth rate. Our findings define relevant constraints that any theoretical description should reproduce, and will help future studies aiming to falsify some of the competing models of the cell cycle existing today in the literature.

Keywords: linear response theory, single-cell growth and division, fluctuation patterns, control of cell division, models, theoretical, data interpretation, statistical

1. INTRODUCTION

Today, dynamically tracked data of many dividing cells offer the possibility to analyze with great precision and detail the decision process leading to cell division (Osella et al., 2017). While such data are starting to be abundant, measurements of many single cells yield a complex tangle of correlation patterns between growth-related variables, which is often difficult to grasp. Consequently, new theoretical and data-analysis tools need to be developed in order to extract from such data the relevant information to understand the control of cell-cycle progression, cell division and their impact on cell proliferation.

Specifically, one can restrict the question to the control of cell size, where some general principles are emerging from empirical data. *E. coli* cells grow with a cell-specific growth rate, they are born with a cell-specific size, and they divide with a cell-specific cell-cycle time. A cell that deviates from average behavior in size or growth rate can correct by a compensatory deviation in doubling time. A "near-adder" control of cell division (Campos et al., 2014; Taheri-Araghi et al., 2015), whereby the differential size extension in a single cell cycle is independent of the initial size of the cell, is an effective principle governing cell division. Additionally, the distributions of interdivision times and cell sizes across conditions (Iyer-Biswas et al., 2014b; Kennard et al., 2016) show a clear link between average values and the variability of these variables from cell to cell, thus suggesting a "universal" shape of the size distribution. However, we still understand relatively little regarding how these simple principles emerge (Ho and Amir, 2015; Harris and Theriot, 2016; Wallden et al., 2016; Osella et al., 2017). For example, there is a debate on whether near-adder behavior could emerge from specific mechanisms or molecular circuits (Taheri-Araghi et al., 2015; Harris and Theriot, 2016), as a byproduct of chekpoint control of cell-cycle progression (Adiciptaningrum et al., 2015; Wallden et al., 2016), or as a consequence of external constraints or trade-offs (Osella et al., 2017).

Comprehensive and precise quantitative methods are necessary in both data analysis and theory to deal with the new data. Broadly, the open question is how much a mechanism can be isolated and specified with available data. The simplest theoretical framework for cell-cycle control describes the cell cycle as a discrete-time process, relating the measured variables (size, interdivision times, etc.) across generations (Amir, 2014; Campos et al., 2014; Taheri-Araghi et al., 2015; Grilli et al., 2017). We have recently shown the general existing equivalence between this simple discrete-time formalism to more detailed continuous-time models describing the division process by a division rate varying with cell state and time (Grilli et al., 2017). In the limit where deviations around the mean initial size (or interdivision time) are small (as first proposed in Amir, 2014), we have shown explicitly how the resulting "linear-response" framework describes a wide range of division control mechanisms and characterizes with remarkable precision the available experimental data. These results make the linear-response framework perfectly suitable to model cell-cycle control in *E. coli* given the available datasets.

However, studies are usually restricted to a subset of the measured variables. Most importantly, while single-cell growth rate clearly fluctuates (Wang et al., 2010), very few studies have addressed the consequences of these fluctuations on cell division control, and the studies that did started from very specific assumptions (Osella et al., 2014; Wallden et al., 2016; Logsdon et al., 2017). More specifically, previous studies adopting the linear-response framework typically neglected cell-to-cell variability in growth rates.

Here, we extend this linear-response framework to incorporate growth rate fluctuations, and we show how it can be used in general to evaluate exhaustively all the possible correlations and fluctuations in the data. We use this systematic approach to evaluate jointly all the homeostatic and noise-generating couplings measured in different experimental studies, and to connect correlation patterns with possible mechanisms underlying cell division.

2. DATA SETS

We tested our theoretical considerations and their implications on the analysis of empirical data with *E. coli* single-cell growth division data from Kiviet et al. (2014), Taheri-Araghi et al. (2015), Kennard et al. (2016), and Wallden et al. (2016). Scripts and formatted data are available with the authors. A detailed list of the growth conditions is available in the Supplementary Information of Cadart et al. (2017).

3. LINEAR-RESPONSE FRAMEWORK FOR CELL GROWTH-DIVISION FLUCTUATIONS

3.1. General Features of Fluctuations of Cell Size and Growth Rate

This section describes the main model assumptions and definitions. We assume exponential growth (Campos et al., 2014; Iyer-Biswas et al., 2014b; Osella et al., 2014; Taheri-Araghi et al., 2015) of cell size $V(t) = V(0)\exp(\alpha t)$, where α is the single-cell growth rate. A cell divides at size $V_f = V_0 \exp(\alpha \tau_d)$, where V_0 is the size at birth and τ_d is the division time. We neglect the fluctuations around binary fission or the process of filamentation and recovery (Osella et al., 2014), thus cell size at division V_f is equally partitioned between the two daughter cells in our description. Given the assumption of exponential growth, it is useful to introduce the logarithmic size $q(t) = \log(V(t)/V^*)$. With this notation, the exponential growth translates into a linear relationship $q(t) = q(0) + \alpha t$. In the definition of q, we introduced a size scale V^*, which in principle can be an arbitrary scale, to make the argument of the logarithm dimensionless. A particularly useful choice is to define $V^* = \langle V_0 \rangle$, so that, at least in the small noise limit, $\langle q(0) \rangle = 0$ independently of the condition (Grilli et al., 2017).

The distribution of cell size at birth gives a static picture of cell-size fluctuations. It is well accepted (Taheri-Araghi et al., 2015; Kennard et al., 2016; Grilli et al., 2017) that the distributions of size at birth (or, equivalently, at division) obtained for different strains and conditions collapse when rescaled by their average. This collapse is the consequence of the existence of a unique size-scale parameter of cell size control that varies across conditions (Kennard et al., 2016; Grilli et al., 2017). A consequence of the collapse is that the expectation of any function of the rescaled size at birth $V_0/\langle V_0 \rangle$ is constant across conditions. **Figure 1** shows that the variance of $\log V_0/\langle V_0 \rangle$ (which is by definition equal to σ_q^2) is constant across conditions, as predicted by the collapse of the size distribution.

Growth rates fluctuations are usually neglected. Contrarily to the well accepted collapse of the size distributions (Giometto et al., 2013; Taheri-Araghi et al., 2015; Kennard et al., 2016; Grilli et al., 2017), there is no consensus on whether the distribution of the single-cell growth rates collapse when rescaled by the average

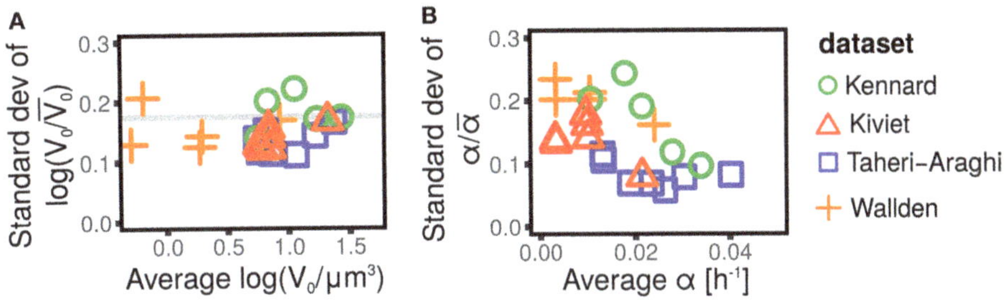

FIGURE 1 | Scaling of size fluctuations, quantified by log-initial size $\log V_0$, and single-cell growth rate α for different datasets (symbols and colors) and conditions. Each symbol represents an average over all single cells available in the dataset (defined by shape and color) for a specific strain in a specific condition. **(A)** Shows that the standard deviation of the rescaled log-initial size $\log(V_0/\langle V_0\rangle)$, is constant across conditions and experiments, and does not depend on the average log-initial size $\langle \log V_0\rangle$. This fact is a consequence of the collapse of initial size distribution (Taheri-Araghi et al., 2015; Kennard et al., 2016; Grilli et al., 2017), since the standard deviation $\log(V_0/\langle V_0\rangle)$ is a function of the coefficient of variation of the initial size only (Grilli et al., 2017). **(B)** Shows the standard deviation of the rescaled growth rate $\alpha/\langle\alpha\rangle$ (which corresponds to the coefficient of variation of α). This quantity is approximately constant for some datasets (e.g., Taheri-Araghi et al., 2015, blue squares), suggesting a collapse of the growth rate distribution (Taheri-Araghi et al., 2015), while it shows a decreasing trend in others (e.g., Kennard et al., 2016, green circles).

growth rate, with some studies reporting the collapse (Taheri-Araghi et al., 2015) and others showing the opposite (Kennard et al., 2016). **Figure 1** shows that the coefficient of variation of the growth rate appears to be constant, independently of the condition and the average growth rate (consistently with the collapse hypothesis) for some datasets (Taheri-Araghi et al., 2015), while it shows a decreasing trend for others (Kiviet et al., 2014; Kennard et al., 2016; Wallden et al., 2016). A possible explanation of these differences could lay in the growth mode of the different experiments (microcolonies in agar or polyacrylamide, wide microfluidics channels, single-file microfluidics channels, etc). Theoretical efforts to provide a rationale of growth rate distributions are available (Iyer-Biswas et al., 2014b; Pugatch, 2015; De Martino et al., 2016; De Martino et al., 2017).

Independently of the controversy about the collapse of the growth rate distributions, it is clear from **Figure 1**, that the fluctuations of the growth rates are not negligible. The coefficients of variation are in the range 0.1–0.3, which is comparable with the typical fluctuations of the sizes at birth, whose CVs are of the order of 0.2 (Amir, 2014; Grilli et al., 2017).

3.2. Linear-Response Framework in Presence of Growth Rate Fluctuations

Since the coefficient of variation of both sizes and growth rates are around 0.2, it is reasonable to assume that the size-control mechanism can be expanded around the average size and average growth rate (Amir, 2014; Grilli et al., 2017). This maps the problem into a discrete-time linear response framework. This section generalizes the usual linear response approach to include single-cell growth rates fluctuations, with the goal of disentangling how size and growth rate fluctuations are connected.

We use a discrete-time description, focusing on a single-cell at a given generation i. This cell has initial size $V_0^{(i)}$, which corresponds to a log-size at birth $q_0^{(i)}$, a growth rate $\alpha^{(i)}$ and divides at a log-size $q_f^{(i)}$ after a time $\tau_d^{(i)}$. These variables are connected by the relation

$$q_f^{(i)} = q_0^{(i)} + \alpha^{(i)}\tau_d^{(i)} \,. \tag{1}$$

We also indicate the net logarithmic multiplicative growth (sometimes referred to as elongation) by

$$G^{(i)} = \alpha^{(i)}\tau_d^{(i)} = \log(V_f^{(i)}/V_b^{(i)}) = q_f^{(i)} - q_0^{(i)} \,.$$

Here, $\alpha^{(i)}$ is a random variable with mean $\langle\alpha\rangle$ (which depends on the condition) and variance σ_α^2. We assume that individual cells maintain the same growth rate for a whole cell cycle (this approximation corresponds to how growth rate distributions are often evaluated in empirical data). The growth rates of subsequent generations are drawn from a common (Gaussian) distribution, but may be correlated across generations, with Pearson correlation coefficient ρ. An alternative approach to model growth fluctuations would consist in modeling the stochastic fluctuations of $\alpha^{(i)}$ in continuous time (Iyer-Biswas et al., 2014a,b). The more minimalistic discrete-time approach that we employ can be seen as a coarse-graining of an appropriate continuous-time model. We also note that the discrete-time modeling approach is equivalent to a continuous-time description for single-exponential growth (Grilli et al., 2017; Ho et al., 2018), but in presence of a growth rate fluctuating in continuous time, this equivalence does not hold anymore. For the small fluctuations observed in the data, the two approaches are qualitatively equivalent and likely quantitatively not distinguishable given the available statistical power.

We introduce the variable fluctuations

$$\begin{aligned}
\delta G^{(i)} &:= G^{(i)} - \langle G\rangle \\
\delta\tau_d^{(i)} &:= \tau_d^{(i)} - \langle\tau\rangle \\
\delta\alpha^{(i)} &:= \alpha^{(i)} - \langle\alpha\rangle \\
\delta q_0^{(i)} &:= q_0^{(i)} - \langle q_0\rangle \,.
\end{aligned} \tag{2}$$

Under the above assumptions, the evolution equations for the fluctuations of the three main variables can be written as

$$\begin{aligned}\frac{\delta G^{(i)}}{\sigma_G} &= -\lambda_{Gq}\frac{\delta q_0^{(i)}}{\sigma_q} - \lambda_{G\alpha}\frac{\delta\alpha^{(i)}}{\sigma_\alpha} + \nu_G^{(i)} \\ \frac{\delta\tau_d^{(i)}}{\sigma_\tau} &= -\lambda_{\tau q}\frac{\delta q_0^{(i)}}{\sigma_q} - \lambda_{\tau\alpha}\frac{\delta\alpha^{(i)}}{\sigma_\alpha} + \nu_\tau^{(i)},\end{aligned} \quad (3)$$

where λ_{XY} are linear coupling parameters (analogous to "susceptibilities") between measured variables. Such linear couplings are central in comparing this theory to experimental data. $\nu_\tau^{(i)}$ and $\nu_G^{(i)}$ are zero-mean random variables representing cell-to-cell variability ("noise"). Previous approaches neglecting cell size fluctuations assume $\delta\alpha = 0$. In order to be solved, Equation 3 requires the specification of how the fluctuations $\delta\alpha^{(i)}$ are related to the fluctuations of logarithmic size $\delta q_0^{(i)}$. We assume that generally the individual growth rate may depend on the initial size, i.e., within a linear-response framework,

$$\frac{\delta\alpha^{(i)}}{\sigma_\alpha} = -\lambda_{\alpha q}\frac{\delta q_0^{(i)}}{\sigma_q} + \nu_\alpha^{(i)}. \quad (4)$$

The values of these coupling constants can be associated to specific mechanisms of size control (Amir, 2014; Grilli et al., 2017). For instance a timer corresponds to $\lambda_{\tau q} = 0$ and $\lambda_{\tau\alpha} = 0$, while a sizer mechanism implies $\lambda_{Gq}\sigma_G/\sigma_q = 1$ and $\lambda_{G\alpha} = 0$. A standard adder corresponds instead to $\lambda_{Gq}\sigma_G/\sigma_q = 1/2$ (Amir, 2014; Grilli et al., 2017).

This general framework can be used to evaluate the direct couplings between fluctuations of birth size, growth rate and division time. In particular, all the couplings λs can be evaluated from the Pearson correlations of these variables or from the slope of conditional averages. The two methods are equivalent if the slopes are estimated using least-squares method. Indeed, assuming a relationship $y = \beta * x + \text{noise}$, and performing a linear regression, the least-square estimate of beta is $\beta = \text{cov}(x, y)/\text{var}(x)$. Any discrepancy of the values of the coupling parameters obtained in these two different ways should be considered as a signature of the violation of the linear-response regime (Cadart et al., 2017). For instance, by multiplying both sides of Equation (4) by $\delta q_0/\sigma_q$ and averaging over the fluctuations, it is easy to obtain (see Supplementary Information)

$$c_{\alpha q} = -\lambda_{\alpha q}, \quad (5)$$

where $c_{\alpha q}$ is the Pearson correlation between the growth rate α and the log-size q. Similarly, multiplying by $\delta q/\sigma_q$ Equation (3) and averaging gives

$$c_{\tau q} = -\lambda_{\tau q} - \lambda_{\tau\alpha}c_{\alpha q} = -\lambda_{\tau q} + \lambda_{\tau\alpha}\lambda_{\alpha q}. \quad (6)$$

Finally, multiplying the same equation by $\delta\alpha/\sigma_\alpha$ and averaging over fluctuations gives

$$c_{\tau\alpha} = -\lambda_{\tau q}c_{\alpha q} - \lambda_{\tau\alpha} = -\lambda_{\tau\alpha} + \lambda_{\tau q}\lambda_{\alpha q}. \quad (7)$$

A detailed comparison between the two ways to evaluate the couplings in empirical data is shown in the Supplementary Information of Cadart et al. (2017). The very small discrepancies found reinforce the ideas that the linear-response framework is effective.

The last two equations clearly show that the presence of a non-zero coupling between single-cell growth rate and birth size induces a non-trivial dependence of the correlations from all the coupling parameters. In other words, when the coupling between division time and cell size is estimated uniquely from $c_{\tau q}$ (Amir, 2014; Grilli et al., 2017), one is measuring a combination of two different effects: the direct coupling between division time and size at birth and an indirect coupling mediated by growth rate fluctuations.

Direct and indirect effects can be disentangled by solving Equations (6) and (7) to calculate $\lambda_{\tau q}$ and $\lambda_{\tau\alpha}$ in terms of measurable covariances,

$$\lambda_{\tau q} = -\frac{c_{\tau q} - c_{\alpha q}c_{\tau\alpha}}{1 - c_{\alpha q}^2}, \quad (8)$$

and

$$\lambda_{\tau\alpha} = -\frac{c_{\tau\alpha} - c_{\alpha q}c_{\tau q}}{1 - c_{\alpha q}^2}. \quad (9)$$

Figure 2 illustrates this procedure over one example dataset.

The main assumption of the above modeling framework is that all the relations between the variables are linear. Deviations from linear response are not captured, but in the data these are fairly small (see **Figure 2**). Additionally, it is difficult to evaluate precisely the non-linear behavior from data because the statistics decreases radically for large fluctuations where such non-linear trends are stronger. Given the statistical power available from existing datasets, it is typically not possible to discriminate non-linear terms from stochastic fluctuations (Grilli et al., 2017).

It has previously been noted that measured Pearson correlations between observed variables in such experiments are very sensitive to (experimental and biological) noise (Eun et al., 2018), so that it is not simple to reconstruct mechanisms from correlations. Our approach, which directly tackles this issue, has two main advantages. First, it subtracts the contributions of indirect correlations and only measures direct couplings. Second, the coupling constants are insensitive to noise, because they are defined as linear slopes of conditional averages, and therefore robust to experimental noise. This is supported by the analysis of Cadart et al. (2017) mentioned above, verifying the equivalence between theory coupling constants measured directly and their expressions in terms of covariances. Indeed, this equivalence not only supports the validity of the linear regime, but also reinforces the idea that scatter in the data (due e.g., to experimental noise) does not bias too much the coupling constants, when they are estimated from the covariances.

We also note that the choice of parameters in Equations (3) and (4) is not unique. There are several alternative (equivalent) choices of fluctuating variables, linked by the condition that $G =$

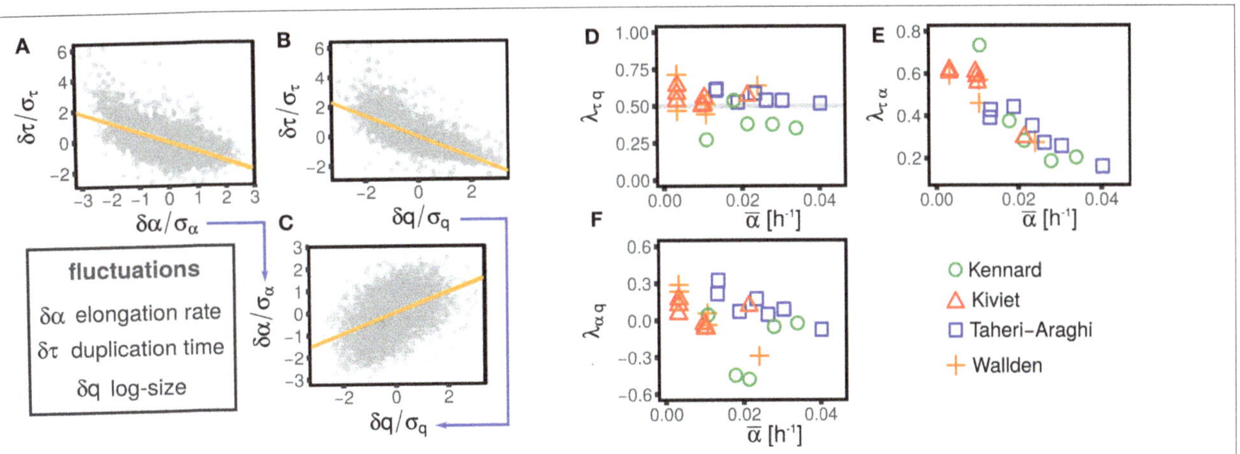

FIGURE 2 | Illustration of the linear-response framework of size control, including growth rate fluctuations. **(A,B)** Show the dependency of the interdivision time fluctuations from fluctuations of logarithmic initial size and growth rate for one dataset (Wallden et al., 2016 intermediate growth rate). Since growth rate and size fluctuations are not independent **(C)**, the slopes observed in **(A,B)** are determined by a combination of the coupling between size fluctuation and interdivision time (with strength $\lambda_{\tau q}$, see Equation 2) and the effect of growth rate fluctuation on interdivision time (with strength $\lambda_{\tau\alpha}$, see Equation 2). The method described in the text disentangles direct and indirect effect on the slopes of **(A–C)** to obtain the direct couplings between τ, α, and q, which are shown in **(D–F)**, for different datasets (symbols and colors) and conditions. Interestingly, all the possible couplings are non-zero. Current models do not describe these nontrivial correlations, which poses a challenge to future models.

$\alpha\tau$: α, τ, q; α, G, q; G, τ, q. Each triplet carries a set of linear-response coupling constants between pairs of variables that can be mapped to the ones defined above.

4. RESULTS

4.1. All the Possible Couplings Between Measured Variables Are Non-zero in Empirical Datasets

One might expect that not all the possible couplings are different from zero, supporting the simplifying assumptions made by most studies. **Figure 2** shows instead that all the independent couplings $\lambda_{\tau q}$, $\lambda_{\tau\alpha}$, and $\lambda_{\alpha q}$ are in fact not negligible.

The existence of nonzero couplings between division time and growth rate $\lambda_{\tau\alpha}$ or, equivalently, elongation and growth rate $\lambda_{G\alpha}$ (see next section and **Figure 3**) suggests that cell size control depends in a non-trivial way on growth rate fluctuations. Moreover, the existence of a coupling between growth rates and cell size (measured by the coefficient $\lambda_{\alpha q}$), affects the observed correlation between division time and size at birth $c_{\tau q}$ (see Equation 6), or, equivalently, the slope of the size-growth plot c_{Gq} (Skotheim, 2013), which is normally used to evaluate the strength of the size control. Different values of these correlations are associated with different strengths of cell size control (Skotheim, 2013; Osella et al., 2014), and, more indirectly, to different cell size control mechanisms (Amir, 2014; Grilli et al., 2017). The presence of non-zero couplings between interdivision time and growth rate and between growth rate and size effectively reduces or increases the observed strength of homeostasis, and might be a signature of the mechanisms effecting such control.

4.2. The Linear-Response Framework Allows to Define Consistency Relations

As discussed above, the framework provides equivalent alternative descriptions for the interdivision time τ or the net growth G as state variables. These two parameter settings are equivalent under the linear-control assumption. This section deals with the mapping between these parametrizations, and reports consistency tests defined by the expected relations using empirical data.

The consistency relations can be obtained from the relation $G^{(i)} = \alpha^{(i)}\tau^{(i)}$, which, under the linearization assumption reduces to (see Supplementary Information)

$$\delta G^{(i)} = \langle\tau\rangle\delta\alpha^{(i)} + \langle\alpha\rangle\delta\tau^{(i)} \, . \qquad (10)$$

Using this relation together with Equation (3), one obtains (see Supplementary Information)

$$\lambda_{Gq} = \langle\alpha\rangle\lambda_{\tau q}\frac{\sigma_\tau}{\sigma_G} \qquad (11)$$

and

$$\lambda_{G\alpha} = \langle\alpha\rangle\lambda_{\tau\alpha}\frac{\sigma_\tau}{\sigma_G} + \langle\tau\rangle\frac{\sigma_\alpha}{\sigma_G} \, . \qquad (12)$$

An important consequence of Equation (12) is that a nonzero $\lambda_{G\alpha}$ (or $\lambda_{\tau\alpha}$) could appear even if the corresponding coupling $\lambda_{\tau\alpha}$ (or $\lambda_{G\alpha}$) was null. For instance, if the coupling $\lambda_{\tau\alpha}$ is negligible, indicating that the interdivision time is independent from the single-cell growth rate, one would still observe a nonzero $\lambda_{G\alpha}$. In fact, in that case, the division time would be independent of the growth rate and therefore, in the same amount of time, cells growing faster would grow more than slow-growing cells. Interestingly, this is not the case. Both $\lambda_{\tau\alpha}$ (see **Figure 2**) and $\lambda_{G\alpha}$

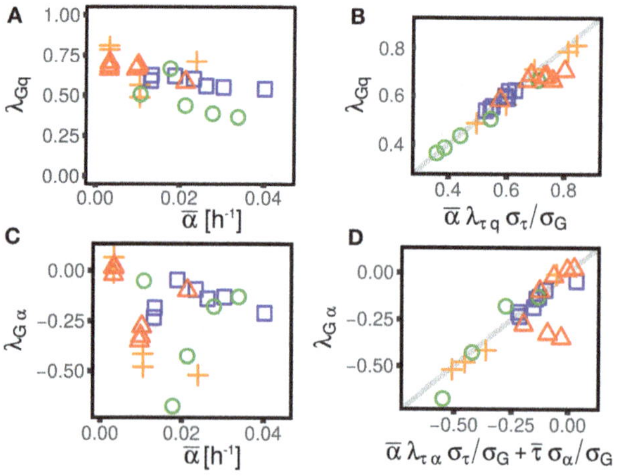

FIGURE 3 | Alternative parameterizations of cell-size control linear response are consistent across different experimental datasets and conditions (symbols and colors are the same as in **Figure 1**). One can parametrize the effect of size and growth rate fluctuations on interdivision time, or, alternatively, their effect on the final size (or the elongation G). **(A,C)** Show the effect of log-size fluctuation on elongation (with strength λ_{Gq}, see Equation 3) and the effect of growth rate fluctuation on interdivision time (with strength $\lambda_{G\alpha}$). The linear framework predicts consistency relations between the alternative parameterizations **(B,D)**, which are verified across datasets and conditions in the plots in the right panel.

are in fact different from zero (see **Figure 2**), suggesting that some specific mechanism must be at play in determining their value.

Figure 3 shows that these consistency relations are verified in the data. This result shows that the generalization of the linear-response framework to include growth rate fluctuations can correctly describe the data. As previously discussed, small deviations from these relations could be due to the presence of non-linearities that cannot be captured by the theoretical framework. However, non-linearities can be appreciated only in presence of large fluctuations (Grilli et al., 2017). Also note that while these deviations could be the signature of not-yet-understood biological mechanisms, they could also come from measurement, segmentation or tracking errors associated to specific experimental and image-analysis protocols.

4.3. Growth Rate Fluctuations Can Confound Standard Tests of Division Control Mechanisms, the Case of "Grower" vs. "Timer"

A standard simple analysis to understand what kind of cell division control is in place is based on size-growth plots (Skotheim, 2013) and more generally on measures of correlation between cell elongation and initial size. Since growth rate fluctuations are typically neglected, a value of $\lambda_{Gq} = 0$, i.e., no correlation between elongation and size at birth (**Figure 4A**), is usually interpreted as the signature of a "timer." However, in presence of growth rate fluctuations the value of this coupling parameter alone is not enough to pinpoint a single mechanism. In fact, $\lambda_{Gq} = 0$ can be the actual result of a timer, i.e., $\lambda_{\tau q} = 0$ which then requires that $\lambda_{\tau\alpha} = 0$ (**Figure 4B**). However, the relevant variable that is actually uncoupled to size could be the net growth (elongation) itself, rather than the doubling time and the growth rate. In this case, one has a "grower" (**Figure 4C**), for which again $\lambda_{Gq} = 0$ and $\lambda_{G\alpha} = 0$. For a grower, $\lambda_{\tau q} = 0$, but $\lambda_{\tau\alpha} = -\langle\tau\rangle\sigma_\alpha/(\langle\alpha\rangle\sigma_\tau)$. Importantly, if $\lambda_{\alpha q} \neq 0$, then the correlation between τ and q ($c_{\tau q}$) is also different from zero in a grower. This illustrative example emphasizes the importance of keeping into account growth rate fluctuations in order to correctly interpret empirical correlations patterns. It also shows how the linear-response framework proposed here can be used to delineate the correct interpretation. A more general decomposition of contribution of growth vs. timing coupling to size control is discussed in Cadart et al. (2017).

We now proceed to describe two situations where our analysis points directly to specific mechanisms contributing to the complexity of the observed correlation patterns.

4.4. Mother-Daughter Growth Rate Correlations Explain the Prevalent Negative Coupling Between Growth Rates and Cell Size Observed in Data

Since the framework describes exhaustively the measured correlations in the datasets, it can be used to explore signatures of mechanisms by which division is coupled to size. This section provides a rationale for the emergence of a non-zero coupling between single-cell growth rate and size at birth $\lambda_{\alpha q}$ observed in most datasets. In particular, we show that a negative $\lambda_{\alpha q}$ may emerge from the presence of a correlation between the growth rates across generations.

Figure 5 shows the Pearson correlation ρ between the mother and the daughter(s) growth rates in several data sets. We computed the correlation using both the daughters (when available in the dataset). Some previous studies have reported, or assumed, small mother-daughter correlations in growth rate (Wang et al., 2010; Eun et al., 2018), while others have found larger ones (Wallden et al., 2016). An important result of the present analysis is that this correlation is never negligible in the analyzed data sets and appears very conserved across conditions and experimental setup. Specifically, all the datasets considered here show a significantly positive correlation, which does not seem to depend strongly on the condition and takes values around 0.3. Variations of single-cell growth rates can be both due to an external origin (e.g., local variation of nutrients in an agar experiment) or inherent to the mechanisms of growth (e.g., from fluctuations in gene expression Iyer-Biswas et al., 2014a). In both cases, one may expect the presence of positive correlation across generations.

We can model mother-daughter correlations in growth rate as a linear constraint on the single-cell growth rates over generations

$$\delta\alpha^{(i+1)} = \rho\delta\alpha^{(i)} + \sigma_\alpha \nu_\alpha^{(i)} . \tag{13}$$

This equation substitutes Equation (4), and must be considered together with Equation 3. The question we address is whether

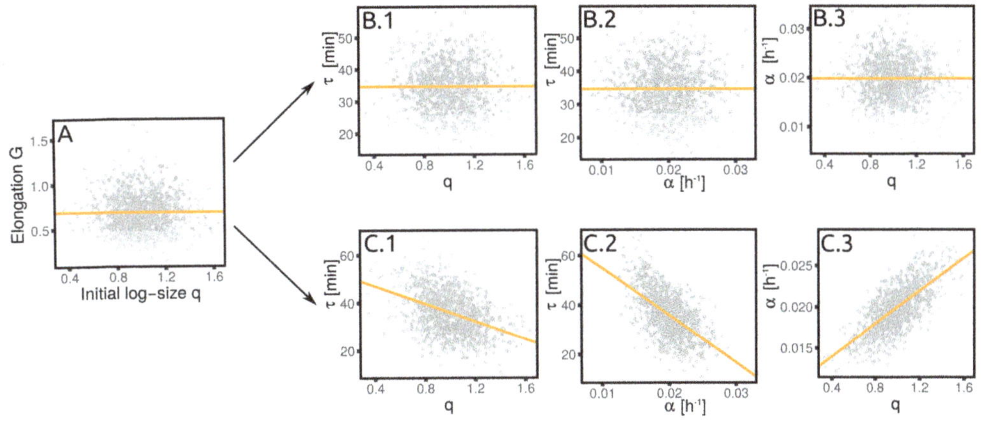

FIGURE 4 | A grower is not a timer. The absence of a correlation between elongation and initial size **(A)** does not necessarily imply a timer **(B)**, where the interdivision time is independent of the initial size. **(C)** Shows the alternative scenario (the grower), where the growth rate depends on the initial size. This correlation, together with the independence of the elongation from the growth rate, implies a dependence of the interdivision time on both initial size and growth rate. Both panels are obtained from numerical simulations of the corresponding linearized model with $\sigma_\alpha/\langle\alpha\rangle = 0.2$ and $\sigma_q^2 = 0.2$.

inter-generational correlations give rise to a non-zero coupling $\lambda_{\alpha q}$ between single-cell growth rate and size.

Since we are considering a fully linear framework, a non-zero coupling $\lambda_{\alpha q}$ can be directly mapped into a non-zero correlation between single-cell growth rate and size $c_{\alpha q}$ via Equation (5). We find that a non-zero correlation $c_{\alpha q}$ emerges in presence of a non-zero correlation of growth rate across generation and a non-zero coupling $\lambda_{G\alpha}$. In the Supplementary Information we obtain the relation

$$\lambda_{\alpha q} = -c_{\alpha q} = \rho \frac{\lambda_{G\alpha}}{1 - \rho\left(1 - \lambda_{Gq}\sqrt{\frac{\sigma_G^2}{\sigma_q^2}}\right)} \sqrt{\frac{\sigma_G^2}{\sigma_q^2}} . \quad (14)$$

From the above equation we observe that, for positive ρ and negative $\lambda_{G\alpha}$ (the case of empirical data) only negative couplings $\lambda_{\alpha q}$ can emerge. Hence, mother-daughter growth rate correlations provide a rationale for this widespread negative coupling.

Figure 5 shows that the prediction of Equation (14) quantitatively reproduces the values of $\lambda_{\alpha q}$ for the datasets that display negative values of this coupling. On the other hand, some datasets have weakly positive couplings $\lambda_{\alpha q}$, which cannot be reproduced via this mechanism. This suggests that some other unknown mechanisms might play a role in coupling growth rates and sizes.

4.5. An Adder Model Captures the Experimental Correlation Patterns Only If Added Size Depends on Single-Cell Growth Rate

This section explores how the observed negative values for the coupling $\lambda_{G\alpha}$ emerge from a simple extension of the adder model that explicitly includes growth rate fluctuations. According to the adder model (Amir, 2014; Taheri-Araghi et al., 2015), the size at division V_f is given by

$$V_f = (V_0 + \Delta)e^{\nu} , \quad (15)$$

where Δ is the added size and ν is a noise term. We also know that the average size $\langle V_0 \rangle$ has an exponential dependence on the average growth rate $\langle \alpha \rangle$, which is often referred to as Schaechter's law (Schaechter et al., 1958; Taheri-Araghi et al., 2015; Kennard et al., 2016; Si et al., 2017). Since for steadily dividing cells the average size at birth is equal to the added size, the exponential dependence of the average size on the average growth rate also implies that the average added size must have an exponential dependence on the average growth rate.

Equation (15) does not take into account fluctuations of the growth rate, and there are several possibilities to extend it. Schaechter's law can be seen as a constraint, as the added size averaged over fluctuations of the growth rate has to scale nearly exponentially with the average growth rate. Two possible extreme scenarios can be disentangled with the data. The first scenario assumes that in a given growth condition the added size does not depend on the fluctuation of the growth rate, i.e.,

$$\Delta = S_0 \exp(\langle \alpha \rangle T) . \quad (16)$$

In such case, two cells with different individual growth rates, but growing in the same condition, will add (on average) the same size. In the opposite extreme scenario, the added size depends on the single-cell growth rate, following Schaecter's law *even for small fluctuations*

$$\Delta = S_0 \exp(\alpha T) . \quad (17)$$

In this second scenario, cells in the same growth conditions growing at different growth rates will add on average different sizes. Conversely, cells growing under different conditions but having the same growth rate due to single-cell fluctuations will

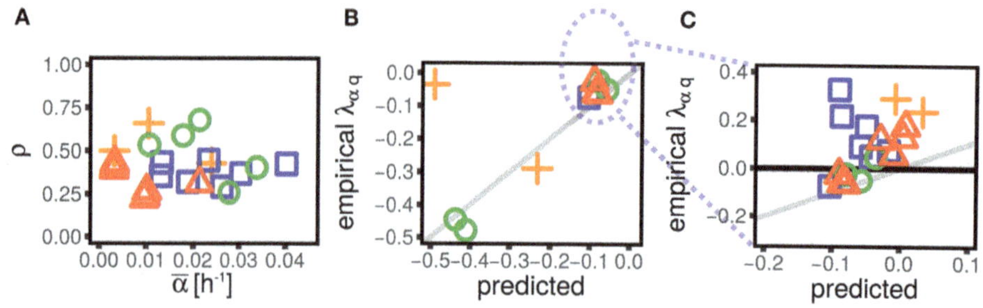

FIGURE 5 | Mother-daughter correlations in growth rate induce negative values of $\lambda_{\alpha q}$. **(A)** Shows the correlation between mother and daughter growth rate across datasets (symbols and colors, see **Figure 1**) and conditions. All the datasets display a significant positive correlation. Panel shows the value of $\lambda_{\alpha q}$ predicted from our theory assuming the empirical mother-daughter growth rate correlations (see Equation 14), compared to the empirical ones for negative values of $\lambda_{\alpha q}$. The assumed mechanism correctly predicts negative values of $\lambda_{\alpha q}$, but the prediction cannot capture the datasets with positive values of $\lambda_{\alpha q}$, which are shown in **(C)** (not visible in **B**).

add (on average) the same size. Since fluctuations of α are small, in both cases the average added size $\langle \Delta \rangle$ will scale exponentially with the average growth rate.

This question was approached in Kennard et al. (2016), but the linear-response theoretical framework gives us the tools to reformulate it more precisely. **Figure 6** shows how we proceed to test the two scenarios. For each dataset we estimated the parameter T (which can be related to the duration of the replication/segregation ("C+D") period Donachie, 1968; Zheng et al., 2016; Si et al., 2017) from $\langle \log \Delta \rangle$ averaged across cells within each condition. Note that for both Equations (16) and (17), the slope of $\langle \log \Delta \rangle$ *vs* $\langle \alpha \rangle$ is T. For each condition, one can then compare the relation between the (logarithmic) added size and the individual growth rate. Under the first scenario, the relation should show no dependency (a flat slope). Under the second scenario, one should instead expect a linear relation with slope T (where T is the mean timing from replication initiation to division, assumed to be nearly constant across conditions).

Under the assumption that fluctuations are small, one can linearize the models and compare the coupling coefficients. In the first case, one obtains $\lambda_{G\alpha} = 0$, while in the second case a non-trivial coupling arises with value

$$\lambda_{G\alpha} = \frac{\sigma_\alpha}{\sigma_G} - \frac{T}{2}. \quad (18)$$

Figure 6 compares this prediction with the empirical values of this coupling parameter measured across datasets, obtaining a remarkable agreement. It is important to stress that the parameter T, which in turn determines the value of $\lambda_{G\alpha}$, has been measured only using the dependence of average added size on the average growth rate across different conditions, without considering single-cell data. This procedure assumes that the parameters S_0 and T are constant within each dataset across different conditions (Donachie, 1968; Si et al., 2017). Variation around Schaecter's law, which could arise for many reasons (Zheng et al., 2016; Si et al., 2017), can affect the values of these parameters. Precise estimates, or measurements of T can improve the prediction of Equation (18). Moreover, the variability around

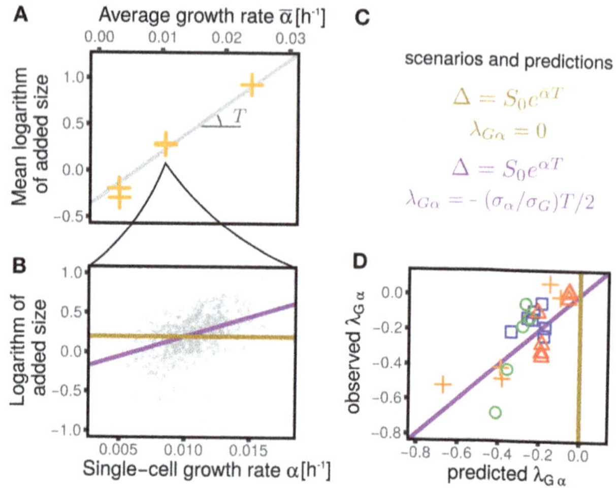

FIGURE 6 | Dependency of added size on single-cell growth rates and effect on the values of $\lambda_{G\alpha}$. **(A)** Shows that the average logarithmic added size is linear when plotted against the average growth rate (for all the conditions from Wallden et al., 2016). We consider two extreme scenarios for the logaritmic added size of individual cells: it could depend only on the average growth rate (gold line) or it could linearly grow with the single-cell growth rate (purple line), with the same slope observed in **(A)**. **(B)** Shows the results of the two scenarios for one condition (intermediate growth rate data set from Wallden et al., 2016). These two scenarios translate into two different predictions for the value of $\lambda_{G\alpha}$, which are reported in panel **(C)** and tested in panel **(D)**. The non-zero values of $\lambda_{G\alpha}$ are therefore a consequences of the dependence of the added size on the the single-cell growth rate.

the adder mechanism (visible in the values of λ_{Gq} plotted in **Figure 3**) also affects the predictions.

5. DISCUSSION AND CONCLUSIONS

With a long list of recent published studies, the current literature on cell division control remains fragmented in terms of conceptual tools and conclusions on the data (Osella et al., 2017).

This work proposes an extension of the linear-response framework to include growth rate fluctuations, and at the same time provides a comparative meta-analysis of different datasets.

As such, we hope it could be useful to the community. The results lead us to propose the linear-response model as a basic common data-analysis framework for the recent (and future) growth division data obtained on bacteria.

Analyzing the experimental data, we found two important results. First, the linear response framework including growth rate fluctuations provides a remarkably good description of the data. Couplings are in all cases only moderately nonlinear (Grilli et al., 2017), and all the predicted self-consistency relations are verified to a satisfactory degree. This carries the important lesson that models where nonlinear couplings are involved might be true, but they would be hard to test with the currently available data. Second, we show that the linear correlation patterns store a wealth of information that is not yet fully appreciated. In particular, single-cell growth rates seem to have a complex effect on cell cycle progression and cell-division control.

The study of the linear coupling parameters between growth rate and size reveals an unexpected variability across experiments. This variability depends on the growth conditions and on the measured mother-daughter correlations in growth rate. Thus, the effect of growth rate fluctuations on cell size varies from one experiment to the other and appears, in some cases, to contribute to size correction. In contrast, the near-adder correlation for the net growth G and the dependency of interdivision time on initial size are robustly observed in all datasets, although quantitatively, they also vary across conditions.

Since all these linear-response effects are clearly testable, we believe that future efforts should be funneled into explaining these patterns comprehensively. Each particular model for a size control mechanism carries a set of constraints for the linear-coupling parameters, which are easily testable in the data. Such testing involves all the measured variables and correlations. Hence, our framework provides a comprehensive and stringent tool to test proposed mechanisms.

The question of specifying the mechanisms for cell division and cell-cycle progression remains largely open, and therefore could benefit from such a systematic tool. One main limitation of current studies is that they discard relevant information and only consider a subset of correlation patterns. Conversely, linear-coupling parameters allow to control all the measured correlations at once. In this work, we provided a proof of principle that this plan of action can be effective by studying two simple examples: (i) the role of mother-daughter correlations in size control, and (ii) the role of growth rate fluctuation in determining the added size.

In particular, we find that growth rate fluctuations are not consistent with an adder model where the added size does not depend on the single-cell growth rate. This could be an important clue on the mechanism underlying the adder behavior and could help selecting one of the currently proposed models (Ho and Amir, 2015; Taheri-Araghi et al., 2015; Harris and Theriot, 2016; Wallden et al., 2016).

Finally, we address the coherence between different experimental data sets. The discrepancies we found could be due to biological factors, but also to possible sources of experimental noise and systematic errors in the data. Current experimental and data-analysis pipelines are different in both the segmentation and tracking steps. This problem is largely disregarded and in particular we know very little about possible induced biases in the measured correlations, which would lead to different conclusions on the mechanistic aspects. As end-users of the data, we propose our pipeline, i.e., comparing all measurable linear correlations as a useful testing ground for different datasets. In our analysis, we have noticed different trends depending on the experimental laboratory, and growth device, as well as outlier points where possibly the growth conditions were not steady or the analysis had problems. Interestingly, while some observations are robust, others are more erratic across datasets. Beyond this, we can just encourage the experimentalists to adopt common protocols and shared comparison pipelines in future studies. It would be sufficient to define a well defined "reference experiment" with fixed strain, growth device, and growth conditions that each data set needs to share, in order to make precise comparisons possible.

AUTHOR CONTRIBUTIONS

MCL and JG planned the study. JG, MCL, and MO developed the modeling framework, JG, CC, and GM, analyzed data. JG and MCL outlined and wrote the paper. MO, CC, and GM contributed with scientific discussion and critical reading.

ACKNOWLEDGMENTS

This work was supported by the International Human Frontier Science Program Organization, grant RGY0070/2014. We acknowledge useful discussions with A. Amir. MO is supported by the Departments of Excellence Grant awarded by the Italian Ministry of Education, University and Research (MIUR) (L. 232/2016). GM was supported by grant no. 31003A 169978 from the Swiss National Science Foundation to Martin Ackermann. JG was supported by an Omidyar Postdoctoral Fellowship at the Santa Fe Institute.

REFERENCES

Adiciptaningrum, A., Osella, M., Moolman, M. C., Cosentino Lagomarsino, M., and Tans, S. J. (2015). Stochasticity and homeostasis in the e. coli replication and division cycle. *Sci. Rep.* 5:18261. doi: 10.1038/srep18261

Amir, A. (2014). Cell size regulation in bacteria. *Phys. Rev. Lett.* 112:208102. doi: 10.1103/PhysRevLett.112.208102

Cadart, C., Monnier, S., Grilli, J., Attia, R., Terriac, E., Baum, B., et al. (2017). Size control in mammalian cells involves modulation of both growth rate and cell cycle duration. *bioRxiv 152728*. doi: 10.1101/152728

Campos, M., Surovtsev, I. V., Kato, S., Paintdakhi, A., Beltran, B., Ebmeier, S. E., et al. (2014). A constant size extension drives bacterial cell size homeostasis. *Cell* 159, 1433–1446. doi: 10.1016/j.cell.2014.11.022

De Martino, D., Capuani, F., and De Martino, A. (2016). Growth against entropy in bacterial metabolism: the phenotypic trade-off behind empirical growth rate distributions in *E. coli*. *Phys. Biol.* 13:036005. doi: 10.1088/1478-3975/13/3/036005

De Martino, D., Capuani, F., and De Martino, A. (2017). Quantifying the entropic cost of cellular growth control. *Phys. Rev. E* 96:010401. doi: 10.1103/PhysRevE.96.010401

Donachie, W. D. (1968). Relationship between cell size and time of initiation of DNA replication. *Nature* 219, 1077–1079. doi: 10.1038/2191077a0

Eun, Y.-J., Ho, P.-Y., Kim, M., LaRussa, S., Robert, L., Renner, L. D., et al. (2018). Archaeal cells share common size control with bacteria despite noisier growth and division. *Nat. Microbiol.* 3, 148–154. doi: 10.1038/s41564-017-0082-6

Giometto, A., Altermatt, F., Carrara, F., Maritan, A., and Rinaldo, A. (2013). Scaling body size fluctuations. *Proc. Natl. Acad. Sci. U.S.A.* 110, 4646–4650. doi: 10.1073/pnas.1301552110

Grilli, J., Osella, M., Kennard, A. S., and Lagomarsino, M. C. (2017). Relevant parameters in models of cell division control. *Phys. Rev. E* 95:032411. doi: 10.1103/PhysRevE.95.032411

Harris, L. K., and Theriot, J. A. (2016). Relative rates of surface and volume synthesis set bacterial cell size. *Cell* 165, 1479–1492. doi: 10.1016/j.cell.2016.05.045

Ho, P.-Y., and Amir, A. (2015). Simultaneous regulation of cell size and chromosome replication in bacteria. *Front. Microbiol.* 6:662. doi: 10.3389/fmicb.2015.00662

Ho, P.-Y., Lin, J., and Amir, A. (2018). Modeling cell size regulation: from single-cell level statistics to molecular mechanisms and population level effects. *Annu. Rev. Biophys.* 47, 251–271. doi: 10.1146/annurev-biophys-070317-032955

Iyer-Biswas, S., Crooks, G. E., Scherer, N. F., and Dinner, A. R. (2014a). Universality in stochastic exponential growth. *Phys. Rev. Lett.* 113:028101. doi: 10.1103/PhysRevLett.113.028101

Iyer-Biswas, S., Wright, C. S., Henry, J. T., Lo, K., Burov, S., Lin, Y., et al. (2014b). Scaling laws governing stochastic growth and division of single bacterial cells. *Proc. Natl. Acad. Sci. U.S.A.* 111, 15912–15917. doi: 10.1073/pnas.1403232111

Kennard, A. S., Osella, M., Javer, A., Grilli, J., Nghe, P., Tans, S. J., et al. (2016). Individuality and universality in the growth-division laws of single E. coli cells. *Phys. Rev. E* 93:012408. doi: 10.1103/PhysRevE.93.012408

Kiviet, D. J., Nghe, P., Walker, N., Boulineau, S., Sunderlikova, V., and Tans, S. J. (2014). Stochasticity of metabolism and growth at the single-cell level. *Nature* 514, 376–379. doi: 10.1038/nature13582

Logsdon, M. M., Ho, P.-Y., Papavinasasundaram, K., Richardson, K., Cokol, M., Sassetti, C. M., et al. (2017). A parallel adder coordinates mycobacterial cell-cycle progression and cell-size homeostasis in the context of asymmetric growth and organization. *Curr. Biol.* 27, 3367–3374.e7. doi: 10.1016/j.cub.2017.09.046

Osella, M., Nugent, E., and Cosentino Lagomarsino, M. (2014). Concerted control of Escherichia coli cell division. *Proc. Natl. Acad. Sci. U.S.A.* 111, 3431–3435. doi: 10.1073/pnas.1313715111

Osella, M., Tans, S. J., and Cosentino Lagomarsino, M. (2017). Step by step, cell by cell: quantification of the bacterial cell cycle. *Trends Microbiol.* 25, 250–256. doi: 10.1016/j.tim.2016.12.005

Pugatch, R. (2015). Greedy scheduling of cellular self-replication leads to optimal doubling times with a log-fréchet distribution. *Proc. Natl. Acad. Sci. U.S.A.* 112, 2611–2616. doi: 10.1073/pnas.1418738112

Schaechter, M., Maaløe, O., and Kjeldgaard, N. O. (1958). Dependency on medium and temperature of cell size and chemical composition during balanced grown of salmonella typhimurium. *J. Gen. Microbiol.* 19, 592–606. doi: 10.1099/00221287-19-3-592

Si, F., Li, D., Cox, S. E., Sauls, J. T., Azizi, O., Sou, C., et al. (2017). Invariance of initiation mass and predictability of cell size in *Escherichia coli*. *Curr. Biol.* 27, 1278–1287. doi: 10.1016/j.cub.2017.03.022

Skotheim, J. M. (2013). Cell growth and cell cycle control. *Mol. Biol. Cell* 24:678. doi: 10.1091/mbc.E13-01-0002

Taheri-Araghi, S., Bradde, S., Sauls, J. T., Hill, N. S., Levin, P. A., Paulsson, J., et al. (2015). Cell-size control and homeostasis in bacteria. *Curr. Biol.* 25, 385–391. doi: 10.1016/j.cub.2014.12.009

Wallden, M., Fange, D., Lundius, E. G., Baltekin, Ö., and Elf, J. (2016). The synchronization of replication and division cycles in individual *E. coli* cells. *Cell* 166, 729–739. doi: 10.1016/j.cell.2016.06.052

Wang, P., Robert, L., Pelletier, J., Dang, W. L., Taddei, F., Wright, A., et al. (2010). Robust growth of *Escherichia coli*. *Curr. Biol.* 20, 1099–1103. doi: 10.1016/j.cub.2010.04.045

Zheng, H., Ho, P.-Y., Jiang, M., Tang, B., Liu, W., Li, D., et al. (2016). Interrogating the escherichia coli cell cycle by cell dimension perturbations. *Proc. Natl. Acad. Sci. U.S.A.* 113, 15000–15005. doi: 10.1073/pnas.1617932114

11

The spatial biology of transcription and translation in rapidly growing *Escherichia coli*

Somenath Bakshi[†]*, Heejun Choi and James C. Weisshaar*[*]

Department of Chemistry and Molecular Biophysics Program, University of Wisconsin-Madison, Madison, WI, USA

***Correspondence:**
James C. Weisshaar, Department of Chemistry and Molecular Biophysics Program, University of Wisconsin-Madison, Madison, WI 53706, USA
weisshaar@chem.wisc.edu

Single-molecule fluorescence provides high resolution spatial distributions of ribosomes and RNA polymerase (RNAP) in live, rapidly growing *Escherichia coli*. Ribosomes are more strongly segregated from the nucleoids (chromosomal DNA) than previous widefield fluorescence studies suggested. While most transcription may be co-translational, the evidence indicates that most translation occurs on free mRNA copies that have diffused from the nucleoids to a ribosome-rich region. Analysis of time-resolved images of the nucleoid spatial distribution after treatment with the transcription-halting drug rifampicin and the translation-halting drug chloramphenicol shows that both drugs cause nucleoid contraction on the 0–3 min timescale. This is consistent with the transertion hypothesis. We suggest that the longer-term (20–30 min) nucleoid expansion after Rif treatment arises from conversion of 70S-polysomes to 30S and 50S subunits, which readily penetrate the nucleoids. Monte Carlo simulations of a polymer bead model built to mimic the chromosomal DNA and ribosomes (either 70S-polysomes or 30S and 50S subunits) explain spatial segregation or mixing of ribosomes and nucleoids in terms of excluded volume and entropic effects alone. A comprehensive model of the transcription-translation-transertion *system* incorporates this new information about the spatial organization of the *E. coli* cytoplasm. We propose that transertion, which radially expands the nucleoids, is essential for recycling of 30S and 50S subunits from ribosome-rich regions back into the nucleoids. There they initiate co-transcriptional translation, which is an important mechanism for maintaining RNAP forward progress and protecting the nascent mRNA chain. Segregation of 70S-polysomes from the nucleoid may facilitate rapid growth by shortening the search time for ribosomes to find free mRNA concentrated outside the nucleoid and the search time for RNAP concentrated within the nucleoid to find transcription initiation sites.

Keywords: single-molecule tracking live cell, *E. coli*, ribosomes, RNA polymerase, nucleoid structure, DNA-ribosome spatial segregation

Introduction

Super-resolution fluorescence methods (Betzig et al., 2006; Hess et al., 2006; Rust et al., 2006) enable detailed exploration of the ways in which the central dogma of molecular biology plays out in *Escherichia coli*. "Photoactivation-localization microscopy" (PALM) can locate and track 1000s of copies of single, specific proteins *in live cells* with ~30 nm spatial resolution and low-ms time resolution. This has enabled detailed, quantitative studies of how ribosomes (Bakshi et al., 2012), chromosomal DNA (Wang et al., 2011), and RNA polymerase (RNAP; Bakshi et al., 2012, 2013; Endesfelder et al., 2013) are distributed in space and move in time within the cytoplasm of single cells. In addition, time-resolved widefield fluorescence imaging after drug treatment has provided new insight into the ways in which transcription and translation determine the internal organization of cytoplasm (Bakshi et al., 2014a). Transcription, translation, and the spatial organization of *E. coli* cytoplasm act as a coupled biochemical–biophysical system. Here, we describe the delicate balance of forces that enables the system to drive rapid cell growth.

In *E. coli*, the chromosomal DNA occupies the region of space called the *nucleoids* (Kellenberger, 1991). During rapid growth, ribosomes are concentrated outside the nucleoids in *ribosome-rich regions* comprising the two polar end-caps, the space between nucleoid lobes, and the thin region proximal to the cytoplasmic membrane (**Figure 1**; Bakshi et al., 2012). The spatial extent of the nucleoids evidently arises from a balance of compacting and expanding forces (Woldringh et al., 1995; Zimmerman, 2006). Likely compacting forces include depletion-attraction of DNA arising from macromolecular crowding by myriad small proteins (Zimmerman and Murphy, 1996); conformational entropy of the confined DNA polymer, which causes the polymer to avoid walls (Mondal et al., 2011; Bakshi et al., 2014a); inter-strand coupling by DNA binding proteins such as H-NS (Dame, 2005; Wang et al., 2011); bending of DNA by IHF (Dame, 2005); and net supercoiling of the DNA by Gyrase and Topoisomerase I (Woldringh et al., 1995). The hypothesized primary expanding force is "*transertion,*" which is the simultaneous co-transcriptional translation *and insertion* of membrane proteins via the translocon machinery (Woldringh, 2002). Transertion implies the existence of DNA-RNAP-mRNA-ribosome-polypeptide-membrane "*transertion chains*" directly linking DNA to the membrane (**Figure 1**). A sufficient number of these chains would radially expand the overall nucleoid. The main evidence for transertion had been the dramatic contraction of the nucleoids on treatment with translation-halting drugs such as chloramphenicol (van Helvoort et al., 1996; Zimmerman, 2002). Treatment with transcription-halting drugs such as rifampicin should have the same effect, but this was not observed on the 30-min timescale studied (Fishov and Woldringh, 1999; Cabrera et al., 2009). Our recent time-dependent imaging study

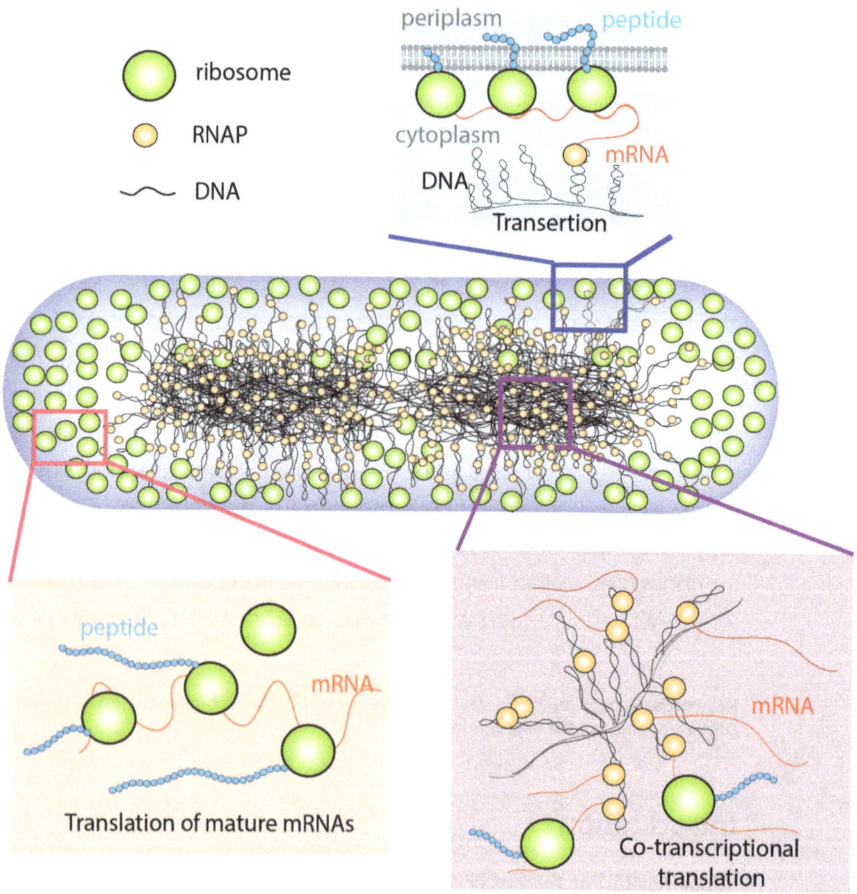

FIGURE 1 | Schematic of the coupled transcription–translation–transertion system in rapidly growing *Escherichia coli*. Adapted from Bakshi et al. (2014a).

discovered nucleoid contraction on a 3-min timescale after rifampicin treatment, placing the transertion hypothesis on solid footing (Bakshi et al., 2014a).

In rapidly growing *E. coli*, only 10–15% of ribosomal subunits lie within the nucleoids (Bakshi et al., 2012). Single-30S subunit tracking studies suggest that most 30S subunits in the ribosome-rich regions are engaged as slowly diffusing 70S-polysomes. The majority of the translation events are presumably carried out on mature, freely diffusing mRNAs within the ribosome-rich regions. While *co-transcriptional* translation (coupled transcription and translation; Miller et al., 1970) is essential for protecting nascent mRNA and ensuring efficient transcription (Burmann et al., 2010, 2012; Proshkin et al., 2010; Svetlov and Nudler, 2012; McGary and Nudler, 2013), it is apparently not the primary means of protein production. The 70S-polysome diffusion coefficient $D_{70S-poly} \sim 0.02\ \mu m^2\ s^{-1}$ is large enough to enable nascent mRNA to diffusively find a ribosome-rich region in ~1 s (Bakshi et al., 2012), a short period compared with the ~6–7 min lifetime of mRNA against degradation (Bernstein et al., 2002).

A number of earlier studies used physical models to explain the formation of compact nucleoids that occupy only a fraction of the entire bacterial cytoplasm. A statistical mechanical model used the osmotic effects of myriad small proteins to explain the apparent phase separation (Odijk, 1998). A recent experimental study of free bacterial nucleoids confined in a microfluidic channel showed that crowding by added polyethylene glycine chains (PEG) led to reversible, first-order "coil-to-globule" collapse of the nucleoids (Pelletier et al., 2012). This study augmented the earlier statistical model to include the entropic spring nature of the nucleoids. A very recent coarse-grained simulation showed that small, spherical crowding agents can induce compaction of a DNA polymer, modeled as a freely jointed chain (Shendruk et al., 2015).

We have found that a simple physical model using Monte Carlo simulations of DNA and ribosome spatial distributions in a confining cytoplasmic space enhances our understanding of the observed ribosome-nucleoid segregation (Mondal et al., 2011). In the model, entropic and excluded volume effects cause strong segregation of the unperturbed nucleoid from 70S-polysomes. The biochemical state of ribosomes (70S-polysomes vs. free 30S and 50S subunits) plays an essential role in ribosome-nucleoid segregation. When the model 70S-polysomes are converted to 30S and 50S subunits, the components mix quite thoroughly with the DNA and the nucleoid expands (Bakshi et al., 2014a). This suggests that the fraction of 70S-polysomes vs. 30S and 50S subunits strongly affects the relative compactness of the nucleoid.

Based on these new experimental and computational results, we are developing a comprehensive model of the spatial organization within the *E. coli* cytoplasm and how it may work to optimize cell growth. Our present view encompasses a variety of inter-related phenomena: (1) In rapidly growing cells, *most translation occurs in the ribosome-rich regions*, not within the nucleoids. (2) Yet there is direct evidence of co-transcriptional translation (Miller et al., 1970), and this is important for protecting nascent mRNA from degradation and for efficient transcription. *Evidently most or all transcription is co-translational, but only ~10–15% of translation is co-transcriptional.* (3) Therefore 30S and 50S subunits must be able to penetrate the nucleoid where they initiate co-transcriptional translation (Sanamrad et al., 2014). This picture implies a *circulation* of 30S and 50S subunits between the nucleoids and the ribosome-rich regions. (4) Transertion expands the nucleoids beyond their relaxed, highly compacted state (Woldringh, 2002). The evidence suggests that transertion plays an essential role, enabling 30S and 50S subunits to move into the nucleoids where they initiate co-transcriptional translation and form nascent 70S-polysomes.

In this view, the transcription and translation machinery and the spatial organization of the cytoplasm act as a coupled biochemical–biophysical system. Effective compartmentalization of the cytoplasm into ribosome-rich regions and DNA- and RNAP-rich nucleoids may enhance the efficiency of protein synthesis and the utilization of RNAP. Segregation of 70S-polysomes from the nucleoid may facilitate rapid growth by minimizing the search time for ribosomes to find the free mRNA concentrated outside the nucleoid and for RNAP concentrated within the nucleoid to find transcription initiation sites. This review describes the evidence behind the comprehensive picture, briefly compares *E. coli* with *Caulobacter crescentus*, and contrasts our "translation-centric" view of nucleoid morphology with a "transcription-centric" view (Jin et al., 2013) presented in another chapter of this volume.

Super-Resolution Imaging of Single Protein Copies in Live Bacterial Cells

Overview

Super-resolution fluorescence microscopy of specific proteins in live cells (**Figure 2**) was enabled by three key technical advances. First, genetic manipulations enable replacement of the gene for the target protein by a gene that appends a fluorescent protein to the target. This "GFP revolution" enables imaging of a specific target protein in a live cell (Zhang et al., 2002). But caution is needed. The fluorescent protein tags may affect the function, aggregation state, spatial distribution, or movement of the target protein (Landgraf et al., 2012).

Second, on a dark background a single fluorescent molecule can produce a punctal, high signal-to-noise image on a sensitive EMCCD camera (Vrljic et al., 2007). However, the full width at half maximum height (FWHM) of this image is broadened to $\sim\lambda/2 = 260$ nm for detection of green light with wavelength $\lambda = 520$ nm. This defines the diffraction limit of light for an optical microscope. Nevertheless, a well-isolated, slowly moving single fluorophore can be located and tracked far more accurately than the diffraction-limited width of its image. In a 1 μm diameter × 3 μm long *E. coli* cell, the localization accuracy for fluorescent proteins is photon- and background-limited to ~30 nm in each of the two projected dimensions (Bakshi et al., 2011).

Third, the super-resolution fluorescence methods use "photoswitchable" fluorescent proteins as the labels to overcome

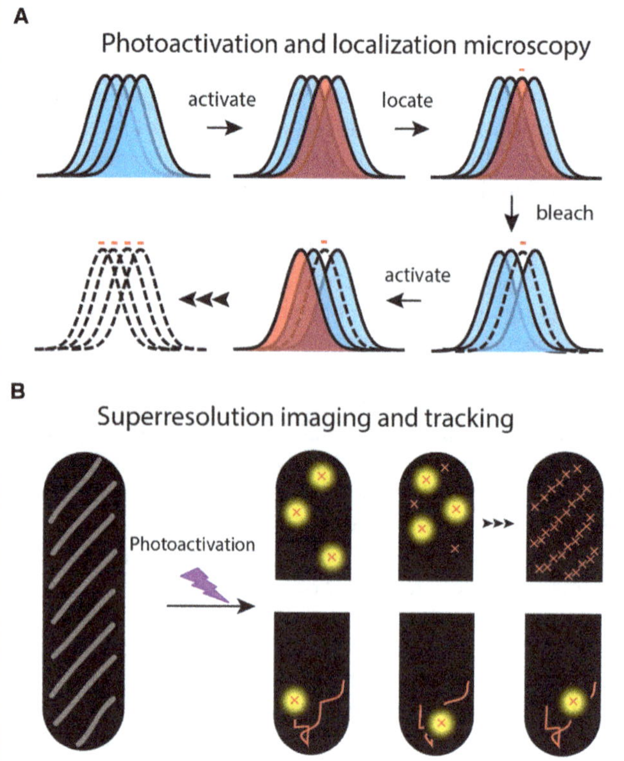

FIGURE 2 | Schematic of the super-resolution imaging method. **(A)** Each fluorescent molecule makes a diffraction-limited, essentially Gaussian image on the camera. A dense set of normal labels (blue images) makes overlapping images. In each imaging cycle, a sparse set of labels is photoactivated (red image) and localized from the well-isolated, single-molecule images. **(B)** An image of the spatial distribution is built up one molecule at a time over 100s or 1000s of optical cycles (upper "half-cells"). Successive images of the same molecule form a diffusive trajectory (lower "half-cells").

the problem of spatial overlap of the myriad images of high copy number fluorescent proteins in small cells (**Figure 2**). For example, PALM and its variations enable sub-diffraction-limit imaging of the ~50,000 labeled 30S-mEos2 ribosomal subunits, essentially by imaging them one or a few copies at a time (Betzig et al., 2006; Rust et al., 2006; Gould et al., 2008). In a widefield fluorescence image, all the 30S images would overlap in space, severely limiting the precision of spatial information (**Figure 2**). Following excitation at shorter wavelengths (typically near 405 nm), photoswitchable proteins undergo photochemistry that shifts their absorption and emission spectra toward longer wavelengths. For the photoswitchable label mEos2, PALM uses weak excitation at 405 nm to create a sparse set of new fluorophores that emit at 584 nm. In each PALM cycle, the few photoswitched copies are located and tracked by excitation with a second probe laser at 561 nm (where the unswitched copies do not absorb). The photoswitched copies bleach all too quickly, after which a new sparse set is photoswitched and tracked. Over several-minutes, literally 1000s of short, 5–15 step trajectories with ~30 nm localization accuracy can be acquired. The stored individual locations are used to reconstruct a PALM image of the two-dimensional spatial distribution of the target protein, averaged over a typical 1–2 min of data acquisition.

The diffusive behavior of each individual copy then provides clues to its biochemical state (Lippincott-Schwartz and Patterson, 2009). For ribosomes, 70S-polysomes diffuse somewhat more slowly than 30S subunits searching for a translation initiation site (Bakshi et al., 2012). Those RNAP copies that are searching for a transcription initiation site diffuse much more rapidly than copies engaged in transcription and thus bound to the DNA polymer (Bakshi et al., 2013). Thus PALM provides not only much more precise spatial distributions than widefield microscopy, but also new information about where specific biochemical processes are carried out within the bacterial cell and the fraction of proteins engaged in each process.

Cell Growth and Preparation

Our preferred *E. coli* strain VH1000 is a modification of MG1655 with the PyrE defect repaired. VH1000 grown in EZRDM ("EZ rich, defined medium"; Neidhardt et al., 1974), MBM (MOPS-buffered minimal medium) using glucose as carbon source, or MBM with glycerol, provides access to a wide range of doubling time (30, 60, and 120 min, respectively, at 37°C). Most of the experiments described here were carried out in EZRDM at 30°C.

We maintain cells at well-defined, constant levels of nutrition and aeration throughout the imaging experiments (Bakshi et al., 2011, 2012). Cells are harvested from exponential growth and plated on a polylysine coated coverslip that forms the base of a microfluidics device providing a continuous flow of fresh, aerated growth medium during imaging. This enables us to study 50–100 cells simultaneously. Length vs. time measurements of single cells shows that they grow at the same rate as in bulk medium to within 5%; evidently there are no harmful effects of the polylysine surface. Visible light can be toxic (Bakshi et al., 2012). It is essential to minimize total photon dosage at 514 or 561 nm. We always ensure that the laser exposure does not alter the property being measured.

Labeling Strategies

The use of λ-Red mediated recombination (Datsenko and Wanner, 2000) to replace wild-type genes on the chromosome with genes for mEos2-labeled proteins has become routine. P1 transduction then transfers the new genes back to the parent strain to eliminate unwanted mutations. Our VH1000 strains expressing the ribosomal protein S2 labeled by mEos2 or the RNAP subunit β′ labeled with mEos2 grow normally. The long time required for mEos2 to achieve its fluorescent state (~130 min) compared to the assembly time of ribosome/RNAP core enzymes (a few min) ensures that we are almost always tracking assembled 30S ribosomal subunits (Bakshi et al., 2012) or complete RNAP enzymes (Bakshi et al., 2013). As a test for possible adverse effects of the labels (Landgraf et al., 2012), we showed that for both β′ and S2, the mEos2 and yGFP labeling schemes produce indistinguishable spatial distributions and diffusive properties.

To image the overall spatial distribution of DNA in growing cells, we strongly prefer the non-perturbative stain SYTOX

Orange over the more standard DAPI imaging (Bakshi et al., 2014a,b). DAPI staining plus UV light perturbs growth rate and nucleoid morphology. SYTOX Orange staining enables normal growth and widefield imaging over 100s of camera frames.

Single-Molecule Localization and Tracking Methodology

The mEos2 constructs yield trajectories of mean length 10 frames and enable control of the time between frames. For each 1000 trajectories, we obtain ~150 trajectories of length >18 frames. We spatially filter the images and fit single-molecule locations using a centroid algorithm (Wang et al., 2010; Bakshi et al., 2011, 2012). Images of cells are rotated so that the long axis is x and the short, transverse axis is y. By plotting a point at the (x, y) coordinates of each centroid, we produce a high-resolution, two-dimensional (2D) projection of the 3D spatial distribution, averaged over the several-minute acquisition period.

A sequence of locations for a particular particle forms a diffusive trajectory. We analyze single-molecule diffusion data in a variety of ways. Plots of mean-square displacement vs. lag time τ, $MSD(\tau)$, are obtained as a running average over each trajectory and over all molecules. To test for sub-diffusion effects due to tethering or caging, $MSD(\tau)$ is compared with Monte Carlo random walk (free diffusion) model calculations within an appropriate confinement volume. To test for heterogeneity of diffusion, we compute histograms of single-molecule diffusion coefficients, estimated as $D_i = \mathrm{msd}_i(\tau)/4\tau$, where $\mathrm{msd}_i(\tau)$ is the single-molecule mean-square displacement averaged over *one trajectory*. The aspect ratio of trajectories can also help distinguish sub-diffusion from free diffusion of single-molecules (Bakshi et al., 2013).

Ribosome-Nucleoid Segregation in Rapid, Exponential Growth

Background

Miller et al. (1970), in their EM study of the contents of the *E. coli* cytoplasm observed long DNA strands to which chains of ribosomes ("70S-polysomes") were attached, direct evidence of translation coupled to transcription ("co-transcriptional translation"). That study inferred that *all* bacterial protein synthesis was co-transcriptional. The total amount of mRNA per cell and the ribosome copy number confirm the importance of polysomes and suggest their typical length to be about ten 70S per mRNA. Co-transcriptional translation evidently assists optimal cell growth. It helps protect nascent mRNA against early termination by Rho and against premature degradation by ribonucleases (Proshkin et al., 2010; McGary and Nudler, 2013). In addition, specific proteins bind simultaneously to both RNAP and the lead ribosome in a polysome chain (Burmann et al., 2012; Svetlov and Nudler, 2012). The translating lead ribosome helps to prevent undesirable RNAP backtracking (Burmann et al., 2010).

However, evidence from EM studies of sections of fixed *E. coli* cells (Kellenberger, 1991) and from widefield immuno-staining of whole, fixed cells (Azam et al., 2000) argued against a model in which all translation is co-transcriptional. The images showed

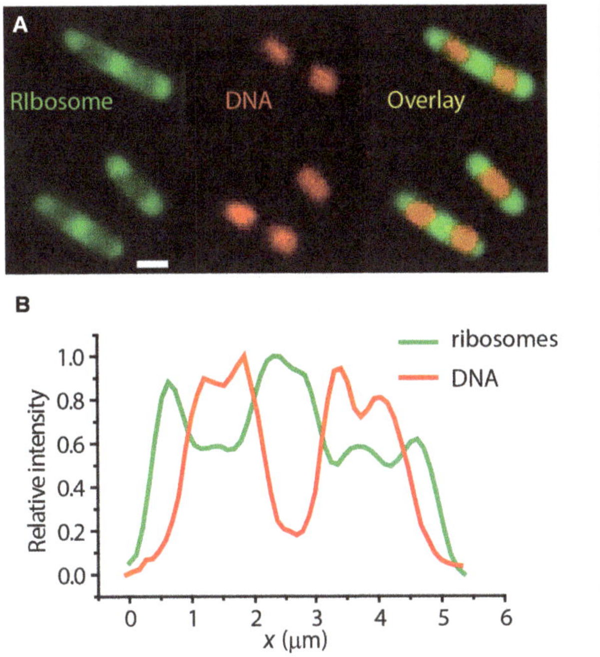

FIGURE 3 | (A) Widefield images of ribosomes (S2-YFP labeling) and DNA (DRAQ5 staining) in single *E. coli* cells. **(B)** Axial linescans of ribosome and DNA intensity vs. the long-axis coordinate x. Anti-correlation is evident. Adapted from Bakshi et al. (2012).

a strong tendency of ribosomes to avoid the nucleoids. Early widefield fluorescence studies in fixed *Bacillus subtilis* co-imaged DNA and ribosomes and again demonstrated strong segregation of ribosomes from DNA (Lewis et al., 2000).

Our own widefield images of live *E. coli* with DNA stained with DRAQ5 and ribsosomes labeled with S2-YFP show clear anti-correlation in the axial spatial distributions of the two species (**Figure 3**). The "peak to valley" ratio of the ribosomes is about 1.5:1. Intriguingly, in the slow-growing species *C. crescentus*, ribosomes and DNA appear to be much more thoroughly mixed (Llopis et al., 2010).

Superresolution Imaging of RNAP and Ribosomes

In Bakshi et al. (2012), we reported super-resolution images of RNAP (β′-yGFP, **Figures 4A–C**) and ribosome (30S-YFP, **Figures 4D–G**) spatial distributions. RNAP spends almost all of its time bound to DNA, either specifically or non-specifically. The experiments revealed a much greater degree of RNAP/ribosome spatial segregation than suggested by the earlier widefield work.

In our study of RNAP motion, we labeled RNAP as β′-mEos2 expressed from the chromosome (Bakshi et al., 2013). Single-molecule diffusive trajectories of 1-s duration cleanly distinguish two states of motion of RNAP. In **Figure 5**, compare the compact purple trajectory with the extended yellow one. Specifically bound copies jiggle in place (sub-diffusion), much like DNA foci, while copies searching for transcription initiation sites undergo apparently free diffusion (a combination of 3D hopping

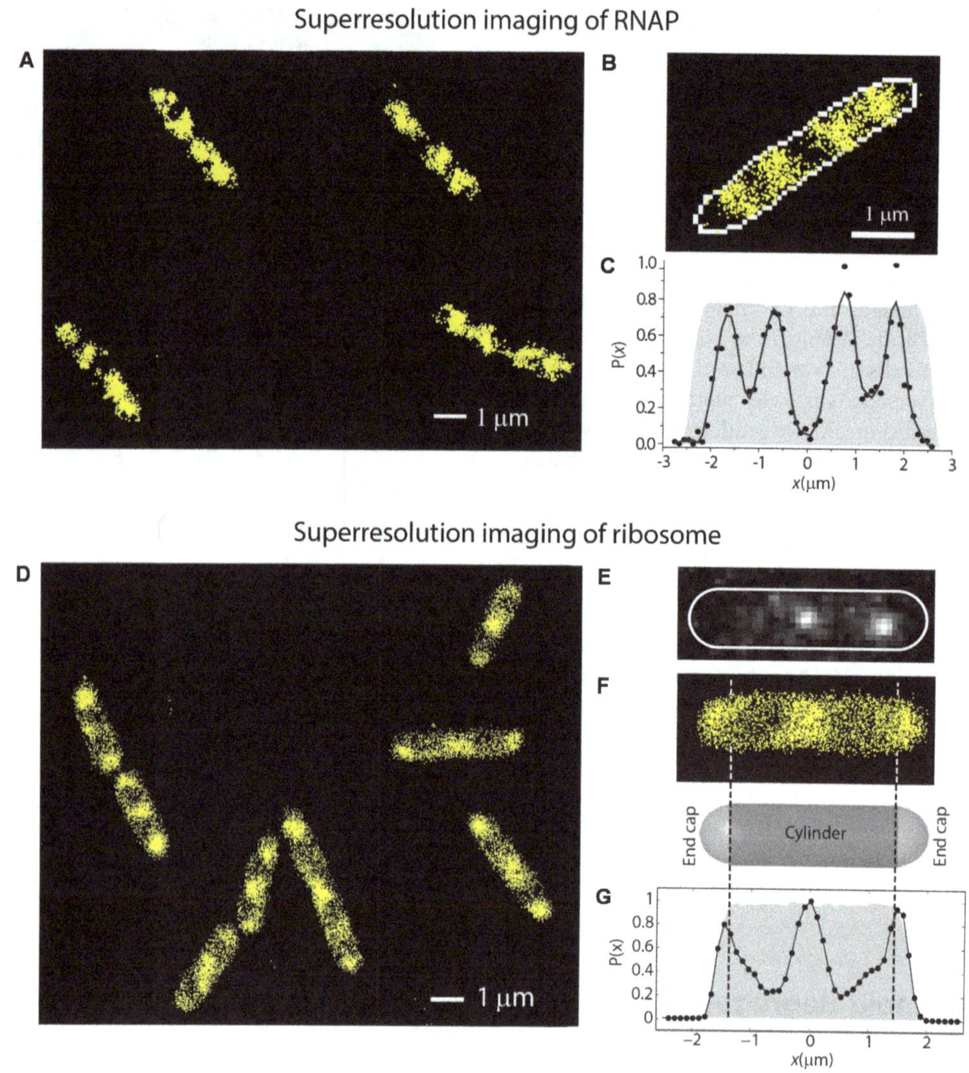

FIGURE 4 | (A–C) Super-resolution imaging of the spatial distributions of RNA polymerase (β′-YFP labeling) in live *E. coli*. **(C)** Shows a histogram of locations projected onto the long axis of the cell in **(B)**. **(D–G)** Super-resolution imaging of the spatial distributions of ribosomes (30S-YFP labeling) in live *E. coli*. **(E)** Shows two single copies imaged in one camera frame. **(G)** Shows an axial histogram of locations for the cell shown in **(F)**. The gray regions in **(C,G)** are simulated histograms for a uniformly filled spherocylinder matching the length and radius of the single cell. Adapted from Bakshi et al. (2012).

and non-specific binding) within the nucleoids. For growth in EZRDM at 37°C, about 50% of RNAP copies are evidently specifically bound to DNA (including all stages of transcription). According to classic estimates (Dennis et al., 2004), about 1/2 to 2/3 of this 50% should be transcribing stable RNA (rRNA and tRNA); the remainder should be transcribing protein genes. In rapidly growing cells, both widefield (Jin and Cabrera, 2006; Bratton et al., 2011) and super-resolution (Endesfelder et al., 2013) fluorescence studies find highly concentrated clusters of RNAP copies. These "transcription foci" presumably comprise RNAP copies engaged in transcription of *rrn* operons.

In the ribosome imaging studies, we found that some 85–90% of the 30S-YFP copies lie in the "ribosome-rich regions" (Bakshi et al., 2012). This suggests that only ~10–15% of translation is co-transcriptional. The diffusive behavior of single 30S-mEos2 copies (**Figure 6**) reveals two distinguishable sub-populations. This enabled us to roughly divide the 30S behavior into 20–30% free 30S with diffusion coefficient $D_{30S} \sim 0.14\ \mu m^2\ s^{-1}$ and 80–70% 70S-polysomes with $D_{70S-poly} \sim 0.02\ \mu m^2\ s^{-1}$. The latter value describes the diffusive motion of mRNA decorated by a variable number of 70S-ribosomes. For comparison, the timescale of transcription of mRNA for a typical protein is ~20 s and the degradation time for messages is ~5 min (Bernstein et al., 2002). These data in *E. coli* are consistent with the view that most transcription of protein genes is co-translational. However, mRNA copies spend most of their lifetime separated from RNAP and DNA. After completion of transcription, $D_{70S-poly}$ is large enough to enable free mRNA decorated with translating 70S ribosomes to diffuse to a ribosome-rich region in ~1 s. There each message can be translated repeatedly before degradation.

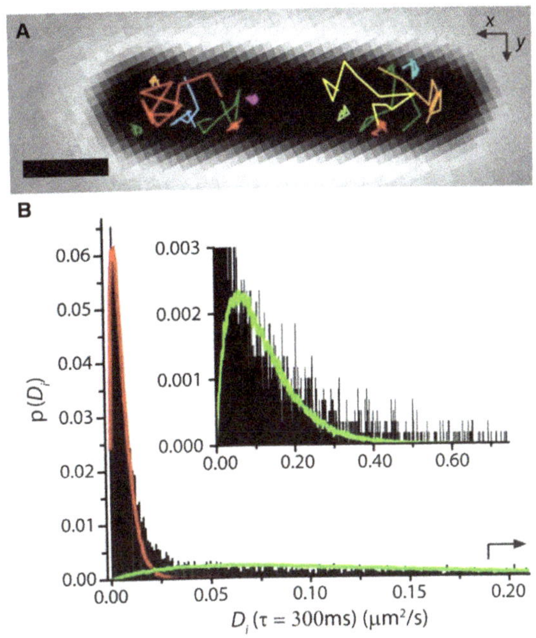

FIGURE 5 | (A) Single-RNAP diffusive trajectories in live *E. coli* growing in EZRDM at 37°C. Labeling was β′-mEos2. Note two types of trajectory. **(B)** Distribution of estimated diffusion coefficients D_i from single-RNAP trajectories. Fast and slow sub-populations are evident. Red and green curves are model distributions for the two components. Adapted from Bakshi et al. (2013).

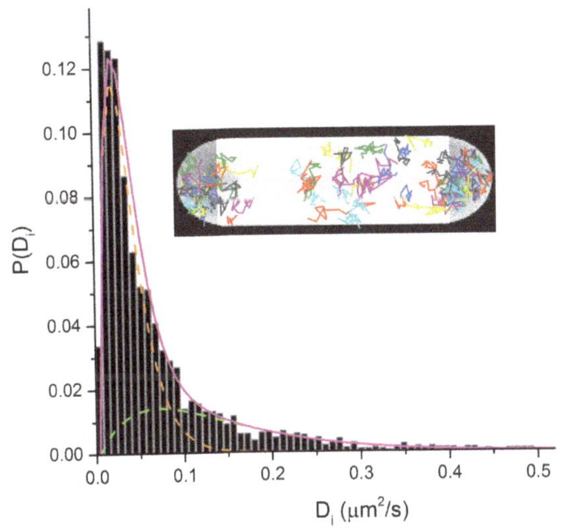

FIGURE 6 | Inset: Single-ribosome diffusive trajectories from a live *E. coli* cell growing in EZRDM at 30°C. Labeling is S2-YFP. **Main figure:** Distribution of estimated single-ribosome diffusion coefficients D_i. S2-mEos2 labeling. Two sub-populations are evident. Model sub-distributions are shown as green and orange dashed lines. Magenta line is their sum.

After complete synthesis of a protein in a ribosome-rich region, the newly free 30S and 50S subunits may engage in repeated rounds of translation in the ribosome-rich region, or "escape" back to the nucleoid region where they can re-initiate co-transcriptional translation (**Figure 7**). Accordingly, the Elf lab recently showed that the more rapidly diffusing 30S and 50S subunits penetrate the nucleoids while the slower 70S-polysomes are largely excluded (Sanamrad et al., 2014). This picture implies a circulation of ribosomal subunits from the nucleoids (where co-transcriptional translation occurs) to the ribosome-rich regions (where most protein synthesis occurs) and back again.

In fast growth conditions, concentration of 70S-polysomes in ribosome-rich regions may enhance the rate of protein synthesis by shortening the search time for translation initiation sites by newly freed 30S and 50S subunits that have just completed synthesis of a protein. Such spatial separation of transcription from most translation in *E. coli* is somewhat reminiscent of eukaryotic cells. However, it occurs without compartmentalization of DNA within a nuclear membrane. The Monte Carlo simulations described below suggest that the underlying segregation mechanism may be primarily physical in nature.

What Driving Forces Induce Nucleoid-Ribosome Mixing or Segregation?

In spite of the 1.5 mm contour length of an *E. coli* chromosome, the nucleoids do not fill the entire volume of the 3–4 μm long, 1 μm diameter cytoplasm (Pettijohn, 1982; Robinow and Kellenberger, 1994). The irregular shape of the *E. coli* nucleoids may be governed by the ring topology of DNA and spatial confinement effects on the ring polymer (Jung et al., 2012; Fisher et al., 2013; Youngren et al., 2014). In *C. crescentus*, chromosome conformation capture (3C) data suggest the nucleoids adopt a "bottle brush" geometry with plectonemes radiating outward from a central spine (Le et al., 2013). Less detail is presently available for the *E. coli* chromosome (Cagliero et al., 2013). Our focus here is on the coarse spatial extent of the nucleoids under different conditions, not the finer details.

Long-Term Effects of Chloramphenicol and Rifampicin on Nucleoid-Ribosome Morphology

It has long been known that transcription- and translation-halting drugs strongly affect nucleoid morphology. Typical imaging experiments have compared nucleoid morphology before and 20–30 min after drug treatment, usually using DAPI as the DNA stain. We used the non-perturbative DNA stain SYTOX Orange (Bakshi et al., 2014b) to study *time-dependent, quantitative* effects of Rif and Cam on nucleoid length and width in live cells over 20 min (Bakshi et al., 2014a). To describe the overall spatial distribution of the chromosomal DNA vs. time, we defined two parameters measured from the SYTOX Orange fluorescence intensity distributions projected along the *x*- and *y*-axes (**Figure 8A**). The axial distribution was characterized by the overall length L_{DNA}, measured as the "outside" full-width at half-maximum height (FWHM) of the projected intensity distribution along *x*. The width W_{DNA}, a rough measure of the mean nucleoid diameter, was defined as

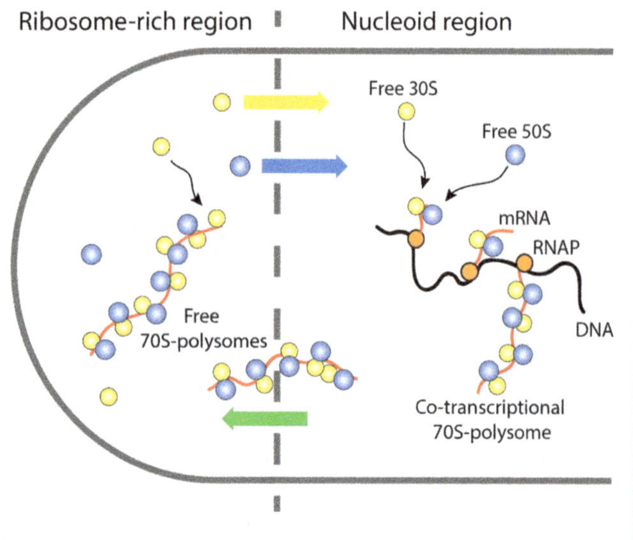

FIGURE 7 | Schematic showing the suggested circulation of ribosomal subunits into and out of the nucleoids and the ribosome-rich regions.

the FWHM of the projection of intensity along the transverse coordinate *y*. We believe this definition of nucleoid length and width is more quantitative than the "relative nucleoid size" used in other work (Jin et al., 2013). We also measured single-ribosome diffusive motion vs. time after drug treatment, using the 30S-mEos2 labeling scheme (Bakshi et al., 2014a). This helps distinguish 70S-polysomes from free 30S subunits after drug treatment.

On the 10–20 min timescale, the nucleoids of Cam-treated cells have become much more compact axially than normally growing cells, while the nucleoids of Rif-treated cells have expanded in both dimensions (**Figures 8B,C**). Similar long-term drug effects were observed in earlier work (Cabrera and Jin, 2006; Zimmerman, 2006). Ribosomes are evidently maintained as 70S-polysomes on a 20-min timescale after Cam treatment, as inferred from the distribution of ribosome diffusion coefficients (Bakshi et al., 2014a). We further suggest that the long-term nucleoid expansion induced by Rif is due to slow degradation of existing mRNA, after transcription initiation is halted by the drug. Imaging of the putative mRNA stain SYTO RNASelect vs. time is consistent with this suggestion. As mRNA is degraded, 70S-polysomes that dissociate to 30S and 50S subunits after completion of translation find fewer and fewer translation initiation sites. The resulting free subunits diffuse much more rapidly than 70S-polysomes. Unlike 70S-polysomes, the free subunits mix thoroughly with the nucleoids.

Nucleoid-Ribosome Mixing Hypothesis

These long-term drug effects have motivated a new nucleoid-ribosome mixing hypothesis (**Figure 9**; Bakshi et al., 2014a). We view the ribosomes and the chromosomal DNA as a *composite biochemical-biophysical system*. Cell physiology and drug treatments dictate the partitioning of ribosomal components between 70S-polysomes and free 30S and 50S subunits. Overall nucleoid spatial extent is then governed by the tendency for nucleoids to segregate from 70S-polysomes and to mix with free 30S and 50S subunits. The Monte Carlo modeling (Mondal et al., 2011; Bakshi et al., 2014a) described next suggests that the underlying driving forces for mixing or segregation are *excluded volume effects* combined with *maximal total entropy* of the composite DNA-ribosome system.

In this view, the drug studies reveal three "states" of the nucleoid-ribosome system. For normally growing, untreated cells, transertion expands the nucleoids, imposing a state of *intermediate DNA density* that prevents 70S polysomes and the chromosomal DNA from mixing while permitting 30S and 50S subunits to penetrate the nucleoids. Transertion may thus be essential for the initiation and maintenance of co-transcriptional translation of nascent mRNA in the dense regions of the nucleoids. Cam freezes ribosomal subunits as 70S-polysomes. Completion of transcription events will break the transertion chains and the 70S-polysomes will then be unable to make new chains. Over 20 min, the chromosomal DNA relaxes to a *fully condensed state* that excludes both 70S-polysomes and 30S and 50S subunits. The fraction of free 30S and 50S subunits becomes even smaller than in normal growth, enabling very strong compaction of the nucleoids. We suggest that this represents the relaxed volume of the nucleoids in the absence of the expanding force of transertion and in the near absence of free ribosomal subunits. Rif treatment also breaks the transertion links on the same timescale as Cam, the time over which transcription events are completed. Again, new transertion chains are prevented from forming. This leads to similar short-term contraction (see below). DNA mixing with 30S and 50S ribosomal subunits eventually occurs on the longer, 10–20 min timescale of mRNA degradation. This leads to the third, *fully expanded* state of the nucleoids, in which few 70S-polysomes exist and DNA and the 30S and 50S ribosomal subunits mix extensively. However, the nucleoids continue to avoid the cylindrical walls and especially the endcaps.

Simple Physical Model of Ribosomes and Nucleoids

In Mondal et al. (2011), we developed a simple physical model of plectonemic DNA and 70S-polysomes confined in a spherocylinder (as pictured in **Figure 10A**). DNA was modeled as a hyperbranched polymer (hard spheres and connecting rods). Based on estimates at the time (Bremer and Dennis, 1996), the model placed two chromosome equivalents of plectonemic DNA (comprising 7000 plectoneme rods) plus 20,000 70S particles organized as freely jointed 70S-polysome 13-mers into a spherocylinder (350 nm radius, 3.0 μm length). DNA-DNA, polysome-DNA, and polysome-polysome excluded volume effects were modeled realistically. There are no attractive interactions between particles. The free energy of the composite system was minimized using Monte Carlo methods. This minimalist, coarse-grained model does not include the effects of transertion, nor does it attempt to describe the level of geometric detail revealed by recent chromosome conformational capture data (Cagliero et al., 2013; Le et al., 2013; Le and Laub, 2014). Instead, it seeks to understand the effects of excluded volume

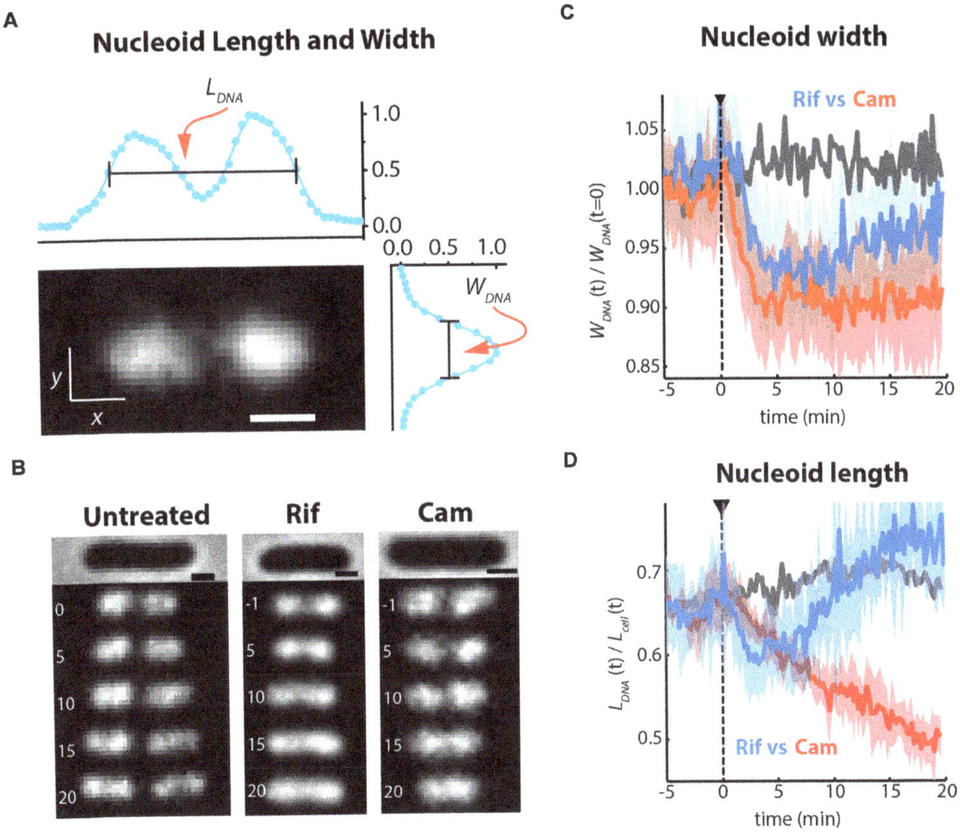

FIGURE 8 | (A) SYTOX orange-stained image of chromosomal DNA in a live *E. coli* cell growing in EZRDM at 30°C. The nucleoid spatial extent is characterized by length L_{DNA} and width W_{DNA} measured as the full width at half-maximum height (FWHM) of intensity distributions projected onto the *x* and *y* axes. **(B)** Time-lapse sequences of images of nucleoids stained by SYTOX Orange. Times in minutes, scale bars are 1 mm. Untreated cells, Rif-treated cells, and Cam-treated cells as indicated. **(C)** Quantitative nucleoid width W_{DNA} vs. time. *Gray:* mean behavior of a set of untreated cells. *Blue:* behavior of Rif-treated cells. *Red:* behavior of Cam-treated cells. For blue and red, heavy lines are averages of traces from many cells; shaded regions show the envelope of single-cell results that were averaged. Dashed line shows time of initiation of flow of drug. **(D)** Same as **(C)**, but with relative nucleoid length plotted as L_{DNA}/L_{cell}. Adapted from Bakshi et al. (2014a).

and entropy on the overall spatial distributions of DNA and ribosomes.

Translational entropy is the entropy of movement in space. It increases with the number of free particles and with the volume available to the particles. DNA conformational entropy is determined by the number of conformational states available to the DNA polymer. *In the model, these entropic plus excluded volume effects lead to strong segregation of DNA from 70S-polysomes* (**Figure 10B**). The model DNA polymer, comprising beads connected by rods, avoids the walls and does not fill the cytoplasm. This is because placing a bead near a wall would eliminate many potential conformations of the polymer and hence decrease conformational entropy.

To help understand why Rif-induced dissociation of 70S-polysomes into 30S and 50S subunits enables mixing of free subunits with the DNA and nucleoid expansion, we have recently performed a new set of simulations based on the same simplified model of DNA (**Figure 10C**; Bakshi et al., 2014a). The 20,000 70S ribosomes engaged as 1538 polysome 13-mers were converted to 20,000 30S spheres and 20,000 50S spheres. Excluded volume effects were adjusted appropriately. *The free 30S and 50S subunits mix with the DNA polymer.* In addition, the nucleoid becomes more expanded, while continuing to avoid the confining walls. Both features are reminiscent of experimental results (Bakshi et al., 2014a; Sanamrad et al., 2014).

The model suggests that the primary driving force for mixing of chromosomal DNA with ribosomal subunits after Rif treatment is *increased translational entropy*. Each 70S-polysome 13-mer has become 26 independent subunits, each "demanding" its own translational entropy. We suggest that the nucleoid expands to provide the 30S and 50S subunits with access to more volume in which to move.

In real cells, the nucleoids adopt their most compact form after Cam treatment, which preserves most ribosomal subunits as 70S-polysomes. Importantly, in this highly compacted form the nucleoids exclude not only 70S-polysomes, but also free 30S subunits (Sanamrad et al., 2014), LacI (Kuhlman and Cox, 2012), and the Kaede tetramer (Bakshi et al., 2011). Accordingly, we believe that it is *transertion* that expands the nucleoids

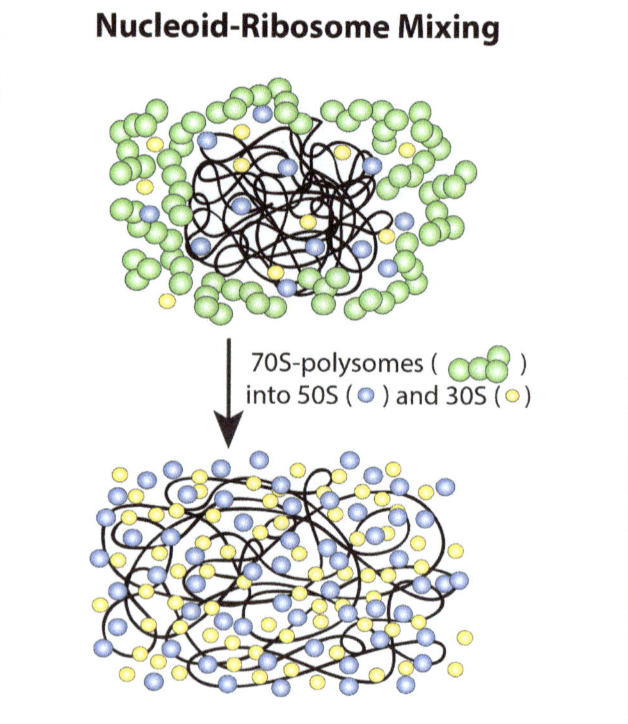

FIGURE 9 | Schematic of ribosome-nucleoid mixing hypothesis. 70S-polysomes and DNA strongly avoid each other, while free 30S and 50S subunits readily penetrate into the nucleoids. Adapted from Bakshi et al. (2014a).

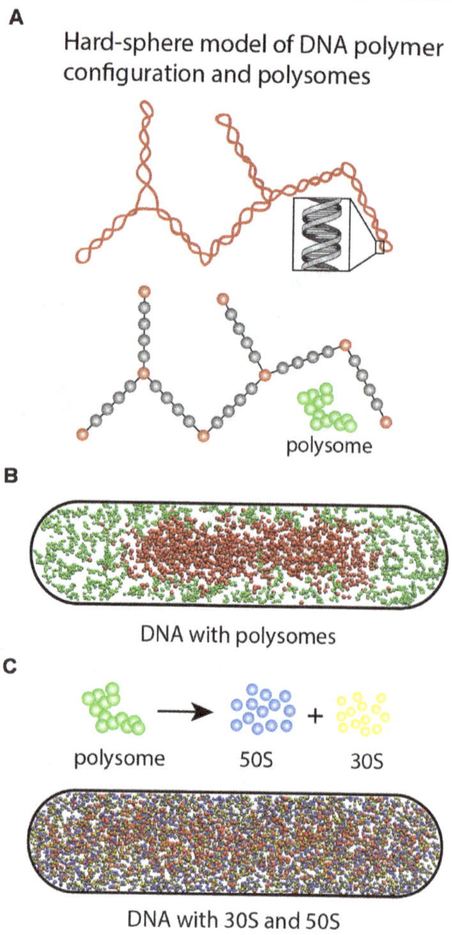

FIGURE 10 | (A) Hyper-branched polymer bead model of plectonemic DNA and 70S-polysomes. The red DNA beads exclude each other; the gray beads are invisible to the DNA beads but act as volume appropriately excluded to the polysomes. Polysomes are represented as freely jointed chains of spheres of appropriate size. **(B)** In Monte Carlo simulations, 70S-polysomes and DNA strongly avoid each other. **(C)** When the 70S-polysomes are dissociated into 50S and 30S monomers, the simulations show strong mixing and nucleoid expansion. Adapted from Bakshi et al. (2014a).

sufficiently to enable 30S and 50S subunits to penetrate and initiate co-transcriptional translation.

Co-transcriptional translation in turn is biologically important for preventing premature degradation of mRNA by endonucleases, termination of transcription by Rho, and excessive backtracking by RNAP. *Thus transertion, whose very existence has long been debated, may be essential to optimal cell growth.*

In our view, the overall physiological state of the cell (slow growth, fast growth, stationary phase, stress response, etc.) determines the number of ribosomal subunits and their partitioning between 70S-polysomes and 30S and 50S subunits. The Monte Carlo modeling then suggests that at least in rapid growth, the corresponding coarse nucleoid morphology is dictated by the tendency of the combined nucleoid-ribosome *system* to maximize total conformational and translational entropy. The modeling further suggests that in the presence of free 30S and 50S subunits, the tendency to maximize translational entropy of ribosomal subunits provides a second expanding force on the nucleoids, in addition to transertion. This tendency is most evident after treatment with the transcription-halting drug Rif. It is less evident, but may still be a significant effect, in normal growth when 30S and 50S subunits are minority ribosomal species. Our model of the factors controlling the overall DNA spatial distribution might be called "translation-centric," because it emphasizes the importance of the DNA-ribosome system.

It is difficult to compare our model of DNA-ribosome interactions with other physical models of DNA confinement and compaction. The two most recent models treat the DNA polymer as a freely jointed chain of beads (Pelletier et al., 2012; Shendruk et al., 2015), whereas we represent DNA as plectonemes modeled as a hyper-branched chain of sticks connecting beads. The other models include small-protein crowders but not ribosomes, whereas we include ribosomes explicitly and neglect small-protein crowders. With its 7000 connecting rods, our hyper-branched chain has many more internal degrees of freedom than a freely jointed chain of several 100 beads, and so may have a greater tendency to avoid walls. We anticipate more detailed physical models in the future, perhaps including crowders of all sizes, the effects of transertion, and new information about specific DNA conformations from chromosome-capture experiments (Cagliero et al., 2013; Le et al., 2013).

New Evidence for Transertion

Background

Transertion is viewed as a dynamic process in which numerous membrane-DNA linkages are constantly forming and breaking during normal transcription and translation events (Norris and Madsen, 1995). The transertion hypothesis implies DNA-RNAP-mRNA-ribosome-polypeptide-membrane "transertion chains" that tether the chromosomal DNA to the cytoplasmic membrane and provide a radially expanding force on the nucleoids (**Figure 1**). The primary evidence for transertion had long been the dramatic contraction of the nucleoids after treatment with drugs that halt translation (e.g., chloramphenicol; van Helvoort et al., 1996; Zimmerman, 2002). Such treatment should break the transertion chains on the timescale of completion of transcription events (~20 s). However, transertion chains should also be broken by drugs such as rifampicin ("Rif"), which prevents transcription initiation. Yet cells observed 20–30 min after Rif treatment showed nucleoid *expansion*, in contradiction of the transertion hypothesis (Jin and Cabrera, 2006).

Recent additional support for transertion came from the Goulian lab. Libby et al. (2012), they discovered a net outward migration of genes encoding cytoplasmic membrane proteins within minutes of induction of transcription. They measured the 2D spatial distribution of fluorescent markers of genes encoding two soluble proteins (*mcherry*, coding the fluorescent protein mCherry; and *aadA*, spectinomycin adenylyltransferase) and of genes encoding two membrane proteins (*lacY*, lactose permease; and *tetA*, the tetracycline efflux pump). For the two membrane-protein genes (but not for the soluble protein genes), induction caused a substantial outward shift in the distribution (toward the membrane), consistent with the transertion hypothesis. The shift occurred within 1–3 min of gene induction.

Short-Time Imaging of Drug Effects on Nucleoid Morphology

Our recent time-resolved drug studies have resolved the Rif dilemma (Bakshi et al., 2014a). The non-perturbative DNA stain SYTOX Orange allows monitoring of nucleoid morphology in growing cells using widefield time-lapse microscopy. We are able follow the size and shape of the nucleoids at 10 s intervals over 25 min after cells are treated with drugs. *On the 0–5 min timescale, both Cam and Rif shrink both the length and width of the nucleoids* (**Figures 8B–D**). This is consistent with the transertion hypothesis. Both Cam and Rif should cause radial contraction of the nucleoids on the timescale of completion of transcription events within the transertion chains. On the 10–20 min timescale following Rif treatment, the nucleoids of the Rif-treated cells expand axially, becoming longer than in the unperturbed state. As described above, we believe the expansion is due to converstion of 70S-polysomes to 30S and 50S subunits, which freely mix with the nucleoids. However, the nucleoid width remains smaller than that of the unperturbed state, consistent with wall avoidance by the DNA polymer, which is no longer tethered to the plasma membrane.

A "Transcription-Centric" Model of *E. coli* Organization

In a separate chapter of this compendium, Ding Jin presents a very different view of the factors governing the overall DNA spatial distribution. The model from the Jin lab might be called "transcription-centric." Jin et al. (2013) labeled RNAP using a GFP tag on the β′ subunit and imaged fixed *E. coli* cells harvested from rapidly growing cultures. They discovered that each cell exhibits several bright puncta of RNAP-GFP intensity, which they dubbed "transcription foci" (Cabrera et al., 2009). Similar foci have been observed in a recent super-resolution study (Endesfelder et al., 2013). The Jin lab studied the presence or absence of transcription foci as well as the overall nucleoid spatial extent in a wide variety of growth and stress conditions (Jin et al., 2013). Transcription foci dissipate rapidly when fast-growing cells are starved or treated with Rif, among other conditions. The foci are not evident in slowly growing cells.

The transcription foci are very likely concentrated centers of transcription of *rrn* operons, the predominant type of transcription in rapidly growing cells. The seven *rrn* operons are distributed broadly within the half of the chromosome closest to *oriC*. Rapidly growing cells containing perhaps four genome equivalents of DNA can easily harbor ~40 *rrn* operons, a number that far exceeds the number of transcription foci observed. This strongly suggests *clustering* of *rrn* operons into hubs of rRNA transcription, perhaps analogous to the nucleolus of eukaryotic cells. The cause of such clustering is unknown. Cook and co-workers developed a statistical model that suggested that a strong depletion-attraction force would arise between two *rrn* operons heavily decorated with RNAP copies (Marenduzzo et al., 2006).

Across a variety of growth conditions, mutant strains, and drug treatments, Jin et al. (2013) observed a strong, positive *correlation* between the presence of transcription foci and the occurrence of relatively compact states of the nucleoids. This led to the suggestion that binding of a large fraction of RNAP copies within transcription foci somehow *causes* the nucleoid to compact.

The most recent concept involves the redistribution of RNAP copies that occurs in conditions that "dissolve" transcription foci. Imaging evidence indicates that the transcription foci tend to lie at or near the nucleoid periphery (Jin et al., 2013). In rapidly growing cells, RNAP may thus be depleted from the bulk of the nucleoids. When the RNAP previously concentrated in peripheral *rrn* clusters are dispersed by Rif treatment or by inducing the stringent response, they bind specifically and non-specifically to numerous sites within the bulk of the DNA. The suggestion is that this expands the nucleoids. However, such expansion cannot be due to insertion of the extra volume of RNAPs into the bulk of the nucleoids. The nucleoids occupy roughly half of the total cytoplasmic volume, or ~2 μm^3 in rapid growth. The RNAP copy number per cell is ~5000 (Bakshi et al., 2012). Even if we place all the RNAP copies within the nucleoids, their total volume is only ~0.003 μm^3. Insertion of RNAP copies into the bulk of the nucleoids in and of itself cannot induce anything like the observed volume expansion induced by Rif. Similarly, the presence in the nucleoid of 10–15% of the ribosomes has only a minor effect on overall nucleoid volume (Bakshi et al., 2012). The total volume of 7,500 70S ribosomes is only $\sim$0.03 μm^3.

Instead, we believe it is the translational entropy of free ribosomal subunits that drives the expansion of the nucleoid (Bakshi et al., 2014a). The fraction of ribosomes involved in active translation as 70S polysomes decreases upon Rif treatment or induction of the stringent response. The increased contribution of translational entropy from the increased fraction of free, non-translating ribosomal subunits causes the nucleoid to expand, making the entire the cytoplasm accessible to the 30S and 50S subunits (**Figures 9** and **10**).

It might be suggested that the gathering of multiple *rrn* operons into a single *rrn* cluster, whatever the underlying cause, constrains the entire nucleoid to be more compact. A critical test of this suggestion would use dual-color fluorescence reporter–operator systems (FROS) to label two different *rrn* operons and test whether *rrn* clusters persist or disperse in slow growth. If they disperse and the nucleoids expand, the idea may have merit. At present, a counter-argument is that the linear contour of one genome measures 1.5 mm in length. Whatever causes *rrn* operons to cluster, it seems likely to us that there is plenty of "slack" in the overall chromosome to accommodate clustering of distant operons without compacting the overall nucleoid morphology. Additional study of the confined DNA polymer model, including pinning together of distant beads to each other, may be informative here.

Also relevant is a recent study of DAPI-stained nucleoid morphology in different growth conditions (Kuhlman and Cox, 2012). From a quantitative comparison of the integrated curvature of 2D images of the nucleoids, they inferred that the nucleoids are substantially *more condensed* (have higher density) in slowly growing cells than in rapidly growing cells. Since transcription foci do not occur in slowly growing cells, this would appear to oppose the trend predicted by the transcription-centric model of Jin et al. (2013).

We plan to extend our ribosome, RNAP, and nucleoid imaging and tracking studies to very slow growth conditions to test the nucleoid-ribosome mixing hypothesis in a very different physiological context.

Does mRNA Co-localize with Ribosomes or with the Gene from Which it was Transcribed?

According to our model of *E. coli* spatial organization, in rapid growth most translation is physically separated from most transcription. The exception is the small fraction of co-transcriptional translation (perhaps 10–15%), which should co-localize with the chromosomal DNA. If this picture is essentially correct, then the spatial distribution of total mRNA should mimic that of the ribosomes more closely than that of the chromosomal DNA.

As reviewed by Amster-Choder and co-workers (Buskila et al., 2014), there are two primary methods for labeling specific mRNAs: fluorescence *in situ* hybridization ("FISH," which requires fixation and permeabilization of the cells) and tagging of mRNA with a sequence to which a fluorescently labeled protein such as MS2-GFP binds specifically (which can be carried out on live cells). That review describes the current evidence on mRNA localization. The resulting picture is complicated and far from complete. It is possible that the picture of strong ribosome-nucleoid segregation and rapid diffusion of mRNA-70S polysomes away from their point of origin after completion of transcription holds only for rapidly growing cells.

Results in Rapidly Growing *E. coli* and *B. subtilis*

Early on, Golding and Cox (2004) used the MS2 scheme to label a very large, artificial mRNA transcribed from a plasmid. The resulting MS2-coated mRNA copies have mass of several MDa, comparable to a ribosome. They localized near the cell poles, much like ribosomes. The diffusive and localization properties of such large objects are not closely related to those of normal gene transcripts.

Amster-Choder and co-workers labeled specific *E. coli* genes in live cells under fast growth conditions using the MS2-GFP procedure (Nevo-Dinur et al., 2011). They found that the mRNA coding for the membrane proteins *lacY* and *bglF* was *localized at the membrane*. This behavior persisted even after treatment with translation-halting drugs such as Cam. The interpretation was that the mRNA itself contains information that targets the transcript to the location where the protein will ultimately be used, analogous to what occurs in eukaryotic cells. However, if the lead ribosome (i.e., the one connected to the RNA polymerase) is indeed necessary to prevent premature intrinsic or Rho-dependent termination of transcription (Burmann et al., 2010), then Cam may prevent efficient transcription as well. We suggest that this result might be interpreted as further evidence of transertion chains. The ribosome-mRNA-membrane linkages would persist after Cam treatment, so the localization of mRNA at the membrane would also persist.

In contrast, the mRNAs coding for the cytoplasmic proteins *cat* and *bglB* were distributed throughout the cytoplasm in a pattern that was described as helical (Nevo-Dinur et al., 2011). However, to our eyes the *cat* and *bglB* mRNA distributions look like the strongly segregated ribosomal spatial distributions in fast growth conditions (**Figure 4**).

We recently used the fluorescent stain SYTO RNASelect to monitor the degradation of total mRNA following Rif treatment (Bakshi et al., 2014a). SYTO RNASelect purportedly stains RNA in preference to DNA. For normally growing cells in EZRDM at 30°C, the resulting images look qualitatively similar to widefield images of ribosomes labeled by S2-YFP in the same growth conditions (**Figure 3**; Bakshi et al., 2012). The *caveat* here is the possibility that SYTO RNASelect is actually staining ribosomal RNA, not mRNA. However, most rRNA is buried in the ribosome interior and not accessible to the stain. In addition, the SYTO RNASelect signal decayed on a 10-min timescale after Rif treatment, consistent with the timescale of degradation of mRNA. Ribosomes are still present after the decay of SYTO RNASelect fluorescence.

Allowing for some re-interpretation of the Amster-Choder results, the available data on mRNA distributions from live, rapidly growing *E. coli* cells seems quite consistent with our overall picture of strong ribosome-DNA spatial segregation in rapidly growing cells.

Rapidly growing *B. subtilis* cells also exhibit strong nucleoid-ribosome segregation (Lewis et al., 2000). The first two-color widefield studies of cells with DNA stained and ribosomes labeled showed two or three ribosome-rich regions with interleaved regions of concentrated DNA. We know of no data on mRNA spatial distribution in *B. subtilis*, but we expect the distribution would closely mimic the ribosome distribution.

Results in Slowly Growing *C. crescentus* and *E. coli*

Using FISH in the slowly growing species *C. crescentus*, the Jacobs-Wagner lab found that six different mRNA messages formed images comprising one or a few *puncta* (Llopis et al., 2010). Remarkably, each of these puncta co-localized with the corresponding gene (also detected by FISH). In addition, the ribosomes and DNA mix extensively in *C. crescentus*. The mixing means that there is no contradiction between retention of mRNA near the site of its transcription and repeated translation of the same message. In slowly growing *E. coli*, similar mRNA and gene co-localization was observed for *lacZ*. This suggests that in slow growth conditions, mRNA diffusion away from the gene from which it was transcribed is very slow, in contrast to our picture of facile mRNA escape from the nucleoids in rapid growth conditions.

In a similar vein, Kuhlman and Cox (2012) used DAPI staining, "FROS" (a fluorescent reporter–operator system), FISH, and Venus labeling in *E. coli* to measure the spatial distributions (averaged over many cells) of the overall nucleoid, the *lacI* gene, its *lacI* mRNA product, and its LacI-Venus protein product. In slow growth, they varied the position of the *lacI* gene, either on an extrachromosomal plasmid, near *oriC*, or near *ter*. The gene and mRNA distributions are non-uniform and quite different in the three cases, exhibiting two, one, and three axial peaks, respectively. In all three cases, the mRNA distribution closely mimicked the gene distribution, again suggesting that the mRNA is not readily diffusing away from the location where it was transcribed. It is perhaps worth noting that the three different *lacI* gene and mRNA spatial distributions in slow growth conditions all have strong peaks in locations that would match our three ribosome-rich regions in rapidly growing cells. However, in rapidly growing *E. coli*, Kuhlman and Cox (2012) found that the *lacI* mRNA spatial distributions from FISH mimic the overall nucleoid distributions, again suggesting that the mRNA does not readily escape the nucleoids. That is, the *lacI* mRNA does not form a pattern mimicking the *ribosome* spatial distributions of **Figures 3** and **4**.

In slowly growing *E. coli* the evidence for two different genes (*lac Z* and *lacI*) indicates that the corresponding mRNA message remains in the vicinity of the gene itself rather than diffusing away (Kuhlman and Cox, 2012). It is possible that the behavior of *lacZ* and *lacI* mRNA in slowly growing cells is somehow not representative of the predominant mRNA behavior. It is also possible that strong DNA-ribosome segregation only occurs in rapidly growing cells. However, facile DNA-ribosome mixing in slow growth would seem contrary to the Kuhlman and Cox (2012) inference of higher DNA density in slowly growing cells than in rapidly growing cells. According to our Monte Carlo modeling, higher DNA density in slowly growing cells would exclude 70S-polysomes to a greater extent, assuming that they can escape the nucleoid interior in the first place.

At present, the mRNA data on slowly growing cells, both *E. coli* and *C. crescentus*, stand in opposition to a picture of strong DNA-ribosome segregation and diffusion of free mRNA to ribosome-rich regions where the bulk of translation occurs. We have not carried out detailed super-resolution studies of ribosomes in slowly growing *E. coli*. We plan to repeat the spatial distribution and diffusion studies to measure the degree of nucleoid-ribosome segregation and the diffusive properties of the ribosomes. Studies of the SYTO RNASelect staining pattern in slow growth are also of interest.

Returning to rapidly growing *E. coli* cells, it seems highly implausible to us that the nucleoids and ribosomes would be strongly segregated while the *overall* mRNA spatial distribution would mimic that of the chromosomal DNA. If that were the case, then what are the ~80% of ribosomes doing while they are so far from the majority of the mRNA that they must translate? The situation could be quite different in slowly growing *E. coli*. Additional work is needed.

Summary

We have presented a comprehensive model whose goal is to explain the coarse spatial organization of the *E. coli* transcription and translation machinery in rapid growth conditions with and without the influence of transcription- or translation-halting drugs. This builds on a great deal of earlier work from other labs (Odijk, 1998; Pelletier et al., 2012). Our model takes account of both the biochemical state of the cell (fraction of ribosomes engaged in translation, presence or absence of transertion chains) and the important underlying physical effects of excluded volume and conformational and translational entropy. By treating the ribosomes and DNA as a coupled biochemical–biophysical *system*, the model explains a wide variety of experimental results.

A wise cell biologist once complained that the trouble with physics-based hypotheses attempting to explain aspects of cellular behavior is that they can never be critically tested. There are no appropriate negative controls—you cannot turn the physics off. Our response would be that neither should the physics be neglected in our thinking. Physical models that make experimentally testable predictions have real value in cell biology. Our physical model can explain why in normal, rapid growth conditions 70S-polysomes and the nucleoids segregate, while free 30S and 50S subunits mix with the nucleoids. The model is oversimplified to be sure, but we believe it captures essential interactions that must not be ignored.

From the time-dependent drug studies, we further infer that transertion maintains the nucleoids in a sufficiently expanded state to enable recycling of 30S and 50S subunits between ribosome-rich regions and the nucleoid interior, where they initiate co-transcriptional translation. This in turn protects the nascent mRNA and prevents undesirable backtracking of RNA polymerase.

These concepts suggest how the strongly coupled transcription-translation-transertion *system* may enhance the maximum growth rate of *E. coli*. Strong segregation of DNA and RNAP from 70S-polysomes in rapid growth may significantly decrease the search times of RNAP for transcription initiation sites and of free 30S and 50S ribosomal subunits for translation initiation sites. Future work will provide additional quantitative detail in rapid growth and seek a better understanding of the situation in slowly growing cells.

Acknowledgments

Different aspects of this research were supported by the grants CHE-1213860 from the US National Science Foundation and R01-GM094510 and R01-GM093265 from the National Institutes of Health (NIGMS). We are deeply indebted to our co-workers and collaborators, including Dr. Jagannath Mondal, Prof. Arun Yethiraj, Prof. Mark Goulian, Dr. Albert Siryaporn, Dr. Benjamin Bratton, Dr. Renee Dalrymple, and Ms. Wenting Li.

References

Azam, T. A., Hiraga, S., and Ishihama, A. (2000). Two types of localization of the DNA-binding proteins within the *Escherichia coli* nucleoid. *Genes Cells* 5, 613–626. doi: 10.1046/j.1365-2443.2000.00350.x

Bakshi, S., Bratton, B. P., and Weisshaar, J. C. (2011). Subdiffraction-limit study of Kaede diffusion and spatial distribution in live *Escherichia coli*. *Biophys. J.* 101, 2535–2544. doi: 10.1016/j.bpj.2011.10.013

Bakshi, S., Choi, H., Mondal, J., and Weisshaar, J. C. (2014a). Time-dependent effects of transcription- and translation-halting drugs on the spatial distributions of the *E. coli*, chromosome and ribosomes. *Mol. Microbiol.* 94, 871–887. doi: 10.1111/mmi.12805

Bakshi, S., Choi, H., Rangarajan, N., Barns, K. J., Bratton, B. P., and Weisshaar, J. C. (2014b). Nonperturbative imaging of nucleoid morphology in live bacterial cells during an antimicrobial peptide attack. *Appl. Environ. Microbiol.* 80, 4977–4986. doi: 10.1128/AEM.00989-14

Bakshi, S., Dalrymple, R. M., Li, W., Choi, H., and Weisshaar, J. C. (2013). Partitioning of RNA polymerase activity in live *Escherichia coli* from analysis of single-molecule diffusive trajectories. *Biophys. J.* 105, 2676–2686. doi: 10.1016/j.bpj.2013.10.024

Bakshi, S., Siryaporn, A., Goulian, M., and Weisshaar, J. C. (2012). Superresolution imaging of ribosomes and RNA polymerase in live *Escherichia coli* cells. *Mol. Microbiol.* 85, 21–38. doi: 10.1111/j.1365-2958.2012.08081.x

Bernstein, J. A., Khodursky, A. B., Lin, P. H., Lin, S.,-C., and Cohen, S. N. (2002). Global analysis of mRNA decay and abundance in *Escherichia coli* at single-gene resolution using two-color fluorescent DNA microarrays. *Proc. Natl. Acad. Sci. U.S.A.* 99, 9697–9702. doi: 10.1073/pnas.112318199

Betzig, E., Patterson, G. H., Sougrat, R., Lindwasser, O. W., Olenych, S., Bonifacino, J. S., et al. (2006). Imaging intracellular fluorescent proteins at nanometer resolution. *Science* 313, 1642–1645. doi: 10.1126/science.1127344

Bratton, B. P., Mooney, R. A., and Weisshaar, J. C. (2011). Spatial distribution and diffusive motion of RNA polymerase in live *Escherichia coli*. *Bacteriol. J.* 193, 5138–5146. doi: 10.1128/JB.00198-11

Bremer, H., and Dennis, P. (1996). "Modulation of chemical composition and other parameters of the cell by growth rate," in *Escherichia coli and Salmonella: Cellular and Molecular Biology*, eds F. C. Neidhardt and R. Curtiss (Washington, DC: ASM Press), 1553–1569.

Burmann, B. M., Knauer, S. H., Sevostyanova, A., Schweimer, K., Mooney, R. A., Landick, R., et al. (2012). An alpha helix to beta barrel domain switch transforms the transcription factor RfaH into a translation factor. *Cell* 150, 291–303. doi: 10.1016/j.cell.2012.05.042

Burmann, B. M., Schweimer, K., Luo, X., Wahl, M. C., Stitt, B. L., Gottesman, M. E., et al. (2010). A NusE:NusG complex links transcription and translation. *Science* 328, 501–504. doi: 10.1126/science.1184953

Buskila, A. A., Kannaiah, S., and Amster, O.-C. (2014). RNA localization in bacteria. *RNA Biol.* 11, 1051–1060. doi: 10.4161/rna.36135

Cabrera, J. E., Cagliero, C., Quan, S., Squires, C. L., and Jin, D. J. (2009). Active transcription of rRNA operons condenses the nucleoid in *Escherichia coli*: examining the effect of transcription on nucleoid structure in the absence of transertion. *J. Bacteriol.* 191, 4180–4185. doi: 10.1128/JB.01707-08

Cabrera, J. E., and Jin, D. J. (2006). Active transcription of rRNA operons is a driving force for the distribution of RNA polymerase in bacteria: effect of extrachromosomal copies of rrnB on the in vivo localization of RNA polymerase. *J. Bacteriol.* 188, 4007–4014. doi: 10.1128/JB.01893-05

Cagliero, C., Grand, R. S., Jones, M. B., Jin, D. J., and O'Sullivan, J. M. (2013). Genome conformation capture reveals that the *Escherichia coli* chromosome is organized by replication and transcription. *Nucleic Acid Res.* 41, 6058–6071. doi: 10.1093/nar/gkt325

Dame, R. T. (2005). The role of nucleoid-associated proteins in the organization and compaction of bacterial chromatin. *Mol. Microbiol.* 56, 858–870. doi: 10.1111/j.1365-2958.2005.04598.x

Datsenko, K. A., and Wanner, B. L. (2000). One-step inactivation of chromosomal genes in *Escherichia coli* K-12 using PCR products. *Proc. Natl. Acad. Sci. U.S.A.* 97, 6640–6645. doi: 10.1073/pnas.120163297

Dennis, P. P., Ehrenberg, M., and Bremer, H. (2004). Control of rRNA synthesis in *Escherichia coli*: a systems biology approach. *Microbiol. Mol. Biol. Rev.* 68, 639–668. doi: 10.1128/MMBR.68.4.639-668.2004

Endesfelder, U., Finan, K., Holden, S. J., Cook, P. R., Kapanidis, A. N., and Heilemann, M. (2013). Multiscale spatial organization of RNA polymerase in *Escherichia coli*. *Biophys. J.* 105, 172–181. doi: 10.1016/j.bpj.2013.05.048

Fisher, J. K., Bourniquel, A., Witz, G., Weiner, B., Prentiss, M., and Kleckner, N. (2013). Four-dimensional imaging of *E. coli*, nucleoid organization and dynamics in living cells. *Cell* 153, 882–895. doi: 10.1016/j.cell.2013.04.006

Fishov, I., and Woldringh, C. L. (1999). Visualization of membrane domains in *Escherichia coli*. *Mol. Microbiol.* 32, 1166–1172. doi: 10.1046/j.1365-2958.1999.01425.x

Golding, I., and Cox, E. C. (2004). RNA dynamics in live *Escherichia coli* cells. *Proc. Natl. Acad. Sci. U.S.A.* 101, 11310–11315. doi: 10.1073/pnas.0404443101

Gould, T. J., Gunewardene, M. S., Gudheti, M. V., Verkhusha, V. V., Yin, S. R., Gosse, J. A., et al. (2008). Nanoscale imaging of molecular positions and anisotropies. *Nat. Methods* 5, 1027–1030. doi: 10.1038/nmeth.1271

Hess, S. T., Girirajan, T. P., and Mason, M. D. (2006). Ultra-high resolution imaging by fluorescence photoactivation localization microscopy. *Biophys. J.* 91, 4258–4272. doi: 10.1529/biophysj.106.091116

Jin, D. J., and Cabrera, J. E. (2006). Coupling the distribution of RNA polymerase to global gene regulation and the dynamic structure of the bacterial nucleoid in *Escherichia coli*. *Struct. J. Biol.* 156, 284–291. doi: 10.1016/j.jsb.2006.07.005

Jin, D. J., Cagliero, C., and Zhou, Y. N. (2013). Role of RNA polymerase and transcription in the organization of the bacterial nucleoid. *Chem. Rev.* 113, 8662–8682. doi: 10.1021/cr4001429

Jung, Y., Jeon, C., Kim, J., Jeong, H., Jun, S., and Ha, B.-Y. (2012). Ring polymers as model bacterial chromosomes: confinement, chain topology, single chain statistics, and how they interact. *Soft Matter* 8, 2095–2102. doi: 10.1039/C1SM05706E

Kellenberger, E. (1991). Functional consequences of improved structural information on bacterial nucleoids. *Res. Microbiol.* 142, 229–238. doi: 10.1016/0923-2508(91)90035-9

Kuhlman, T. E., and Cox, E. C. (2012). Gene location and DNA density determine transcription factor distributions in *Escherichia coli*. *Mol Syst. Biol.* 8:610. doi: 10.1038/msb.2012.42

Landgraf, D., Okumus, B., Chien, P., Baker, T. A., and Paulsson, J. (2012). Segregation of molecules at cell division reveals native protein localization. *Nat. Methods* 9, 480–498. doi: 10.1038/nmeth.1955

Le, T. B., Imakaev, M. V., Mirny, L. A., and Laub, M. T. (2013). High-resolution mapping of the spatial organization of a bacterial chromosome. *Science* 342, 731–734. doi: 10.1126/science.1242059

Le, T. B., and Laub, M. T. (2014). New approaches to understanding the spatial organization of bacterial genomes. *Curr. Opin. Microbiol.* 22C, 15–21. doi: 10.1016/j.mib.2014.09.014

Lewis, P. J., Thaker, S. D., and Errington, J. (2000). Compartmentalization of transcription and translation in *Bacillus subtilis*. *EMBO J.* 19, 710–718. doi: 10.1093/emboj/19.4.710

Libby, E. A., Roggiani, M., and Goulian, M. (2012). Membrane protein expression triggers chromosomal locus repositioning in bacteria. *Proc. Natl. Acad. Sci. U.S.A.* 109, 7445–7450. doi: 10.1073/pnas.1109479109

Lippincott-Schwartz, J., and Patterson, G. H. (2009). Photoactivatable fluorescent proteins for diffraction-limited and super-resolution imaging. *Trends Cell Biol.* 19, 555–565. doi: 10.1016/j.tcb.2009.09.003

Llopis, P. M., Jackson, A. F., Sliusarenko, O., Surovtsev, I., Heinritz, J., Emonet, T., et al. (2010). Spatial organization of the flow of genetic information in bacteria. *Nature* 466, U77–U90. doi: 10.1038/nature09152

Marenduzzo, D., Micheletti, C., and Cook, P. R. (2006). Entropy-driven genome organization. *Biophys. J.* 90, 3712–3721. doi: 10.1529/biophysj.105.077685

McGary, K., and Nudler, E. (2013). RNA polymerase and the ribosome: the close relationship. *Curr. Opin. Microbiol.* 16, 112–117. doi: 10.1016/j.mib.2013.01.010

Miller, O. L., Hamkalo, B. A., and Thomas, C. A. (1970). Visualization of bacterial genes in action. *Science* 169, 392–395. doi: 10.1126/science.169.3943.392

Mondal, J., Bratton, B. P., Li, Y., Yethiraj, A., and Weisshaar, J. C. (2011). Entropy-based mechanism of ribosome-nucleoid segregation in *E. coli*, cells. *Biophys. J.* 100, 2605–2613. doi: 10.1016/j.bpj.2011.04.030

Neidhardt, F. C., Bloch, P. L., and Smith, D. F. (1974). Culture medium for enterobacteria. *J. Bacteriol.* 119, 736–747.

Nevo-Dinur, K., Nussbaum, A.,-Shochat, Ben, S.,-Y., and Amster, O.-C. (2011). Translation-independent localization of mRNA in *E. coli*, *Science* 331, 1081–1084. doi: 10.1126/science.1195691

Norris, V., and Madsen, M. S. (1995). Autocatalytic gene expression occurs via transertion and membrane domain formation and underlies differentiation in bacteria: a model. *Mol. J. Biol.* 253, 739–748. doi: 10.1006/jmbi.1995.0587

Odijk, T. (1998). Osmotic compaction of supercoiled DNA into a bacterial nucleoid. *Biophys. Chem.* 73, 23–29. doi: 10.1016/S0301-4622(98)00115-X

Pelletier, J., Halvorsen, K., Ha, B. Y., Paparcone, R., Sandler, S. J., Woldringh, C. L., et al. (2012). Physical manipulation of the *Escherichia coli* chromosome reveals its soft nature. *Proc. Natl. Acad. Sci. U.S.A.* 109, E2649–E2656. doi: 10.1073/pnas.1208689109

Pettijohn, D. E. (1982). Structure and properties of the bacterial nucleoid. *Cell* 30, 667–669. doi: 10.1016/0092-8674(82)90269-0

Proshkin, S., Rahmouni, A. R., Mironov, A., and Nudler, E. (2010). Cooperation between translating ribosomes and RNA polymerase in transcription elongation. *Science* 328, 504–508. doi: 10.1126/science.1184939

Robinow, C., and Kellenberger, E. (1994). The bacterial nucleoid revisited. *Microbiol. Rev.* 58, 211–232.

Rust, M. J., Bates, M., and Zhuang, X. (2006). Sub-diffraction-limit imaging by stochastic optical reconstruction microscopy (STORM). *Nat. Methods* 3, 793–795. doi: 10.1038/nmeth929

Sanamrad, A., Persson, F., Lundius, E. G., Fange, D., Gynna, A. H., and Elf, J. (2014). Single-particle tracking reveals that free ribosomal subunits are not excluded from the *Escherichia coli* nucleoid. *Proc. Natl. Acad. Sci. U.S.A.* 111, 11413–11418. doi: 10.1073/pnas.1411558111

Shendruk, T. N., Bertrand, M., de Haan, H. W., Harden, J. L., and Slater, G. W. (2015). Simulating the entropic collapse of coarse-grained chromosomes. *Biophys. J.* 108, 810–820. doi: 10.1016/j.bpj.2014.11.3487

Svetlov, V., and Nudler, E. (2012). Unfolding the bridge between transcription and translation. *Cell* 150, 243–245. doi: 10.1016/j.cell.2012.06.025

van Helvoort, J. M., Kool, J., and Woldringh, C. L. (1996). Chloramphenicol causes fusion of separated nucleoids in *Escherichia coli* K-12 cells and filaments. *J. Bacteriol.* 178, 4289–4293.

Vrljic, M., Nishimura, S. Y., and Moerner, W. E. (2007). Single-molecule tracking. *Methods Mol. Biol.* 398, 193–219. doi: 10.1007/978-1-59745-513-8_14

Wang, T., Ingram, C., and Weisshaar, J. C. (2010). Model lipid bilayer with facile diffusion of lipids and integral membrane proteins. *Langmuir* 26, 11157–11164. doi: 10.1021/la101046r

Wang, W. Q., Li, G. W., Chen, C. Y., Xie, X. S., and Zhuang, X. W. (2011). Chromosome organization by a nucleoid-associated protein in live bacteria. *Science* 333, 1445–1449. doi: 10.1126/science.1204697

Woldringh, C. L. (2002). The role of co-transcriptional translation and protein translocation (transertion) in bacterial chromosome segregation. *Mol. Microbiol.* 45, 17–29. doi: 10.1046/j.1365-2958.2002.02993.x

Woldringh, C. L., Jensen, P. R., and Westerhoff, H. V. (1995). Structure and partitioning of bacterial DNA: determined by a balance of compaction and expansion forces? *FEMS Microbiol. Lett.* 131, 235–242. doi: 10.1111/j.1574-6968.1995.tb07782.x

Youngren, B., Nielsen, H. J., Jun, S., and Austin, S. (2014). The multifork *Escherichia coli* chromosome is a self-duplicating and self-segregating thermodynamic ring polymer. *Genes Dev.* 28, 71–84. doi: 10.1101/gad.231050.113

Zhang, J., Campbell, R. E., Ting, A. Y., and Tsien, R. Y. (2002). Creating new fluorescent probes for cell biology. *Nat. Rev. Mol. Cell Biol.* 3, 906–918. doi: 10.1038/nrm976

Zimmerman, S. B. (2002). Toroidal nucleoids in *Escherichia coli* exposed to chloramphenicol. *Struct. J. Biol.* 138, 199–206. doi: 10.1016/S1047-8477(02)00036-9

Zimmerman, S. B. (2006). Shape and compaction of *Escherichia coli* nucleoids. *J. Struct. Biol.* 156, 255–261. doi: 10.1016/j.jsb.2006.03.022

Zimmerman, S. B., and Murphy, L. D. (1996). Macromolecular crowding and the mandatory condensation of DNA in bacteria. *FEBS Lett.* 390, 245–248. doi: 10.1016/0014-5793(96)00725-9

12

The DnaA cycle in *Escherichia coli*: Activation, function and inactivation of the initiator protein

Tsutomu Katayama*, Kazutoshi Kasho and Hironori Kawakami

Department of Molecular Biology, Graduate School of Pharmaceutical Sciences, Kyushu University, Fukuoka, Japan

***Correspondence:**
Tsutomu Katayama
katayama@phar.kyushu-u.ac.jp

This review summarizes the mechanisms of the initiator protein DnaA in replication initiation and its regulation in *Escherichia coli*. The chromosomal origin (*oriC*) DNA is unwound by the replication initiation complex to allow loading of DnaB helicases and replisome formation. The initiation complex consists of the DnaA protein, DnaA-initiator-associating protein DiaA, integration host factor (IHF), and *oriC*, which contains a duplex-unwinding element (DUE) and a DnaA-oligomerization region (DOR) containing DnaA-binding sites (DnaA boxes) and a single IHF-binding site that induces sharp DNA bending. DiaA binds to DnaA and stimulates DnaA assembly at the DOR. DnaA binds tightly to ATP and ADP. ATP-DnaA constructs functionally different sub-complexes at DOR, and the DUE-proximal DnaA sub-complex contains IHF and promotes DUE unwinding. The first part of this review presents the structures and mechanisms of *oriC*-DnaA complexes involved in the regulation of replication initiation. During the cell cycle, the level of ATP-DnaA level, the active form for initiation, is strictly regulated by multiple systems, resulting in timely replication initiation. After initiation, regulatory inactivation of DnaA (RIDA) intervenes to reduce ATP-DnaA level by hydrolyzing the DnaA-bound ATP to ADP to yield ADP-DnaA, the inactive form. RIDA involves the binding of the DNA polymerase clamp on newly synthesized DNA to the DnaA-inactivator Hda protein. In *datA*-dependent DnaA-ATP hydrolysis (DDAH), binding of IHF at the chromosomal locus *datA*, which contains a cluster of DnaA boxes, results in further hydrolysis of DnaA-bound ATP. SeqA protein inhibits untimely initiation at *oriC* by binding to newly synthesized *oriC* DNA and represses *dnaA* transcription in a cell cycle dependent manner. To reinitiate DNA replication, ADP-DnaA forms oligomers at DnaA-reactivating sequences (*DARS1* and *DARS2*), resulting in the dissociation of ADP and the release of nucleotide-free apo-DnaA, which then binds ATP to regenerate ATP-DnaA. *In vivo*, *DARS2* plays an important role in this process and its activation is regulated by timely binding of IHF to *DARS2* in the cell cycle. Chromosomal locations of *DARS* sites are optimized for the strict regulation for timely replication initiation. The last part of this review describes how DDAH and DARS regulate DnaA activity.

Keywords: DnaA, *oriC*, *DARS*, *datA*, chromosome replication

INTRODUCTION

The genome of *Escherichia coli* consists of a single circular 4.6 Mb chromosome, with a unique replication origin called *oriC*. Replication initiation at *oriC* results in construction of a pair of replisomes, which migrate bi-directionally to replicate the entire chromosome. Replication initiation at *oriC* is regulated to occur only once during each cell cycle, and the timing of initiation is coordinated with the cellular growth rate. Even when cells grow rapidly and the copy number of *oriC* increases to more than two per cell, initiation occurs at sister *oriC* regions simultaneously only once at a specific time during the cell cycle. As such, the time of initiation at *oriC* is regulated and re-initiation during the same cell cycle is strictly repressed (Skarstad and Katayama, 2013; Wolański et al., 2015; Riber et al., 2016).

The 245 bp minimal *oriC* region has multiple binding sites for the chromosomal replication initiator protein DnaA (DnaA boxes), and a single binding site for the integration host factor (IHF), in addition to an AT-rich duplex-unwinding element (DUE) (**Figures 1A, 2A**; Kaguni, 2011; Leonard and Grimwade, 2015; Wolański et al., 2015; Shimizu et al., 2016). The 9-mer DnaA box consensus sequence is 5′-TTATnCACA-3′. DnaA-initiator-associating protein DiaA is a DnaA-binding protein that stimulates ATP-bound DnaA (ATP-DnaA) assembly on *oriC*, and that is required for timely replication initiation (Ishida et al., 2004; Keyamura et al., 2007). Binding of IHF to DNA causes a sharp (120–180°) bend in the double helix (Swinger and Rice, 2004). A complex consisting of *oriC*, IHF, DiaA, and oligomeric ATP-DnaA is considered to make up the initiation complex in *E. coli* (Keyamura et al., 2007, 2009). This complex unwinds the *oriC* DUE, enabling loading of DnaB helicases onto single-stranded DNA (ssDNA) by specific protein–protein interactions and dissociations, which in turn leads to formation of replisomes (for a review, see Bell and Kaguni, 2013).

Multiple negative and positive regulatory systems target the *oriC*, the *dnaA* gene, and DnaA and work harmoniously to ensure that initiation occurs in a timely manner, in some cases via negative feedback from DNA replication. As for *oriC*, the minimal region contains 11 sites with 5′-GATC-3′ sequences that are specific targets of DNA adenine methylase (Dam) (Waldminghaus and Skarstad, 2009). As GATC is palindromic in duplex DNA, A residues in both strands can be methylated. In newly replicated DNA, only the sites on the parental strand are methylated, whereas those on the daughter strand are unmethylated. The hemimethylated state of *oriC* persists for ~10 min in cells with a doubling time of 30 min depending on SeqA protein (Lu et al., 1994). SeqA protein has an N-terminal self-oligomerization domain and a C-terminal DNA-binding domain, and binds to the hemimethylated sites, forming self-oligomers. *oriC*–SeqA complexes inhibit binding of DnaA to *oriC*, blocking re-initiation from newly replicated DNA. SeqA sequestration is, therefore, a negative-feedback system coupled to DNA replication (Waldminghaus and Skarstad, 2009; Skarstad and Katayama, 2013). In another negative-feedback system, transcription of *dnaA* is autoregulated by DnaA, and is also repressed by SeqA–Dam-dependent post-replicative regulation (Campbell and Kleckner, 1990; Bogan and Helmstetter, 1997; Speck et al., 1999; Waldminghaus and Skarstad, 2009). These SeqA mechanisms have been well documented elsewhere (Waldminghaus and Skarstad, 2009), and this review will focus instead on regulation of DnaA protein.

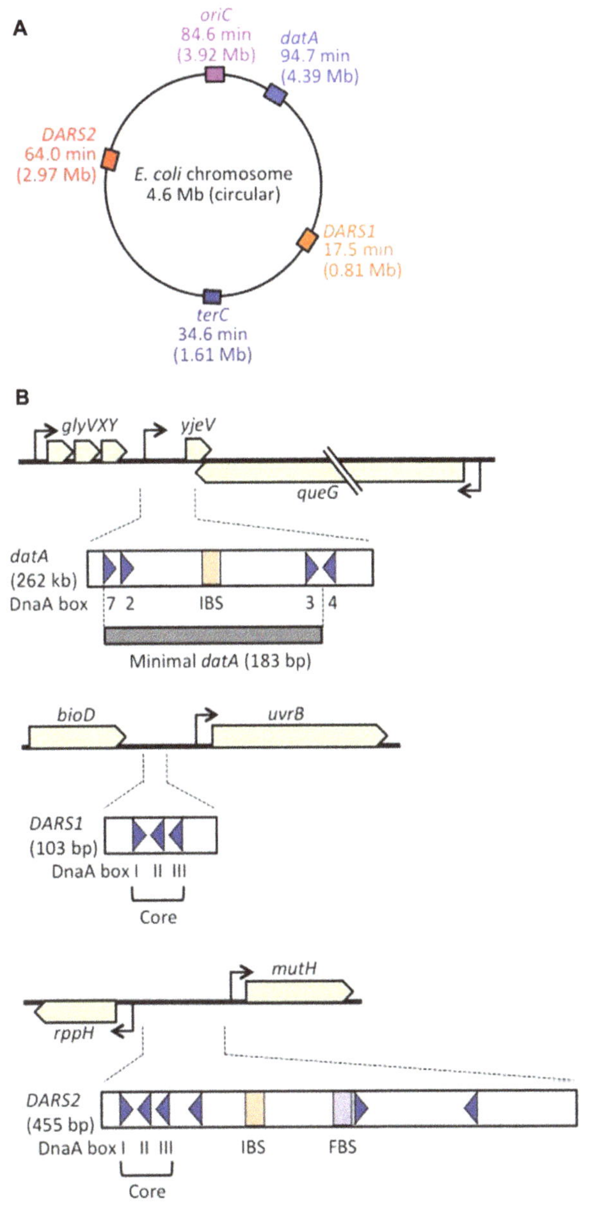

FIGURE 1 | Regulatory DNA elements involved in replication initiation on the *E. coli* genome. **(A)** Schematic representation of the genomic loci of *oriC*, *datA*, *DARS1*, and *DARS2* (and *terC*) in the 4.6 Mb circular *E. coli* genome, with positions also indicated on the scale of 0–100 min. **(B)** Structures of *datA*, *DARS1*, and *DARS2*. Open or gray bars indicate minimal regions. Triangles represent 9 bp DnaA-binding sites (DnaA boxes). Filled squares represent IHF-binding sites (IBS; shown in orange) and a Fis-binding site (FBS; shown in purple). Minimal *datA* consists of DnaA boxes 2, 3, and 7 and a single IBS. DnaA box 4 stimulates DDAH *in vitro*. *DARS1* and *DARS2* both have core regions containing DnaA boxes I–III. *DARS2* also contains additional DnaA boxes and regulatory IBS and FBS.

Escherichia coli cells have at least three regulatory systems for DnaA activity (Skarstad and Katayama, 2013; Riber et al., 2016). In regulatory inactivation of DnaA (RIDA), ATP-DnaA is inactivated in a negative-feedback manner coupled to DNA replication (Katayama et al., 2010). In this system, the clamp subunit of DNA polymerase III holoenzyme has a key role; it remains on the nascent DNA strand after Okazaki-fragment completion, and the clamp–DNA complex binds to the ADP form of DNA regulatory inactivator Hda protein, an 'ATPases associated with various cellular activities' (AAA+) protein with an N-terminal clamp-binding site (Katayama et al., 1998; Kato and Katayama, 2001; Su'etsugu et al., 2008; Baxter and Sutton, 2012; Kim et al., 2017). The resultant ADP-Hda–clamp–DNA complex interacts with ATP-DnaA molecules catalytically, simulating ATP hydrolysis to yield ADP-DnaA. This system is predominant in the inactivation of DnaA after replication initiation, and strongly represses over-initiation of replication. RIDA has been well documented elsewhere (Katayama et al., 2010; Skarstad and Katayama, 2013; Riber et al., 2016), and this review will focus on the two other DnaA regulatory systems.

DDAH (*datA*-dependent DnaA-ATP hydrolysis) regulates the inactivation of DnaA (Kasho and Katayama, 2013; Riber et al., 2016) and requires the non-coding, chromosomal DNA element *datA* (Kitagawa et al., 1996, 1998; **Figure 1**). The 262 bp minimal *datA* region has a specific DnaA-box cluster and a single IHF-binding site (IBS) (Kasho and Katayama, 2013; Kasho et al., 2017; **Figure 1B**). A *datA*–IHF complex forms after replication initiation, and stimulates formation of

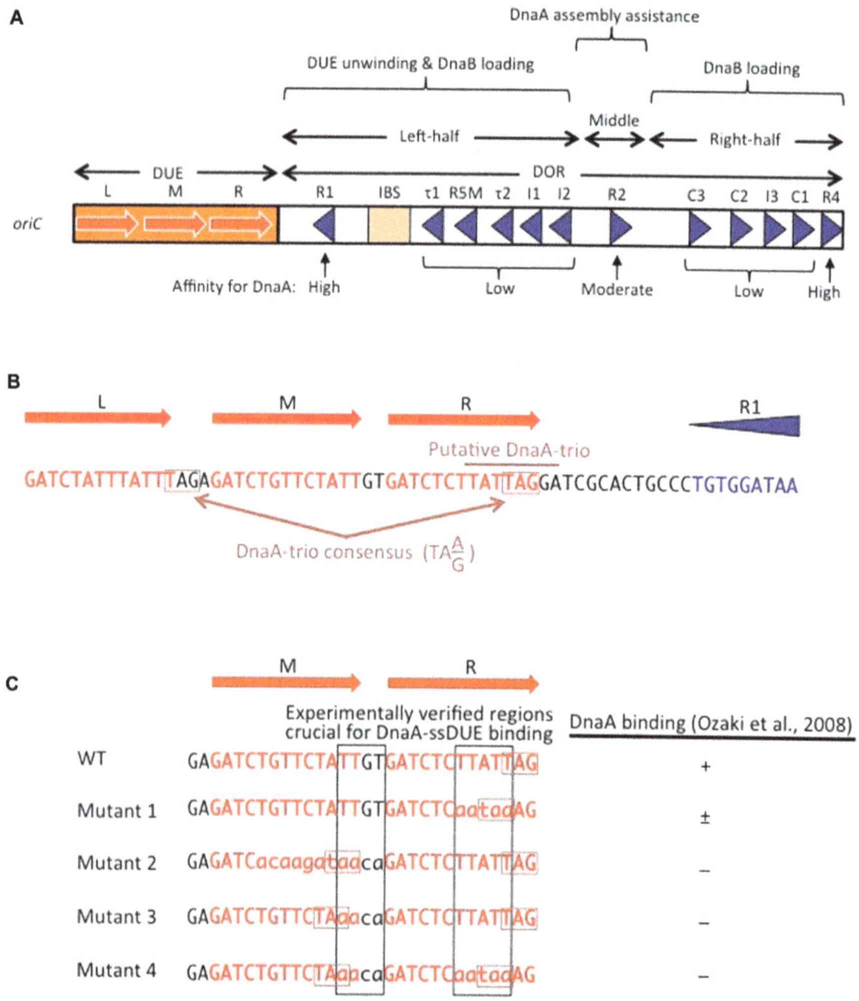

FIGURE 2 | Basic features of *oriC*. **(A)** Structure of *oriC*. Features of the minimal 245 bp *oriC* sequence are shown, including DnaA boxes (triangles), IHF-binding site (IBS; rectangle), and duplex-unwinding element (DUE) AT-rich 13 bp elements (red arrows). Boxes with high, moderate, and low affinity for DnaA are indicated. **(B)** A portion of the *oriC* sequence including the DUE and R1 is shown in detail, with the putative DnaA-trio indicated (Richardson et al., 2016). Perfect DnaA-trio consensus sequences are boxed. **(C)** Effect of mutations on single-stranded DUE (ssDUE) DnaA binding. A portion of the DUE including the M and R 13-mers is indicated in red. Wild-type sequences and mutations are in uppercase and lowercase, respectively. Perfect DnaA-trio consensus sequences are boxed in red. DnaA binding with the indicated ssDUE sequences is summarized (Ozaki et al., 2008). Experimentally verified regions essential for DnaA-ssDUE binding are boxed in black (i.e., TTGT and TTATT). Note that even if the DnaA-trio consensus is preserved, mutations in the boxed sequences abolish DnaA binding.

specific ATP-DnaA oligomers, and hydrolysis of DnaA-bound ATP, independently of RIDA.

DnaA-reactivating sequence (DARS) sites promote reactivation of DnaA (Fujimitsu et al., 2009). The *E. coli* genome has at least two DARS sites (*DARS1* and *DARS2*) (**Figure 1**), which contain specific DnaA-box clusters (**Figure 1B**), and promote oligomerization of ADP-DnaA, leading to the release of ADP (Fujimitsu et al., 2009; Kasho et al., 2014). The resultant nucleotide-free apo-DnaA preferentially dissociates from the complex, and binds ATP. Both *DARS1* and *DARS2* are required to sustain timely initiation *in vivo*. In particular, *DARS2* is activated by binding of IHF and Fis (a nucleoid-associated factor); IHF binding occurs temporarily before replication initiation (Kasho et al., 2014). The RIDA, DDAH, and DARS DnaA-regulating systems are all required for timely initiation of chromosomal replication.

This review provides an up-to-date detailed synopsis of the various factors and mechanisms involved, particularly from the perspective of the mechanisms involved in the regulation of the timely activation and inactivation of the initiator protein DnaA. Sections "Basic Features of *oriC* and DnaA and DnaA Complex on *oriC*" describe mainly the principal features of *oriC*, DnaA and *oriC*-DnaA complexes, which are based on progress gained over the last 30 years. Sections "DDAH System and DARS System" mainly describe recent developments regarding DDAH and DARS systems and the salient features of DnaA complexes at *datA* and DARS sites. These latter sections provide information on the mechanisms of functionally different DnaA-DNA complexes. Readers who are interested only in DDAH and DARS systems might skip the Section "DnaA Complex on *oriC*," while readers who are interested only in *oriC*-DnaA complexes might skip the Sections "DDAH System and DARS System."

BASIC FEATURES OF *oriC* AND DnaA

oriC

The *E. coli oriC* is a 245 bp DNA element located at 84.6 min of the circular chromosome, and composed of two functionally distinct regions: the DUE and the DnaA-oligomerization region (DOR) (**Figures 1A, 2A**). The DUE has three AT-rich 13-mer sequences (L, M, and R) with the consensus 5′-GATCTnTTnTTTT-3′, where the Watson and the Crick strands are T-rich and A-rich, respectively (Bramhill and Kornberg, 1988; Messer, 2002). An additional AT-rich region flanking the L sequence assists in DUE unwinding (Messer, 2002). When the DUE is unwound, the single-stranded T-rich DUE strand binds to DnaA oligomers that are bound to the DOR (Ozaki et al., 2008) (also see below). Stable binding of the single-stranded DUE (ssDUE) requires the presence of a region spanning at least the M and R sequences; 5′-TTGT-3′ and 5′-TTATT-3′ sequences are specifically required for the binding (Ozaki et al., 2008; **Figure 2B**). In *Bacillus subtilis oriC*, a repeating trinucleotide motif (DnaA-trio) with the consensus 5′-TA(A/G)-3′ is present in the AT-rich DUE and has been proposed to sustain ssDNA binding by direct interaction with DnaA (**Figure 2**; Richardson et al., 2016). As requirements of specific sequences in DUE are different, *E. coli* and *B. subtilis* may differ in their mechanisms for recognition of ssDUE by DnaA (**Figure 2B**).

The DOR is directly connected to the right edge of the DUE, and contains 12 DnaA boxes, which are present in both orientations (**Figure 2A**). The DOR can be subdivided into three structurally distinct sub-regions: left-half, middle, and right-half (Leonard and Grimwade, 2015; Shimizu et al., 2016). Of these, the left-half sub-region contains six DnaA boxes (R1, τ1, R5M, τ2, I1, I2) in one orientation, and the right-half sub-region contains five boxes (C3, C2, I3, C1, R4) in the opposite orientation, whereas the middle region contains one box (R2) in the right-half orientation (McGarry et al., 2004; Rozgaja et al., 2011; Ozaki and Katayama, 2012; Ozaki et al., 2012a; Shimizu et al., 2016; **Figure 2A**). DnaA complexes assembled on these different sub-regions have specific roles in DUE unwinding and DnaB-helicase loading.

DnaA boxes R1 and R4 (at the left and right ends of the DOR, respectively) match the consensus 5′-TTATnCACA-3′ and have the highest affinity among DOR boxes for DnaA (in both its ATP and ADP bound forms) (Messer, 2002; McGarry et al., 2004; Kawakami et al., 2005; **Figure 2A**). DnaA box R2 has basically moderate affinity, and ATP-DnaA more efficiently binds to this box present in DOR than ADP-DnaA (Messer, 2002; McGarry et al., 2004; Kawakami et al., 2005; Keyamura et al., 2009). All the other DnaA boxes (τ1–I2 and R4–C1 in the left and right halves, respectively) are clustered, and each individual DnaA box only has a low affinity (McGarry et al., 2004; Kawakami et al., 2005; Rozgaja et al., 2011). Although these low-affinity boxes have only moderate similarities to the DnaA-box consensus, ATP-DnaA-specific cooperative binding does occur. This binding depends on DnaA Arg285 for recognition of DnaA-bound ATP (Kawakami et al., 2005; Keyamura et al., 2007, 2009; Kaur et al., 2014; Sakiyama et al., 2017) (for details, see below). ATP-DnaA binding to box τ1 occurs in the absence of IHF binding, but not in the presence of IHF binding, probably because of steric interference (Kawakami et al., 2005; Sakiyama et al., 2017) (see below).

The left-half DOR has a specific binding motif for the nucleoid-associated, DNA-bending protein IHF (**Figure 2A**). The consensus of the IBS is 5′-(A/T)ATCAAnnnnTT(A/G)-3′ (Swinger and Rice, 2004). *In vitro*, IHF binding stimulates DUE unwinding and replication from *oriC*; another nucleoid-associated protein (HU) can substitute for IHF in these reactions (Hwang and Kornberg, 1992). IHF stimulates DnaA binding to moderate-affinity (R2) and low-affinity (R5M and I1-3) sites (Grimwade et al., 2000; McGarry et al., 2004), and promotes binding to ssDUE by DnaA that is complexed with the DOR (Ozaki and Katayama, 2012). Double mutations resulting in deficiency of both IHF and HU severely inhibit cell growth, and cause synthetic lethality at high temperatures (Kano and Imamoto, 1990). Although a Fis binding consensus sequence is present between the R2 and C3 boxes, mutant cells bearing a sequence defective in Fis binding and the results of *in vitro* replication systems suggest that binding of Fis to *oriC* is not an important element in the regulation

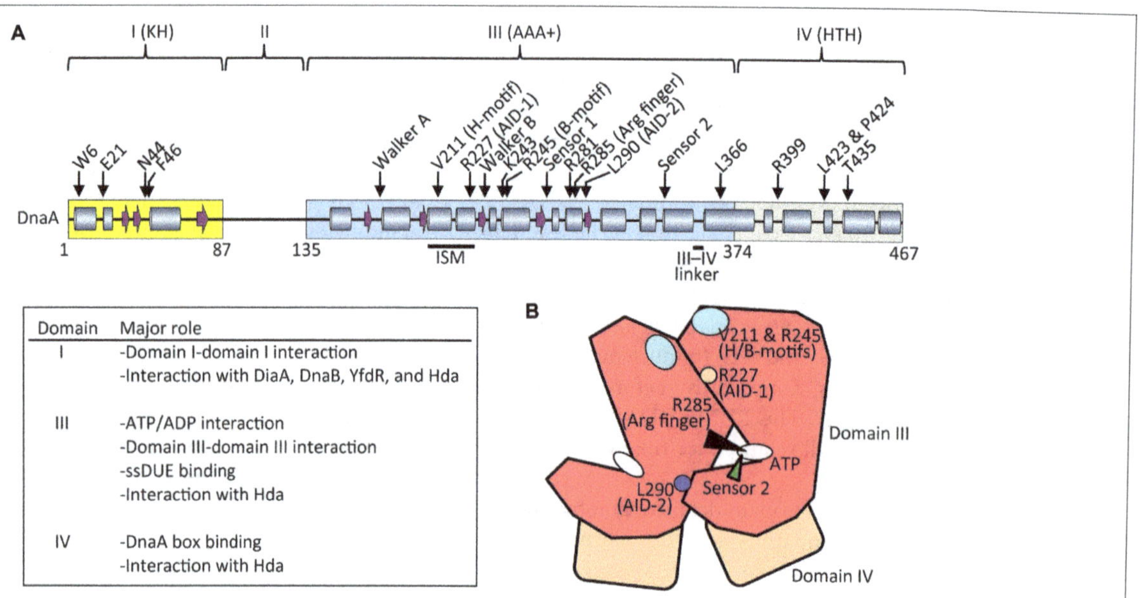

FIGURE 3 | Basic features of DnaA. **(A)** Possible secondary structure of full length DnaA. Domains I, III, and IV are shown in yellow, blue, and gray boxes, respectively. α helices (cylinders) and β strands (arrows) were determined from the NMR structure of DnaA domain I–II (PDB: 2E0G; Abe et al., 2007) and the crystal structure of DnaA domain IV (PDB: 1J1V; Fujikawa et al., 2003), combined with a homology model of DnaA domain III generated using the crystal structure of *Thermotoga maritima* DnaA (PDB: 2Z4S; Ozaki et al., 2008) (SWISS-MODEL Repository: P03004). For clarity, extremely short helices in the homology model are omitted according to the crystal structure of *A. aeolicus* DnaA domain III (Erzberger et al., 2002). Representative motifs in the AAA+ fold and residues crucial for DnaA functions are indicated. Major roles of domains I, II, and IV are summarized. Arg finger, arginine finger (Arg285). **(B)** Schematic presentation of domain III (*red*) and IV (*brown*). Important motifs and residues for DnaA–DnaA interaction and ATP recognition/hydrolysis in DnaA complexes are indicated. ATP, white oval; arginine finger Arg285, black arrowhead; AID-1 Arg227, orange circle; AID-2 Leu270, blue circle; sensor 2, green arrowhead; and H/B-motifs Val211 and Arg245, cyan oval.

of initiation (Margulies and Kaguni, 1998; Weigel et al., 2001).

DnaA

DnaA is a 52.5 kDa basic protein composed of 473 amino acids in four domains (Ozaki and Katayama, 2009; Kaguni, 2011; **Figure 3A**). The largest (domain III) has an AAA+ ATPase fold that contains Walker A and B ATP/ADP-binding/hydrolysis motifs, motifs for domain III–domain III head-to-tail interaction, and motifs for ssDUE binding. Domain III AAA+ sensor 1 motifs and N-linker motif contribute to an extraordinarily high affinity for ATP/ADP (K_d = 10–100 nM) (Kawakami et al., 2006; Ozaki et al., 2012b). ATP binding causes conformational changes in DnaA, and structural differences between ATP-DnaA and ADP-DnaA are thought to occur in domains II and III (Saxena et al., 2015). Lys178, an essential residue within the Walker A motif of DnaA, is acetylated specifically in stationary-phase cells (Zhang et al., 2016).

ATP-DnaA forms oligomers at *oriC* more efficiently than ADP-DnaA. The arginine finger motif (Arg285), ATP-DnaA-specific interactive locus for DUE unwinding (AID) motifs (Arg227, Leu290), box VII motif (Arg281), and Lys243 are required for functional domain III–domain III interactions (Felczak and Kaguni, 2004; Kawakami et al., 2005; Ozaki et al., 2008, 2012a; **Figure 3**) (see section Structure of the DnaA–IHF–*oriC* Complex for DUE Unwinding and DnaB Loading). In particular, Arg285 Arg finger is important for the ATP-dependent activation of DnaA complexes on *oriC* (Kawakami et al., 2005) and is exposed on the side opposite to that of the ATP/ADP-binding site of domain III (**Figure 3B**). Whereas DNA-free DnaA is monomeric in solution, domain III enables DnaA to self-oligomerize in an ATP-dependent, head-to-tail manner via interactions between bound ATP and Arg285 of adjacent protomers, resulting in the binding of multiple DnaA protomers to *oriC* (Kawakami et al., 2005; Ozaki et al., 2008).

During unwinding of DUE by DnaA, hydrophobic (H) and basic (B) motifs (Val211 and Arg245, respectively), which are highly conserved in DnaA orthologs, have a crucial role in binding to the T-rich ssDUE strand (Ozaki et al., 2008; Ozaki and Katayama, 2012; Sakiyama et al., 2017; **Figure 3**) (see Section Structure of the DnaA–IHF–*oriC* Complex for DUE Unwinding and DnaB Loading). The H and B motifs are located in helices between the Walker A and B motifs and the Walker B and sensor 1 motifs, respectively (Ozaki and Katayama, 2009; **Figure 3A**). Whereas typical AAA+ proteins have only a single α helix between Walker A and B motifs, DnaA-related proteins (collectively termed the replication initiator clade) have two α helices in tandem, and the H motif is located on one of the two, giving a structure called an initiator/loader-specific motif (Iyer et al., 2004; Dueber et al., 2007; Mott et al., 2008; Ozaki et al., 2008; Ozaki and Katayama, 2009).

Certain residues in DnaA have distinct roles in its ATPase activity. Sensor 2 Arg334 is important for the intrinsic ATPase activity of DnaA, RIDA, and DDAH, probably via an interaction between the γ phosphate of ATP and Arg334. However, this residue is dispensable for the formation of functional initiation complexes on *oriC* (Sekimizu et al., 1987; Nishida et al., 2002; Kasho and Katayama, 2013; **Figure 3**). Arginine finger Arg285 is essential for intrinsic ATPase activity and for regulation by DDAH, but not by RIDA (Kawakami et al., 2005; Kasho and Katayama, 2013). Box VII Arg281 represses intrinsic ATPase activity and is dispensable for RIDA, but is required for DDAH (Kawakami et al., 2005; Kasho and Katayama, 2013) (see below).

The DnaA C-terminal domain IV has a typical helix-turn-helix (HTH) motif and binds with sequence specificity to DnaA boxes (Erzberger et al., 2002; Obita et al., 2002; Fujikawa et al., 2003; Yoshida et al., 2003; **Figure 3A**). This binding causes DNA bending by about 20°. An α helix consisting of amino-acid residues 434–451 is part of the HTH motif that is inserted in the major groove of the 9 bp DnaA box, recognizing the 3′ portion (5′-TnCACA-3′) of the consensus sequence (Fujikawa et al., 2003). Arg399, which is located in the turn (or loop) region of the HTH motif, is inserted in the minor groove of the DnaA box, recognizing the 5′ portion (5′-TTA-3′) of the consensus sequence (Fujikawa et al., 2003). In consistent, DnaA T435M and DnaA R399A are defective in DnaA-box binding (Sutton and Kaguni, 1997; Blaesing et al., 2000). The short linker connecting domains III and IV is a structurally flexible loop that enables swiveling of domain IV (Erzberger et al., 2002), and the results of molecular dynamics simulation indicate that it is important in the assembly of DnaA oligomers on *oriC* (Shimizu et al., 2016; **Figure 3A**). Leu366 located in an α helix downstream of the short linker is required for a conformational change in DnaA complexes constructed on *oriC* (Garner and Crooke, 1996; Saxena et al., 2011; **Figure 3A**). In addition, Leu423 and Pro424 contribute to the Hda binding that is crucial for RIDA (Keyamura and Katayama, 2011).

The N-terminal domain I of DnaA has important roles in protein–protein interaction. Weak domain I–domain I binding depends on Trp6, which contributes to DnaA self-oligomerization (Sutton et al., 1998; Simmons et al., 2003; Felczak et al., 2005; **Figure 3A**). Glu21 and Phe46 form a binding site for both DnaB helicase and DiaA protein (Abe et al., 2007; Keyamura et al., 2009). Phe46 contributes more to DiaA binding than Glu21, and is also required for binding to YfdR protein, a potential inhibitor of DnaA–DiaA or DnaA–DnaB interactions that is encoded by a cryptic prophage (Noguchi and Katayama, 2016). Asn44 is thought to interact with Hda to facilitate functional interaction between DnaA domain III and Hda during RIDA (Su'etsugu et al., 2013). Domain I also binds to HU, Dps (DNA protection during starvation protein), and ribosomal L2 protein, but the residues involved in these interactions have not yet been identified (Chodavarapu et al., 2008a,b, 2011). In addition, DnaA domain I contains a K-homology domain, which enables binding to ssDNA and RNA (Abe et al., 2007). However, domain I has only slight activity in ssDUE binding *in vitro* (Abe et al., 2007), and notably, DnaA with deletion of domains I and II retains full DUE-unwinding activity *in vitro* (Sutton et al., 1998), which indicates that domain III plays a predominant role in ssDUE binding sustaining DUE unwinding (Ozaki et al., 2008; Ozaki and Katayama, 2012; Sakiyama et al., 2017). Domain II is an unstructured linker and the least-conserved domain among DnaA orthologs in eubacterial species (Abe et al., 2007; Nozaki and Ogawa, 2008; **Figure 3A**).

DnaA COMPLEX ON *oriC*

The Role of DiaA

The 196 amino acid DiaA protein forms a homotetramer in which each protomer has a DnaA-binding site (Ishida et al., 2004; Keyamura et al., 2007, 2009). DiaA stimulates ATP-DnaA assembly on *oriC* and the unwinding of DUE. A single DiaA tetramer can bind multiple DnaA molecules (theoretically up to four; experimentally at least three), potentially leading to the linkage effect for stimulating cooperative binding. With the linkage effect, it has been demonstrated that, if two proteins each with a weak affinity for DNA are stably connected by a linker molecule, the affinity of the linked molecule for DNA is increases drastically (theoretically by up to 10^3- to 10^5-fold) because it permits binding between the protein and multiple points on the DNA (Zhou, 2001; Stauffer and Chazin, 2004). A mutant that encodes a variant of DiaA with constitutive binding of just two DnaA protomers is defective in stimulation of DnaA assembly on *oriC* and DUE unwinding (Keyamura et al., 2007). The sequence of the *diaA* gene is widely conserved in the genomes of eubacterial species (Keyamura et al., 2007).

The DiaA-binding site is located in domain I of DnaA (and includes residues Glu21 and Phe46), and this site also binds DnaB helicase, as described in Section DnaA (Abe et al., 2007; Keyamura et al., 2009; **Figure 3A**). Both overproduction of DiaA and deletion of the *diaA* gene moderately inhibit initiation *in vivo* (Ishida et al., 2004; Flåtten et al., 2015). DiaA therefore has a positive effect on initiation through stimulation of DnaA assembly, and a negative effect probably via its ability to inhibit DnaA–DnaB binding. DiaA that is bound to the DnaA assembly on *oriC* is thought to dissociate (enabling DnaB binding) at (or after) the time of DUE unwinding, by an unknown mechanism (Keyamura et al., 2009). Regulation of the inhibitory activity of DiaA has not yet been determined.

Structure of the DnaA–IHF–*oriC* Complex for DUE Unwinding and DnaB Loading

The question of the regularity and directionality of DnaA boxes in *oriC* has been resolved by the identification of low-affinity DnaA boxes, and by the correction of previously obtained information. The left-half DOR is now known to contain only DnaA boxes with a leftward directionality (R1, τ1, R5M, τ2, I1, I2), whereas the middle region and right-half DOR contain only DnaA boxes with a rightward directionality (R2, C3, C2, I3, C1, R4) (**Figure 2A**). As explained above, head-to-tail binding of domain III leads to ATP-DnaA-dependent assembly of sub-complexes with opposite orientations in the left-half and right-half DORs (Kawakami et al., 2005; Noguchi et al., 2015; Shimizu et al., 2016; Sakiyama et al., 2017; **Figures 3, 4**).

In DnaA sub-complexes, nucleotides that are associated with R1 box- and R4 box-bound DnaA protomers are located at the

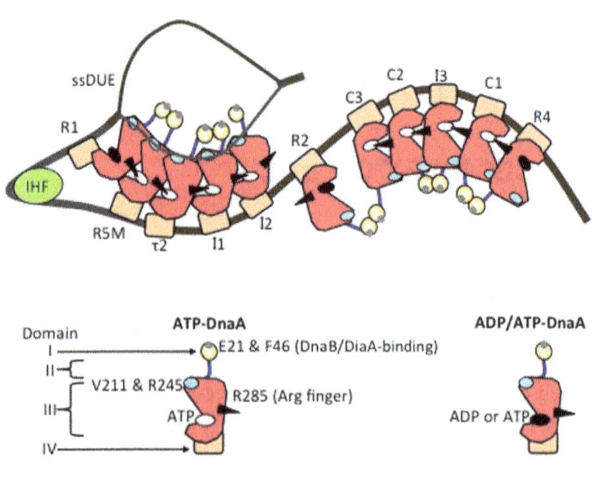

FIGURE 4 | Model of an unwound state of the initiation complex. DnaA domains I (*orange*), II (*blue*), III (*red*), and IV (*brown*) are schematically presented. A small gray circle on domain I shows a patch including Glu21 and Phe46 (for binding of DiaA and DnaB helicase). A black arrowhead, cyan oval, white oval, and black oval on domain III represent the arginine finger Arg285, H/B motifs, ATP, and either ADP or ATP, respectively. IHF is shown as a light green circle. DnaA-box positions are labeled. On the left-half and right-half DnaA-oligomerization regions, head-to-tail DnaA-oligomer sub-complexes assemble, with interactions between ATP and Domain III Arg285. The T-rich ssDUE strand interacts with the DnaA H/B motifs in the left-half sub-complex.

outer edges of the DOR, so that those cannot interact with DnaA protomers bound to other DOR sites (**Figures 2A**, **4**). Thus, even when ADP-DnaA is bound to R1, DUE-unwinding activity is fully sustained (Noguchi et al., 2015). ATP-DnaA assembly at the left-half DOR fundamentally depends on ATP–DnaA binding to the low-affinity R5M box (but not to the R1 box) and is stimulated by R1-bound DnaA (Sakiyama et al., 2017). Cooperative binding of ATP-DnaA molecules would operate effectively in the region from the R5M box to the I2 box (Sakiyama et al., 2017). Also, when ADP-DnaA is bound to the R4 box, assembly of the DnaA sub-complex on the right-half DOR is fully sustained with ATP-DnaA (Noguchi et al., 2015). ATP-DnaA assembly to the low-affinity sites in the right-half DOR depends on DnaA binding to the R4 box (Noguchi et al., 2015).

Evidence suggests that domain III of R2-bound DnaA does not interact with DnaA protomers bound to I2 or C3 (Rozgaja et al., 2011; Shimizu et al., 2016), as expected by the long spacing between R2 and I2 (9 bp) and between R2 and C3 (20 bp) (**Figure 2A**). Each box in the τ1-R5M-τ2-I1-I2 region and the R4-C1-I3-C2-C3 region is separated by 2–4 bp. However, domain I of R2-bound DnaA might interact with that of I2-bound or C3-bound DnaA, simulating cooperative DnaA binding (Miller et al., 2009; Ozaki and Katayama, 2012; Shimizu et al., 2016; **Figure 4**).

The existence of nucleotide-independent activities of DnaA bound to R1, R4, and R2 is consistent with data from *in vivo* dimethyl-sulfate footprint analysis, which suggest that DnaA binding to these boxes is stable throughout the cell cycle (Samitt et al., 1989; McGarry et al., 2004; Kawakami et al., 2005). By contrast, in *in vivo* experiments with DnaA fused to enhanced yellow fluorescent protein, results from photobleaching and high-resolution microscopy suggest that, overall, DnaA complexes bound to *oriC* are rapidly turned over with a time scale of a few seconds (Schenk et al., 2017). These results are consistent with the unstable binding of DnaA to low-affinity sites. Such dynamic features might be important for timely assembly of ATP-DnaA on *oriC* after the DARS system has increased in total level of ATP-DnaA.

The left-half ATP-DnaA sub-complex possesses full activity for DUE unwinding and ssDUE binding (Ozaki and Katayama, 2012; **Figure 4**). In addition to the Arg finger Arg285, AID1 (Arg227) and AID2 (Leu290) motifs, which are highly conserved in DnaA orthologs, are thought to be located at the domain III–domain III interface and to contribute to ATP–DnaA-specific interactions and the establishment of DnaA oligomers within the left-half DOR that are competent for DUE unwinding (Ozaki et al., 2012a; **Figure 3B**). Lys243 also contributes to oligomerization of ATP-DnaA on *oriC* and DUE unwinding (Ozaki et al., 2008). Also, Lys243 can be acetylated *in vivo*. Although the significance of this is unknown, it moderately affects DnaA binding to *oriC in vitro* (Li et al., 2017). Molecular dynamics simulation shows that sharp DNA bending (>120°) caused by IHF binding enables domain III–domain III interaction between DnaA protomers bound to R1 and R5M in the absence of DnaA–τ1 binding (Shimizu et al., 2016; **Figure 4**). Consistently, *in vitro* experiments indicate that DnaA–τ1 interaction is inhibited by IHF binding (Sakiyama et al., 2017).

DUE unwinding depends on temperature and superhelicity of *oriC* DNA, which contribute to destabilization of the duplex state of the DUE (Baker and Kornberg, 1988; Sekimizu et al., 1988). In addition, the unstable ssDUE binds to, and is stabilized by, DnaA domain III, enabling loading of DnaB helicase to the unwound region (Ozaki et al., 2008). In *E. coli* and *B. subtilis*, only a specific strand (the T-rich strand in *E. coli*) binds to DnaA (Ozaki et al., 2008; Richardson et al., 2016). Also, in *E. coli*, the H/B-motif residues of domain III are required for ssDUE binding (Ozaki et al., 2008; Sakiyama et al., 2017). This binding mode is consistent with that seen in the co-crystal structure of the poly-A ssDNA-bound DnaA domain III ortholog from the hyperthermophilic bacterium *Aquifex aeolicus* (Duderstadt et al., 2011).

Notably, when multiple DnaA molecules bind to the DOR, ssDUE binding by DnaA increases dramatically and is much higher than that of DOR-unbound DnaA molecules (Ozaki et al., 2008; Ozaki and Katayama, 2012; Sakiyama et al., 2017). ssDUE directly binds to domain III of DnaA protomers that have associations through domain IV with the left-half DOR (Ozaki et al., 2008; Ozaki and Katayama, 2012; Sakiyama et al., 2017; **Figure 4**). In particular, domain III of the R1- and R5M-bound DnaA protomers play predominant roles in ssDUE binding for replication initiation (Sakiyama et al., 2017). These and the other observations mentioned above support the idea that ssDUE directly binds to R1- and R5M-bound DnaA protomers via sharp DNA bending by IHF, which stimulates ssDUE recruitment to those DnaA molecules. This is referred to as the ssDUE-recruitment mechanism (**Figure 4**; Ozaki et al., 2008; Ozaki and Katayama, 2012; Noguchi et al., 2015; Shimizu et al., 2016; Sakiyama et al., 2017). This reasonably explains the role of IHF and dispensability of ATP of R1-bound DnaA in DUE unwinding, and the striking stimulation of ssDUE binding by DnaA-oligomer assembly on the DOR, as well as the strict requirements in DUE unwinding for the spacing between the DUE and the DnaA R1 box and between the DnaA R1 box and the IBS, and the requirement for H/B motifs of the DOR-bound DnaA for ssDUE binding (**Figure 4**) (for details, see Sakiyama et al., 2017).

Multiple DnaA molecules bound to the DOR are required for binding of DnaB helicase which is a homohexamer (Abe et al., 2007; Keyamura et al., 2009; Ozaki and Katayama, 2009). Multiple DnaA domain I molecules of the DOR-bound DnaA complexes would be allied so as to stably bind DnaB helicases by using multiple binding points (**Figure 4**). In consistent, Box VII Arg281 stabilizes DnaA oligomers assembled on the DOR, and is required for DnaB-helicase loading, but not for DUE unwinding (Felczak and Kaguni, 2004). This suggests that stability of DnaA complexes is more important for DnaB-helicase loading than for DUE unwinding.

In addition to the basic DnaB-binding/loading activity of the left-half-DOR DnaA sub-complex, the right-half-DOR DnaA sub-complex stimulates DnaB binding/loading (Ozaki and Katayama, 2012; **Figures 2A, 4**). This activity might result in binding of two DnaB helicases, one to each of the DnaA sub-complexes, enabling bi-directional replication (Ozaki and Katayama, 2012; Noguchi et al., 2015; **Figure 4**). The spacing between the R2 and C3 boxes is important for efficient DnaB loading, suggesting coordination of the left and right DnaA subcomplexes in a DnaB loading process (Shimizu et al., 2016; **Figure 4**).

DDAH SYSTEM

Function

datA-dependent DnaA inactivation has been termed DDAH. Three hundred 9-mer DnaA box consensus sequences are widely distributed throughout the *E. coli* chromosome (Tesfa-Selase and Drabble, 1992; Roth and Messer, 1998; Hansen et al., 2007). Some of these sites regulate the activity of DnaA for replication initiation, whereas some of others function for transcriptional regulation by DnaA (Kitagawa et al., 1996; Messer and Weigel, 1997; Speck et al., 1999; Olliver et al., 2010). The *datA* locus includes a DnaA-box cluster and was originally identified as a potent DnaA-binding locus that can repress untimely replication initiation (Kitagawa et al., 1996; Ogawa et al., 2002; Nozaki et al., 2009; **Figure 1A**). DnaA binding at *datA* is stimulated by IHF (Nozaki et al., 2009); the *datA*–IHF complex stimulates hydrolysis of DnaA-bound ATP independently of RIDA to produce ADP-DnaA (Kasho and Katayama, 2013; Kasho et al., 2017; **Figure 5A**).

Several lines of *in vivo* evidence point to the important role of the *datA*–IHF complex in the repression of untimely replication initiation. Deletion of *datA* or IHF genes (*ihfA* or *ihfB*) increases cellular levels of ATP-DnaA, resulting in higher levels of ATP-DnaA than in wild-type cells and untimely replication initiation without cell growth inhibition (Kitagawa et al., 1996; Nozaki et al., 2009; Kasho and Katayama, 2013). As inhibition of RIDA causes severe over-initiation of replication and arrest of cell growth (Kato and Katayama, 2001; Nishida et al., 2002; Fujimitsu et al., 2008), DDAH is thought to play a supportive role in RIDA. Also, in DDAH-deficient mutants, replication initiation is not completely repressed in the presence of rifampicin, an inhibitor of transcription and replication initiation (von Freiesleben et al., 2000; Morigen, Skarstad and Molina, 2005), probably due to elevated levels of ATP-DnaA.

Structure of *datA*

The minimal *datA* for DDAH function is 183 bp containing high-affinity DnaA boxes 2 and 3, low-affinity DnaA box 7 and a single IBS, which are all essential for efficient ATP–DnaA binding and DDAH activity (Kitagawa et al., 1998; Ogawa et al., 2002; Nozaki et al., 2009; Kasho and Katayama, 2013; Kasho et al., 2017; **Figure 1B**). All the essential DnaA boxes have the same directionality, suggesting the importance of head-to-tail ATP–DnaA interactions. The results of mutation analyses indicate that, as in the *oriC* DORs, the orientations and interval lengths of the *datA* DnaA boxes 2 and 3 and the IBS are optimized for repression of untimely initiations through DDAH (Nozaki et al., 2009; Kasho and Katayama, 2013). The *datA* DnaA box 4 stimulates DnaA assembly and DnaA-ATP hydrolysis *in vitro*, but is not essential for initiation regulation *in vivo* (Ogawa et al., 2002; Kasho and Katayama, 2013; Kasho et al., 2017; **Figure 1B**). Unidentified factor(s) might stimulate DnaA assembly *in vivo* even in the absence of DnaA box 4 (see below).

Mechanism

In the *datA* region, oligomers containing three ATP-DnaA molecules form on and around the core DnaA boxes and higher oligomers form depending on IHF (Kasho and Katayama, 2013; Kasho et al., 2017; **Figure 6A**). IHF binding causes DNA looping, which would promote interaction between the box 2-bound DnaA and the box 3-bound DnaA, thereby stabilizing DnaA complexes in a cooperative manner for the activation of DnaA-ATP hydrolysis. The results of biochemical analyses using mutant DnaA proteins provide evidence that, in *datA*–DnaA complexes, DnaA–DnaA interactions depend on AAA+ domain III arginine finger Arg285, box VII Arg281, and AID2 Leu290, and stimulation of ATP hydrolysis depends on AAA+ sensor 2 motif Arg334 (Kasho and Katayama, 2013; Kasho et al., 2017). DnaA AID-2 Leu290 might facilitate dissociation of the resultant ADP-DnaA, which is unstable in *datA*, thereby enabling catalytic reaction of DDAH (Kasho et al., 2017). Overall, these requirements for specific residues are similar to those of the initiation complex, except for the crucial role of sensor 2 Arg334 in ATP hydrolysis. This seems reasonable because both DDAH complexes and the initiation complex are constructed mainly from head-to-tail DnaA–DnaA interactions (**Figures 3B, 6A**).

The stimulatory role of IHF in DDAH cannot be replaced by HU, which is a structural homolog of IHF and a widely conserved nucleoid-associated protein in most bacteria, although IHF can be functionally replaced by HU in an *in vitro* reconstituted system

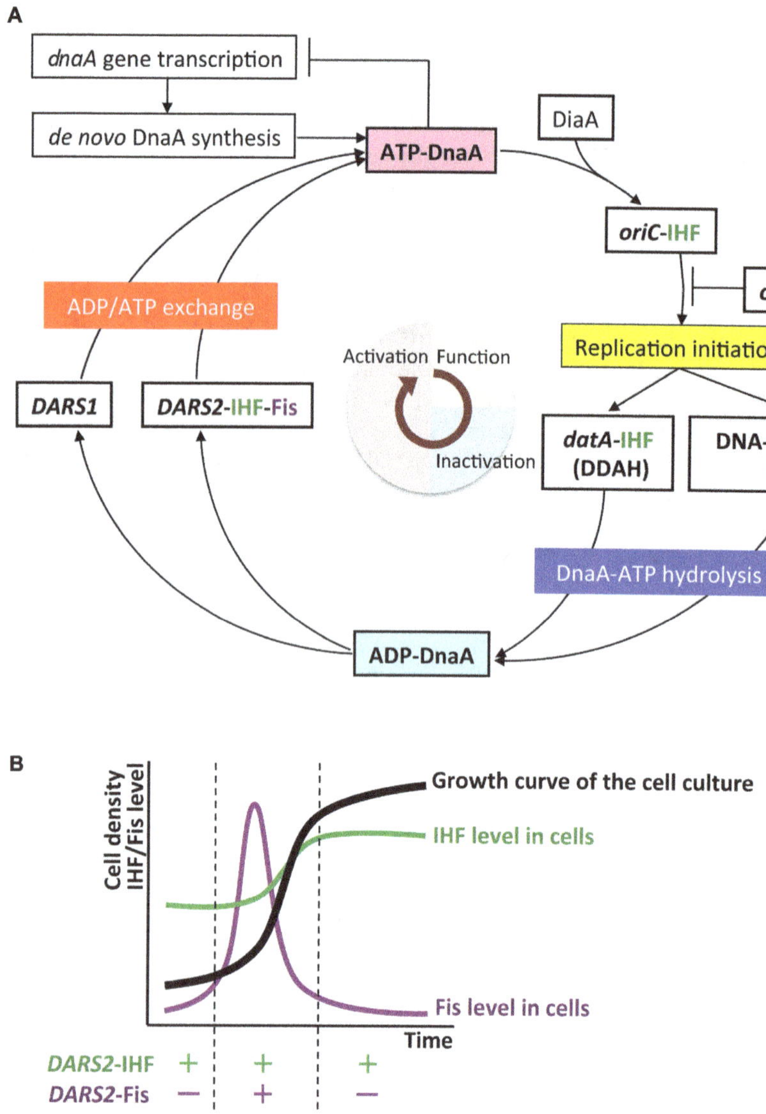

FIGURE 5 | The regulatory cycle of DnaA. **(A)** Schematic presentation of the regulatory cycle of DnaA. ATP-DnaA forms oligomers on *oriC* with the aid of integration host factor (IHF) and DnaA interacting protein DiaA, and initiates replication (period in light yellow). After initiation, ATP-DnaA is converted to replication-inactive ADP-DnaA (RIDA) and *datA*-dependent DnaA-ATP hydrolysis (DDAH) systems (period in light blue). RIDA involves Hda protein and the clamp subunit of DNA polymerase III holoenzyme, and DDAH involves the *datA* locus and IHF. DnaA-reactivating sequence (*DARS1* and *DARS2*) loci stimulate nucleotide exchange of ADP-DnaA and regenerate ATP-DnaA (period in light red). IHF binds to specific sites in *oriC*, *datA*, and *DARS2* in a cell-cycle-coordinated manner. Fis binds to *DARS2* in log-phase cells. **(B)** Growth-phase coordination of regulation of Fis expression. The cellular level of Fis varies through the growth phases, and is much higher in exponential-phase cells (enabling Fis binding to *DARS2*) than in stationary-phase cells. By contrast, the cellular level of IHF is highest in stationary-phase cells.

of replication initiation from *oriC* (Hwang and Kornberg, 1992; Dillon and Dorman, 2010; Kasho and Katayama, 2013). Like IHF, HU induces sharp DNA bending, but unlike IHF, HU binds to DNA without sequence specificity. At *datA* and *oriC*, the IBS is located between DnaA boxes, but the inter-DnaA box region in *datA* (i.e., between box 2 and box 3) is longer than that (i.e., between R1 box and R5M box) at *oriC* (**Figures 1B, 2A**), suggesting that a functional conformational change in *datA* for activating DDAH is strongly dependent on specific DNA bending at the *datA* IBS induced by IHF binding.

Regulation

Coupling of *datA* Dosage (Copy Number) with Replication

datA is located at 94.7 min (4.39 Mb position) in the *E. coli* genome, near to *oriC* (84.6 min, 3.92 Mb position), which means that *datA* is duplicated soon after replication initiation, resulting in a temporary higher dosage (copy number) of *datA* per cell. This is thought to be important for the repression of untimely initiations under certain growth conditions (Kitagawa et al., 1996; **Figure 1A**). In agreement with these observations, increasing the

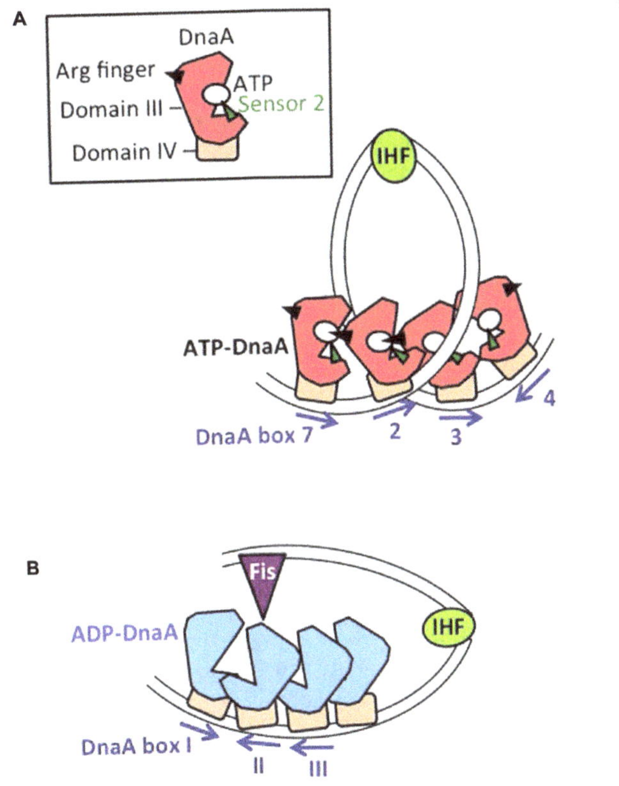

FIGURE 6 | Mechanistic view of DnaA regulation by *datA* and *DARS2*. **(A)** Mechanistic model of *datA*-dependent DnaA-ATP hydrolysis (DDAH). Oligomers of ATP-DnaA (with domain III shown as a red polygon, and domain IV as an brown square) assemble on the *datA* minimal region containing DnaA boxes 7, 2, and 3. Sharp DNA bending by IHF could stimulate contact between DnaA bound at boxes 2 and 3. Cooperative ATP-DnaA binding on *datA* DnaA box 7 might induce conformational change of the DnaA to activate DDAH. **(B)** Mechanistic model of *DARS2*-mediated DnaA activation. Oligomers of ADP-DnaA (with domain III shown as a blue polygon, and domain IV as an brown square) assemble on the *DARS2* core region. IHF and Fis bind to specific sites in the *DARS2* regulatory region, and the resultant complex stimulates conformational changes in the core-bound DnaA oligomer, to induce structural changes in DnaA and ADP dissociation. Nucleotide-free apo-DnaA dissociates from the core region and binds ATP (which is abundant in the cytosol), yielding ATP-DnaA.

datA copy number via a multi-copy plasmid negatively impacts replication initiation and cell growth (Morigen et al., 2003; Kasho et al., 2017). Conversely, translocation of *datA* to the replication-termination locus *terC* at 34.6 min, which is >2 Mb apart from *oriC* (**Figure 1A**), does not result in a temporary increase in its copy number until replication is nearly completed, allowing untimely initiations in nutrition-rich medium, similar to those observed in *datA*-deleted cells (Kitagawa et al., 1998; Frimodt-Møller et al., 2015). The *datA* copy number may also be important for the regulation of cell division (Morigen et al., 2014). Notably, in *B. subtilis* and *Streptomyces coelicolor*, chromosomal DnaA-box clusters analogous to *datA* repress untimely replication initiation (Smulczyk-Krawczyszyn et al., 2006; Okumura et al., 2012), suggesting that the DnaA regulation by *datA* is a conserved feature of most eubacterial species.

Cell-Cycle-Dependent Binding of IHF

The activation of DDAH and RIDA reduces the ATP-DnaA level and represses untimely initiations (Kasho and Katayama, 2013). Regulation of IHF binding to *datA* is essential for the timely activation of DDAH. Cell-cycle analysis of *datA*–IHF binding indicates that IHF dissociates from *datA* before replication initiation and temporarily binds to *datA* after initiation (Kasho and Katayama, 2013). As both IHF binding to and dissociation from *datA* occur during on-going chromosome replication, DDAH is suggested to be regulated by a mechanism that couples certain cell cycle events to *datA*–IHF binding. Conversely, RIDA is activated by DNA replication-coupled feedback (Katayama et al., 2010; Kasho and Katayama, 2013)

The transcription inhibitor rifampicin might inhibit timely dissociation of *datA*–IHF complexes (Kasho and Katayama, 2013). As IHF is abundant in exponentially growing cells (Ali Azam et al., 1999), a specific factor inhibiting *datA*–IHF binding might be required to be transcribed at a particular moment in the cell cycle. DDAH activity is inhibited by translocation of the *datA* sequence to a highly transcribed region (Frimodt-Møller et al., 2016), suggesting that transcription through *datA* might inhibit *datA*–IHF binding. Consistently, the essential core region of *datA* (i.e., box 7-box 2-IBS-box 3) is located at an intergenic region downstream of the *glyVXY* operon and *queG* gene (Kitagawa et al., 1998; Nozaki et al., 2009; Kasho et al., 2017; **Figure 1B**). It remains to be determined whether transcription of these genes oscillates in a coordinated manner during the cell cycle.

Effects of DNA Superhelicity and Nutrition

DDAH is suggested to regulated also by DNA supercoiling. *In vitro* experiments indicate that negative DNA supercoiling, which is modulated by DNA topoisomerases such as DNA gyrase and by nucleoid protein binding (Travers and Muskhelishvili, 2005), stabilizes ATP-DnaA oligomerization and IHF binding on *datA* and increases DDAH activity twofold compared with *datA* activity on relaxed DNA (Kasho et al., 2017). Modulation of negative supercoiling by novobiocin, a DNA-gyrase inhibitor, decreases the function of *datA* on plasmid DNA for repressing replication initiation (Kasho et al., 2017), suggesting supercoiling-dependent regulation of DDAH *in vivo*. Negative supercoiling is predicted to increase under osmotic-stress conditions (Higgins et al., 1988), which might affect *datA* activity for repressing initiation.

Nutrition-dependent regulation of DDAH has been suggested. Untimely initiations in *datA*-null cells are more severely exhibited in nutrient-poor medium containing acetate or glucose than in rich medium containing glucose-casamino acid or LB (Kitagawa et al., 1998; Flåtten et al., 2015). Chromosome-conformation-capture analysis has revealed that *oriC* can physically interact with *datA* under conditions of replication stress induced by serine hydroxamate, which causes amino-acid starvation and inhibits replication initiation (Cagliero et al., 2013). Free DnaA molecules present around *oriC* could efficiently interact with *datA* under poor nutrition conditions. Conversely, in rich medium, the *datA* dosage could be important for the regulation of DDAH. As described above, *datA* translocation to *terC* causes excess initiations in rich medium, but not in poor medium (Kitagawa et al., 1998).

DARS SYSTEM

Function

In *E. coli*, ATP-DnaA is produced by binding of apo-DnaA to ATP, which is abundant in the cytosol (Bochner and Ames, 1982). Replication initiation occurs when the cellular level of ATP-DnaA reaches a peak that is high enough to enable assembly on *oriC* (Kurokawa et al., 1999). At least three mechanisms for production of apo-DnaA have been characterized: *de novo* DnaA synthesis; nucleotide dissociation from ADP-DnaA by acidic phospholipids in the cell membrane; and a mechanism involving specific chromosomal DNA elements termed DARS sites (**Figure 1**).

In *de novo* synthesis, newly translated apo-DnaA binds to ATP in the cytosol (**Figure 5A**). In addition to autoregulation by DnaA (Speck et al., 1999), transcription of *dnaA* is down-regulated by the SeqA-Dam system immediately after initiation (Campbell and Kleckner, 1990; Bogan and Helmstetter, 1997). This would contribute to the reduction in *de novo* DnaA synthesis (i.e., production of ATP-DnaA) immediately after initiation and maintain coordination between replication initiation and the cell cycle (Riber and Løbner-Olesen, 2005).

The *E. coli* cell membrane contains acidic phospholipids such as cardiolipin and phosphatidylglycerol that can promote dissociation of ADP or ATP from DnaA by interaction with DnaA domain III. These phospholipids can convert ADP-DnaA to ATP-DnaA *in vitro* (Saxena et al., 2013). In mutant cells with reduction of levels of acidic phospholipids, initiation from *oriC* is repressed (Fingland et al., 2012). The possibility that acidic phospholipids regulate DnaA activity during the cell cycle deserves to be further investigated.

The *DARS1* and *DARS2* chromosomal DNA sequences promote dissociation of ADP from ADP-DnaA, stimulating nucleotide exchange and increasing levels of ATP-DnaA, thereby stimulating replication initiation (Fujimitsu et al., 2009; Kasho et al., 2014; **Figures 1, 5A**). *In vitro*, *DARS1* can promote ADP dissociation of ADP-DnaA without additional factors, whereas *DARS2* has little activity by itself, but is substantially activated by IHF and Fis, which binds site-specifically and bends DNA by <60° (Fujimitsu et al., 2009; Kasho et al., 2014; **Figure 1B**). In cells, *DARS1* and *DARS2* both have stimulatory roles in ATP-DnaA production and replication initiation, and the deletion of either or both delays initiation (Fujimitsu et al., 2009; Kasho et al., 2014). Deletion of *DARS2* inhibits timely initiation more severely than deletion of *DARS1*, causing asynchronous initiations, and consistently, oversupply of *DARS2* has a greater effect than oversupply of *DARS1* on stimulation of extra initiations, which can cause inhibition of cell growth (Fujimitsu et al., 2009; Kasho et al., 2014). Deletion of *DARS2,* substitution of the IBS or Fis-binding site (FBS), or deletion of Fis gene decreases the cellular ATP-DnaA level more than deletion of *DARS1* under RIDA-deficient conditions (Fujimitsu et al., 2009; Kasho et al., 2014; **Figure 1B**). These observations suggest that the complex formed by *DARS2*, IHF, and Fis is important for increasing the level of ATP-DnaA and determining initiation timing, and that *DARS1* might maintain basal levels of ATP-DnaA.

Structure of *DARS1* and *DARS2*

A common structural feature of *DARS1* (101 bp) and *DARS2* (455 bp) is the presence of a core region containing three DnaA boxes (DnaA boxes I, II, and III): the spacing between these boxes is the same at both *DARS1* and *DARS2*, and box I has the opposite orientation to boxes II and III (**Figure 1B**). In addition to the core regions, *DARS1* and *DARS2* have regulatory regions of different lengths. The ~50 bp regulatory region of *DARS1* enhances DnaA-ADP-dissociation activity *in vitro*, but its role *in vivo* has not been determined (Fujimitsu et al., 2009). The ~400 bp *DARS2* regulatory region contains the IBS and FBS, which stimulate ADP-DnaA assembly on *DARS2* and are essential for ADP dissociation from DnaA (Kasho et al., 2014).

Mechanism

Unlike *oriC* or *datA*, DARS sites preferentially bind ADP-DnaA rather than ATP-DnaA. Notably, head-to-head (but not head-to-tail) interactions would occur between the box I-bound DnaA and the box II-bound DnaA, which is a prominent feature at DARS sites (**Figures 1B, 6B**). Formation of oligomers of four or more ADP-DnaA molecules on the *DARS2* core region depends on DnaA AID-2 Leu290, and results in dissociation of ADP (**Figure 6B**). Formation of similar oligomers on the *DARS1* core region promotes dissociation of ADP and is dependent on DnaA AAA+ sensor 1 Asp269, which would participate in probable heat-dependent conformational change of the DnaA nucleotide-binding pocket (Fujimitsu et al., 2009). DnaA sensor 1 Asp269 is dispensable for ADP dissociation at *DARS2* (Kasho et al., 2014), possibly because of still unknown functions of the activator proteins IHF and Fis (**Figure 6B**). Dissociation of ADP produces apo-DnaA molecules that are unstable in binding to the DARS core region, enabling ATP binding and the production of ATP-DnaA (Fujimitsu et al., 2009; Kasho et al., 2014).

Binding of IHF is essential for *DARS2* activity, and, as with DDAH, IHF cannot be replaced by HU. The complex of *DARS2*–IHF–Fis promotes ~10 times as much ADP dissociation as *DARS1 in vitro* (Kasho et al., 2014). The results of mutational analyses indicate that the spacing between the core-region DnaA boxes, IBS, and FBS is optimized for efficient activation of *DARS2* (Kasho et al., 2014). The sequences and lengths of the spaces between the *DARS2* core region, IBS, and FBS are highly conserved in *E. coli*-proximal β- or γ-proteobacterial species in which IHF and Fis homologs are conserved, suggesting that the regulatory mechanism for *DARS2* is shared among proteobacterial species, including pathogenic species (Kasho et al., 2014).

Regulation

Cell-Cycle-Dependent IHF Binding

IHF binding to *DARS2* is dynamically regulated in a cell-cycle-coordinated manner (Kasho et al., 2014). In contrast to *datA*–IHF binding, *DARS2*–IHF binding occurs specifically in the pre-initiation period, consistent with its role in timely production of ATP-DnaA for replication initiation (Kasho et al., 2014). Regulation of *DARS2*–IHF binding and dissociation is resistant to rifampicin (Kasho et al., 2014), suggesting that it is independent of transcription. The timing of *DARS2*–IHF binding is also independent of replication initiation at *oriC* or replication fork progression (Kasho et al., 2014), supporting the suggestion that functional activation of *DARS2* might be controlled by specific cell cycle (but not DNA replication)-coupled events.

Growth-Phase-Dependent Fis Binding

In addition to cell-cycle regulation, growth-phase-dependent regulation contributes to timely activation of *DARS2*. The cellular level of Fis increases specifically in the exponential-growth phase (Ali Azam et al., 1999; Dillon and Dorman, 2010; **Figure 5B**). Fis binding to *DARS2* occurs in the exponential phase, but not in the stationary phase (Kasho et al., 2014). The activation of *DARS2* only in exponential phase is consistent with the requirement for ATP-DnaA for replication initiation.

Chromosomal-Position Effect

Escherichia coli DARS1 (101 bp) and *DARS2* (455 bp) are both located in intergenic regions, at 17.5 min (0.81 Mb position) and 64.0 min (2.97 Mb position), respectively, between *oriC* and *terC* (**Figure 1A**), and these positions are well conserved among related bacteria (Fujimitsu et al., 2009; Kasho et al., 2014; Frimodt-Møller et al., 2015). Translocation of *DARS2* to a *terC*-proximal locus moderately inhibits the timing of initiation, as does its translocation to an *oriC*-proximal locus (Frimodt-Møller et al., 2016; Inoue et al., 2016). Inhibition at the *terC*-proximal locus is sustained in cells lacking MatP, a DNA-binding protein involved in formation of *terC*-specific subchromosomal structure (Mercier et al., 2008; Inoue et al., 2016). The results of 3C (chromosome-conformation-capture) analysis suggest that *oriC* can physically interact with the wild-type *DARS2* locus (Cagliero et al., 2013). These results suggest that the chromosomal location of *DARS2* might be important for efficient interaction with *oriC*. In addition, a specific local chromosomal structure might be important for DARS function.

DnaA activation by *DARS1* and *DARS2* stimulates efficient cell growth under nutrient-poor conditions, and colonization of the large intestine of streptomycin-treated mice (Frimodt-Møller et al., 2015), suggesting a requirement for *DARS1* and *DARS2* for environmental adaptation of *E. coli*. Stimulation of replication initiation by oversupply of *DARS2* has more severe effects on viability under aerobic conditions than under anaerobic conditions (Charbon et al., 2014), suggesting that over-initiation in aerobic conditions leads to lethal levels of encounters between replication forks and sites of oxidative damage (Charbon et al., 2017).

PERSPECTIVES FOR COUPLING BETWEEN DnaA REGULATION AND CELL GROWTH IN *E. coli*

During the *E. coli* cell cycle, levels of ATP-DnaA rise to a peak that induces replication initiation, after which they fall, predominantly as a result of RIDA, a replication-coupled negative-feedback mechanism (Katayama et al., 2010). In the pre-initiation period, *DARS2* is predominant in the conversion of ADP-DnaA to ATP-DnaA. IHF binding to *DARS2* promotes this activity in a timely manner, but IHF is abundant throughout the cell cycle, and the mechanism responsible for its timely binding and dissociation is not yet known. Similarly, the mechanism responsible for the regulation of *datA*–IHF binding, which occurs at a specific time after replication initiation, has not yet been determined. Further studies are required to determine which mechanisms contribute to coupling between these processes and cell-cycle progression. In addition, further studies are required to understand DnaA complex formation/dissociation, ADP dissociation, and ATP hydrolysis during DDAH and in the formation of protein complexes at DARS sites.

Chromosomal positioning of *DARS1*, *DARS2*, and *datA* is important for regulation of the timing of initiation under specific growth conditions. Chromosomal structural determinants such as superhelicity, co-localization of specific loci in the 3D chromosomal structure, or the timing of changes in copy number during replication that is related to the distance from *oriC* might all contribute to the importance of chromosomal location. Further studies can help to determine which of these mechanisms is relevant for the correct timing of replication initiation.

How DnaB helicase binding to *oriC*-DnaA complexes *in vivo* is regulated is an open question. Even in the presence of a high ATP-DnaA level and *oriC* DUE unwinding, DnaB helicase might still not be able to be bind to *oriC*-DnaA complexes because of the stable interaction between DiaA and DnaA. Further studies of the regulatory mechanisms governing the DiaA binding/dissociation to/from DnaA and DnaB helicase loading *in vivo* are required to understand fully how initiation is timed correctly during the cell cycle.

Intensive research over the last 40 years has increased knowledge of the factors and regulatory mechanisms involved in chromosome replication initiation in *E. coli*. Our knowledge has exploded, and it is possible today to provide a considerably detailed outline about the factors involved and how replication initiation is controlled. At the same time, such progress creates further important mysteries, too. Further work in this area will undoubtedly continue to surprise and inform us about how replication is controlled in this erstwhile considered simple organism and advance knowledge about how this process is controlled in other organisms as well.

AUTHOR CONTRIBUTIONS

All authors listed have made a substantial, direct and intellectual contribution to the work, and approved it for publication.

FUNDING

This work was supported by MEXT/JSPS KAKENHI Grant numbers 17H03656, 16H00775, 15K18479 and 17K07338.

REFERENCES

Abe, Y., Jo, T., Matsuda, Y., Matsunaga, C., Katayama, T., and Ueda, T. (2007). Structure and function of DnaA N-terminal domains: specific sites and mechanisms in inter-DnaA interaction and in DnaB helicase loading on *oriC*. *J. Biol. Chem.* 282, 17816–17827. doi: 10.1074/jbc.M701841200

Ali Azam, T., Iwata, A., Nishimura, A., Ueda, S., and Ishihama, A. (1999). Growth phase-dependent variation in protein composition of the *Escherichia coli* nucleoid. *J. Bacteriol.* 181, 6361–6370.

Baker, T. A., and Kornberg, A. (1988). Transcriptional activation of initiation of replication from the *E. coli* chromosomal origin: an RNA-DNA hybrid near *oriC*. *Cell* 55, 113–123. doi: 10.1016/0092-8674(88)90014-1

Baxter, J. C., and Sutton, M. D. (2012). Evidence for roles of the *Escherichia coli* Hda protein beyond regulatory inactivation of DnaA. *Mol. Microbiol.* 85, 648–668. doi: 10.1111/j.1365-2958.2012.08129.x

Bell, S. P., and Kaguni, J. M. (2013). Helicase loading at chromosomal origins of replication. *Cold Spring Harb. Perspect. Biol.* 5:a010124. doi: 10.1101/cshperspect.a010124

Blaesing, F., Weigel, C., Welzeck, M., and Messer, W. (2000). Analysis of the DNA-binding domain of *Escherichia coli* DnaA protein. *Mol. Microbiol.* 36, 557–569. doi: 10.1046/j.1365-2958.2000.01881.x

Bochner, B. R., and Ames, B. N. (1982). Complete analysis of cellular nucleotides by two-dimensional thin layer chromatography. *J. Biol. Chem.* 257, 9759–9769.

Bogan, J. A., and Helmstetter, C. E. (1997). DNA sequestration and transcription in the *oriC* region of *Escherichia coli*. *Mol. Microbiol.* 26, 889–896. doi: 10.1046/j.1365-2958.1997.6221989.x

Bramhill, D., and Kornberg, A. (1988). A model for initiation at origins of DNA replication. *Cell* 54, 915–918. doi: 10.1016/0092-8674(88)90102-X

Cagliero, C., Grand, R. S., Jones, M. B., Jin, D. J., and O'Sullivan, J. M. (2013). Genome conformation capture reveals that the *Escherichia coli* chromosome is organized by replication and transcription. *Nucleic Acids Res.* 41, 6058–6071. doi: 10.1093/nar/gkt325

Campbell, J. L., and Kleckner, N. (1990). *E. coli oriC* and the *dnaA* gene promoter are sequestered from dam methyltransferase following the passage of the chromosomal replication fork. *Cell* 62, 967–979. doi: 10.1016/0092-8674(90)90271-F

Charbon, G., Bjørn, L., Mendoza-Chamizo, B., Frimodt-Møller, J., and Løbner-Olesen, A. (2014). Oxidative DNA damage is instrumental in hyperreplication stress-induced inviability of *Escherichia coli*. *Nucleic Acids Res.* 42, 13228–13241. doi: 10.1093/nar/gku1149

Charbon, G., Campion, C., Chan, S. H. J., Bjørn, L., Weimann, A., da Silva, L. C. N., et al. (2017). Rewiring of energy metabolism promotes viability during hyperreplication stress in *E. coli*. *PLOS Genet.* 13:e1006590. doi: 10.1371/journal.pgen.1006590

Chodavarapu, S., Felczak, M. M., and Kaguni, J. M. (2011). Two forms of ribosomal protein L2 of *Escherichia coli* that inhibit DnaA in DNA replication. *Nucleic Acids Res.* 39, 4180–4191. doi: 10.1093/nar/gkq120

Chodavarapu, S., Felczak, M. M., Rouvière-Yaniv, J., and Kaguni, J. M. (2008a). *Escherichia coli* DnaA interacts with HU in initiation at the *E. coli* replication origin. *Mol. Microbiol.* 67, 781–792. doi: 10.1111/j.1365-2958.2007.06094.x

Chodavarapu, S., Gomez, R., Vicente, M., and Kaguni, J. M. (2008b). *Escherichia coli* Dps interacts with DnaA protein to impede initiation: a model of adaptive mutation. *Mol. Microbiol.* 67, 1331–1346. doi: 10.1111/j.1365-2958.2008.06127.x

Dillon, S. C., and Dorman, C. J. (2010). Bacterial nucleoid-associated proteins, nucleoid structure and gene expression. *Nat. Rev. Microbiol.* 8, 185–195. doi: 10.1038/nrmicro2261

Duderstadt, K. E., Chuang, K., and Berger, J. M. (2011). DNA stretching by bacterial initiators promotes replication origin opening. *Nature* 478, 209–213. doi: 10.1038/nature10455

Dueber, E. L. C., Corn, J. E., Bell, S. D., and Berger, J. M. (2007). Replication origin recognition and deformation by a heterodimeric archaeal Orc1 complex. *Science* 317, 1210–1213. doi: 10.1126/science.1143690

Erzberger, J. P., Pirruccello, M. M., and Berger, J. M. (2002). The structure of bacterial DnaA: implications for general mechanisms underlying DNA replication initiation. *EMBO J.* 21, 4763–4773. doi: 10.1093/emboj/cdf496

Felczak, M. M., and Kaguni, J. M. (2004). The box VII motif of *Escherichia coli* DnaA protein is required for DnaA oligomerization at the *E. coli* replication origin. *J. Biol. Chem.* 279, 51156–51162. doi: 10.1074/jbc.M409695200

Felczak, M. M., Simmons, L. A., and Kaguni, J. M. (2005). An essential tryptophan of *Escherichia coli* DnaA protein functions in oligomerization at the *E. coli* replication origin. *J. Biol. Chem.* 280, 24627–24633. doi: 10.1074/jbc.M503684200

Fingland, N., Flåtten, I., Downey, C. D., Fossum-Raunehaug, S., Skarstad, K., and Crooke, E. (2012). Depletion of acidic phospholipids influences chromosomal replication in *Escherichia coli*. *Microbiology* 1, 450–466. doi: 10.1002/mbo3.46

Flåtten, I., Fossum-Raunehaug, S., Taipale, R., Martinsen, S., and Skarstad, K. (2015). The DnaA protein is not the limiting factor for initiation of replication in *Escherichia coli*. *PLOS Genet.* 11:e1005276. doi: 10.1371/journal.pgen.1005276

Frimodt-Møller, J., Charbon, G., Krogfelt, K. A., and Løbner-Olesen, A. (2015). Control regions for chromosome replication are conserved with respect to sequence and location among *Escherichia coli* strains. *Front. Microbiol.* 6:1011. doi: 10.3389/fmicb.2015.01011

Frimodt-Møller, J., Charbon, G., Krogfelt, K. A., and Løbner-Olesen, A. (2016). DNA replication control is linked to genomic positioning of control regions in *Escherichia coli*. *PLOS Genet.* 12:e1006286. doi: 10.1371/journal.pgen.1006286

Fujikawa, N., Kurumizaka, H., Nureki, O., Terada, T., Shirouzu, M., Katayama, T., et al. (2003). Structural basis of replication origin recognition by the DnaA protein. *Nucleic Acids Res.* 31, 2077–2086. doi: 10.1093/nar/gkg309

Fujimitsu, K., Senriuchi, T., and Katayama, T. (2009). Specific genomic sequences of *E. coli* promote replicational initiation by directly reactivating ADP-DnaA. *Genes Dev.* 23, 1221–1233. doi: 10.1101/gad.1775809

Fujimitsu, K., Su'etsugu, M., Yamaguchi, Y., Mazda, K., Fu, N., Kawakami, H., et al. (2008). Modes of overinitiation, *dnaA* gene expression, and inhibition of cell division in a novel cold-sensitive *hda* mutant of *Escherichia coli*. *J. Bacteriol.* 190, 5368–5381. doi: 10.1128/JB.00044-08

Garner, J., and Crooke, E. (1996). Membrane regulation of the chromosomal replication activity of *E. coli* DnaA requires a discrete site on the protein. *EMBO J.* 15, 2313–2321.

Grimwade, J. E., Ryan, V. T., and Leonard, A. C. (2000). IHF redistributes bound initiator protein, DnaA, on supercoiled *oriC* of *Escherichia coli*. *Mol. Microbiol.* 35, 835–844. doi: 10.1046/j.1365-2958.2000.01755.x

Hansen, F. G., Christensen, B. B., and Atlung, T. (2007). Sequence characteristics required for cooperative binding and efficient *in vivo* titration of the replication initiator protein DnaA in *E. coli*. *J. Mol. Biol.* 367, 942–952. doi: 10.1016/j.jmb.2007.01.056

Higgins, C. F., Dorman, C. J., Stirling, D. A., Waddell, L., Booth, I. R., May, G., et al. (1988). A physiological role for DNA supercoiling in the osmotic regulation of gene expression in *S. typhimurium* and *E. coli*. *Cell* 52, 569–584.

Hwang, D. S., and Kornberg, A. (1992). Opening of the replication origin of *Escherichia coli* by DnaA protein with protein HU or IHF. *J. Biol. Chem.* 267, 23083–23086.

Inoue, Y., Tanaka, H., Kasho, K., Fujimitsu, K., Oshima, T., and Katayama, T. (2016). Chromosomal location of the DnaA-reactivating sequence *DARS2* is important to regulate timely initiation of DNA replication in *Escherichia coli*. *Genes Cells* 21, 1015–1023. doi: 10.1111/gtc.12395

Ishida, T., Akimitsu, N., Kashioka, T., Hatano, M., Kubota, T., Ogata, Y., et al. (2004). DiaA, a novel DnaA-binding protein, ensures the initiation timing of *E. coli* chromosome replication. *J. Biol. Chem.* 279, 45546–45555. doi: 10.1074/jbc.M402762200

Iyer, L. M., Leipe, D. D., Koonin, E. V., and Aravind, L. (2004). Evolutionary history and higher order classification of AAA+ ATPases. *J. Struct. Biol.* 146, 11–31. doi: 10.1016/j.jsb.2003.10.010

Kaguni, J. M. (2011). Replication initiation at the *Escherichia coli* chromosomal origin. *Curr. Opin. Chem. Biol.* 15, 606–613. doi: 10.1016/j.cbpa.2011.07.016

Kano, Y., and Imamoto, F. (1990). Requirement of integration host factor (IHF) for growth of *Escherichia coli* deficient in HU protein. *Gene* 89, 133–137. doi: 10.1016/0378-1119(90)90216-E

Kasho, K., Fujimitsu, K., Matoba, T., Oshima, T., and Katayama, T. (2014). Timely binding of IHF and Fis to *DARS2* regulates ATP-DnaA production and replication initiation. *Nucleic Acids Res.* 42, 13134–13149. doi: 10.1093/nar/gku1051

Kasho, K., and Katayama, T. (2013). DnaA binding locus *datA* promotes DnaA-ATP hydrolysis to enable cell cycle-coordinated replication initiation. *Proc. Natl. Acad. Sci. U.S.A.* 110, 936–941. doi: 10.1073/pnas.1212070110

Kasho, K., Tanaka, H., Sakai, R., and Katayama, T. (2017). Cooperative DnaA binding to the negatively supercoiled *datA* locus stimulates DnaA-ATP hydrolysis. *J. Biol. Chem.* 292, 1251–1266. doi: 10.1074/jbc.M116.762815

Katayama, T., Kubota, T., Kurokawa, K., Crooke, E., and Sekimizu, K. (1998). The initiator function of DnaA protein is negatively regulated by the sliding clamp of the *E. coli* chromosomal replicase. *Cell* 94, 61–71.

Katayama, T., Ozaki, S., Keyamura, K., and Fujimitsu, K. (2010). Regulation of the replication cycle: conserved and diverse regulatory systems for DnaA and *oriC*. *Nat. Rev. Microbiol.* 8, 163–170. doi: 10.1038/nrmicro2314

Kato, J., and Katayama, T. (2001). Hda, a novel DnaA-related protein, regulates the replication cycle in *Escherichia coli*. *EMBO J.* 20, 4253–4262. doi: 10.1093/emboj/20.15.4253

Kaur, G., Vora, M. P., Czerwonka, C. A., Rozgaja, T. A., Grimwade, J. E., and Leonard, A. C. (2014). Building the bacterial orisome: high-affinity DnaA recognition plays a role in setting the conformation of *oriC* DNA. *Mol. Microbiol.* 91, 1148–1163. doi: 10.1111/mmi.12525

Kawakami, H., Keyamura, K., and Katayama, T. (2005). Formation of an ATP-DnaA-specific initiation complex requires DnaA Arginine 285, a conserved motif in the AAA+ protein family. *J. Biol. Chem.* 280, 27420–27430. doi: 10.1074/jbc.M502764200

Kawakami, H., Su'etsugu, M., and Katayama, T. (2006). An isolated Hda-clamp complex is functional in the regulatory inactivation of DnaA and DNA replication. *J. Struct. Biol.* 156, 220–229. doi: 10.1016/j.jsb.2006.02.007

Keyamura, K., Abe, Y., Higashi, M., Ueda, T., and Katayama, T. (2009). DiaA dynamics are coupled with changes in initial origin complexes leading to helicase loading. *J. Biol. Chem.* 284, 25038–25050. doi: 10.1074/jbc.M109.002717

Keyamura, K., Fujikawa, N., Ishida, T., Ozaki, S., Su'etsugu, M., Fujimitsu, K., et al. (2007). The interaction of DiaA and DnaA regulates the replication cycle in *E. coli* by directly promoting ATP-DnaA-specific initiation complexes. *Genes Dev.* 21, 2083–2099. doi: 10.1101/gad.1561207

Keyamura, K., and Katayama, T. (2011). DnaA protein DNA-binding domain binds to Hda protein to promote inter-AAA+ domain interaction involved in regulatory inactivation of DnaA. *J. Biol. Chem.* 286, 29336–29346. doi: 10.1074/jbc.M111.233403

Kim, J. S., Nanfara, M. T., Chodavarapu, S., Jin, K. S., Babu, V. M. P., Ghazy, M. A., et al. (2017). Dynamic assembly of Hda and the sliding clamp in the regulation of replication licensing. *Nucleic Acids Res.* 45, 3888–3905. doi: 10.1093/nar/gkx081

Kitagawa, R., Mitsuki, H., Okazaki, T., and Ogawa, T. (1996). A novel DnaA protein-binding site at 94.7 min on the *Escherichia coli* chromosome. *Mol. Microbiol.* 19, 1137–1147. doi: 10.1046/j.1365-2958.1996.453983.x

Kitagawa, R., Ozaki, T., Moriya, S., and Ogawa, T. (1998). Negative control of replication initiation by a novel chromosomal locus exhibiting exceptional affinity for *Escherichia coli* DnaA protein. *Genes Dev.* 12, 3032–3043. doi: 10.1101/gad.12.19.3032

Kurokawa, K., Nishida, S., Emoto, A., Sekimizu, K., and Katayama, T. (1999) Replication cycle-coordinated change of the adenine nucleotide-bound forms of DnaA protein in *Escherichia coli*. *EMBO J.* 18, 6642–6652. doi: 10.1093/emboj/18.23.6642

Leonard, A. C., and Grimwade, J. E. (2015). The orisome: structure and function. *Front. Microbiol.* 6:545. doi: 10.3389/fmicb.2015.00545

Li, S., Zhang, Q., Xu, Z., and Yao, Y. (2017). Acetylation of lysine 243 inhibits the *oriC* binding ability of DnaA in *Escherichia coli*. *Front. Microbiol.* 8:699. doi: 10.3389/fmicb.2017.00699

Lu, M., Campbell, J. L., Boye, E., and Kleckner, N. (1994). SeqA: a negative modulator of replication initiation in *E. coli*. *Cell* 77, 413–426. doi: 10.1016/0092-8674(94)90156-2

Margulies, C., and Kaguni, J. M. (1998). The FIS protein fails to block the binding of DnaA protein to *oriC*, the *Escherichia coli* chromosomal origin. *Nucleic Acids Res.* 26, 5170–5175. doi: 10.1093/nar/26.22.5170

McGarry, K. C., Ryan, V. T., Grimwade, J. E., and Leonard, A. C. (2004). Two discriminatory binding sites in the *Escherichia coli* replication origin are required for DNA strand opening by initiator DnaA-ATP. *Proc. Natl. Acad. Sci. U.S.A.* 101, 2811–2816. doi: 10.1073/pnas.0400340101

Mercier, R., Petit, M. A., Schbath, S., Robin, S., El Karoui, M., Boccard, F., et al. (2008). The MatP/*matS* site-specific system organizes the terminus region of the *E. coli* chromosome into a macrodomain. *Cell* 135, 475–485. doi: 10.1016/j.cell.2008.08.031

Messer, W. (2002). The bacterial replication initiator DnaA. DnaA and oriC, the bacterial mode to initiate DNA replication. *FEMS Microbiol. Rev.* 26, 355–374.

Messer, W., and Weigel, C. (1997). DnaA initiator–also a transcription factor. *Mol. Microbiol.* 24, 1–6. doi: 10.1046/j.1365-2958.1997.3171678.x

Miller, D. T., Grimwade, J. E., Betteridge, T., Rozgaja, T., Torgue, J. J., and Leonard, A. C. (2009). Bacterial origin recognition complexes direct assembly of higher-order DnaA oligomeric structures. *Proc. Natl. Acad. Sci. U.S.A.* 106, 18479–18484. doi: 10.1073/pnas.0909472106

Morigen, Løbner-Olesen, A., and Skarstad, K. (2003). Titration of the *Escherichia coli* DnaA protein to excess *datA* sites causes destabilization of replication forks, delayed replication initiation and delayed cell division. *Mol. Microbiol.* 50, 349–362. doi: 10.1046/j.1365-2958.2003.03695.x

Morigen, Skarstad, K., and Molina, F. (2005). Deletion of the *datA* site does not affect once-per-cell-cycle timing but induces rifampin-resistant replication. *J. Bacteriol.* 187, 3913–3920. doi: 10.1128/JB.187.12.3913-3920.2005

Morigen, M., Flåtten, I., and Skarstad, K. (2014). The *Escherichia coli datA* site promotes proper regulation of cell division. *Microbiology* 160, 703–710. doi: 10.1099/mic.0.074898-0

Mott, M. L., Erzberger, J. P., Coons, M. M., and Berger, J. M. (2008). Structural synergy and molecular crosstalk between bacterial helicase loaders and replication initiators. *Cell* 135, 623–634. doi: 10.1016/j.cell.2008.09.058

Nishida, S., Fujimitsu, K., Sekimizu, K., Ohmura, T., Ueda, T., and Katayama, T. (2002). A nucleotide switch in the *Escherichia coli* DnaA protein initiates chromosomal replication: evidence from a mutant DnaA protein defective in regulatory ATP hydrolysis *in vitro* and *in vivo*. *J. Biol. Chem.* 277, 14986–14995. doi: 10.1074/jbc.M108303200

Noguchi, Y., and Katayama, T. (2016). The *Escherichia coli* cryptic prophage protein YfdR binds to DnaA and initiation of chromosomal replication is inhibited by overexpression of the gene cluster *yfdQ-yfdR-yfdS-yfdT*. *Front. Microbiol.* 7:239. doi: 10.3389/fmicb.2016.00239

Noguchi, Y., Sakiyama, Y., Kawakami, H., and Katayama, T. (2015). The Arg fingers of key DnaA protomers are oriented inward within the replication origin *oriC* and stimulate DnaA subcomplexes in the initiation complex. *J. Biol. Chem.* 290, 20295–20312. doi: 10.1074/jbc.M115.662601

Nozaki, S., and Ogawa, T. (2008). Determination of the minimum domain II size of *Escherichia coli* DnaA protein essential for cell viability. *Microbiology* 154, 3379–3384. doi: 10.1099/mic.0.2008/019745-0

Nozaki, S., Yamada, Y., and Ogawa, T. (2009). Initiator titration complex formed at *datA* with the aid of IHF regulates replication timing in *Escherichia coli*. *Genes Cells* 14, 329–341. doi: 10.1111/j.1365-2443.2008.01269.x

Obita, T., Iwura, T., Su'etsugu, M., Yoshida, Y., Tanaka, Y., Katayama, T., et al. (2002). Determination of the secondary structure in solution of the *Escherichia coli* DnaA DNA-binding domain. *Biochem. Biophys. Res. Commun.* 299, 42–48. doi: 10.1016/S0006-291X(02)02590-1

Ogawa, T., Yamada, Y., Kuroda, T., Kishi, T., and Moriya, S. (2002). The *datA* locus predominantly contributes to the initiator titration mechanism in the control of replication initiation in *Escherichia coli*. *Mol. Microbiol.* 44, 1367–1375. doi: 10.1046/j.1365-2958.2002.02969.x

Okumura, H., Yoshimura, M., Ueki, M., Oshima, T., Ogasawara, N., and Ishikawa, S. (2012). Regulation of chromosomal replication initiation by *oriC*-proximal DnaA-box clusters in *Bacillus subtilis*. *Nucleic Acids Res.* 40, 220–234. doi: 10.1093/nar/gkr716

Olliver, A., Saggioro, C., Herrick, J., and Sclavi, B. (2010). DnaA-ATP acts as a molecular switch to control levels of ribonucleotide reductase expression in *Escherichia coli*. *Mol. Microbiol.* 76, 1555–1571. doi: 10.1111/j.1365-2958.2010.07185.x

Ozaki, S., and Katayama, T. (2009). DnaA structure, function, and dynamics in the initiation at the chromosomal origin. *Plasmid* 62, 71–82. doi: 10.1016/j.plasmid.2009.06.003

Ozaki, S., and Katayama, T. (2012). Highly organized DnaA-*oriC* complexes recruit the single-stranded DNA for replication initiation. *Nucleic Acids Res.* 40, 1648–1665. doi: 10.1093/nar/gkr832

Ozaki, S., Kawakami, H., Nakamura, K., Fujikawa, N., Kagawa, W., Park, S.-Y., et al. (2008). A common mechanism for the ATP-DnaA-dependent formation of open complexes at the replication origin. *J. Biol. Chem.* 283, 8351–8362. doi: 10.1074/jbc.M708684200

Ozaki, S., Noguchi, Y., Hayashi, Y., Miyazaki, E., and Katayama, T. (2012a). Differentiation of the DnaA-*oriC* subcomplex for DNA unwinding in a replication initiation complex. *J. Biol. Chem.* 287, 37458–37471. doi: 10.1074/jbc.M112.372052

Ozaki, S., Noguchi, Y., Nishimura, M., and Katayama, T. (2012b). Stable nucleotide binding to DnaA requires a specific glutamic acid residue within the AAA+ box II motif. *J. Struct. Biol.* 179, 242–250. doi: 10.1016/j.jsb.2012.05.001

Riber, L., Frimodt-Møller, J., Charbon, G., and Løbner-Olesen, A. (2016). Multiple DNA binding proteins contribute to timing of chromosome replication in *E. coli*. *Front. Mol. Biosci.* 3:29. doi: 10.3389/fmolb.2016.00029

Riber, L., and Løbner-Olesen, A. (2005). Coordinated replication and sequestration of oriC and dnaA are required for maintaining controlled once-per-cell-cycle initiation in *Escherichia coli*. *J. Bacteriol.* 187, 5605–5613. doi: 10.1128/JB.187.16.5605-5613.2005

Richardson, T. T., Harran, O., and Murray, H. (2016). The bacterial DnaA-trio replication origin element specifies single-stranded DNA initiator binding. *Nature* 534, 412–416. doi: 10.1038/nature17962

Roth, A., and Messer, W. (1998). High-affinity binding sites for the initiator protein DnaA on the chromosome of *Escherichia coli*. *Mol. Microbiol.* 28, 395–401. doi: 10.1046/j.1365-2958.1998.00813.x

Rozgaja, T. A., Grimwade, J. E., Iqbal, M., Czerwonka, C., Vora, M., and Leonard, A. C. (2011). Two oppositely oriented arrays of low-affinity recognition sites in *oriC* guide progressive binding of DnaA during *Escherichia coli* pre-RC assembly. *Mol. Microbiol.* 82, 475–488. doi: 10.1111/j.1365-2958.2011.07827.x

Sakiyama, Y., Kasho, K., Noguchi, Y., Kawakami, H., and Katayama, T. (2017). Regulatory dynamics in the ternary DnaA complex for initiation of chromosomal replication in *Escherichia coli*. *Nucleic Acids Res.* 45, 12354–12373. doi: 10.1093/nar/gkx914

Samitt, C. E., Hansen, F. G., Miller, J. F., and Schaechter, M. (1989). *In vivo* studies of DnaA binding to the origin of replication of *Escherichia coli*. *EMBO J.* 8, 989–993.

Saxena, R., Fingland, N., Patil, D., Sharma, A. K., and Crooke, E. (2013). Crosstalk between DnaA protein, the initiator of *Escherichia coli* chromosomal replication, and acidic phospholipids present in bacterial membranes. *Int. J. Mol. Sci.* 14, 8517–8537. doi: 10.3390/ijms14048517

Saxena, R., Rozgaja, T., Grimwade, J., and Crooke, E. (2011). Remodeling of nucleoprotein complexes is independent of the nucleotide state of a mutant AAA+ protein. *J. Biol. Chem.* 286, 33770–33777. doi: 10.1074/jbc.M111.223495

Saxena, R., Vasudevan, S., Patil, D., Ashoura, N., Grimwade, J. E., and Crooke, E. (2015). Nucleotide-induced conformational changes in *Escherichia coli* DnaA protein are required for bacterial ORC to pre-RC conversion at the chromosomal origin. *Int. J. Mol. Sci.* 16, 27897–27911. doi: 10.3390/ijms161126064

Schenk, K., Hervás, A. B., Rösch, T. C., Eisemann, M., Schmitt, B. A., Dahlke, S., et al. (2017). Rapid turnover of DnaA at replication origin regions contributes to initiation control of DNA replication. *PLOS Genet.* 13:e1006561. doi: 10.1371/journal.pgen.1006561

Sekimizu, K., Bramhill, D., and Kornberg, A. (1987). ATP activates dnaA protein in initiating replication of plasmids bearing the origin of the *E. coli* chromosome. *Cell* 50, 259–265. doi: 10.1016/0092-8674(87)90221-2

Sekimizu, K., Bramhill, D., and Kornberg, A. (1988). Sequential early stages in the in vitro initiation of replication at the origin of the *Escherichia coli* chromosome. *J. Biol. Chem.* 263, 7124–7130.

Shimizu, M., Noguchi, Y., Sakiyama, Y., Kawakami, H., Katayama, T., and Takada, S. (2016). Near-atomic structural model for bacterial DNA replication initiation complex and its functional insights. *Proc. Natl. Acad. Sci. U.S.A.* 113, E8021–E8030. doi: 10.1073/pnas.1609649113

Simmons, L. A., Felczak, M., and Kaguni, J. M. (2003). DnaA Protein of *Escherichia coli*: oligomerization at the *E. coli* chromosomal origin is required for initiation and involves specific N-terminal amino acids. *Mol. Microbiol.* 49, 849–858. doi: 10.1046/j.1365-2958.2003.03603.x

Skarstad, K., and Katayama, T. (2013). Regulating DNA replication in bacteria. *Cold Spring Harb. Perspect. Biol.* 5:a012922. doi: 10.1101/cshperspect.a012922

Smulczyk-Krawczyszyn, A., Jakimowicz, D., Ruban-Osmialowska, B., Zawilak-Pawlik, A., Majka, J., Chater, K., et al. (2006). Cluster of DnaA boxes involved in regulation of *Streptomyces* chromosome replication: from *in silico* to *in vivo* studies. *J. Bacteriol.* 188, 6184–6194. doi: 10.1128/JB.00528-06

Speck, C., Weigel, C., and Messer, W. (1999). ATP- and ADP-dnaA protein, a molecular switch in gene regulation. *EMBO J.* 18, 6169–6176. doi: 10.1093/emboj/18.21.6169

Stauffer, M. E., and Chazin, W. J. (2004). Structural mechanisms of DNA replication, repair, and recombination. *J. Biol. Chem.* 279, 30915–30918. doi: 10.1074/jbc.R400015200

Su'etsugu, M., Harada, Y., Keyamura, K., Matsunaga, C., Kasho, K., Abe, Y., et al. (2013). The DnaA N-terminal domain interacts with Hda to facilitate replicase clamp-mediated inactivation of DnaA. *Environ. Microbiol.* 15, 3183–3195. doi: 10.1111/1462-2920.12147

Su'etsugu, M., Nakamura, K., Keyamura, K., Kudo, Y., and Katayama, T. (2008). Hda monomerization by ADP binding promotes replicase clamp-mediated DnaA-ATP hydrolysis. *J. Biol. Chem.* 283, 36118–36131. doi: 10.1074/jbc.M803158200

Sutton, M. D., Carr, K. M., Vicente, M., and Kaguni, J. M. (1998). *Escherichia coli* DnaA protein. The N-terminal domain and loading of DnaB helicase at the *E. coli* chromosomal origin. *J. Biol. Chem.* 273, 34255–34262. doi: 10.1074/jbc.273.51.34255

Sutton, M. D., and Kaguni, J. M. (1997). Threonine 435 of *Escherichia coli* DnaA protein confers sequence-specific DNA binding activity. *J. Biol. Chem.* 272, 23017–23024. doi: 10.1074/jbc.272.37.23017

Swinger, K. K., and Rice, P. A. (2004). IHF and HU: flexible architects of bent DNA. *Curr. Opin. Struct. Biol.* 14, 28–35. doi: 10.1016/j.sbi.2003.12.003

Tesfa-Selase, F., and Drabble, W. T. (1992). Regulation of the *gua* operon of *Escherichia coli* by the DnaA protein. *Mol. Gen. Genet.* 231, 256–264.

Travers, A., and Muskhelishvili, G. (2005). DNA supercoiling - a global transcriptional regulator for enterobacterial growth? *Nat. Rev. Microbiol.* 3, 157–169.

von Freiesleben, U., Rasmussen, K. V., Atlung, T., and Hansen, F. G. (2000). Rifampicin-resistant initiation of chromosome replication from *oriC* in *ihf* mutants. *Mol. Microbiol.* 37, 1087–1093. doi: 10.1046/j.1365-2958.2000.02060.x

Waldminghaus, T., and Skarstad, K. (2009). The *Escherichia coli* SeqA protein. *Plasmid* 61, 141–150. doi: 10.1016/j.plasmid.2009.02.004

Weigel, C., Messer, W., Preiss, S., Welzeck, M., Morigen, and Boye, E. (2001). The sequence requirements for a functional *Escherichia coli* replication origin are different for the chromosome and a minichromosome. *Mol. Microbiol.* 40, 498–507. doi: 10.1046/j.1365-2958.2001.02409.x

Wolański, M., Donczew, R., Zawilak-Pawlik, A., and Zakrzewska-Czerwińska, J. (2015). *oriC*-encoded instructions for the initiation of bacterial chromosome replication. *Front. Microbiol.* 5:735. doi: 10.3389/fmicb.2014.00735

Yoshida, Y., Obita, T., Kokusho, Y., Ohmura, T., Katayama, T., Ueda, T., et al. (2003). Identification of the region in *Escherichia coli* DnaA protein required for specific recognition of the DnaA box. *Cell. Mol. Life Sci.* 60, 1998–2008. doi: 10.1007/s00018-003-3176-7

Zhang, Q., Zhou, A., Li, S., Ni, J., Tao, J., Lu, J., et al. (2016). Reversible lysine acetylation is involved in DNA replication initiation by regulating activities of initiator DnaA in *Escherichia coli*. *Sci. Rep.* 6:30837. doi: 10.1038/srep30837

Zhou, H. X. (2001). The affinity-enhancing roles of flexible linkers in two-domain DNA-binding proteins. *Biochemistry* 40, 15069–15073. doi: 10.1021/bi015795g

13

A balanced perspective on unbalanced growth and thymineless death

Philip C. Hanawalt*

Department of Biology, Stanford University, Stanford, CA, USA

***Correspondence:**
*Philip C. Hanawalt,
Department of Biology, Stanford
University, 371 Serra Mall, Stanford,
CA 94305-5020, USA
hanawalt@stanford.edu*

The early history of the esoteric phenomenon of thymineless death (TLD) is recounted, from the pioneering discovery by Seymour Cohen and Hazel Barner, through my graduate studies at Yale and postdoctoral research in Copenhagen. My principal contribution was the discovery that restricted synthesis of protein and RNA permits cultures of *Escherichia coli* to complete their DNA replication cycles without initiating new ones, and that cells held in this physiological state are immune to the lethality of thymine deprivation; unbalanced growth is not the fundamental cause of TLD. The successful synchronization of the DNA replication cycle contributed to formulation of the replicon concept. Studies at Stanford revealed a specific requirement for transcription and led to the discovery of a TLD-resistant mutant in a new gene, termed *recQ*, with important homologs in humans and most other organisms. The lessons learned from research on TLD underscore the value of basic research in bacterial systems that can have profound implications for human health.

Keywords: DNA replication cycle, DNA degradation, replication origin, DNA repair, thymine starvation, transcription, genomic uracil, RecQ

Introduction

This review constitutes a reflection upon my early experiences in biomedical research, with particular focus upon the enigmatic phenomenon of thymineless death (TLD) that began during my graduate study and that still commands my intrigue. I came to appreciate how advances in science and their promotion are coupled to the personalities and aspirations of the investigators. While a discovery may incorporate novel insights, it always builds upon the contributions from predecessors and usually includes elements of serendipity. I will candidly relate the collaborative interactions among some pioneering investigators who have strived to achieve a mechanistic understanding of TLD, while the ultimate comprehension still eludes their grasp.

Upon graduation from Oberlin College with a major in physics, I joined the biophysics program at Yale for graduate study in 1954. That was an exciting era; the Watson–Crick DNA structure and a plausible model for DNA replication had just emerged, but we did not have a clue about the functions of RNA or its relationships to DNA. My first research mentor, Harold Morowitz, suggested that I should synchronize the division cycle in cultures of the bacterium, *Escherichia coli* (*E. coli*), so that we could study the time-course of biochemical events from one division to the next. Unfortunately, a published synchronization procedure involving glucose starvation was not reproducible by us or by anyone else. Although I retained my interest in the cell cycle, I then joined the laboratory of Richard Setlow, with whom I had just completed an outstanding spectroscopy course. I wanted to explore the application of short-wavelength ultraviolet light (UV) to probe macromolecular activities in bacterial cells through action spectra. I developed an approach to resolve the cellular components of DNA, RNA, and phospholipids so that I could follow their respective syntheses by labeling with ^{32}P orthophosphate (Hanawalt, 1959). However, I also wanted to follow DNA synthesis at higher specificity with ^{14}C-labeled thymine, so I needed to improve the efficiency of its exogenous incorporation. In a literature search I found a timely paper describing a thymine-requiring bacterium (Barner and Cohen, 1954).

Seymour Cohen was characterizing *E. coli* 15T−, later shown to lack functional thymidylate synthetase. This auxotrophic mutant, originally isolated by Raymond Roepke from UV-irradiated *E. coli* strain 15, had been acquired from Joseph Gots at the University of Pennsylvania. Unlike the biostatic consequences of starvation for most required growth factors, such as uracil or cytosine, *E. coli* 15T− suffered a striking loss of viability when thymine was omitted from the defined growth medium, which may explain the scarcity of this type of mutant. The lethality was initially attributed to "unbalanced growth," since the bacteria continued to grow in mass while cell division was curtailed and hardly any DNA synthesis was evident (Cohen and Barner, 1954); however, the cells in minimal medium survived if the sole carbon source, glucose, was omitted, indicating that metabolism was required for TLD. The suggestion that the inhibition of DNA synthesis by UV might also kill cells by unbalanced growth sparked my interest. Balanced growth, in contrast to unbalanced growth, was defined as the physiological condition over a time interval during which every extensive property of the growing system increased by the same factor (Campbell, 1957).

TLD Studies at Yale

Cohen generously provided *E. coli* 15T−, and my research followed upon his work, with an eventual thesis entitled: "Macromolecular synthesis in *Escherichia coli* during conditions of unbalanced growth," in which I compared the effects of thymine starvation to those caused by short-wavelength UV irradiation (Hanawalt, 1958). I confirmed that during thymine starvation the bacteria stopped dividing and formed long filaments, similar to those observed following UV irradiation (Deering and Setlow, 1957). The optical density (turbidity) continued to follow that of cultures in balanced growth for one division period or a little longer, while viability remained constant for about 30 min before it decreased exponentially at roughly 90% per division period. RNA synthesis in the absence of thymine was linear for two division periods before declining, while protein synthesis (followed by labeling with ^{14}C-proline or $^{35}SO_4$) leveled off earlier. Unlike the case for thymine starvation, DNA synthesis eventually recovered following low doses of UV (Hanawalt and Setlow, 1960) and the lag in recovery could be shortened by exposing the cells to photoreactivating light (Hanawalt and Buehler, 1960). Of course, the ^{32}P labeling protocol was needed to confirm that there was very little DNA synthesis in the thymine-starved cells (Hanawalt, 1959). Cohen and Barner had also established that the thymine analog, 5-bromouracil (5BU), could partially fulfill the thymine requirement and prolong the viability of *E. coli* 15T− (Cohen and Barner, 1956). The incorporation of deoxyuridine or 5BU during DNA synthesis *in vitro* was documented by Bessman et al. (1958). At that point in history we were unaware of any effects of thymine starvation on DNA structure or function and we did not yet know the nature of the DNA damage produced by UV.

TLD Studies in Denmark

Upon completion of my Ph.D., I wanted to return to my earlier interest in synchronous growth, so I obtained an NIH postdoctoral fellowship to join Ole Maaløe in Denmark, where Gordon Lark, Elio Schaechter, and others had collaborated in studies of bacterial growth and division patterns upon changing media or shifting temperature. Maaløe was considered the international expert in bacterial synchronous growth, and he had just been appointed Professor/Chair of a newly established Department of Microbiology at the University of Copenhagen. Shortly before leaving Yale I received from Cohen a triple-mutant derivative of *E. coli* 15T− called "TAU," auxotrophic for arginine (A) and uracil (U), as well as for thymine (T; Barner and Cohen, 1958). Maaløe offered no immediate suggestions for a research project upon my arrival in his lab, so I simply continued with the TLD studies I had begun at Yale. Although I recorded some interesting effects of incubation temperature on TLD, the most exciting results were those obtained when A and U were withheld (henceforth, "AU," since both requirements were due to the same mutation in carbamoyl phosphate synthase). The first thing I noted was that although mass (turbidity) did not increase in the absence of AU, nearly all of the cells (~97%) still suffered TLD. Thus, unbalanced growth at the level of cell mass in relation to DNA content could not be responsible for the lethality. In the presence of T but absence of AU, DNA synthesis continued for over an hour to achieve roughly a 40% increase before it leveled off. Interestingly, the fraction of the population resistant to TLD in the absence of AU also increased during that period of allowed DNA synthesis. By the time DNA synthesis had ceased, the entire population of cells had become immune to TLD, unless protein and RNA syntheses were again permitted (**Figure 1**). These results prompted me to suggest that the cellular DNA replication cycle could be completed in the absence of protein and RNA synthesis, but that protein and/or RNA synthesis were required for re-initiation of the cycle; and furthermore, that cells are only susceptible to TLD if they are actively replicating DNA. (This made logical sense: why would they need thymine if they were not replicating DNA? Of course, we did not yet know about *repair replication*!) I had inadvertently discovered a simple approach for synchronizing the DNA replication cycle in bacteria. At this point Maaløe became intensely interested in my project and I enjoyed more frequent discussions with him. We decided that we would need to use autoradiography to learn the status of DNA replication in individual cells in order to verify the model. Elio Schaechter had previously employed single cell autoradiography in Maaløe's group to demonstrate that *E. coli* normally replicates DNA throughout most of the cell cycle (Schaechter et al., 1959). Unfortunately, the technology had not been retained in Maaløe's lab when Schaechter departed. However, my graduate colleague from Yale, Bob Van Tubergen, was an expert in single cell autoradiography and had just joined Roy Markham's group in

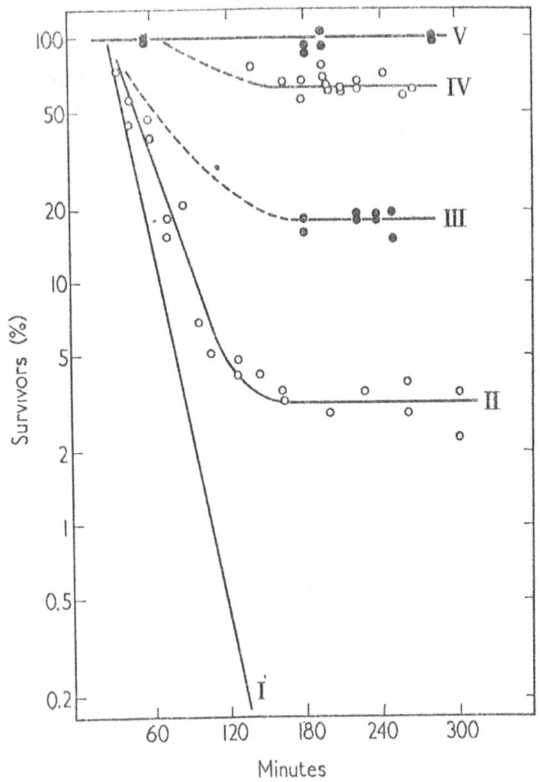

FIGURE 1 | Effect of pre-incubation with T but without AU on survival of *E. coli* TAU during subsequent incubation, −T − AU. II; −T − AU (no pre-incubation): III; 30 min pre-incubation +T − AU: IV; 60 min pre-incubation +T − AU: V; 90 min pre-incubation +T − AU: I; Comparison curve for −T + AU, from another experiment. From Maaløe and Hanawalt (1961).

Cambridge (Van Tubergen, 1961). That afforded an opportunity for me to visit England, to learn the autoradiographic procedures from Bob, and to carry out several preliminary experiments under his guidance. I then completed the single cell autoradiographic studies to confirm our model that individual cells terminated DNA synthesis at different times during incubation with T in the absence of AU. A new postdoc, Donald Cummings, soon joined Maaløe's group and contributed experience in density-labeling DNA with 5BU to follow DNA synthesis. I was pleased to have Don as a collaborator because I appreciated the unique advantages of density-labeling for analysis of DNA replication (Meselson and Stahl, 1958). Density-labeling nascent DNA could tell us whether the DNA that had replicated in the absence of AU was able to begin a new round of synthesis before AU was again supplied. The studies of TLD in Copenhagen provided early insights into control of the normal DNA replication cycle in bacteria and contributed to the formulation of the concept of the replicon (Jacob et al., 1963).

Other researchers were also attracted to the TLD field, while Barner and Cohen (1956) continued their leadership with many additional contributions, including evidence for at least one synchronous division when thymine was re-added 30 min after starvation, the report that a single amino acid deficiency does not completely block TLD although it inhibits growth (Barner and Cohen, 1957), and the finding that deprivation of uracil (in a double-mutant for T and U) prolongs the lag period for TLD and induces RNA turnover (Barner and Cohen, 1958). The UV irradiation of cells prior to thymine starvation was shown to shorten the lag period before the onset of TLD, suggesting some similarities in the "cellular units" affected by the respective treatments (Gallant and Suskind, 1961). However, there was still no information on the nature of DNA damage due to either UV or thymine starvation.

TLD Studies at Caltech

With an interest in learning more about biophysical approaches for characterizing nucleic acids, and hoping to discover what thymine starvation actually did to DNA, I joined the group of Robert Sinsheimer at the California Institute of Technology for a third-year postdoc. Shortly after my arrival at Caltech in late 1960 I heard about an upcoming Cold Spring Harbor Symposium on "Cellular Regulatory Mechanisms," for which Maaløe was an invited speaker. Unfortunately, the conference had been oversubscribed so it was too late for me to attend. I learned later that my mentor had featured our TLD project in his lecture and in the published proceedings as sole author, and had acknowledged his research assistant, Jens Ole Rostock, for the autoradiography (Maaløe, 1961). The results were simultaneously published in two journal articles (Hanawalt et al., 1961; Maaløe and Hanawalt, 1961). The authorship of the second paper included not only Don Cummings for the 5BU studies but also Elio Schaechter, who had not participated in the project and whom I had not even met. (That was my first encounter with "courtesy authorship.") Although these were disappointing experiences, I learned important lessons from them about appropriate attribution of contributions in the dissemination of scientific advances. Seven years later, at the 1968 Cold Spring Harbor Symposium on "Replication of DNA in Microorganisms," two of my graduate students gave invited talks on their own work, and I included four additional graduate students and two postdocs as co-authors for my presentation (Hanawalt et al., 1968). I also convened a reunion of alumni from Maaløe's group for beer and informal discussions during the conference (**Figure 2**).

Although I learned a lot at Caltech from productive interactions with Sinsheimer, Max Delbrück, Matt Meselson, Jerry Vinograd, and their students, my search for DNA alterations during TLD did not reveal anything significant. The most exciting new information that year was the report that UV irradiation caused thymine to form covalent dimers (Beukers and Berends, 1960), and these were soon shown to be the predominant photoproducts in cellular DNA; so there was finally a specific lesion to study in relation to DNA synthesis inhibition and damage processing in UV-irradiated bacteria.

TLD Studies at Stanford

I joined the Biophysics Laboratory at Stanford University in September 1961 with a primary focus on DNA replication in

FIGURE 2 | Photos from Cold Spring Harbor Symposium Vol. XXXIII on "Replication of DNA in Micro-organisms" (1968). (A) Professor Ole Maaløe, with his ever-present cigar. **(B)** Reunion of alumni from Ole Maaløe's research group Standing: P. Hanawalt, J. Clark, C. Levinthal, J. Watson, P. Kuempel, C. Helmstetter, D. Glaser Seated: K. G. Lark, M. Schaechter, S. Cooper, O. Maaløe, with cigar.

UV-irradiated *E. coli*, but also with ongoing TLD studies. I then acquired a new multiple auxotroph of *E. coli* 15T− from Richard Wax that required methionine, tryptophan, proline, and arginine as well as thymine, and had a more stringent uracil requirement than that in strain TAU. (I named it TAU-bar, simply to avoid confusion with TAU). The removal of uracil greatly reduced TLD and implicated mRNA transcription in the lethality; however, I emphasized that one must distinguish between the event that causes killing and an event that converts the cell from the immune state to the susceptible state (Hanawalt, 1963). Gallant and Suskind (1962), in studies of *E. coli B3,* had concluded that TLD is correlated with unbalanced synthesis of RNA. In later studies we showed that the transcription inhibitor, rifampicin, essentially eliminates TLD, while also preventing the DNA strand breaks we had observed during thymine deprivation (Nakayama and Hanawalt, 1975). The accumulation of strand breaks had been seen, or not, by a number of other researchers using different bacterial strains and different conditions, but it was difficult to assess whether this could be a cause or an effect of TLD, in part because some of the strains carried inducible prophage or colicin factors (Freifelder, 1969; Ahmad et al., 1998 for review). Following upon our discovery of repair replication in UV-irradiated *E. coli* (Pettijohn and Hanawalt, 1964), we documented the non-conservative repair mode of DNA replication in thymine-starved bacteria (Pauling and Hanawalt, 1965), and we speculated that transcription was somehow responsible for introducing strand breaks or gaps that required thymine for their repair. An obvious hypothesis was that, in the absence of thymidine, DNA polymerases incorporated deoxyuridine and this would lead to a futile cycle of removal and resynthesis, as the uracil in DNA is recognized by glycosylases (e.g., Ung) for repair. The pool of deoxyuridylate was reported to increase markedly during thymine starvation (Goulian and Beck, 1966). Meanwhile, adding to the complexity of models, it was shown that initiation of replication from the chromosomal origin occurred before completion of the previous replication cycle during thymine starvation, and that amino acid starvation prevented what was called "*premature* initiation" (Pritchard and Lark, 1964). Cells growing in very low thymine concentrations exhibited increased *thymineless mutagenesis* and required an extended period for completion of chromosomal DNA synthesis (Zaritsky and Pritchard, 1971). Since thymine limitation can be seen as the *approach* to thymine exhaustion, some features of thymine deficiency may already be apparent, as exemplified by the more recent (40 years later!) results of Zaritsky et al. (2011), in which delayed initiations during thymine limitation are ascribed to possible "collapsing replisomes" and double strand breaks in the origin region.

Donachie and Hobbs (1967) had proposed that the lethality during TLD could result from sensitivity to plating on nutrient agar, but Nakayama and Couch (1973) established that the lethal event must occur *during the period of thymine starvation*, as they assayed viability entirely in the liquid phase, through application of the Poisson distribution to dilute suspensions of the bacteria.

Thymineless death was demonstrated in a member of the simplest class of living cells, *Mycoplasma laidlawii B* (Smith and Hanawalt, 1968) and in the most UV-resistant of known bacteria, *Deinococcus radiodurans*, which suffered TLD with roughly the same kinetics as *E. coli*, thereby challenging the hypothesis that the DNA damage due to thymine deprivation was similar to that due to UV (Little and Hanawalt, 1973). TLD was also documented in the eukaryote, *Saccharomyces cerevisiae* (Barclay and Little, 1978; Little, 1985), and of course, eventually in cultured mammalian cells. Goulian et al. (1985) studied mammalian cells treated with methotrexate, which inhibits dihydrofolate reductase to result in thymidine deficiency, and found a greater than 1000-fold increase in dUTP (normally <0.3 nM) associated with the drop in intracellular dTTP. They documented the incorporation of uracil and fragmentation of newly replicated DNA together with lethality, supporting the hypothesis that an elevated concentration of dUTP in mammalian cells resulted in TLD through a mechanism of uracil incorporation and removal during DNA repair (Goulian et al., 1985). For current reviews on uracil in DNA and its processing, see Krokan et al. (2014)and Williams and Kunkel (2014).

An undergraduate from Cornell University, Jack Nunberg, spent the summer of 1971 in our laboratory to study the DNA degradation associated with thymine deprivation, following the report of selective degradation of nascent DNA during TLD in *Bacillis subtilis* (Reiter and Ramareddy, 1970). Breitman et al. (1972) had also documented loss of thymine from DNA and an early increase in deoxyribose during thymine starvation in *E. coli* TAU-bar. We had previously documented preferential loss of nascent DNA in UV-irradiated *E. coli* (Hanawalt and Brempelis, 1967), and many years later we learned that the RecQ helicase and RecJ 5′–3′ exonuclease were responsible for that specific degradation of the lagging DNA strand (Courcelle and Hanawalt, 1999). Nunberg found that the *E. coli* CR34/74 temperature-sensitive DNA synthesis mutant (deficient in the DnaB helicase that operates at the replication fork), exhibited nascent DNA degradation at the restrictive temperature but without loss of viability. However, TLD at the restrictive temperature in CR34/74 did not differ significantly from that at the normal growth temperature; thus, active DNA replication forks are not required for TLD! Using *E. coli* TAU-bar prelabeled with ^{14}C-thymine and pulse labeled with ^{3}H-thymine, Nunberg also found that "bulk" DNA degradation but not nascent DNA degradation was eliminated by rifampicin, in parallel with the protection from TLD (Nunberg, 1971). We had previously documented turnover of deoxyribonucleotides in DNA at ~0.02% per generation period at the restrictive temperature in *E. coli* CR34/74, and this was markedly suppressed by rifampicin, in support of our hypothesis that transcription may sometimes result in repairable single-strand breaks (Grivell et al., 1975).

Hiroaki Nakayama selected survivors from a thymine-starved *E. coli K12* strain to isolate a mutant that was remarkably resistant to TLD. Upon return to Kyushu University from Stanford, he named the gene *recQ* (for *Kyushu*!) and identified the product as a helicase (Nakayama et al., 1984). RecQ participates in the RecF recombination pathway, and it turns out that mutants in other genes in this pathway are also resistant to TLD, whereas *recBC* mutants are unusually sensitive (Nakayama et al., 1982, 1988). RecQ is the prototype for the large family of RecQ helicases found in many bacterial strains and most eukaryotes, and its discovery is one of many examples that validate the importance of basic research on seemingly esoteric topics for future advances in biomedical science. Human cells have 5 *RecQ* homologs; mutation in one causes premature aging, mutation in another causes high levels of sister-chromatid exchanges and mutation in a third also has a cancer prone phenotype. For reviews of RecQ, see Nakayama (2005) and Larsen and Hickson (2013).

Structural changes in DNA from thymine-starved *E. coli* were revealed by pulsed field gel electrophoresis following cell lysis in agarose gels and post-lysis treatment with restriction endonucleases. Non-migrating DNA fragments accumulated in cells during thymine deprivation. Interestingly, the amounts of specific migrating DNA fragments were reduced inversely in relation to their map distances from the replication origin (*oriC*). Electron microscopy of the non-migrating DNA revealed convoluted structures, with branching and single-strand tails. The accumulation of non-migrating DNA was largely dependent upon RecA and *recF*-family genes, including *recQ*, implying links between the non-migrating DNA and TLD (Nakayama et al., 1994). However, the question remained as to whether these observations were revealing causes or consequences of TLD.

In the early 1990s the TLD field went into "eclipse" for a decade, with sparse publication of bacterial work, although the number of clinically relevant papers began to increase (for examples, see Canman et al., 1992; Houghton et al., 1997; Ladner, 2001; Liao et al., 2005). The *stalemate* in mechanistic comprehension is highlighted by the fact that a drafted review on TLD by Shamim Ahmad, Raymond Devoret, Hiroaki Nakayama, and myself in 1988, was eventually completed and published by Ahmad and several other co-authors 10 years later (Ahmad et al., 1998).

Re-Awakened Interest in TLD

Periodically, I would suggest to a student that there were many interesting things yet to be learned from the study of TLD, but very few took the "bait" until an undergraduate, Pam Morganroth, joined my group in her junior year to study genetic control of TLD, for graduation with Honors (Morganroth and Hanawalt, 2006). TLD was characterized in *E. coli* mutants deficient in DNA synthesis (*dnaA, dnaB, dnaC,* and *dnaE*) and in cells treated with hydroxyurea (HU) to inhibit replication. Substantial resistance to TLD was observed when replication initiation was inhibited (*dnaA* and *dnaC*), but little resistance was noted when only replication elongation was inhibited (*dnaE* and HU). Morganroth's results confirmed those of Bouvier and Sicard (1975), who had reported TLD resistance in several other temperature sensitive *dnaA* and *dnaC* mutants at the non-permissive temperature. It was concluded that active replication elongation is not required for the major TLD pathway, consistent with the findings of Nunberg (1971). A *uvrA* mutant, deficient in both global and transcription coupled nucleotide excision repair, exhibited normal TLD, ruling out participation of that ubiquitous pathway. In addition, the *lexA3* mutation, causing deficient activation of the SOS genomic stress response, did not affect TLD. However, the later studies of Fonville et al. (2010) indicated that part of the sensitivity to thymine starvation is dependent upon the SOS response, through activation of *sulA* to block cell division, in contrast to Morganroth's results. Further study resolved the discrepancy by establishing different genotypes in the respective strains employed and by showing inexplicably that TLD is inhibited by CsrA, a carbon storage regulator, in cells lacking the SOS response (Hamilton et al., 2013). Reflecting upon the roles of the *recF*-family genes, it is provocative to consider the paradoxical observation that while RecF proteins operate at arrested replication forks in UV-irradiated cells to enhance survival, their *absence* enhances survival in thymine-starved cells (Courcelle, 2005).

A major breakthrough in our understanding of TLD resulted from the application of genome wide DNA microarrays by Sangurdekar et al. (2010), who revealed a striking loss of genetic material, spreading outward from the origin (*oriC*), that correlated with lethality during thymine starvation. TLD resistant mutants in the RecFORJQ pathway had correspondingly less damage in the origin region. Using the same approach, the loss of origin-specific DNA during TLD has been confirmed by Kuong and Kuzminov (2012). Fonville et al. (2010) employed FISH analyses to record loss of DNA near the origin during thymine deprivation as well as eventual loss near the replication terminus at late times. Martín and Guzmán (2011) have confirmed the initiation of new replication initiations at the origin during thymine deprivation that correlate with TLD, and that are inhibited by rifampicin. Chapters from the groups of Guzmân and Khodursky follow in this special collection of articles on The Bacterial Cell (Guzmân and Martîn, 2015; Ostrer et al., 2015). It is exciting as we appear to be closing in on the prime culprit for TLD at the replication origin. The ultimate answer to the mechanism of TLD will surely implicate the essential role of RNA polymerase in initiating the DNA replication cycle (Lark, 1972; Martîn et al., 2014) as well as the genomic instability due to incorporation of uracil into DNA during the initiation of DNA synthesis at *oriC*. TLD underlies the mechanisms of action of a number of clinically important drugs such as methotrexate, trimethoprim, and fluorouracil. It is more than likely that the insights derived from the continuing studies of TLD in bacteria will have translational impact for chemotherapies that depend upon TLD-like responses.

Acknowledgments

I wish to dedicate this review to the memory of my graduate mentor, Richard (Dick) Setlow, who passed away April 6, 2015 at age 94. Dick was largely responsible for my graduate training and ultimate scientific career in the fields of UV photobiology and DNA repair. We are all indebted to Seymour Cohen and Hazel Barner for initiating the TLD field, and for the ideas and efforts of many talented students and their mentors over the past 60 years. I apologize to those whose contributions were overlooked or not cited for lack of space and time. The work in my laboratory has been supported by the National Institutes of Health, most recently by a grant CA077712 from the National Cancer Institute.

References

Ahmad, S. I., Kirk, S. H., and Eisenstark, A. (1998). Thymine metabolism and thymineless death in prokaryotes and eukaryotes. *Annu. Rev. Microbiol.* 52, 591–625. doi: 10.1146/annurev.micro.52.1.591

Barclay, B. J., and Little, J. G. (1978). Genetic damage during thymidylate starvation in *Saccharomyces cerevisiae*. *Mol. Gen. Genet.* 160, 33–40.

Barner, H. D., and Cohen, S. S. (1954). The induction of thymine synthesis by T2 infection of a thymine requiring mutant of *Escherichia coli*. *J. Bacteriol.* 68, 80–88.

Barner, H. D., and Cohen, S. S. (1956). Synchronization of division of a thymineless mutant of *Escherichia coli*. *J. Bacteriol.* 72, 115–123.

Barner, H. D., and Cohen, S. S. (1957). The isolation and properties of amino acid requiring mutants of a thymineless bacterium. *J. Bacteriol.* 74, 350–355.

Barner, H. D., and Cohen, S. S. (1958). Protein synthesis and RNA turnover in a pyrimidine-deficient bacterium. *Biochim. Biophys. Acta* 30, 12–20. doi: 10.1016/0006-3002(58)90234-8

Bessman, M. J., Lehman, I. R., Adler, J., Zimmerman, S. B., Simms, E. S., and Kornberg, A. (1958). Enzymatic synthesis of deoxoribonucleic acide III. The incorporation of pyrimidine and purine analogues into deoxyribonucleic acid. *Proc. Natl. Acad. Sci. U.S.A.* 44, 633–640. doi: 10.1073/pnas.44.7.633

Beukers, R., and Berends, W. (1960). Isolation and identification of the irradiation product of thymine. *Biochim. Biophys. Acta* 41, 550–551. doi: 10.1016/0006-3002(60)90063-9

Bouvier, F., and Sicard, N. (1975). Interference of dna-ts mutations of *Escherichia coli* with thymineless death. *J. Bacteriol.* 124, 1198–1204.

Breitman, T. R., Maury, P. B., and Toal, J. N. (1972). Loss of deoxyribonucleic acid-thymine during thymine starvation of *Escherichia coli*. *J. Bacteriol.* 112, 646–648.

Campbell, A. (1957). Synchronization of cell division. *Bacteriol. Rev.* 21, 263–272.

Canman, C. E., Tang, H.-Y., Normolle, D. P., Lawrence, T. S., and Maybaum, J. (1992). Variations in patterns of DNA damage induced in human colorectal tumor cells by 5-fluorodeoxyuridine: implications for mechanisms of resistance and cytotoxicity. *Proc. Natl. Acad. Sci. U.S.A.* 89, 10474–10478. doi: 10.1073/pnas.89.21.10474

Cohen, S. S., and Barner, H. D. (1954). Studies on unbalanced growth in *E. coli*. *Proc. Natl. Acad. Sci. U.S.A.* 40, 885–893. doi: 10.1073/pnas.40.10.885

Cohen, S. S., and Barner, H. D. (1956). Studies on the induction of thymine deficiency and on the effects of thymine and thymidine analogues in *Escherichia coli*. *J. Bacteriol.* 71, 588–597.

Courcelle, J. (2005). Recs preventing wrecks. *Mutat. Res.* 577, 217–227. doi: 10.1016/j.mrfmmm.2005.03.019

Courcelle, J., and Hanawalt, P. C. (1999). RecQ and RecJ process blocked replication forks prior to the resumption of replication in UV-irradiated *Escherichia coli*. *Mol. Gen. Genet.* 262, 543–551. doi: 10.1007/s004380051116

Deering, R. A., and Setlow, R. B. (1957). Inhibition of cell division of *Escherichia coli* by low doses of ultraviolet light. *Science* 126, 397–398. doi: 10.1126/science.126.3270.397

Donachie, W. D., and Hobbs, D. G. (1967). Recovery from "thymineless death" in *Escherichia coli* 15T-. *Biochem. Biophys. Res. Commun.* 29, 172–177. doi: 10.1016/0006-291X(67)90582-7

Fonville, N., Bates, D., Hastings, P. J., Hanawalt, P. C., and Rosenberg, S. M. (2010). Role of RecA and the SOS response in thymineless death in *Escherichia coli*. *PLoS Genet.* 6:e1000865. doi: 10.1371/journal.pgen.1000865

Freifelder, D. (1969). Single-strand breaks in bacterial DNA associated with thymine starvation. *J. Mol. Biol.* 45, 1–7. doi: 10.1016/0022-2836(69)90205-8

Gallant, J., and Suskind, S. R. (1961). Relationship between thymineless death and ultraviolet inactivation in *Escherichia coli*. *J. Bacteriol.* 82, 187–194.

Gallant, J., and Suskind, S. R. (1962). Ribonucleic acid synthesis and thymineless death. *Biochim. Biophys.* 55, 627–638. doi: 10.1016/0006-3002(62)90841-7

Goulian, M., and Beck, W. (1966). Variations of intracellular deoxyribosyl compounds in deficiencies of vitamin B12, folic acid and thymine. *Biochim. Biophys. Acta* 129, 336–349. doi: 10.1016/0005-2787(66)90376-5

Goulian, M., Ingraham, H., Tseng, B., Dickey, L., Bleile, B., and Neynaber, S. (1985). Mechanisms of thymineless death. *Pediatr. Res.* 19, 756–762. doi: 10.1203/00006450-198507000-00093

Grivell, A. R., Grivell, M. B., and Hanawalt, P. C. (1975). Turnover in bacterial DNA containing thymine or 5-bromouracil. *J. Mol. Biol.* 98, 219–233. doi: 10.1016/S0022-2836(75)80110-0

Guzmân, E. C., and Martîn, C. M. (2015). Death at the origin. *Front. Microbiol.* 6:499. doi: 10.3389/fmicb.2015.00499

Hamilton, H. M., Wilson, R., Blythe, M., Nehring, R. B., Fonville, N. C., Louis, E. J., et al. (2013). Thymineless death is inhibited by CsrA in *Escherichia coli* lacking the SOS response. *DNA Repair* 12, 993–999. doi: 10.1016/j.dnarep.2013.08.011

Hanawalt, P. C. (1958). *Macromolecular Synthesis in E. coli Under Conditions of Unbalanced Growth*. Ph.D. thesis, Yale University.

Hanawalt, P. C. (1959). Use of phosphorus-32 in microassay for nucleic acid synthesis in *E. coli*. *Science* 130, 386–387. doi: 10.1126/science.130.3372.386

Hanawalt, P. C. (1963). Involvement of synthesis of RNA in thymineless death. *Nature* 198, 286. doi: 10.1038/198286a0

Hanawalt, P. C., and Brempelis, I. (1967). "Selective degradation of newly-replicated DNA after inhibition of DNA synthesis in *E. coli*," in *Proceedings of Seventh International Congress of Biochemistry*, Japan, 650.

Hanawalt, P. C., and Buehler, J. (1960). Photoreactivation of macromolecular synthesis in *E. coli*. *Biochim. Biophys. Acta* 37, 141–143. doi: 10.1016/0006-3002(60)90088-3

Hanawalt, P. C., Maaløe, O., Cummings, O., and Schaechter, M. (1961). The normal DNA replication cycle II. *J. Mol. Biol.* 3, 156–165. doi: 10.1016/S0022-2836(61)80042-9

Hanawalt, P. C., Pettijohn, D. E., Pauling, C. E., Brunk, C. F., Smith, D. W., Kanner, L. C., et al. (1968). Repair replication of DNA in vivo. *Cold Spring Harb. Symp. Quant. Biol.* 36, 187–194. doi: 10.1101/SQB.1968.033.01.022

Hanawalt, P. C., and Setlow, R. B. (1960). Effect of monochromatic UV on macromolecular synthesis in *E. coli*. *Biochim. Biophys. Acta* 41, 283–294. doi: 10.1016/0006-3002(60)90011-1

Houghton, J. A., Harwood, F. G., and Tillman, D. M. (1997). Thymineless death in colon carcinoma cells is mediated by fas signaling. *Proc. Natl. Acad. Sci. U.S.A.* 94, 8144–8149. doi: 10.1073/pnas.94.15.8144

Jacob, F., Brenner, S., and Cuzin, F. (1963). On the regulation of DNA replication in bacteria. *Cold Spring Harb. Symp. Quant. Biol.* 28, 329–438. doi: 10.1101/SQB.1963.028.01.048

Krokan, H. E., Sætrom, P., Aas, P. A., Pettersen, H. S., Kavli, B., and Slupphaug, G. (2014). Error-free versus mutagenic processing of genomic uracil-relevance to cancer. *DNA Repair* 19, 38–47. doi: 10.1016/j.dnarep.2014.03.028

Kuong, K. J., and Kuzminov, A. (2012). Disintegration of nascent replication bubbles during thymine starvation triggers RecA- and RecBCD-dependent replication origin destruction. *J. Biol. Chem.* 287, 23958–23970. doi: 10.1074/jbc.M112.359687

Ladner, R. D. (2001). The role of dUTPase and uracil-DNA repair in cancer chemotherapy. *Curr. Protein Peptide Sci.* 2, 361–370. doi: 10.2174/1389203013380991

Lark, K. G. (1972). Evidence for the direct involvement of RNA in the initiation of DNA replication in *E. coli* 15T-. *J. Mol. Biol.* 64, 47–60. doi: 10.1016/0022-2836(72)90320-8

Larsen, N. B., and Hickson, I. D. (2013). RecQ helicases: conserved guardians of genomic integrity. *Adv. Exp. Med. Biol.* 767, 161–184. doi: 10.1007/978-1-4614-5037-5_8

Liao, Z. Y., Sordet, O., Zhang, H. L., Kohlhagen, G., Antony, S., Gmeiner, W. H., et al. (2005). A novel polypyrimidine antitumor agent FdUMP[10] induces thymineless death with topoisomerase I-DNA complexes. *Cancer Res.* 65, 4844–4851. doi: 10.1158/0008-5472.CAN-04-1302

Little, J. G. (1985). Genetic and biochemical effects of thymidylate stress in yeast. *Basic Life Sci.* 31, 211–231. doi: 10.1007/978-1-4613-2449-2_13

Little, J. G., and Hanawalt, P. C. (1973). Thymineless death and ultraviolet sensitivity in *Micrococcus radiodurans*. *J. Bacteriol.* 113, 233–240.

Maaløe, O. (1961). The control of normal DNA replication in bacteria. *Cold Spring Harb. Symp. (Quant.) Biol.* 26, 45–52. doi: 10.1101/SQB.1961.026.01.010

Maaløe, O., and Hanawalt, P. C. (1961). Thymine deficiency and the normal DNA replication cycle I. *J. Mol. Biol.* 3, 144–155. doi: 10.1016/S0022-2836(61)80041-7

Martín, C. M., and Guzmán, E. C. (2011). DNA replication initiation as a key element in thymineless death. *DNA Repair* 10, 94–101. doi: 10.1016/j.dnarep.2010.10.005

Martîn, C. M., Viguera, E., and Guzmân, E. C. (2014). Rifampicin suppresses thymineless death by blocking the transcription-dependent step of chromosome initiation. *DNA Repair* 18, 10–17. doi: 10.1016/j.dnarep.2014.03.004

Meselson, M., and Stahl, F. (1958). The replication of DNA in *Escherichia coli*. *Proc. Natl. Acad. Sci. U.S.A.* 44, 671–682. doi: 10.1073/pnas.44.7.671

Morganroth, P., and Hanawalt, P. (2006). Role of DNA replication and repair in thymineless death in *Escherichia coli*. *J. Bacteriol.* 188, 5286–5288. doi: 10.1128/JB.00543-06

Nakayama, H. (2005). *Escherichia coli* RecQ helicase: a player in thymineless death. *Mutat. Res.* 577, 228–236. doi: 10.1016/j.mrfmmm.2005.02.015

Nakayama, H., and Couch, J. (1973). Effects of thymine deprivation in thymine auxotrophs of *E. coli*; thymineless death in liquid assay system. *J. Bacteriol.* 114, 228–232.

Nakayama, H., and Hanawalt, P. C. (1975). Sedimentation analysis of deoxyribonucleic acid from thymine-starved *Escherichia coli*. *J. Bacteriol.* 121, 537–547.

Nakayama, H., Nakayama, K., Nakayama, R., Irino, N., Nakayama, Y., and Hanawalt, P. C. (1984). Isolation and genetic characterization of a thymineless death-resistant mutant of *Escherichia coli* K-12: identification of a new mutation (recQ1) that blocks the RecF recombination pathway. *Mol. Gen. Genet.* 195, 474–480. doi: 10.1007/BF00341449

Nakayama, H., Nakayama, K., Nakayama, R., and Nakayama, Y. (1982). Recombination-deficient mutations and thymineless death in *Escherichia coli* K12: reciprocal effects of recBC and recF and indifference of recA mutations. *Can. J. Microbiol.* 28, 425–430. doi: 10.1139/m82-064

Nakayama, K., Kusano, K., Irino, N., and Nakayama, H. (1994). Thymine starvation-induced structural changes in *Escherichia coli* DNA. Detection by pulsed field gel electrophoresis and evidence for involvement of homologous recombination. *J. Mol. Biol.* 243, 611–620. doi: 10.1016/0022-2836(94)90036-1

Nakayama, K., Shiota, S., and Nakayama, H. (1988). Thymineless death in *Escherichia coli* mutants deficient in the RecF recombination pathway. *Can. J. Microbiol.* 34, 905–907. doi: 10.1139/m88-157

Nunberg, J. (1971). *On the Correlation of DNA Degradation with Thymineless Death*. Undergraduate Honors Thesis, Division of Biological Sciences, Cornell University.

Ostrer, L., Hamann, B. L., and Khodursky, A. (2015). Perturbed states of the bacterial chromosome: a thymineless death case study. *Front. Microbiol.* 6:363. doi: 10.3389/fmicb.2015.00363

Pauling, C., and Hanawalt, P. C. (1965). Nonconservative DNA replication in bacteria after thymine starvation. *Proc. Natl. Acad. Sci. U.S.A.* 54, 1728–1735. doi: 10.1073/pnas.54.6.1728

Pettijohn, D., and Hanawalt, P. C. (1964). Evidence for repair-replication of ultraviolet damaged DNA in bacteria. *J. Mol. Biol.* 9, 395–410. doi: 10.1016/S0022-2836(64)80216-3

Pritchard, R. H., and Lark, K. G. (1964). Induction of replication by thymine starvation at the chromosome origin in *Escherichia coli*. *J. Mol. Biol.* 9, 288–307. doi: 10.1016/S0022-2836(64)80208-4

Reiter, H., and Ramareddy, G. (1970). Loss of DNA behind the growing point of thymine-starved *Bacillus subtilis* 168. *J. Mol. Biol.* 50, 533–548. doi: 10.1016/0022-2836(70)90210-X

Sangurdekar, D. P., Hamann, B. L., Smirnov, D., Srienc, F., Hanawalt, P. C., and Khodursky, A. B. (2010). Thymineless death is associated with loss of essential genetic information from the replication origin. *Mol. Microbiol.* 6, 1455–1467. doi: 10.1111/j.1365-2958.2010.07072.x

Schaechter, M., Bentson, M. W., and Maaløe, O. (1959). Synthesis of deoxyribonucleic acid during the division cycle of bacteria. *Nature* 183, 1207–1208. doi: 10.1038/1831207a0

Smith, D., and Hanawalt, P. C. (1968). Macromolecular synthesis and thymineless death in *Mycoplasma laidlawii* B. *J. Bacteriol.* 96, 2066–2076.

Van Tubergen, R. P. (1961). The use of radioautography and electron microscopy for the localization of tritium label in bacteria. *J. Biophys. Biochem. Cytol.* 9, 219–222. doi: 10.1083/jcb.9.1.219

Williams, J. S., and Kunkel, T. A. (2014). Ribonucleotides in DNA: origins, repair and consequences. *DNA Repair* 19, 27–37. doi: 10.1016/j.dnarep.2014.03.029

Zaritsky, A., and Pritchard, R. (1971). Replication time of the chromosome in thymineles mutants of *Escherichia coli*. *J. Mol. Biol.* 60, 65–74. doi: 10.1016/0022-2836(71)90447-5

Zaritsky, A., Wang, P., and Vischer, N. O. E. (2011). Instructive simulation of the bacterial division cycle. *Microbiology* 157, 1876–1885. doi: 10.1099/mic.0.049403-0

14

Turgor pressure and possible constriction mechanisms in bacterial division

Masaki Osawa and Harold P. Erickson*

Department of Cell Biology, Duke University Medical Center, Durham, NC, United States

***Correspondence:**
Masaki Osawa
masaki.osawa@duke.edu

Bacterial cytokinesis begins with the assembly of FtsZ into a Z ring at the center of the cell. The Z-ring constriction in Gram-negative bacteria may occur in an environment where the periplasm and the cytoplasm are isoosmotic, but in Gram-positive bacteria the constriction may have to overcome a substantial turgor pressure. We address three potential sources of invagination force. (1) FtsZ itself may generate force by curved protofilaments bending the attached membrane. This is sufficient to constrict liposomes *in vitro*. However, this force is on the order of a few pN, and would not be enough to overcome turgor. (2) Cell wall (CW) synthesis may generate force by pushing the plasma membrane from the outside. However, this would probably require some kind of Brownian ratchet to separate the CW and membrane sufficiently to allow a glycan strand to slip in. The elastic element is not obvious. (3) Excess membrane production has the potential to contribute significantly to the invagination force. If the excess membrane is produced under the CW, it would force the membrane to bleb inward. We propose here that a combination of FtsZ pulling from the inside, and excess membrane pushing membrane inward may generate a substantial constriction force at the division site. This combined force generation mechanism may be sufficient to overcome turgor pressure. This would abolish the need for a Brownian ratchet for CW growth, and would permit CW to operate by reinforcing the constrictions generated by FtsZ and excess membrane.

Keywords: ftsZ, bacterial division, tubulin, cytokinesis, turgor pressure

INTRODUCTION

The cytokinesis system of almost all bacteria and many archaea is based on a ring of the tubulin homolog FtsZ (Bi and Lutkenhaus, 1991), termed the Z ring, which recruits up to a dozen downstream proteins to achieve constriction of the membranes and cell wall (CW). FtsZ polymerizes into short protofilaments (~150–300 nm) which further associate into the Z ring. In many bacteria the protofilaments are tethered to the plasma membrane (PM) by FtsA, which binds the C-terminal peptide of FtsZ and inserts its own C-terminal amphipathic helix into the PM (Pichoff and Lutkenhaus, 2005). FtsA also plays a role in recruitment of downstream molecules that remodel the peptidoglycan CW (Weiss et al., 1999).

An important question for cytokinesis is whether the invagination of the PM needs to overcome the turgor force. The turgor force is considerable, with estimates ranging from 0.3 atm (Deng et al., 2011) to 3 atm (Cayley et al., 2000) for *Escherichia coli*, and up to 20 atm for *Bacillus subtilis* (Whatmore and Reed, 1990). For comparison a racing bicycle tire has a pressure of about 10 atm. If cytokinesis has to overcome this turgor force it will need very high forces. Whether it does has been a question of controversy in the field, and even between the two present authors.

Erickson has suggested that bacterial cytokinesis might "cheat turgor pressure" by taking place completely within the high pressure environment (Erickson, 2017). This seems well-established for Gram-negative bacteria, where the turgor pressure is generated and sustained by the outer membrane plus the CW, and the periplasm is isoosmotic with the cytoplasm. Then both PM ingression and adding peptidoglycan to the inner surface of the CW to grow the septum take place without fighting turgor. Erickson argued that this is likely also true for Gram-positive bacteria. The strongest argument for this is that cryo EM shows a periplasmic space of ~20 nm between the PM and CW for Gram-positive bacteria. If the periplasm were not isoosmotic, the PM should be pressed against the CW. Isoosmolarity would be achieved if the thicker CW itself functions as a semipermeable membrane, with a porosity excluding proteins and molecules larger than ~1,000 Da.

On the contrary, Osawa notes several lines of evidence that the CW of Gram-positive bacteria is leakier than the CW plus outer membrane of Gram-negative bacteria. First, at least one study of the CW of *B. subtilis* proposed a sieving size of 2 nm, which corresponds to the diameter of a 20 kDa globular protein (Demchick and Koch, 1996).

Second, the Sec and Tat secretion systems of *B. subtilis* have been well-delineated for transport across the PM, but no dedicated system has been described for transport across the CW. The simplest mechanism would be that once transported across the PM, small proteins such as beta-lactamase (29 kDa), cutinase (24 kDa), and GFP (28 kDa) diffusely pass through the CW into the extracellular medium (Palva et al., 1982; Tjalsma et al., 2004; Brockmeier et al., 2006; Vrancken et al., 2007). In contrast Gram-negative bacteria have a "main terminal branch" for transport of secretory proteins across the CW and outer membrane. In fact, an *E. coli* strain K 12 that does not express this transporter virtually does not secrete the periplasmic proteins to the media (Pugsley et al., 1997).

Third, many bacteriocins, which are peptides and proteins up to ~20 kDa, are known to attack the PM of various Gram-positive bacteria, which means they are able to cross the CW. For example cyclic antibacterial peptide Enterocin AS-48 (7 kDa), which has a compact globular structure, kills a wide variety of Gram-positive bacteria by damaging the PM (Grande Burgos et al., 2014). Phospholipase A2 (14 kDa) also kills actively growing Gram-positive bacteria by extensive degradation of phospholipid (Foreman-Wykert et al., 1999). Dysgalacticin (21.5 kDa), a heat-labile bacteriocin, can kill both growing and non-growing *Streptococcus pyogenes* (Gram-positive bacteria) (Swe et al., 2009).

These results suggest the possibility that ~20 kDa molecules can diffusively pass through the CW of Gram-positive bacteria. If this is the case, how could Gram-positive bacteria maintain a periplasm isoosmolar with cytoplasm? One possibility is that negatively charged lipoteichoic acids tethered to the PM create a Donnan-like equilibrium that mitigates the difference of osmotic pressure between cytoplasm and periplasm. This has been suggested by two studies (Oku et al., 2009; Percy and Grundling, 2014). Quantitative analysis is needed here. It remains possible that cell division might cheat the turgor force in Gram-negative bacteria, but would have to constrict against the full and much larger turgor force in Gram-positive bacteria.

Here we would like to discuss three potential sources for constriction force, and how they might function even in the presence of turgor force: (A) FtsZ bending the PM from the inside; (B) peptidoglycan assembly pushing the PM from the outside; and (C) excess membrane synthesis pushing the membrane to expand inward. The idea of membrane pushing suggests a new force mechanism in which the combined force of FtsZ pulling and membrane pushing could divide bacteria even against turgor pressure.

(A) FtsZ bending the PM from the inside: A number of mechanisms have been proposed for how FtsZ might generate a constriction force, which we have discussed previously (Erickson, 2009; Erickson et al., 2010; Erickson and Osawa, 2017). We favor a model where curved protofilaments generate a bending force on the membrane (Osawa et al., 2009). The strongest evidence came from experiments where FtsZ was tethered to liposomes by an artificial membrane tether. With the tether [(mts) membrane targeting sequence] on the normal C-terminal side, FtsZ-mts assembled Z rings inside tubular liposomes that constricted from the inside (Osawa et al., 2008). When the tether was switched to the opposite side, mts-FtsZ assembled Z rings on the outside, and these constricted the liposomes by squeezing (Osawa and Erickson, 2011). See Supplementary Figure S1 in Supplemental Material for images and diagram. These results are consistent with the FtsZ protofilaments having a defined curvature, and the direction of force determined by whether the tether was on the outside or inside of the curve.

A curved protofilament conformation was first observed for *E. coli* FtsZ adsorbed to cationic lipid monolayers (**Figure 1a**) (Erickson et al., 1996). These 24 nm diameter minirings were favored by GDP, suggesting a hydrolysis-linked conformational change. However, they have only been found for FtsZ from a few bacterial species (discussed in Erickson and Osawa, 2017). Also, GTPase deficient FtsZ can function in suppressor strains, suggesting that GTP hydrolysis is not essential for force generation and cell division (Bi and Lutkenhaus, 1992; Osawa and Erickson, 2006).

There is an alternative curved conformation that we have called "intermediate curved" (Erickson et al., 2010; Erickson and Osawa, 2017) with a diameter 100–200 nm. In many cases, these curved filaments appear mixed with straight protofilaments by negative stain EM, but they were more prominently developed after adsorption to a mica surface (Mingorance et al., 2005; Hamon et al., 2009; Mateos-Gil et al., 2012) or in crowding conditions (Popp et al., 2009) (**Figure 1b**). Intermediate curved

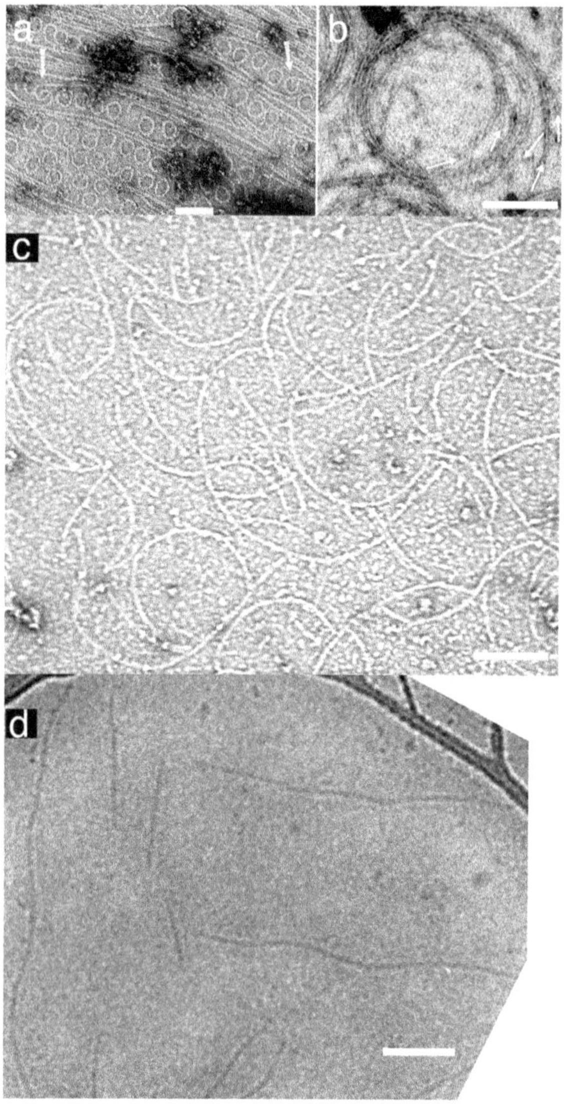

FIGURE 1 | Straight and curved protofilaments. **(a)** FtsZ protofilaments were absorbed on a cationic lipid membrane and then negatively stained for observation by EM. Many minirings approximately 24 nm diameter were observed. These minirings were probably stabilized with GDP-FtsZ (reprinted from Erickson et al., 1996). **(b)** FtsZ was polymerized into protofilaments in crowding buffer solution and then negatively stained. The protofilaments attached loosely to each other and formed curved bundles (inner diameter of the toroid is approximately 200 nm) (reprinted from Popp et al., 2009). **(c)** Protofilament assembly was observed without Mg^{2+}. In this condition all subunits in protofilaments were FtsZ-GTP. *Pseudomonas aeruginosa* FtsZ was used in this image (micrograph courtesy of Dr. Yaodong Chen, Northwest University, Xi'an, China). **(d)** FtsZ was polymerized into protofilaments in solution and then observed by cryo EM. These protofilaments, which apparently made no lateral contact with each other or a substrate, appear straight with some wavy structure (reprinted from Turner et al., 2012). Bars are 100 nm.

protofilaments could also be assembled in GTP plus EDTA, where the absence of Mg completely blocks GTP hydrolysis (Chen et al., 2005) (**Figure 1c**). Intermediate curved filaments were observed in a reconstitution system *in vitro* with FtsZ-mts (Osawa et al., 2009) or FtsZ/FtsA on a supported lipid bilayer (Szwedziak et al., 2014). Furthermore, the diameter of this curvature fits ~200–300 nm minimum diameter of the Z ring *in vivo* (Soderstrom et al., 2014; Coltharp et al., 2016). We consider this intermediate curvature the likely conformation that generates bending force on the membrane (Erickson and Osawa, 2017).

Two recent studies have shown FtsI moving in circular paths around the Z ring, driven by patches of treadmilling FtsZ (Bisson-Filho et al., 2017; Yang et al., 2017). This may be important for distribution of the CW remodeling molecules, but may not be essential for constriction, because the treadmilling is governed by GTP hydrolysis which is not essential for division.

If GTP hydrolysis is not involved, then what drives the conformational change from straight to curved? One possibility is that the protofilaments are intrinsically curved and may bend the membrane like Bar domains in eukaryotic systems (Saarikangas et al., 2009). Another possibility is a transition from twist to curvature. Although FtsZ protofilaments with no attachment to any surfaces appear mostly straight in cryo EM (**Figure 1d**) (Turner et al., 2012), those filaments might be naturally twisted. Previous studies have suggested that the protofilament conformation might be a balance of twist and curvature (Arumugam et al., 2012; Gonzalez de Prado Salas et al., 2014). When a twisted filament binds the membrane the twist must straighten (so that all subunits face the membrane) and the energy is transferred to a curvature. Lateral interactions of several types seem to induce the curved conformation, perhaps by forcing untwisting (Mingorance et al., 2005; Hamon et al., 2009; Popp et al., 2009; Mateos-Gil et al., 2012). This hypothesis seems to be contradicted by the existence of straight, untwisted protofilaments in crystals of *Staphylococcus aureus* FtsZ (Matsui et al., 2012), but it remains an interesting idea.

(A1) Estimation of energy and force generated by Z rings: Knowing that reconstituted Z rings can divide unilamellar liposomes (**Figure 2A**), a minimum requirement of energy and force for liposome division can be estimated. Certain geometries, such as that shown in **Figure 2B** (Supplementary Material), generate daughters with surface area and volume equal to the mother. We assume that successful divisions of liposomes *in vitro* (**Figure 2A**) (Osawa and Erickson, 2013) approximate this geometry.

To estimate the minimum energy needed to achieve liposome division, we used Helfrich's equation (Helfrich, 1973), and calculated the bending energy of the membrane before and after division (**Figure 2B**). The derivation is given in Supplementary Material. A second calculation gave the minimum initial constriction force that could be generated by a Z ring based on the bending energy/force of isotropic rods (Lan et al., 2009). The minimum energy for full constriction of the liposome shown in **Figure 2B** is 8.0×10^{-19} J, which also estimates 0.35 pN as the minimum initial constriction force for a liposome of 1 μm diameter. The second calculation gives 5.6×10^{-18} J and 2.45 pN as the potential energy and force that could be generated by FtsZ filaments for initial constriction of a 1 μm diameter bacterium. These are similar to values described previously with different parameters (Lan et al., 2009). Both the energy and force for FtsZ bending are about 10 times larger than the minimum

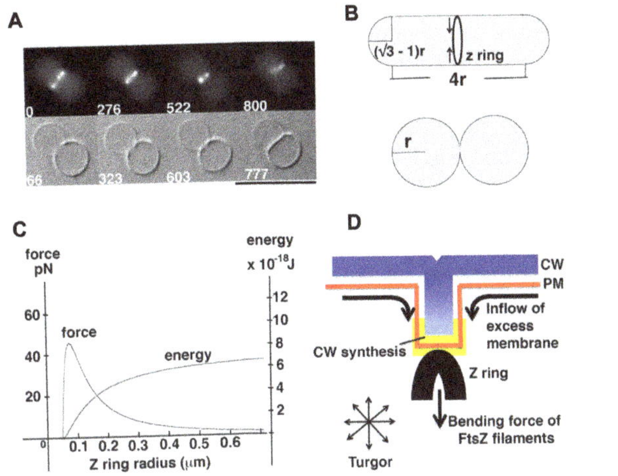

FIGURE 2 | Model for force generation by the Z ring *in vitro* and *in vivo*. **(A)** FtsZ formed Z rings in unilamellar liposomes with FtsA* (R286W), and the reconstituted Z rings divided liposomes with a clear septum (reprinted from Osawa and Erickson, 2013). A constraint on liposome division is posed by the fact that the membrane cannot be stretched more than 4% of the original area (Needham and Nunn, 1990). Because the membrane is impermeable to ions, reducing the volume by removing water would lead to higher salt inside and consequent osmotic pressure. With these constraints one cannot divide a spherical liposome, because the two daughter liposomes would have to shed 20% of the water of the mother. **(B)** Schematic view of liposome division by Z rings. In this particular geometry surface area and volume of the liposome does not change after division, in which case the Z ring only needs to bend the membrane. The successful division of the liposome in **(A)** probably approximates this geometry. **(C)** The relationship between force and energy with different Z ring radius. The graph was made from the equations for bending an isotropic rod (Supplementary Material) (Lan et al., 2009). **(D)** A model showing how excess phospholipid production can play a key role in Z ring constriction.

requirements for liposome division, consistent with the observed constriction *in vitro* (Osawa and Erickson, 2013). As shown in **Figure 2C**, the constriction force actually increases up to 45 pN at 200 nm diameter, which is approximately where FtsZ leaves the Z ring (Soderstrom et al., 2014; Coltharp et al., 2016).

In the absence of turgor, this force would be sufficient to divide a liposome of appropriate geometry. FtsZ bending membranes is almost certainly the major source of constriction force in mycoplasma and some archaea, which lack a CW and have minimal turgor force. If excess membrane were provided it could divide almost any geometry. However, it would probably not be sufficient to constrict the PM in the face of strong turgor.

(B) Peptidoglycan assembly pushing the PM from the outside: Another source of constriction force could be inward growth of the CW. Because the CW is a rigid structure, it should be capable of supporting the PM even if it is pressing against it with full turgor pressure. Coltharp and colleagues have suggested recently that the inward growth of the CW should be the major source of constriction force, while FtsZ serves primarily as a scaffold to define the location and orientation of the CW remodeling proteins (Coltharp et al., 2016; Coltharp and Xiao, 2017). In this case the curvature of FtsZ filaments may just work for fitting to the diameter of constriction site.

A detailed mechanism of pushing by CW ingrowth has not yet been developed. If the CW is to push the PM, it would seem necessary for it to contact the PM directly or indirectly through a membrane protein. This would need to eliminate the periplasmic space at the site of the site of contact. The invaginating CW must then insert glycan strands against the force on the PM. This would likely require a Brownian ratchet mechanism, in which either the CW or the PM were moving with sufficient distance and frequency to permit the glycan strands to slip in. In models of actin pushing membranes it was found that a simple Brownian ratchet would not work. The key was an elastic Brownian ratchet, in which flexing of the actin filaments provided the gaps for insertion (Mogilner and Oster, 1996). It is not clear where the Brownian motion would occur for CW pushing the PM against turgor. This may not work.

Plant and yeast cells are in a similar situation where the PM is surrounded by a rigid CW and is exposed to turgor pressure (Tepfer and Taylor, 1981; De Nobel et al., 1989). A recent paper proposed that the actin-myosin contractile ring for yeast division cannot constrict against this high turgor pressure (Proctor et al., 2012). This study found that constriction often continued after disruption of the actin-myosin ring, and concluded that the major constriction force is generated by the CW invagination.

Another consideration is that both plant and yeast cells can achieve endocytosis with an intracellular apparatus made of protein molecules (Samaj et al., 2005; Goode et al., 2015). This indicates that intracellular proteins can deform the membrane against high turgor pressure, without any contribution from CW invagination. It is not known how this works, but the mechanism may apply also to how bacteria divide even under turgor pressure. We suggest that the mechanism may utilize force from excess membrane synthesis, discussed next.

(C) Excess membrane synthesis pushing the membrane to expand inward: During cell elongation, membrane synthesis must match the expansion of the CW. If, however, membrane is synthesized in excess, it could only be accommodated by blebbing of the membrane. For a bacterial, plant or yeast cell, where the membrane is trapped by the rigid CW, the membrane can only bleb inward. Importantly, if a bleb is initiated by even a small force from a Z ring (or an endocytic apparatus), the excess membrane synthesis may contribute to further extension, perhaps even matching the turgor pressure (**Figure 2D**).

The extension of the rod during bacterial elongation is constant in both *E. coli* and *B. subtilis*, but this extension is reduced during Z ring constriction in both species (Burdett et al., 1986; Coltharp et al., 2016). If membrane synthesis continues at the same rate, it would be produced in excess during the constriction phase. One study actually showed that phospholipid production increased in both species during cross wall formation (Carty and Ingram, 1981), although another study of *E. coli* suggested that the productions of CW and phospholipid are synchronized (Gally et al., 1993). These studies were done 20–30 years ago and there has been no update. Because phospholipids are produced in the PM in bacteria (Zhang and Rock, 2008), phospholipid synthesis directly increases the area of PM and may largely cancel the effect of turgor pressure pressing the PM against the CW, especially once the PM initiates bending.

Because of the constraining CW, excess phospholipid production must push the membrane inward. A possible example is in Gram-positive *Streptomyces* species that grow as vegetative hyphae, where the PM can invaginate and form membrane cross walls without Z ring or CW support (Celler et al., 2016). This suggests that membrane production may be sufficient for membrane invagination against turgor pressure [However, the existence of cross walls in *Streptomyces* lacking FtsZ has been questioned by a recent study (Santos-Beneit et al., 2017)]. Membrane septa without CW were also observed in *B. subtilis* when one peptidoglycan synthase, PBP2B, was deleted (Daniel et al., 2000). Similar membrane septa were observed in *E. coli* with deletion of several lytic hydrolases (Heidrich et al., 2002), or with mutations in FtsK (Berezuk et al., 2014). Internalized PM was also observed in *B. subtilis* when membrane was overproduced by overexpression of AccDA, the enzyme for fatty acid synthesis (Mercier et al., 2013). That study also made a strong case that excess membrane synthesis was necessary and sufficient for division of *B. subtilis* L forms, lacking the CW and FtsZ. We suggest that septum formation without CW invagination is probably driven by excess membrane pushing the membrane inward, in some cases directed and assisted by the constriction force of the Z ring.

Combined Forces of FtsZ Pulling and Membrane Pushing May Divide Bacteria Even under Turgor Pressure

Assuming a substantial contribution from excess membrane production, the requirement of force generation by Z rings can probably be much smaller than expected previously (Lan et al., 2007). If excess membrane is exerting a lateral compressive force on the PM, a bleb may be initiated with only a small invagination force from FtsZ. A force of 2.45 pN or even 0.35 pN may suffice, depending on factors such as the pushing force of membrane and the initial bending curvature of membrane. Continued membrane synthesis would add substantially to any invagination force.

Excess membrane production may also facilitate CW ingrowth by canceling the effect of turgor pressure that pushes the PM against the ingressing CW. This could facilitate or eliminate the need for a Brownian ratchet to insert new glycan strands. In this case CW production might contribute to constriction by working as a ratchet to stabilize the newly constricted membrane. This could also explain how CW synthesis could be rate limiting for constriction (Coltharp et al., 2016). Bisson-Filho et al. (2017) suggested that dynamic FtsZ filaments may deform the membrane, and PG synthesis could reinforce the deformation on the other side. The PG reinforcement could be rate limiting.

The force generated by membrane synthesis could be sufficient to power invagination against a turgor force. Even in Gram-negative bacteria, whose PM is thought to be in an isoosmotic balance between cytoplasm and periplasm, membrane synthesis is needed to keep up with the invagination, and to avoid developing turgor pressure as cytoplasmic volume is reduced by septal ingrowth. This proposal might be also extend to yeast division and to endocytosis in yeast and plant cells, with excess PM providing an essential part of the force for membrane invagination.

Future Directions

A major problem is lack of information about the osmolarity of the periplasmic space in Gram-positive bacteria. Without this we don't even know whether cell division needs to overcome turgor pressure. Important new tools would be bacterial strains where FtsZ bending, CW synthesis and membrane synthesis can be regulated up and down. Biosensors that can measure the force generated by FtsZ and CW would be especially useful.

AUTHOR CONTRIBUTIONS

MO wrote and performed the theoretical analysis; HE assisted writing and deeply discussed a wide range of problems.

FUNDING

This work was supported by NIH grant R01-GM066014.

REFERENCES

Arumugam, S., Chwastek, G., Fischer-Friedrich, E., Ehrig, C., Monch, I., and Schwille, P. (2012). Surface topology engineering of membranes for the mechanical investigation of the tubulin homologue FtsZ. *Angew. Chem. Int. Ed. Engl.* 51, 11858–11862. doi: 10.1002/anie.201204332

Berezuk, A. M., Goodyear, M., and Khursigara, C. M. (2014). Site-directed fluorescence labeling reveals a revised N-terminal membrane topology and functional periplasmic residues in the *Escherichia coli* cell division protein FtsK. *J. Biol. Chem.* 289, 23287–23301. doi: 10.1074/jbc.M114.569624

Bi, E., and Lutkenhaus, J. (1991). FtsZ ring structure associated with division in *Escherichia coli*. *Nature* 354, 161–164. doi: 10.1038/354161a0

Bi, E., and Lutkenhaus, J. (1992). Isolation and characterization of *ftsZ* alleles that affect septal morphology. *J. Bacteriol.* 174, 5414–5423. doi: 10.1128/jb.174.16.5414-5423.1992

Bisson-Filho, A. W., Hsu, Y. P., Squyres, G. R., Kuru, E., Wu, F., Jukes, C., et al. (2017). Treadmilling by FtsZ filaments drives peptidoglycan synthesis and bacterial cell division. *Science* 355, 739–743. doi: 10.1126/science.aak9973

Brockmeier, U., Wendorff, M., and Eggert, T. (2006). Versatile expression and secretion vectors for *Bacillus subtilis*. *Curr. Microbiol.* 52, 143–148. doi: 10.1007/s00284-005-0231-7

Burdett, I. D., Kirkwood, T. B., and Whalley, J. B. (1986). Growth kinetics of individual *Bacillus subtilis* cells and correlation with nucleoid extension. *J. Bacteriol.* 167, 219–230. doi: 10.1128/jb.167.1.219-230.1986

Carty, C. E., and Ingram, L. O. (1981). Lipid synthesis during the *Escherichia coli* cell cycle. *J. Bacteriol.* 145, 472–478.

Cayley, D. S., Guttman, H. J., and Record, M. T. Jr. (2000). Biophysical characterization of changes in amounts and activity of *Escherichia coli* cell and compartment water and turgor pressure in response to osmotic stress. *Biophys. J.* 78, 1748–1764. doi: 10.1016/S0006-3495(00)76726-9

Celler, K., Koning, R. I., Willemse, J., Koster, A. J., and Van Wezel, G. P. (2016). Cross-membranes orchestrate compartmentalization and morphogenesis in *Streptomyces*. *Nat. Commun.* 7:11836. doi: 10.1038/ncomms 11836

Chen, Y., Bjornson, K., Redick, S. D., and Erickson, H. P. (2005). A rapid fluorescence assay for FtsZ assembly indicates cooperative assembly with a dimer nucleus. *Biophys. J.* 88, 505–514. doi: 10.1529/biophysj.104. 044149

Coltharp, C., Buss, J., Plumer, T. M., and Xiao, J. (2016). Defining the rate-limiting processes of bacterial cytokinesis. *Proc. Natl. Acad. Sci. U.S.A.* 113, E1044–E1053. doi: 10.1073/pnas.1514296113

Coltharp, C., and Xiao, J. (2017). Beyond force generation: why is a dynamic ring of FtsZ polymers essential for bacterial cytokinesis. *Bioessays* 39, 1–11. doi: 10.1002/bies.201600179

Daniel, R. A., Harry, E. J., and Errington, J. (2000). Role of penicillin-binding protein PBP 2B in assembly and functioning of the division machinery of *Bacillus subtilis*. *Mol. Microbiol.* 35, 299–311. doi: 10.1046/j.1365-2958.2000. 01724.x

Demchick, P., and Koch, A. L. (1996). The permeability of the wall fabric of *Escherichia coli* and *Bacillus subtilis*. *J. Bacteriol.* 178, 768–773. doi: 10.1128/ jb.178.3.768-773.1996

Deng, Y., Sun, M., and Shaevitz, J. W. (2011). Direct measurement of cell wall stress stiffening and turgor pressure in live bacterial cells. *Phys. Rev. Lett.* 107:158101. doi: 10.1103/PhysRevLett.107.158101

De Nobel, J. G., Dijkers, C., Hooijberg, E., and Klis, F. M. (1989). Increased cell-wall porosity in *Saccharomyces cerevisiae* after treatment with Dithiothreitol or EDTA. *J. Gen. Microbiol.* 135, 2077–2084. doi: 10.1099/00221287-135-7-2077

Erickson, H. P. (2009). Modeling the physics of FtsZ assembly and force generation. *Proc. Natl. Acad. Sci. U.S.A.* 106, 9238–9243. doi: 10.1073/pnas.0902258106

Erickson, H. P. (2017). How bacterial cell division might cheat turgor pressure - a unified mechanism of septal division in Gram-positive and Gram-negative bacteria. *Bioessays* 39:1700045. doi: 10.1002/bies.201700045

Erickson, H. P., Anderson, D. E., and Osawa, M. (2010). FtsZ in Bacterial cytokinesis: cytoskeleton and force generator all in one. *Microbiol. Mol. Biol. Rev.* 74, 504–528. doi: 10.1128/MMBR.00021-10

Erickson, H. P., and Osawa, M. (2017). FtsZ constriction force - curved protofilaments bending membranes. *Subcell. Biochem.* 84, 139–160. doi: 10.1007/978-3-319-53047-5_5

Erickson, H. P., Taylor, D. W., Taylor, K. A., and Bramhill, D. (1996). Bacterial cell division protein FtsZ assembles into protofilament sheets and minirings, structural homologs of tubulin polymers. *Proc. Natl. Acad. Sci. U.S.A.* 93, 519–523. doi: 10.1073/pnas.93.1.519

Foreman-Wykert, A. K., Weinrauch, Y., Elsbach, P., and Weiss, J. (1999). Cell-wall determinants of the bactericidal action of group IIA phospholipase A2 against Gram-positive bacteria. *J. Clin. Invest.* 103, 715–721. doi: 10.1172/JCI5468

Gally, D., Bray, K., and Cooper, S. (1993). Synthesis of peptidoglycan and membrane during the division cycle of rod-shaped, gram-negative bacteria. *J. Bacteriol.* 175, 3121–3130. doi: 10.1128/jb.175.10.3121-3130.1993

Gonzalez de Prado Salas, P., Horger, I., Martin-Garcia, F., Mendieta, J., Alonso, A., Encinar, M., et al. (2014). Torsion and curvature of FtsZ filaments. *Soft Matter* 10, 1977–1986. doi: 10.1039/c3sm52516c

Goode, B. L., Eskin, J. A., and Wendland, B. (2015). Actin and endocytosis in budding yeast. *Genetics* 199, 315–358. doi: 10.1534/genetics.112. 145540

Grande Burgos, M. J., Pulido, R. P., Del Carmen Lopez Aguayo, M., Galvez, A., and Lucas, R. (2014). The cyclic antibacterial peptide enterocin AS-48: isolation, mode of action, and possible food applications. *Int. J. Mol. Sci.* 15, 22706–22727. doi: 10.3390/ijms151222706

Hamon, L., Panda, D., Savarin, P., Joshi, V., Bernhard, J., Mucher, E., et al. (2009). Mica surface promotes the assembly of cytoskeletal proteins. *Langmuir* 25, 3331–3335. doi: 10.1021/la8035743

Heidrich, C., Ursinus, A., Berger, J., Schwarz, H., and Holtje, J. V. (2002). Effects of multiple deletions of murein hydrolases on viability, septum cleavage, and sensitivity to large toxic molecules in *Escherichia coli*. *J. Bacteriol.* 184, 6093–6099. doi: 10.1128/JB.184.22.6093-6099.2002

Helfrich, W. (1973). Elastic properties of lipid bilayers: theory and possible experiments. *Z. Naturforsch. C* 28, 693–703. doi: 10.1515/znc-1973-11-1209

Lan, G., Daniels, B. R., Dobrowsky, T. M., Wirtz, D., and Sun, S. X. (2009). Condensation of FtsZ filaments can drive bacterial cell division. *Proc. Natl. Acad. Sci. U.S.A.* 106, 121–126. doi: 10.1073/pnas.0807963106

Lan, G., Wolgemuth, C. W., and Sun, S. X. (2007). Z-ring force and cell shape during division in rod-like bacteria. *Proc. Natl. Acad. Sci. U.S.A.* 104, 16110–16115. doi: 10.1073/pnas.0702925104

Mateos-Gil, P., Paez, A., Horger, I., Rivas, G., Vicente, M., Tarazona, P., et al. (2012). Depolymerization dynamics of individual filaments of bacterial cytoskeletal protein FtsZ. *Proc. Natl. Acad. Sci. U.S.A.* 109, 8133–8138. doi: 10.1073/pnas. 1204844109

Matsui, T., Yamane, J., Mogi, N., Yamaguchi, H., Takemoto, H., Yao, M., et al. (2012). Structural reorganization of the bacterial cell-division protein FtsZ from *Staphylococcus aureus*. *Acta Crystallogr. D Biol. Crystallogr.* 68, 1175–1188. doi: 10.1107/S0907444912022640

Mercier, R., Kawai, Y., and Errington, J. (2013). Excess membrane synthesis drives a primitive mode of cell proliferation. *Cell* 152, 997–1007. doi: 10.1016/j.cell. 2013.01.043

Mingorance, J., Tadros, M., Vicente, M., Gonzalez, J. M., Rivas, G., and Velez, M. (2005). Visualization of single *Escherichia coli* FtsZ filament dynamics with atomic force microscopy. *J. Biol. Chem.* 280, 20909–20914. doi: 10.1074/jbc. M503059200

Mogilner, A., and Oster, G. (1996). Cell motility driven by actin polymerization. *Biophys. J.* 71, 3030–3045. doi: 10.1016/S0006-3495(96) 79496-1

Needham, D., and Nunn, R. S. (1990). Elastic deformation and failure of lipid bilayer membranes containing cholesterol. *Biophys. J.* 58, 997–1009. doi: 10.1016/S0006-3495(90)82444-9

Oku, Y., Kurokawa, K., Matsuo, M., Yamada, S., Lee, B. L., and Sekimizu, K. (2009). Pleiotropic roles of polyglycerolphosphate synthase of lipoteichoic acid in growth of *Staphylococcus aureus* cells. *J. Bacteriol.* 191, 141–151. doi: 10.1128/ JB.01221-08

Osawa, M., Anderson, D. E., and Erickson, H. P. (2008). Reconstitution of contractile FtsZ rings in liposomes. *Science* 320, 792–794. doi: 10.1126/science. 1154520

Osawa, M., Anderson, D. E., and Erickson, H. P. (2009). Curved FtsZ protofilaments generate bending forces on liposome membranes. *EMBO J.* 28, 3476–3484. doi: 10.1038/emboj.2009.277

Osawa, M., and Erickson, H. P. (2006). FtsZ from divergent foreign bacteria can function for cell division in *Escherichia coli*. *J. Bacteriol.* 188, 7132–7140. doi: 10.1128/JB.00647-06

Osawa, M., and Erickson, H. P. (2011). Inside-out Z rings - constriction with and without GTP hydrolysis. *Mol. Microbiol.* 81, 571–579. doi: 10.1111/j.1365-2958. 2011.07716.x

Osawa, M., and Erickson, H. P. (2013). Liposome division by a simple bacterial division machinery. *Proc. Natl. Acad. Sci. U.S.A.* 110, 11000–11004. doi: 10. 1073/pnas.1222254110

Palva, I., Sarvas, M., Lehtovaara, P., Sibakov, M., and Kaariainen, L. (1982). Secretion of *Escherichia coli* beta-lactamase from *Bacillus subtilis* by the aid of alpha-amylase signal sequence. *Proc. Natl. Acad. Sci. U.S.A.* 79, 5582–5586. doi: 10.1073/pnas.79.18.5582

Percy, M. G., and Grundling, A. (2014). Lipoteichoic acid synthesis and function in gram-positive bacteria. *Annu. Rev. Microbiol.* 68, 81–100. doi: 10.1146/ annurev-micro-091213-112949

Pichoff, S., and Lutkenhaus, J. (2005). Tethering the Z ring to the membrane through a conserved membrane targeting sequence in FtsA. *Mol. Microbiol.* 55, 1722–1734. doi: 10.1111/j.1365-2958.2005.04522.x

Popp, D., Iwasa, M., Narita, A., Erickson, H. P., and Maeda, Y. (2009). FtsZ condensates: an in vitro electron microscopy study. *Biopolymers* 91, 340–350. doi: 10.1002/bip.21136

Proctor, S. A., Minc, N., Boudaoud, A., and Chang, F. (2012). Contributions of turgor pressure, the contractile ring, and septum assembly to forces in cytokinesis in fission yeast. *Curr. Biol.* 22, 1601–1608. doi: 10.1016/j.cub.2012.06.042

Pugsley, A. P., Francetic, O., Possot, O. M., Sauvonnet, N., and Hardie, K. R. (1997). Recent progress and future directions in studies of the main terminal branch of the general secretory pathway in Gram-negative bacteria–a review. *Gene* 192, 13–19. doi: 10.1016/S0378-1119(96)00803-7

Saarikangas, J., Zhao, H., Pykalainen, A., Laurinmaki, P., Mattila, P. K., Kinnunen, P. K., et al. (2009). Molecular mechanisms of membrane deformation by I-BAR domain proteins. *Curr. Biol.* 19, 95–107. doi: 10.1016/j.cub.2008.12.029

Samaj, J., Read, N. D., Volkmann, D., Menzel, D., and Baluska, F. (2005). The endocytic network in plants. *Trends Cell Biol.* 15, 425–433. doi: 10.1016/j.tcb.2005.06.006

Santos-Beneit, F., Roberts, D. M., Cantlay, S., Mccormick, J. R., and Errington, J. (2017). A mechanism for FtsZ-independent proliferation in *Streptomyces. Nat. Commun.* 8:1378. doi: 10.1038/s41467-017-01596-z

Soderstrom, B., Skoog, K., Blom, H., Weiss, D. S., Von Heijne, G., and Daley, D. O. (2014). Disassembly of the divisome in *Escherichia coli*: evidence that FtsZ dissociates before compartmentalization. *Mol. Microbiol.* 92, 1–9. doi: 10.1111/mmi.12534

Swe, P. M., Cook, G. M., Tagg, J. R., and Jack, R. W. (2009). Mode of action of dysgalacticin: a large heat-labile bacteriocin. *J. Antimicrob. Chemother.* 63, 679–686. doi: 10.1093/jac/dkn552

Szwedziak, P., Wang, Q., Bharat, T. A., Tsim, M., and Lowe, J. (2014). Architecture of the ring formed by the tubulin homologue FtsZ in bacterial cell division. *Elife* 3:e04601. doi: 10.7554/eLife.04601

Tepfer, M., and Taylor, I. E. (1981). The permeability of plant cell walls as measured by gel filtration chromatography. *Science* 213, 761–763. doi: 10.1126/science.213.4509.761

Tjalsma, H., Antelmann, H., Jongbloed, J. D., Braun, P. G., Darmon, E., Dorenbos, R., et al. (2004). Proteomics of protein secretion by *Bacillus subtilis*: separating the "secrets" of the secretome. *Microbiol. Mol. Biol. Rev.* 68, 207–233. doi: 10.1128/MMBR.68.2.207-233.2004

Turner, D. J., Portman, I., Dafforn, T. R., Rodger, A., Roper, D. I., Smith, C. J., et al. (2012). The mechanics of FtsZ fibers. *Biophys. J.* 102, 731–738. doi: 10.1016/j.bpj.2012.01.015

Vrancken, K., De Keersmaeker, S., Geukens, N., Lammertyn, E., Anne, J., and Van Mellaert, L. (2007). pspA overexpression in *Streptomyces lividans* improves both Sec- and Tat-dependent protein secretion. *Appl. Microbiol. Biotechnol.* 73, 1150–1157. doi: 10.1007/s00253-006-0571-7

Weiss, D. S., Chen, J. C., Ghigo, J. M., Boyd, D., and Beckwith, J. (1999). Localization of FtsI (PBP3) to the septal ring requires its membrane anchor, the Z ring, FtsA, FtsQ, and FtsL. *J. Bacteriol.* 181, 508–520.

Whatmore, A. M., and Reed, R. H. (1990). Determination of turgor pressure in *Bacillus subtilis*: a possible role for K+ in turgor regulation. *J. Gen. Microbiol.* 136, 2521–2526. doi: 10.1099/00221287-136-12-2521

Yang, X., Lyu, Z., Miguel, A., Mcquillen, R., Huang, K. C., and Xiao, J. (2017). GTPase activity-coupled treadmilling of the bacterial tubulin FtsZ organizes septal cell wall synthesis. *Science* 355, 744–747. doi: 10.1126/science.aak9995

Zhang, Y. M., and Rock, C. O. (2008). Membrane lipid homeostasis in bacteria. *Nat. Rev. Microbiol.* 6, 222–233. doi: 10.1038/nrmicro1839

15

Perturbed states of the bacterial chromosome: A thymineless death case study

Lev Ostrer, Bree L. Hamann[†] *and Arkady Khodursky*[*]

Department of Biochemistry, Molecular Biology and Biophysics, Biotechnology Institute, University of Minnesota, St. Paul, MN, USA

***Correspondence:** *Arkady Khodursky, Department of Biochemistry, Molecular Biology and Biophysics, Biotechnology Institute, University of Minnesota, 244 Gortner Lab, 1479 Gortner Avenue, St. Paul, MN 55108, USA khodu001@umn.edu*

[†]*Formerly associated with Biotechnology Institute, University of Minnesota, St. Paul, MN, USA*

Spatial patterns of transcriptional activity in the living genome of *Escherichia coli* represent one of the more peculiar aspects of the *E. coli* chromosome biology. Spatial transcriptional correlations can be observed throughout the chromosome, and their formation depends on the state of replication in the cell. The condition of thymine starvation leading to thymineless death (TLD) is at the "cross-roads" of replication and transcription. According to a current view, e.g., (Cagliero et al., 2014), one of the cellular objectives is to segregate the processes of transcription and replication in time and space. An ultimate segregation would take place when one process is inhibited and another is not, as it happens during thymine starvation, which results in numerous molecular and physiological abnormalities associated with TLD. One of such abnormalities is the loss of spatial correlations in the vicinity of the origin of replication. We review the transcriptional consequences of replication inhibition by thymine starvation in a context of the state of DNA template in the starved cells and opine about a possible significance of normal physiological coupling between the processes of replication and transcription.

Keywords: transcription, replication, spatial correlations, thymine starvation, thymineless death

Introduction

Replication and transcription are intertwined in a number of ways, as molecular reactions and biological processes. In the cell, the two polymerization reactions utilize, and compete for, the same DNA template. That may force local interactions between DNA and RNA polymerase complexes (Liu and Alberts, 1995; Felipe-Abrio et al., 2014), which may be a part of the normal replication process (French, 1992; Merrikh et al., 2011) or, under special circumstances, may be a source of chromosome pathology (Merrikh et al., 2012). The reactions are also linked at a deep evolutionary level via the activity of the enzyme ribonucleoside di(tri)phosphate reductase which converts ribonucleotides, precursors of substrates (or substrates) for the reaction of transcription, into deoxyribonucleotides, substrates for DNA polymerization reaction (Reichard, 2010).

The biological processes of replication and transcription are also coupled via a number of regulatory mechanisms, some of which are still unexplained or/and whose physiological significance is not yet understood. First, DNA replication determines intrinsic levels of gene expression by establishing gene dosage gradients along the chromosome (Chandler and Pritchard, 1975; Schmid and Roth, 1987; Couturier and Rocha, 2006; Slager et al., 2014). Second, transcription is required for initiation of chromosomal DNA replication (Lark, 1972; Messer, 1972). Third, DNA replication is required for establishing normal spatial patterns of chromosomal transcription (Jeong et al., 2006). Fourth, interference with DNA replication results in a specialized DNA damage transcriptional response (Simmons et al., 2008; Kreuzer, 2013). Fifth, the replication initiator protein DnaA

(Kaguni, 2006) controls, as an activator and repressor, transcription of the ribonucleotide reductase operon (Olliver et al., 2010). Sixth, activity of both processes can be modulated by (p)ppGpp (Chiaramello and Zyskind, 1990; Levine et al., 1991; Denapoli et al., 2013) and by changes in DNA supercoiling (Kowalski and Eddy, 1989; Crooke et al., 1991; von Freiesleben and Rasmussen, 1992; Peter et al., 2004; Rovinskiy et al., 2012), two molecular sensors that can relay information about critical fluctuations in the cell's environment to the parts of the replication and transcription machineries (Dorman, 2006; Jin et al., 2012; Sobetzko et al., 2012). Last and perhaps more of a fortuitous link is a genetic one, exemplified by the evolutionarily conserved arrangement of genes encoding the DNA primase DnaG and the main sigma factor RpoD in one operon (Burton et al., 1983; Versalovic et al., 1993).

All these interactions have been shaped and calibrated by the evolutionary forces to accommodate species specific differences in genome organization and physiology. Disruption of normal temporal and spatial relationships between replication and transcription may have detrimental consequences for the cell. And nowhere such consequences are more pronounced than during thymine starvation.

Thymine Starvation and Transcription

Thymine is one of the five common nucleobases and it is found primarily in DNA where it makes a Watson-Crick pair with Adenine. Active thymine compound in the cell is TTP: it is a precursor of DNA as well as of thymidine nucleotide sugars that serve as intermediates in O-antigen biosynthesis (Samuel and Reeves, 2003). In most organisms, cells synthesize thymine on the level of a nucleotide as thymidine monophosphate from deoxyuridine monophosphate (www.kegg.jp, www.ecocyc.org). Thymine dependency, or auxotrophy, can be established by mutating a gene encoding thymidylate synthase (Belfort et al., 1983). Thymine requiring mutants can only grow in presence of the exogenous base or nucleoside (Barner and Cohen, 1954). Starving cells for thymine results in a rapid decay in colony counts (Barner and Cohen, 1954). The decay is preceded by a near complete cessation of DNA synthesis (Barner and Cohen, 1958; Kuong and Kuzminov, 2010) and it is paralleled by increase in the cell size (Barner and Cohen, 1954), which is indicative of division arrest. However, inhibition of RNA and protein synthesis across a population of starved cells lags significantly behind the drop in colony formation (Barner and Cohen, 1958; McFall and Magasanik, 1960). Thus, cells deprived of thymine go through temporally ordered events: 1-inhibition of bulk DNA synthesis; 2-cell elongation and division arrest; 3-decrease in colony formation; 4-inhibition of protein and RNA synthesis. The overall phenomenon of thymine starvation leading to an exponential decrease in colony counts is known as thymineless death (Cohen and Barner, 1954), or TLD. It was postulated that the cause of death is "inhibition of DNA synthesis under conditions of continued cytoplasmic synthesis" (Barner and Cohen, 1957), including RNA, protein and overall biomass. Such uncoupling between biosynthetic processes was termed an "unbalanced growth" (Cohen and Barner, 1954) and it was proposed to be the underlying macro-mechanism of the bactericidal effect of not only thymine starvation, but also of a number of antibiotics (Barner and Cohen, 1956). However, the pathology of TLD has been associated with the transcriptional activity of the starved cells and not with other aspects of the unbalanced growth (Gallant and Suskind, 1962; Hanawalt, 1963; Rolfe, 1967; Morganroth and Hanawalt, 2006; Martin et al., 2014).

Transcriptional activity of a population of bacterial cells is a sum of the activities of individual cells. Even when growth rate of a population is kept constant, cells making up that population may have different microenvironments, may come from different stages of the cell cycle, and may be of different sizes and ages. As a result, transcriptional activity across a population of cells in a steady state can be viewed as a type of noise that must be deconvoluted on the basis of physiological parameters in order to make sense of the activity. In part because transcriptional activity is noisy, even under controlled conditions, transcriptional states of the cell have been traditionally characterized in terms of dominant differences in transcript abundances, or in promoter activities, which could be associated with physiological or/and environmental changes. Activation of the DNA damage sensitive promoter of the *recA* gene was originally shown to be one of such dominant changes elicited by thymine deprivation (Casaregola et al., 1982).

Introduction of whole-genome microarray technology (Schena et al., 1995) made monitoring transcriptional activity less biased and more quantitative. Genome-wide surveys of changes in transcript abundances elicited by thymine deprivation confirmed that the condition induces the SOS regulon (Sangurdekar et al., 2010, 2011), a collection of genetically unlinked genes whose transcription is primarily controlled by the LexA repressor and whose activities allow the cell to repair or bypass DNA lesions in time before the delayed cell division (Simmons et al., 2008). The surveys also found that activity of the SOS regulon contrasts thymine limitation and thymine deprivation: whereas both limitation and starvation result in transcriptional upregulation of genes of the deoxyribose salvage pathway and downregulation of genes whose products are involved in pyrimidine biosynthesis, only the starvation induces transcription of the SOS regulon (Sangurdekar et al., 2010, 2011).

Relative transcript abundances measured for nearly every gene in a genome can be used to correlate transcriptional activities of genes over time or conditions (Eisen et al., 1998; Tamayo et al., 1999).The resulting correlation patterns can be explained in terms of underlying regulatory mechanisms, e.g., (van Helden et al., 1998). Co-transcriptional patterns in prokaryotic genomes result primarily from organization of genes in operons (Gutierrez-Rios et al., 2003) and regulons (Khodursky et al., 2000a; Courcelle et al., 2001). However, an operonal organization can be inferred not only from multidimensional comparative profiles of transcriptional activity of genomes, but also from one dimensional spatial series of transcript abundances, strongly suggesting that transcriptional signals recorded across a population of cells, despite being a product of multiple sources of biological variation, contain biologically relevant structural information (Jeong et al., 2004). In fact, the spatial transcriptional signal can be used to model geometric features of the chromosomal DNA superstructure, assuming juxtaposition of co-regulated genes in 3-D space (Xiao et al., 2011). Mapping of contacts between linearly non-adjacent DNA segments in the *Caulobacter crescentus* chromosome using the high-resolution chromosome capture methodology (Lieberman-Aiden et al., 2009) revealed the general properties of physical organization of chromosomal DNA inside the cell (Le et al., 2013). These properties, including the dimensions of short- and medium-range features, were consistent with the estimates obtained from the geometric model which was based on expression data alone. This concordance bolsters the notion that spatial transcriptional profiles contain information about underlying structure of the chromosomal DNA inside the cell (Jeong et al., 2004).

Phenomenologically, transcriptional activity of any two genes can be up or down relative to some baseline level. If activities of multiple pairs of genes vary in the same direction, and genes in the pairs are situated the same or nearly the same distance apart, then the transcriptional signal can be viewed as spatially regular with a characteristic distance between genes with co-varying activity. Such characteristic distance, or spatial frequency, can be evaluated using conventional signal processing techniques, including autocorrelation or the Fourier transform. Using these methods, it was demonstrated that, independently of growth conditions, transcriptional signal from the *E. coli* chromosome can be characterized by several statistically significant spatial frequencies: 1 kbp, 4 ± 2 kbp, 16 ± 4 kbp, 32 ± 6 kbp, 128 ± 12 kbp, and 570 ± 50 kbp (Jeong et al., 2004). However, co-variations in transcriptional activities do not have to be spatially regular or have the same regularity across the entire chromosome. To account for that the signal can be analyzed using the wavelet transform which provides space-frequency representation of original spatial series (Torrence and Compo, 1998). Moreover, it may be useful to know what spectral components occur in which chromosomal intervals, especially for the purposes of comparative analysis (Allen et al., 2006). As expected, chromosome wide transcriptional signals are not stationary; all spatial frequency components, from short- (up to 16 kbp) to long-range are found in certain regions of the chromosome but not in others (Jeong et al., 2004; Allen et al., 2006). Furthermore, different parts of the chromosome may contain different amount of spectral information. For example, nearly all spatial frequency components can be found in a region of the chromosome proximal to the origin of replication in the counter-clockwise replichore (Jeong et al., 2004). In fact, the presence of multiple frequency components, up to 128 ± 12 kbp, in this region is a characteristic feature of spatial transcriptional profiles obtained under dozens of conditions, including growth phase transitions, nutritional shifts, recoveries, antibiotic treatments, etc. (Jeong et al., 2004, 2006; Sangurdekar et al., 2006; Xiao et al., 2011). The exceptions were DNA damaging conditions causing inhibition of DNA replication, including treatment with a quinolone antibiotic, norfloxacin (Jeong et al., 2004, 2006), and thymine starvation leading to TLD (**Figure 1**). A study by Jeong et al. (2006) demonstrated, using mutants with conditionally defective initiation of chromosomal DNA replication, that ongoing replication is required for observing spatial patterns across scales. Both the quinolone treatment (Goss et al., 1965; Benbrook and Miller, 1986; Khodursky and Cozzarelli, 1998) and thymine starvation (Barner and Cohen, 1958; Kuong and Kuzminov, 2010) result in rapid inhibition of DNA replication, which in turn likely

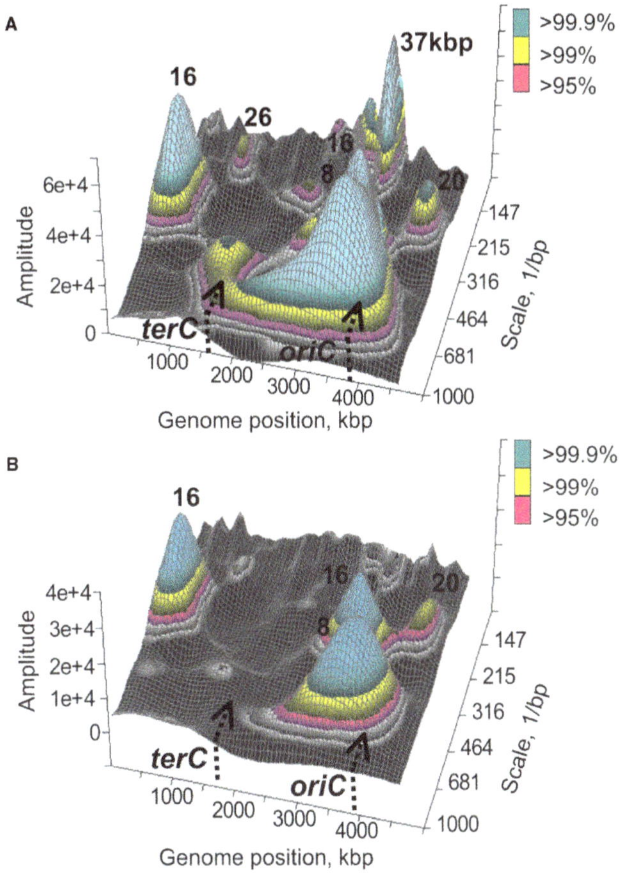

FIGURE 1 | Spatial correlations in transcript abundances along the *E. coli* chromosome. Significant wavelets with corresponding main frequencies are shown for a genome-wide transcriptional signal recorded prior to thymine starvation **(A)** and 30 min into starvation **(B)**. The figure was obtained using publicly available data as previously described (Jeong et al., 2004).

brings about structural changes in the chromosome that disfavor spatial pattern formation.

State of DNA in Thymine Starved Cells

Observations of the dissolution of spatial transcriptional patterns may be confounded in part by the fact that the loss of the patterns occurs under conditions that induce DNA damage. The damaged DNA may serve as a poor template for transcription, thereby providing an almost trivial reason for the loss of the spatial correlations in transcriptional activity.

Indeed, several types of DNA damage have been observed in cells starved for thymine, some of which may culminate in the template destruction: single strand breaks in DNA (Freifelder, 1969; Hill and Fangman, 1973; Nakayama and Hanawalt, 1975); double strand breaks (Yoshinaga, 1973; Guarino et al., 2007; Kuong and Kuzminov, 2010, 2012); geographically limited degradation of *Bacillus subtilis* chromosomal DNA behind the replication fork (Ramareddy and Reiter, 1970; Reiter and Ramareddy, 1970) and on one side of the origin of replication

(Regamey et al., 2000); partial unstructuring of the *E. coli* chromosome in a region around the origin of replication (Nakayama et al., 1994). It was shown that DNA from thymine starved cells is indeed susceptible to a single strand endonuclease activity *in vitro* and that thymine starvation induces transcription of the ssb gene, encoding the single strand DNA binding protein, which is consistent with an increase in the fraction of single strand DNA *in vivo* (Sangurdekar et al., 2010). Furthermore, investigations of DNA metabolism indicated that the starvation not only inhibits normal elongation of DNA replication (Maaloe and Hanawalt, 1961) but also results in a residual DNA synthesis (Kuong and Kuzminov, 2010) and small but quantifiable increase in the amount of bulk DNA (Barner and Cohen, 1958; Breitman et al., 1972; Kuong and Kuzminov, 2012), which is concomitant with the loss of thymine from DNA (Breitman et al., 1972) and is followed by contained DNA degradation (Kuong and Kuzminov, 2012).

Thus DNA degradation appears to be the main confounding factor in interpreting the loss of spatial transcriptional patterns in the vicinity of the origin of replication. In their attempt to map single strand DNA regions and/or regions of unstructured DNA, Sangurdekar et al. used the array comparative genomic hybridization (aCGH) method (Sangurdekar et al., 2010), which was earlier adapted (Khodursky et al., 2000b) for genetic marker frequency analysis (Sueoka and Yoshikawa, 1965) with a single gene resolution. Hybridization of labeled genomic DNA to microarray probes representing nearly every single gene and intergenic sequence in the *E. coli* genome (Khodursky et al., 2003) results in a characteristic profile of average abundances of sequences along the chromosome in a population of cells. It was shown that thymine starvation triggers gradual loss of DNA in the 500 kbp region surrounding the origin of replication (Sangurdekar et al., 2010): in the course of starvation the average dosage of DNA at and in the immediate vicinity of the origin of replication is reduced to the dosage of the replication terminus, the area of the chromosome present at the lowest frequency in an exponentially growing bacterial population (**Figure 2A**). Comparable localized variations in gene dosage have been observed by Rosenberg and colleagues (Fonville et al., 2010) and by Kuong and Kuzminov (2012). These results are consistent with a model according to which DNA template collapses at the replication origin and its neighborhood either as a result of degradation of nascent leading and lagging DNA strands in both chromatids or from a random, relatively short patch degradation of parental and nascent DNA, also in both chromatids (Sangurdekar et al., 2010). Both interpretations imply that DNA in the region may be degenerated to a point where it can no longer be used as a template in the reaction of transcription, explaining why any transcriptional patterns in that region might also be degraded. This view however, is complicated by an observation that thymine deprived cells continue transcribing genes in the region of the lesion well into starvation and even during the killing phase of TLD (Sangurdekar et al., 2010): half-life of a representative *E. coli* mRNA is about 60 times shorter than the timeline of a typical thymine starvation experiment (Bernstein et al., 2002) and there is no indication that mRNA stability is affected under conditions of thymine starvation, and the average abundance of

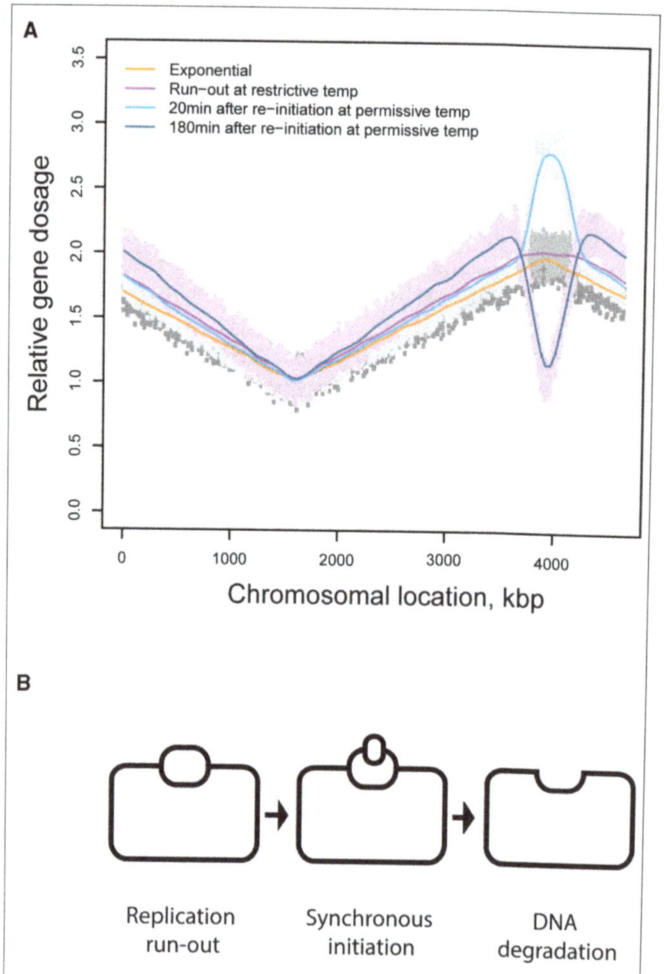

FIGURE 2 | (A) DNA dosage profiles in chromosomes of cells synchronized for replication initiation by using a temperature sensitive replication initiation mutant (Hamann, 2013). **(B)** A model of DNA loss in thymine starved cells synchronized for initiation of DNA replication.

transcripts from the region in question changes less than 50% in the course of the starvation (Sangurdekar et al., 2010). Thus an alternative model of the formation of the origin-centric lesion must be entertained, in which at least one DNA template remains intact.

Figure 2A depicts DNA dosage variations in a population of the thymine starved cells synchronized for initiation of DNA replication using a dnaC^{ts} mutant (Hamann, 2013); a similar profile with analogous interpretation was independently obtained by Kuong and Kuzminov (2012) in an asynchronous population of starved cells. Three conclusions can be drawn from this observation. First, following thymine withdrawal from a population of cells that can no longer initiate new rounds of replication, replication continues at least in a fraction of cells for a short period of time and then stops, resulting in a characteristic plateau spanning the stretch of the chromosome that underwent residual replication. Second, upon return to the permissive temperature, starved cells can initiate new rounds of replication even though the cells were deprived of thymine and could only support limited extension of the ongoing replication rounds. Third, new rounds of replication are initiated only on one of two partially replicated sister chromatids followed by complete destruction of the newly established replication bubble. (Although the increase in DNA dosage may be interpreted as a variation in a fraction of the population, the decrease cannot be explained in fractional terms.). Such sequence of events would result in one intact DNA template (**Figure 2B**). Although this template is transcriptionally active, its activity is spatially disorganized.

Concluding Remarks

Spatial regularity is one of the attributes of transcriptional activity of living genomes. The nature of spatial correlations at linear distances greater than the average size of an operon is poorly understood. The patterns somewhat coincide with the purported structural features of the chromosomal DNA and may in small part be explained by co-regulation of genes that are spaced with some periodicity on the chromosome (Kepes, 2004). The correlations are particularly sensitive to the state of DNA. Conditions that interfere with either DNA supercoiling or DNA replication result in diminishment or dissolution of the patterns, suggesting that the correlations are set, or modulated, by the moving replication fork.

Thymine starvation is one of the conditions that inhibit spatial pattern formation in the region of the chromosome adjacent to the origin of replication. However, this condition not merely inhibits DNA replication but also results in structural, copy number variations in the same region of the chromosome that loses the spatial pattern and yet does not become transcriptionally silent. It raises a formal possibility that, if spatial transcriptional correlations are the result of chromosomal DNA folding behind the replication fork, the normal folding of the chromosome in a region may also depend on the regional gene dosage or ploidy of the region. Moreover, the behind-the-fork organization of the chromosomal DNA into supercoiled loops (Postow et al., 2004) or rosettes of similar size loops (Kavenoff and Bowen, 1976) may result in a coordinated, basal transcriptional activity along the newly replicated stretch of DNA. This activity may facilitate further chromosomal folding which in turn may be required for the following round of replication initiation and nucleoid segregation. Consistent with this view are observations that global inhibition of transcription results in chromosome decondensation (Pettijohn and Hecht, 1974) and precludes initiation of DNA replication without a need for locus-specific transcription (Bates et al., 1997).

Acknowledgments

AK would like to thank Anrei Kuzminov, Pat Higgins and Georgi Muskhelishvili for positive reviews and critical reading of the manuscript.

References

Allen, T. E., Price, N. D., Joyce, A. R., and Palsson, B. O. (2006). Long-range periodic patterns in microbial genomes indicate significant multi-scale chromosomal organization. *PLoS Comput. Biol.* 2:e2. doi: 10.1371/journal.pcbi.0020002

Barner, H. D., and Cohen, S. S. (1954). The induction of thymine synthesis by T2 infection of a thymine requiring mutant of *Escherichia coli*. *J. Bacteriol.* 68, 80–88.

Barner, H. D., and Cohen, S. S. (1956). The relation of growth to the lethal damage induced by ultraviolet irradiation in *Escherichia coli*. *J. Bacteriol.* 71, 149–157.

Barner, H. D., and Cohen, S. S. (1957). The isolation and properties of amino acid requiring mutants of a thymineless bacterium. *J. Bacteriol.* 74, 350–355.

Barner, H. D., and Cohen, S. S. (1958). Protein synthesis and RNA turnover in a pyrimidine-deficient bacterium. *Biochim. Biophys. Acta* 30, 12–20. doi: 10.1016/0006-3002(58)90234-8

Bates, D. B., Boye, E., Asai, T., and Kogoma, T. (1997). The absence of effect of gid or mioC transcription on the initiation of chromosomal replication in *Escherichia coli*. *Proc. Natl. Acad. Sci. U.S.A.* 94, 12497–12502. doi: 10.1073/pnas.94.23.12497

Belfort, M., Maley, G., Pedersen-Lane, J., and Maley, F. (1983). Primary structure of the *Escherichia coli* thyA gene and its thymidylate synthase product. *Proc. Natl. Acad. Sci. U.S.A.* 80, 4914–4918. doi: 10.1073/pnas.80.16.4914

Benbrook, D. M., and Miller, R. V. (1986). Effects of norfloxacin on DNA metabolism in *Pseudomonas aeruginosa*. *Antimicrob. Agents Chemother.* 29, 1–6. doi: 10.1128/AAC.29.1.1

Bernstein, J. A., Khodursky, A. B., Lin, P. H., Lin-Chao, S., and Cohen, S. N. (2002). Global analysis of mRNA decay and abundance in *Escherichia coli* at single-gene resolution using two-color fluorescent DNA microarrays. *Proc. Natl. Acad. Sci. U.S.A.* 99, 9697–9702. doi: 10.1073/pnas.112318199

Breitman, T. R., Maury, P. B., and Toal, J. N. (1972). Loss of deoxyribonucleic acid-thymine during thymine starvation of *Escherichia coli*. *J. Bacteriol.* 112, 646–648.

Burton, Z. F., Gross, C. A., Watanabe, K. K., and Burgess, R. R. (1983). The operon that encodes the sigma subunit of RNA polymerase also encodes ribosomal protein S21 and DNA primase in *E. coli* K12. *Cell* 32, 335–349. doi: 10.1016/0092-8674(83)90453-1

Cagliero, C., Zhou, Y. N., and Jin, D. J. (2014). Spatial organization of transcription machinery and its segregation from the replisome in fast-growing bacterial cells. *Nucleic Acids Res.* 42, 13696–13705. doi: 10.1093/nar/gku1103

Casaregola, S., D'ari, R., and Huisman, O. (1982). Role of DNA replication in the induction and turn-off of the SOS response in *Escherichia coli*. *Mol. Gen. Genet.* 185, 440–444. doi: 10.1007/BF00334136

Chandler, M. G., and Pritchard, R. H. (1975). The effect of gene concentration and relative gene dosage on gene output in *Escherichia coli*. *Mol. Gen. Genet.* 138, 127–141. doi: 10.1007/BF02428117

Chiaramello, A. E., and Zyskind, J. W. (1990). Coupling of DNA replication to growth rate in *Escherichia coli*: a possible role for guanosine tetraphosphate. *J. Bacteriol.* 172, 2013–2019.

Cohen, S. S., and Barner, H. D. (1954). Studies on Unbalanced Growth in *Escherichia Coli*. *Proc. Natl. Acad. Sci. U.S.A.* 40, 885–893. doi: 10.1073/pnas.40.10.885

Courcelle, J., Khodursky, A., Peter, B., Brown, P. O., and Hanawalt, P. C. (2001). Comparative gene expression profiles following UV exposure in wild-type and SOS-deficient *Escherichia coli*. *Genetics* 158, 41–64.

Couturier, E., and Rocha, E. P. (2006). Replication-associated gene dosage effects shape the genomes of fast-growing bacteria but only for transcription and translation genes. *Mol. Microbiol.* 59, 1506–1518. doi: 10.1111/j.1365-2958.2006.05046.x

Crooke, E., Hwang, D. S., Skarstad, K., Thony, B., and Kornberg, A. (1991). *E. coli* minichromosome replication: regulation of initiation at *oriC*. *Res. Microbiol.* 142, 127–130. doi: 10.1016/0923-2508(91)90019-7

Denapoli, J., Tehranchi, A. K., and Wang, J. D. (2013). Dose-dependent reduction of replication elongation rate by (p)ppGpp in *Escherichia coli* and *Bacillus subtilis*. *Mol. Microbiol.* 88, 93–104. doi: 10.1111/mmi.12172

Dorman, C. J. (2006). DNA supercoiling and bacterial gene expression. *Sci. Prog.* 89, 151–166. doi: 10.3184/003685006783238317

Eisen, M. B., Spellman, P. T., Brown, P. O., and Botstein, D. (1998). Cluster analysis and display of genome-wide expression patterns. *Proc. Natl. Acad. Sci. U.S.A.* 95, 14863–14868. doi: 10.1073/pnas.95.25.14863

Felipe-Abrio, I., Lafuente-Barquero, J., Garcia-Rubio, M. L., and Aguilera, A. (2014). RNA polymerase II contributes to preventing transcription-mediated replication fork stalls. *EMBO J.* 34, 236–250. doi: 10.15252/embj.201488544

Fonville, N. C., Bates, D., Hastings, P. J., Hanawalt, P. C., and Rosenberg, S. M. (2010). Role of RecA and the SOS response in thymineless death in *Escherichia coli*. *PLoS Genet.* 6:e1000865. doi: 10.1371/journal.pgen.1000865

Freifelder, D. (1969). Single-strand breaks in bacterial DNA associated with thymine starvation. *J. Mol. Biol.* 45, 1–7. doi: 10.1016/0022-2836(69)90205-8

French, S. (1992). Consequences of replication fork movement through transcription units *in vivo*. *Science* 258, 1362–1365. doi: 10.1126/science.1455232

Gallant, J., and Suskind, S. R. (1962). Ribonucleic acid synthesis and thymineless death. *Biochim. Biophys. Acta* 55, 627–638. doi: 10.1016/0006-3002(62)90841-7

Goss, W. A., Deitz, W. H., and Cook, T. M. (1965). Mechanism of action of nalidixic acid on *Escherichia Coli*. ii. inhibition of deoxyribonucleic acid synthesis. *J. Bacteriol.* 89, 1068–1074.

Guarino, E., Salguero, I., Jimenez-Sanchez, A., and Guzman, E. C. (2007). Double-strand break generation under deoxyribonucleotide starvation in *Escherichia coli*. *J. Bacteriol.* 189, 5782–5786. doi: 10.1128/JB.00411-07

Gutierrez-Rios, R. M., Rosenblueth, D. A., Loza, J. A., Huerta, A. M., Glasner, J. D., Blattner, F. R., et al. (2003). Regulatory network of *Escherichia coli*: consistency between literature knowledge and microarray profiles. *Genome Res.* 13, 2435–2443. doi: 10.1101/gr.1387003

Hamann, B. L. (2013). *Elucidating the Mechanism of Thymineless Death in Escherichia coli*. Ph.D. thesis, University of Minnesota, Minnesota.

Hanawalt, P. C. (1963). Involvement of synthesis of RNA in thymineless death. *Nature* 198, 286. doi: 10.1038/198286a0

Hill, W. E., and Fangman, W. L. (1973). Single-strand breaks in deoxyribonucleic acid and viability loss during deoxyribonucleic acid synthesis inhibition in *Escherichia coli*. *J. Bacteriol.* 116, 1329–1335.

Jeong, K. S., Ahn, J., and Khodursky, A. B. (2004). Spatial patterns of transcriptional activity in the chromosome of *Escherichia coli*. *Genome Biol.* 5, R86. doi: 10.1186/gb-2004-5-11-r86

Jeong, K. S., Xie, Y., Hiasa, H., and Khodursky, A. B. (2006). Analysis of pleiotropic transcriptional profiles: a case study of DNA gyrase inhibition. *PLoS Genet.* 2:e152. doi: 10.1371/journal.pgen.0020152

Jin, D. J., Cagliero, C., and Zhou, Y. N. (2012). Growth rate regulation in *Escherichia coli*. *FEMS Microbiol. Rev.* 36, 269–287. doi: 10.1111/j.1574-6976.2011.00279.x

Kaguni, J. M. (2006). DnaA: controlling the initiation of bacterial DNA replication and more. *Annu. Rev. Microbiol.* 60, 351–375. doi: 10.1146/annurev.micro.60.080805.142111

Kavenoff, R., and Bowen, B. C. (1976). Electron microscopy of membrane-free folded chromosomes from *Escherichia coli*. *Chromosoma* 59, 89–101. doi: 10.1007/BF00328479

Kepes, F. (2004). Periodic transcriptional organization of the *E.coli* genome. *J. Mol. Biol.* 340, 957–964. doi: 10.1016/j.jmb.2004.05.039

Khodursky, A. B., Bernstein, J. A., Peter, B. J., Rhodius, V., Wendisch, V. F., and Zimmer, D. P. (2003). *Escherichia coli* spotted double-strand DNA microarrays: RNA extraction, labeling, hybridization, quality control, and data management. *Methods Mol. Biol.* 224, 61–78. doi: 10.1385/1-59259-364-X:61

Khodursky, A. B., and Cozzarelli, N. R. (1998). The mechanism of inhibition of topoisomerase IV by quinolone antibacterials. *J. Biol. Chem.* 273, 27668–27677. doi: 10.1074/jbc.273.42.27668

Khodursky, A. B., Peter, B. J., Cozzarelli, N. R., Botstein, D., Brown, P. O., and Yanofsky, C. (2000a). DNA microarray analysis of gene expression in response to physiological and genetic changes that affect tryptophan metabolism in *Escherichia coli*. *Proc. Natl. Acad. Sci. U.S.A.* 97, 12170–12175. doi: 10.1073/pnas.220414297

Khodursky, A. B., Peter, B. J., Schmid, M. B., Derisi, J., Botstein, D., Brown, P. O., et al. (2000b). Analysis of topoisomerase function in bacterial replication fork movement: use of DNA microarrays. *Proc. Natl. Acad. Sci. U.S.A.* 97, 9419–9424. doi: 10.1073/pnas.97.17.9419

Kowalski, D., and Eddy, M. J. (1989). The DNA unwinding element: a novel, *cis*-acting component that facilitates opening of the *Escherichia coli* replication origin. *EMBO J.* 8, 4335–4344.

Kreuzer, K. N. (2013). DNA damage responses in prokaryotes: regulating gene expression, modulating growth patterns, and manipulating replication forks. *Cold Spring Harb. Perspect. Biol.* 5, a012674. doi: 10.1101/cshperspect.a012674

Kuong, K. J., and Kuzminov, A. (2010). Stalled replication fork repair and misrepair during thymineless death in *Escherichia coli*. *Genes Cells* 15, 619–634. doi: 10.1111/j.1365-2443.2010.01405.x

Kuong, K. J., and Kuzminov, A. (2012). Disintegration of nascent replication bubbles during thymine starvation triggers RecA- and RecBCD-dependent replication origin destruction. *J. Biol. Chem.* 287, 23958–23970. doi: 10.1074/jbc.M112.359687

Lark, K. G. (1972). Evidence for the direct involvement of RNA in the initiation of DNA replication in *Escherichia coli* 15T. *J. Mol. Biol.* 64, 47–60. doi: 10.1016/0022-2836(72)90320-8

Le, T. B., Imakaev, M. V., Mirny, L. A., and Laub, M. T. (2013). High-resolution mapping of the spatial organization of a bacterial chromosome. *Science* 342, 731–734. doi: 10.1126/science.1242059

Levine, A., Vannier, F., Dehbi, M., Henckes, G., and Seror, S. J. (1991). The stringent response blocks DNA replication outside the ori region in *Bacillus subtilis* and at the origin in *Escherichia coli*. *J. Mol. Biol.* 219, 605–613. doi: 10.1016/0022-2836(91)90657-R

Lieberman-Aiden, E., Van Berkum, N. L., Williams, L., Imakaev, M., Ragoczy, T., Telling, A., et al. (2009). Comprehensive mapping of long-range interactions reveals folding principles of the human genome. *Science* 326, 289–293. doi: 10.1126/science.1181369

Liu, B., and Alberts, B. M. (1995). Head-on collision between a DNA replication apparatus and RNA polymerase transcription complex. *Science* 267, 1131–1137. doi: 10.1126/science.7855590

Maaloe, O., and Hanawalt, P. C. (1961). Thymine deficiency and the normal DNA replication cycle. I. *J. Mol. Biol.* 3, 144–155. doi: 10.1016/S0022-2836(61)80041-7

Martin, C. M., Viguera, E., and Guzman, E. C. (2014). Rifampicin suppresses thymineless death by blocking the transcription-dependent step of chromosome initiation. *DNA Repair (Amst)* 18, 10–17. doi: 10.1016/j.dnarep.2014.03.004

McFall, E., and Magasanik, B. (1960). Thymine starvation and enzyme synthesis. *Biochim. Biophys. Acta* 45, 610–612. doi: 10.1016/0006-3002(60)91505-5

Merrikh, H., Machon, C., Grainger, W. H., Grossman, A. D., and Soultanas, P. (2011). Co-directional replication-transcription conflicts lead to replication restart. *Nature* 470, 554–557. doi: 10.1038/nature09758

Merrikh, H., Zhang, Y., Grossman, A. D., and Wang, J. D. (2012). Replication-transcription conflicts in bacteria. *Nat. Rev. Microbiol.* 10, 449–458. doi: 10.1038/nrmicro2800

Messer, W. (1972). Initiation of deoxyribonucleic acid replication in *Escherichia coli* B-r: chronology of events and transcriptional control of initiation. *J. Bacteriol.* 112, 7–12.

Morganroth, P. A., and Hanawalt, P. C. (2006). Role of DNA replication and repair in thymineless death in *Escherichia coli*. *J. Bacteriol.* 188, 5286–5288. doi: 10.1128/JB.00543-06

Nakayama, H., and Hanawalt, P. (1975). Sedimentation analysis of deoxyribonucleic acid from thymine-starved *Escherichia coli*. *J. Bacteriol.* 121, 537–547.

Nakayama, K., Kusano, K., Irino, N., and Nakayama, H. (1994). Thymine starvation-induced structural changes in *Escherichia coli* DNA. Detection by pulsed field gel electrophoresis and evidence for involvement of homologous recombination. *J. Mol. Biol.* 243, 611–620. doi: 10.1016/0022-2836(94)90036-1

Olliver, A., Saggioro, C., Herrick, J., and Sclavi, B. (2010). DnaA-ATP acts as a molecular switch to control levels of ribonucleotide reductase expression in *Escherichia coli*. *Mol. Microbiol.* 76, 1555–1571. doi: 10.1111/j.1365-2958.2010.07185.x

Peter, B. J., Arsuaga, J., Breier, A. M., Khodursky, A. B., Brown, P. O., and Cozzarelli, N. R. (2004). Genomic transcriptional response to loss of chromosomal supercoiling in *Escherichia coli*. *Genome Biol.* 5, R87. doi: 10.1186/gb-2004-5-11-r87

Pettijohn, D. E., and Hecht, R. (1974). RNA molecules bound to the folded bacterial genome stabilize DNA folds and segregate domains of supercoiling. *Cold Spring Harb. Symp. Quant. Biol.* 38, 31–41. doi: 10.1101/SQB.1974.038.01.006

Postow, L., Hardy, C. D., Arsuaga, J., and Cozzarelli, N. R. (2004). Topological domain structure of the *Escherichia coli* chromosome. *Genes Dev.* 18, 1766–1779. doi: 10.1101/gad.1207504

Ramareddy, G., and Reiter, H. (1970). Sequential loss of loci in thymine-starved *Bacillus subtilis* 168 cells. Evidence for a circular chromosome. *J. Mol. Biol.* 50, 525–532. doi: 10.1016/0022-2836(70)90209-3

Regamey, A., Harry, E. J., and Wake, R. G. (2000). Mid-cell Z ring assembly in the absence of entry into the elongation phase of the round of replication in bacteria: co-ordinating chromosome replication with cell division. *Mol. Microbiol.* 38, 423–434. doi: 10.1046/j.1365-2958.2000.02130.x

Reichard, P. (2010). Ribonucleotide reductases: substrate specificity by allostery. *Biochem. Biophys. Res. Commun.* 396, 19–23. doi: 10.1016/j.bbrc.2010.02.108

Reiter, H., and Ramareddy, G. (1970). Loss of DNA behind the growing point of thymine-starved *Bacillus subtilis* 168. *J. Mol. Biol.* 50, 533–548. doi: 10.1016/0022-2836(70)90210-X

Rolfe, R. (1967). On the mechanism of thymineless death in *Bacillus subtilis*. *Proc. Natl. Acad. Sci. U.S.A.* 57, 114–121. doi: 10.1073/pnas.57.1.114

Rovinskiy, N., Agbleke, A. A., Chesnokova, O., Pang, Z., and Higgins, N. P. (2012). Rates of gyrase supercoiling and transcription elongation control supercoil density in a bacterial chromosome. *PLoS Genet.* 8:e1002845. doi: 10.1371/journal.pgen.1002845

Samuel, G., and Reeves, P. (2003). Biosynthesis of O-antigens: genes and pathways involved in nucleotide sugar precursor synthesis and O-antigen assembly. *Carbohydr. Res.* 338, 2503–2519. doi: 10.1016/j.carres.2003.07.009

Sangurdekar, D. P., Hamann, B. L., Smirnov, D., Srienc, F., Hanawalt, P. C., and Khodursky, A. B. (2010). Thymineless death is associated with loss of essential genetic information from the replication origin. *Mol. Microbiol.* 75, 1455–1467. doi: 10.1111/j.1365-2958.2010.07072.x

Sangurdekar, D. P., Srienc, F., and Khodursky, A. B. (2006). A classification based framework for quantitative description of large-scale microarray data. *Genome Biol.* 7, R32. doi: 10.1186/gb-2006-7-4-r32

Sangurdekar, D. P., Zhang, Z., and Khodursky, A. B. (2011). The association of DNA damage response and nucleotide level modulation with the antibacterial mechanism of the anti-folate drug trimethoprim. *BMC Genomics* 12:583. doi: 10.1186/1471-2164-12-583

Schena, M., Shalon, D., Davis, R. W., and Brown, P. O. (1995). Quantitative monitoring of gene expression patterns with a complementary DNA microarray. *Science* 270, 467–470. doi: 10.1126/science.270.5235.467

Schmid, M. B., and Roth, J. R. (1987). Gene location affects expression level in *Salmonella typhimurium*. *J. Bacteriol.* 169, 2872–2875.

Simmons, L. A., Foti, J. J., Cohen, S. E., and Walker, G. C. (2008). The SOS regulatory network. *Ecosal Plus* 2008, 1–48. doi: 10.1128/ecosalplus.5.4.3

Slager, J., Kjos, M., Attaiech, L., and Veening, J. W. (2014). Antibiotic-induced replication stress triggers bacterial competence by increasing gene dosage near the origin. *Cell* 157, 395–406. doi: 10.1016/j.cell.2014.01.068

Sobetzko, P., Travers, A., and Muskhelishvili, G. (2012). Gene order and chromosome dynamics coordinate spatiotemporal gene expression during the bacterial growth cycle. *Proc. Natl. Acad. Sci. U.S.A.* 109, E42–E50. doi: 10.1073/pnas.1108229109

Sueoka, N., and Yoshikawa, H. (1965). The chromosome of *Bacillus subtilis*. I. Theory of marker frequency analysis. *Genetics* 52, 747–757.

Tamayo, P., Slonim, D., Mesirov, J., Zhu, Q., Kitareewan, S., Dmitrovsky, E., et al. (1999). Interpreting patterns of gene expression with self-organizing maps: methods and application to hematopoietic differentiation. *Proc. Natl. Acad. Sci. U.S.A.* 96, 2907–2912. doi: 10.1073/pnas.96.6.2907

Torrence, C., and Compo, G. (1998). A practical guide to wavelet analysis. *Bull. Amer. Met. Soc.* 79, 61–78. doi: 10.1175/1520-0477(1998)079<0061:APGTWA>2.0.CO;2

van Helden, J., Andre, B., and Collado-Vides, J. (1998). Extracting regulatory sites from the upstream region of yeast genes by computational analysis of oligonucleotide frequencies. *J. Mol. Biol.* 281, 827–842. doi: 10.1006/jmbi.1998.1947

Versalovic, J., Koeuth, T., Britton, R., Geszvain, K., and Lupski, J. R. (1993). Conservation and evolution of the rpsU-dnaG-rpoD macromolecular synthesis operon in bacteria. *Mol. Microbiol.* 8, 343–355. doi: 10.1111/j.1365-2958.1993.tb01578.x

von Freiesleben, U., and Rasmussen, K. V. (1992). The level of supercoiling affects the regulation of DNA replication in *Escherichia coli*. *Res. Microbiol.* 143, 655–663. doi: 10.1016/0923-2508(92)90060-2

Xiao, G., Wang, X., and Khodursky, A. B. (2011). Modeling three-dimensional chromosome structures using gene expression data. *J. Am. Stat. Assoc.* 106, 61–72. doi: 10.1198/jasa.2010.ap09504

Yoshinaga, K. (1973). Double-strand scission of DNA involved in thymineless death of *Escherichia coli* 15 TAU. *Biochim. Biophys. Acta* 294, 204–213. doi: 10.1016/0005-2787(73)90293-1

16

Low affinity DnaA-ATP recognition sites in *E. coli oriC* make non-equivalent and growth rate-dependent contributions to the regulated timing of chromosome replication

Prassanna Rao[1], Tania A. Rozgaja[2], Abdulaziz Alqahtani[1], Julia E. Grimwade[1] and Alan C. Leonard[1]*

[1] Department of Biological Sciences, Florida Institute of Technology, Melbourne, FL, United States, [2] AREVA Med, Plano, TX, United States

***Correspondence:**
Alan C. Leonard
aleonard@fit.edu

Although the mechanisms that precisely time initiation of chromosome replication in bacteria remain unclear, most clock models are based on accumulation of the active initiator protein, DnaA-ATP. During each cell division cycle, sufficient DnaA-ATP must become available to interact with a distinct set of low affinity recognition sites in the unique chromosomal replication origin, *oriC*, and assemble the pre-replicative complex (orisome) that unwinds origin DNA and helps load the replicative helicase. The low affinity *oriC*-DnaA-ATP interactions are required for the orisome's mechanical functions, and may also play a role in timing of new rounds of DNA synthesis. To further examine this possibility, we constructed chromosomal *oriC*s with equal preference for DnaA-ADP or DnaA-ATP at one or more low affinity recognition sites, thereby lowering the DnaA-ATP requirement for orisome assembly, and measured the effect of the mutations on cell cycle timing of DNA synthesis. Under slow growth conditions, mutation of any one of the six low affinity DnaA-ATP sites in chromosomal *oriC* resulted in initiation earlier in the cell cycle, but the shift was not equivalent for every recognition site. Mutation of τ2 caused a greater change in initiation age, suggesting its occupation by DnaA-ATP is a temporal bottleneck during orisome assembly. In contrast, during rapid growth, all origins with a single mutated site displayed wild-type initiation timing. Based on these observations, we propose that *E. coli* uses two different, DnaA-ATP-dependent initiation timing mechanisms; a slow growth timer that is directly coupled to individual site occupation, and a fast growth timer comprising DnaA-ATP and additional factors that regulate DnaA access to *oriC*. Analysis of origins with paired mutated sites suggests that Fis is an important component of the fast growth timing mechanism.

Keywords: *oriC*, DnaA, DNA replication, replication origin, orisomes, pre-replicative complexes, DNA binding proteins, cell cycle

INTRODUCTION

All cells must initiate new rounds of chromosomal DNA synthesis with temporal precision and only once per cell cycle to avoid genome instability. Bacteria accomplish this by using a triggering mechanism that requires the AAA+ (ATPases Associated with various cellular Activities) initiator protein, DnaA. In cells, DnaA is bound to either ATP or ADP, and DnaA-ATP is considered to be the active form of the initiator based on *in vitro* unwinding and replication assays (Sekimizu et al., 1987). Multiple DnaA molecules interact with distinct high and low affinity recognition sites in the unique chromosomal replication origin (*oriC*) to assemble a pre-replicative complex (orisome) that unwinds origin DNA and then assists with loading the replicative helicase onto the single strands (reviewed in Bell and Kaguni, 2013; Leonard and Grimwade, 2015; Chodavarapu and Kaguni, 2016).

In *E. coli oriC*, there are 11 DnaA binding sites within the 190 base pairs that are adjacent to the DNA Unwinding Element (DUE) (**Figure 1A**; Bramhill and Kornberg, 1988; Kowalski and Eddy, 1989; Rozgaja et al., 2011). Three of these sites (R1, R2, and R4), with consensus sequence, 5′-TTA/TTNCACA-3′, bind DnaA with high affinity. The remaining eight sites deviate at two or more positions and bind DnaA with low affinity (**Figure 1A**), reviewed in Leonard and Grimwade (2011), Leonard and Méchali (2013), and Wolański et al. (2014). These weak binding sites lie in the gap regions between R1 and R2, and R2 and R4, with each gap containing a cluster of four sites separated by 2 bp (**Figure 1A**). All low affinity DnaA recognition sites in *oriC* are incapable of binding DnaA directly (Schaper and Messer, 1995); instead, their occupation requires the DnaA occupying high affinity sites (usually R1 and R4) to recruit and donate DnaA to the nearest low affinity site (usually R5M and C1), followed by cooperative DnaA binding between the arrayed low affinity sites, so that DnaA occupation progresses toward the center of *oriC* (Simmons et al., 2003; Miller et al., 2009; Rozgaja et al., 2011).

Two DNA bending proteins, IHF and Fis, modulate the low affinity DnaA interactions in *oriC*. IHF-catalyzed bending of the DNA between R1 and R5M places these two sites into proximity, facilitating cooperative binding in *trans*, and thus beginning the progressive DnaA occupation of the left half of *oriC* (Grimwade et al., 2000; Rozgaja et al., 2011). However, IHF binding to its recognition site is prohibited as long as Fis

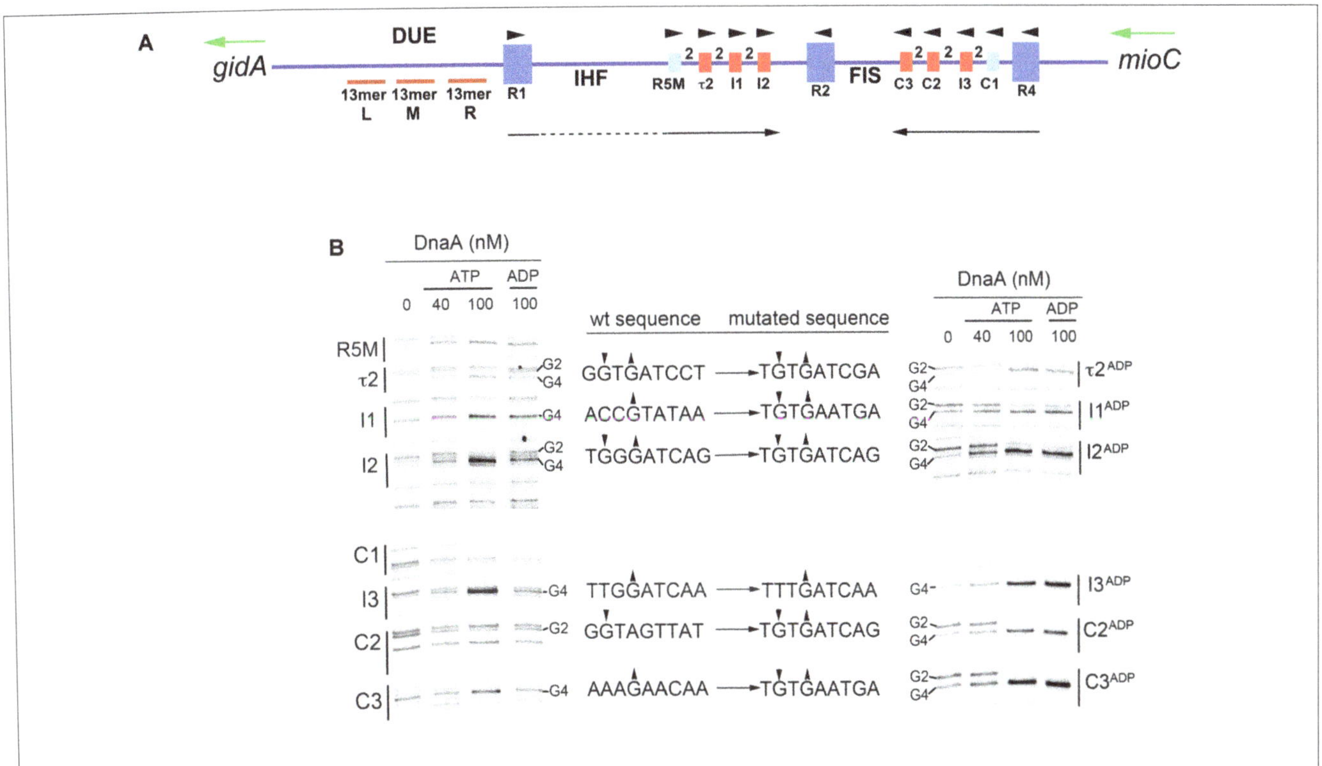

FIGURE 1 | Six low affinity DnaA-ATP sites in *E. coli oriC* can be converted to sites that bind DnaA-ADP. **(A)** Map of *E. coli oriC*, showing low affinity DnaA recognition sites that preferentially bind DnaA-ATP (red boxes), and high and low affinity sites that bind both nucleotide forms (dark and light blue boxes, respectively). The number 2 between boxes indicates the spacing between sites. The positions of the binding sites for Fis and IHF are shown, as well as the left (L), middle (M), and right (R) 13mer repeats in the DNA Unwinding Element (DUE) are indicated. Arrowheads indicate the orientation of the binding sites. Black arrows indicate the direction of progressive DnaA binding. Green arrows indicate the direction of transcription of the genes flanking *oriC* (*gidA* and *mioC*). Figure is from Leonard and Grimwade (2015). **(B)** Comparison of DnaA-ATP and DnaA-ADP binding to wild type *oriC* and *oriC* containing the indicated mutations. DMS footprints, showing binding of purified his-tagged DnaA-ATP (40 and 100 nM) and DnaA-ADP (100 nM) are shown. All the binding sites are marked, with positions of the guanosine residues in the second and fourth positions (if present) indicated. Only the footprint of the mutated site on each mutated plasmid is shown in the right panel. Small arrowheads indicate enhancement (up arrow) or suppressions (down arrow) of DMS modification. Intervening lanes on the gels have been removed in the figure, and regions between R1 and R5M, and C3 and R2 were also removed.

occupies its recognition site between R2 and C3, and higher levels of DnaA are required for orisome assembly if the *oriC* template is occupied by Fis (Ryan et al., 2004). Fis is bound to *oriC* throughout most of the cell cycle, but is displaced by the progressive DnaA binding extending from R4 (Ryan et al., 2004); loss of Fis relieves the repression of IHF binding and allows rapid DnaA occupation of any unbound low affinity sites. This dynamic interplay among DnaA, Fis, and IHF ensures synchronous initiations from the multiple *oriC* copies that exist during rapid growth (Weigel et al., 2001; Ryan et al., 2004). However, because Fis is a growth rate-regulated protein (Nilsson et al., 1992; Mallik et al., 2006) it is not likely to be an orisome component in slowly growing cells, and without Fis-mediated inhibition, IHF is likely to be bound thought the cell cycle.

Despite our ever-increasing understanding of orisome assembly and regulation (Leonard and Grimwade, 2011; Skarstad and Katayama, 2013; Katayama et al., 2017; Schenk et al., 2017), it remains unclear how replication initiation is precisely coupled to the bacterial cell division cycle. Most models for initiation timing (Mahaffy and Zyskind, 1989; Hansen et al., 1991b; Browning et al., 2004; Grant et al., 2011; Zhang and Shi, 2012) are based on accumulation of DnaA-ATP, whose levels fluctuate during the cell cycle, peaking near the time of initiation (Kurokawa et al., 1999). There is extensive experimental evidence in support of these models (Atlung et al., 1987; Pierucci et al., 1987; Løbner-Olesen et al., 1989; Skarstad et al., 1989; Atlung and Hansen, 1993), but some controversy remains (Xu and Bremer, 1988; Flåtten et al., 2015). In addition, most models have yet to take into account the manner in which DnaA-ATP interacts with *oriC*. The three high affinity sites bind DnaA-ATP and DnaA-ADP equivalently (McGarry et al., 2004), and remain occupied throughout the cell cycle (Samitt et al., 1989; Nievera et al., 2006). For this reason, occupation of these sites is unlikely to be the rate-limiting step in triggering chromosome replication. Instead, it is more probable that initiation timing is determined by the filling of the low affinity DnaA recognition sites because occupation of these sites is transient and takes place immediately before initiation (Nievera et al., 2006; Rozgaja et al., 2011). Further, although sites C1 and R5M have equal preference for DnaA-ATP and DnaA–ADP, all remaining low affinity sites (τ2, I1, I2, and C3, C2, and I3) exhibit a 4-fold binding preference for DnaA-ATP (McGarry et al., 2004; Grimwade et al., 2018), suggesting that occupation of the sites with DnaA-ATP could couple initiation to levels of available DnaA-ATP.

Site preference or non-preference for DnaA-ATP is determined by nucleotide sequence, and previous work demonstrates that *oriC* remains functional when some or all low affinity sites are altered by mutagenesis to allow both DnaA-ATP and -ADP forms to bind equivalently without changing overall site affinity (McGarry et al., 2004; Grimwade et al., 2018). DnaA is not normally degraded after bound ATP is hydrolyzed, and DnaA-ADP is prevalent in cells, fluctuating between 30 and 70% of total DnaA (Kurokawa et al., 1999). Since mixed complexes of DnaA-ATP and DnaA-ADP can activate *oriC* (Yung et al., 1990; Grimwade et al., 2007, 2018), if DnaA-ADP can access *oriC* at mutated sites, this binding should reduce the requirement for DnaA-ATP in orisome formation, and shift the timing of initiation earlier in the cell cycle. Previous studies support this idea. For example, although cloned wild-type *oriC*, initiates synchronously with the wild-type chromosomal copy (Helmstetter and Leonard, 1987; Koppes, 1987; Løbner-Olesen, 1999), *oriC* plasmids carrying loss of preference mutations at I2 and I3 rapidly integrate and replace chromosomal *oriC*, consistent with the cloned origin initiating earlier and out-competing the chromosomal origin (Grimwade et al., 2007). Further, changing all the DnaA-ATP sites in chromosomal *oriC* into sites that bind both DnaA-ATP and DnaA-ADP causes over-initiation and loss of initiation synchrony (Grimwade et al., 2018), similar to the phenotype of cells in which there is a 3–4-fold excess of DnaA (Atlung and Hansen, 1993). However, it remains unclear whether all DnaA-ATP sites play equivalent timing roles (i.e., each site requiring a fixed amount of time to become occupied as newly synthesized DnaA-ATP becomes available) or if additional factors might contribute to how DnaA accesses the origin sites.

To better understand the relationship between low affinity site occupation and the initiation clock, we measured the timing of DNA synthesis in cells whose only chromosomal *oriC* copy carried mutations resulting in loss of DnaA-ATP preference at individual or pairs of low affinity recognition sites. We observed that during slow growth, the onset of chromosomal DNA replication was shifted to earlier cell cycle times for all single site mutations, but the shift was greater in cells carrying mutations at the τ2 site, revealing a possible bottleneck step during orisome assembly. Initiation timing in cells where pairs of sites were mutated was consistent with this idea. However, altering DnaA-ATP preference at single sites did not affect timing when assays were performed under rapid growth conditions. These observations are consistent with the existence of two distinct, growth-rate dependent timing mechanisms in *E. coli*, and we propose that the occupation of *oriC* low affinity DnaA-ATP recognition sites directly regulates initiation timing only during slow growth. In rapidly growing cells, although initiation is not triggered until all the low affinity sites are occupied, other factors modulate binding of DnaA-ATP to these sites. Examination of double site mutations during fast-growth suggests that one of these factors is the growth rate-regulated protein Fis, making this architectural protein an important component of the initiation timing mechanism.

MATERIALS AND METHODS

Chemicals, Proteins, Enzymes, and Oligonucleotides

Reagent grade chemicals were purchased from IBI, ThermoFisher, or Sigma. Media components were from IBI or Difco. All enzymes were from New England Biolabs. Oligonucleotides were purchased from ThermoFisher. His-tagged DnaA was isolated as described (Li and Crooke, 1999), including a urea wash step to remove bound nucleotide and lipids.

Construction of *oriCs* With Loss of DnaA-ATP Preference

The positions of mutated DnaA recognition sites are shown in **Figure 1A**. Specific changes in nucleotide sequence are listed in **Figure 1B**. Site-directed mutagenesis was performed as described in McGarry et al. (2004). The starting template for all *oriC* mutations was pOC170 (3,852 bp) (Langer et al., 1996), which carries both the pBR322 replication origin and *oriC*. Overlapping 30–35 bp primers carrying the base alterations were flanked by 10 bp of DNA homologous to the template. To make mutations in multiple DnaA binding sites in *oriC*, a single site was mutated per reaction, and the resulting plasmid used as template to make additional mutations. Supercoiled plasmid was isolated using a plasmid preparation kit (IBI). Mutations in *oriC* were confirmed by sequence analysis (sequencing primer was 5′-CTCAACTTTGTCGGCTTGAG) and plasmids were introduced into DH5α by transformation selecting for ampicillin resistance.

Recombineering

PCR fragments carrying the mutant *oriC* were amplified from plasmids using primers near *gidA* (5′-CACGGCCA CCGCTGTAATTAT) and *mioC* (5′-ATCCCATACTTTT CCACAGG). The fragments were introduced by electroporation into ACL402 (*asnB32, relA1, spoT1, thi-1, fuc-1, lysA, ilv-192, zia*::pKN500, Δ*dnaA, mad-1*, λcI857Δ(*cro-bioA*), *nad*::Tn10, Δ*oriC*::*cat-sacB*) following induction of RED gene expression, as described in Kaur et al. (2014). ACL402 was constructed by inserting the cat-*sacB* cassette from pK03 (Link et al., 1997) into ACL401 (Kaur et al., 2014). Transformed cells were grown in LB overnight at 32°C without selection, and then diluted and plated onto LB agar plates containing 5% sucrose and incubated at 32°C overnight. Replacement of the Δ*oriC*::*cat sac* with the mutated *oriC* was verified in sucrose resistant cells by testing for loss of chloramphenicol resistance, and nucleotide sequence analysis. The *oriC* region from verified mutant cells was transduced using P1 phage into an JEG22, an MG1655 derivative which is *asnA*::cat and *asnB*::Tn10 and transductants were selected for growth in the absence of asparagine on minimal salts agar (Helmstetter and Krajewski, 1982), supplemented with 0.1% glucose as described in Kaur et al. (2014).

Dimethyl Sulfate Footprinting

DMS modification of DNA (0.75 μg) *in vitro* was performed as described (Ryan et al., 2004). DnaA was pre-incubated in reaction buffer with 5 mM ATP or ADP for 5 min before addition to reactions at the concentrations indicated in **Figure 1B**. DMS-treated samples were extended with radiolabeled primer as described (Ryan et al., 2004). Primer sequences were 5′-GTATACAGATCGTGCGATC, for revealing R1-I2, and 5′-GGATCATTAACTGTGAATG, for revealing R4-R2. Extension products were resolved on 6% polyacrylamide sequencing gels, and dried gels were scanned on a Bio-Rad Molecular Imager FX PhosphorImager. Representative scans are shown in **Figure 1B**.

Western Blots

MG1655 was grown for at least five generations in exponential growth, in minimal media containing 0.1% uracil supplemented with 0.1% glucose plus 0.1% casamino acids; 0.1% glucose; 0.2% glycerol; or 0.4% succinate. Samples were removed from the cultures, pelleted, and flash frozen in liquid nitrogen three times to lyse the cells. Protein concentrations in each sample were quantified using the MicroBCA assay (Pierce) and then equal amounts of protein (5 mg) were resolved on 12% PAGE/SDS gels. The proteins were transferred to nitrocellulose, and DnaA and Fis on the blots were revealed using anti-DnaA and anti-Fis antibodies with detection by chemiluminescence (Bio-Rad ImunoStar kit). Assays were done in triplicate, with representative scans shown in Supplementary Figure S1.

Flow Cytometry

Wild-type or mutant *oriC* strains (MG1655 derivatives) were grown in minimal salts media (Helmstetter and Krajewski, 1982; Løbner-Olesen et al., 1989) supplemented with 20 μg/ml uracil and either 0.4% succinate or 0.1% glucose, 0.1% casamino acids, and 0.1% uracil, and liquid cultures were grown to exponential phase at 37°C with vigorous shaking. Cells were then treated with 300 μg/ml rifampicin and 15 μg/ml cephalexin (Skarstad et al., 1986; Løbner-Olesen et al., 1989), and incubation was continued at 37°C for at least 4 h. After drug treatment, 1 ml of cells was fixed with 9 ml of 70% ethanol and stored at 4°C. Prior to flow cytometric analysis, 1 ml of the fixed cell suspension was pelleted, washed with 50 mM Tris-Cl, pH 7.5, 150 mM NaCl (TBS), and resuspended in 1 ml of TBS containing 0.5 μl/ml Vybrant DyeCycle Green (ThermoFisher) (final concentration of dye is 2.5 μM). Stained cells (3,000–5,000 cells/ml) were analyzed using an Accuri C6 personal flow cytometer, and data from 10,000 cells were collected. Forward scatter was used for cell mass measurement. CFlow software was used to calculate the percentage of cells in each chromosomal DNA peak.

RESULTS

Conversion of a DnaA-ATP Site Into One That Binds DnaA-ADP Alters Initiation Timing in Slowly Growing *E. coli* Cells

DnaA-ADP is produced in bacterial cells by hydrolysis of DnaA-ATP, which, in *E. coli*, is stimulated by the Hda protein associated with replication forks, as well as by the *datA* locus, reviewed in Katayama et al. (2017). Since *E. coli* cells do not degrade DnaA-ADP, this form is generally available for *oriC* interactions at any recognition sites capable of binding it. Therefore, mutation of a low affinity DnaA-ATP recognition site to allow interaction with DnaA-ADP should decrease the amount of DnaA-ATP required for orisome formation, and, if the mutation is carried in chromosomal *oriC*, it should shift initiation timing to earlier in the cell cycle. To examine this possibility, we first expanded our collection of site mutants (McGarry et al., 2004; Grimwade et al., 2007, 2018) by altering each of the six individual DnaA-ATP recognition sites (see **Figures 1A,B**), as well as pairs of sites in *oriC* specifically chosen for their location (two sites on the right, two on the left, and one in each half). Since we wished to retain the low affinity attribute while allowing DnaA-ADP binding, the I3, C2, C3, I2 I1, and τ2 sites were each individually mutated to sequences identical or similar to R5M. The sequence of the mutated sites is shown in **Figure 1B**. To verify that the mutations relieved the DnaA-ATP preference, but did not alter overall site affinity, binding of DnaA-ATP or DnaA-ADP to each site was evaluated using dimethyl sulfate (DMS) footprinting (**Figure 1B**). DnaA was pre-incubated with nucleotide, and then bound to supercoiled *oriC* plasmid prior to DMS treatment and primer extension. Sites that are occupied by DnaA are revealed by distinctive changes to the DMS modification pattern resolved on sequencing gels. Specifically, the G4 of the 9 mer consensus 5′-TGTGGATAA (or variations of this sequence) becomes hypersensitive, and the G2 less sensitive to DMS (**Figure 1B**), causing darker and lighter bands, respectively. The changes in band intensity increase as all of the plasmid molecules in the treatment population become occupied. It should be noted that not all binding sites have guanosines at both positions. For evaluation of site affinity, two concentrations of DnaA-ATP were used: 40 nM, which is insufficient for full occupation of all low affinity sites, and 100 nM, which previous studies have shown to be a saturating concentration, resulting in complete DnaA-ATP binding to all *oriC* copies in the reaction mix (McGarry et al., 2004; Grimwade et al., 2007). In wild type *oriC* (**Figure 1B**, left panel), the pattern of DnaA-ATP binding is consistent with previous studies showing R1, R2, and R4 as higher affinity sites fully occupied at 40 nM DnaA, and the remaining sites being fully occupied only at the higher concentration. The DnaA-ATP binding pattern is similar for origins containing the mutated sites, indicating that the mutations do not alter any sites' affinity for DnaA-ATP. However, when DnaA-ADP is used in the reaction, only the mutant origin is fully occupied by 100 nM DnaA-ADP (**Figure 1B**); in contrast, binding of 100 nM DnaA-ADP binding to I3, C2, C3, I2 I1, and τ2 in wild type *oriC* gives a DMS modification pattern similar to that seen with 40 nM DnaA-ATP (McGarry et al., 2004).

We next evaluated the effect of single site mutations on initiation timing. A previous study reported that mutations may affect chromosomal *oriC* differently than *oriC*'s harbored on a plasmid (Weigel et al., 2001). Therefore, recombineering (homologous recombination-mediated genetic engineering) was used to delete wild-type *oriC* on the chromosome (Court et al., 2002), and replace it at the native location with one of the mutant *oriC*s carrying an altered low affinity site. After verifying successful origin transplantation by sequence analysis, P1 phage was then used to transfer the mutated origin from the recombineering strain into MG1655 for analysis of cell cycle timing. In the initial experiments, *E. coli* containing wild type or mutant *oriC* were grown in minimal media supplemented with succinate as a carbon source (generation time 101–103 min). This growth rate allows us to examine the simplest initiation scenario, in which only one *oriC* copy per cell must be activated, since the generation time of *E. coli* growing in this media should be greater than the time needed to complete chromosome replication and divide (~60 min), so exponentially growing cells will contain only one copy of *oriC* prior to the time of initiation (Cooper and Helmstetter, 1968). Further, since previous studies by our lab revealed that Fis, when bound to *oriC*, reduces binding of DnaA to low affinity sites, and because Fis is a growth-rate regulated protein, orisome assembly in cells growing in succinate media should take place without Fis inhibition. To confirm this, equal amounts of total protein from cells with wild type *oriC*, growing exponentially at four different growth rates were separated on denaturing polyacrylamide gels, transferred to nitrocellulose, and the blots were incubated with both anti-DnaA and anti-Fis antibodies (Supplementary Figure S1). The blots confirm that Fis levels decrease as a function of growth rate (Flåtten and Skarstad, 2013), and are nearly undetectable in cells using succinate as a carbon source, while DnaA concentrations remain constant among the growth rates tested, as has been reported previously (Hansen et al., 1991a).

To evaluate initiation timing, cells carrying the mutated *oriC*s were grown to mid-exponential phase in minimal media supplemented with succinate. Then, cells were treated with rifampicin to inhibit new rounds of chromosome replication, and cephalexin, which prevents cell division, followed by incubation to allow completion of ongoing rounds of replication. After staining cellular DNA with a fluorescent dye, the number of chromosomes in the cell, which reflects the number of *oriC* copies present at the time of drug addition, was detected by flow cytometry. The origin number doubles at the time of initiation, and at all growth rates, only 2^n origins should be in these cells. Accordingly, cultures of MG1655 cells contained 1 or 2 origins (**Figure 2A**), and initiated ~36 min after cell division (**Table 1**). Although previous studies report that cells growing slowly with acetate as the carbon source display rifampicin resistant initiations (Flåtten et al., 2009), we do not see evidence of this in cells growing in succinate media, since the percentage of cells that have not yet started chromosome replication (1 chromosome peak) does not decrease after addition of rifampicin (Supplementary Figure S2). Interestingly, converting a single DnaA-ATP site into one that binds both of DnaA's nucleotide forms changed initiation timing in all the mutant cells, resulting in an increased number of origins per cell, and an earlier initiation time in the cell cycle (**Figures 2B–G** and **Table 1**). The conversion of the τ2 site to $\tau 2^{ADP}$ caused a greater change in timing (**Table 1**), suggesting that filling this site might represent a rate-limiting step in slow growth orisome formation.

The change in initiation time caused by converting one DnaA-ATP site to the non-discriminatory form raised the possibility that changing more than one site might have an additive effect. To evaluate this, we examined the effect of mutation pairs on initiation during slow growth. Mutation pairs chosen for study were located either in *oriC*'s right half (I3/C3), left half (τ2/I2) or each half (τ2/I3; I2/I3), because there is evidence that DnaA assembled onto the left and right halves of *oriC* play different roles in initiation (Ozaki et al., 2012), and we wished to examine if the two sub-assemblies made equivalent contributions to setting the time of initiation. Accordingly, cells carrying the double mutations were grown in minimal media supplemented with succinate, treated with rifampicin and cephalexin, and the number of chromosome equivalents, corresponding the number

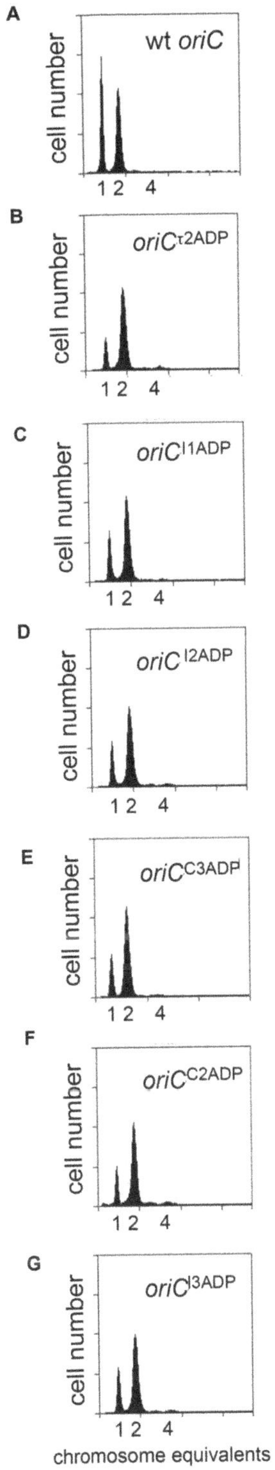

FIGURE 2 | Mutation of individual DnaA-ATP sites into sequences that equivalently bind DnaA-ADP causes a shift in initiation timing in slowly growing *E. coli* cells. **(A–G)** DNA histograms of cells with wild type *oriC* or with *oriC*s carrying the indicated single site mutations, growing in minimal media using succinate as the carbon source were treated with cephalexin and rifampicin, and analyzed by flow cytometry. The number of chromosome equivalents, corresponding to the number of origins at the time of drug treatment, is shown.

of origins at the time of drug treatment, was determined by flow cytometry (**Figure 3**). The effect of mutating two sites in the right sub-assembly ($I3^{ADP}$ and $C3^{ADP}$) did not change timing appreciably from that obtained for cells containing these mutations individually (**Figure 3D** and **Table 1**). In contrast, placing two mutations ($\tau2^{ADP}$ and $I2^{ADP}$) in the left sub-assembly shifted the time of initiation much more than either of the single mutations (**Figure 3B** and **Table 1**). Initiation timing in cells harboring a mutated site in both left and right assemblies appeared to be determined largely by the left half mutation, with the $\tau2^{ADP}/I3^{ADP}$ mutant cells initiating slightly earlier that the $I2^{ADP}/I3^{ADP}$ cells (**Figures 3C,E** and **Table 1**), consistent with the idea that τ2 plays a greater timing role than I2. These results remain consistent with models suggesting that the right and left sub-assemblies are not equivalent (Ozaki et al., 2012).

Individual Low Affinity DnaA-ATP Sites Do Not Play an Initiation Timing Role in Rapidly Growing Cells

Although the studies described above show that mutations causing loss of preference for DnaA-ATP shift initiations to earlier cell cycle times in slow growing cells, Riber et al. previously reported that the same mutations in I2 or I3 had no effect on initiation timing in rapidly growing cells (Riber et al., 2009), suggesting that the timing role of individual DnaA-ATP sites might be dependent on growth rate. To examine this possibility, we evaluated initiation timing of the six single DnaA-ATP site mutations in cells growing in minimal media supplemented with glucose and Casamino acids, with doubling times of ~33 min. Exponentially growing cells were treated with rifampicin and cephalexin, and DNA content was measured by flow cytometry after incubation to allow completion of ongoing rounds of DNA replication. Under these growth conditions, the *E. coli* generation time is less than the time needed to complete chromosome replication, and exponentially growing cells normally contain more than one copy of *oriC*, and all origin copies initiate chromosome replication synchronously, once per cell cycle, on partially duplicated chromosomes (Cooper and Helmstetter, 1968; Skarstad et al., 1986). Accordingly, cultures of MG1655 cells contained 2, 4, or 8 origins (**Figure 4A**), and initiated ~3 min after division. When cells containing origins carrying single site mutations were examined, we found that initiation timing was identical to that observed in cells with wild type *oriC* (**Figures 4B–G** and **Table 2**), in contrast to the shift in timing seen at the slower growth rate. These results confirm previous reports (Riber et al., 2009) and suggest that another factor may contribute to initiation timing regulation in rapidly growing cells.

We also used flow cytometry to examine DNA content in cells containing two site mutations to determine if this might alter timing (**Figure 5**). As was done in the studies of slow growing cells, the double mutants were selected to allow examination of left and right sub-assemblies (e.g., τ2/I2, I3/C3, τ2/I3, and I2/I3). Interestingly, changing both τ2 and I2 in the left half of *oriC* was sufficient to perturb initiation (**Figure 5B**), but the alteration in timing was more complex than the simple shift to earlier

TABLE 1 | Doubling times, number of origins, and cell mass of strains carrying mutated sites growing in minimal media supplemented with succinate.

Site mutated	Doubling time (min)	Number of origins/cell	Relative cell mass[a]	Origins/cell mass	Ai[b]
None	102	1.57	1.0	1.57	0.35 (36 min)
$\tau2^{ADP}$	101	2.00	1.0	2.00	0.13 (13 min)
$I1^{ADP}$	102	1.81	1.0	1.81	0.18 (18 min)
$I2^{ADP}$	103	1.82	1.0	1.82	0.17 (18 min)
$C2^{ADP}$	102	1.81	1.0	1.81	0.18 (18 min)
$C3^{ADP}$	102	1.84	1.0	1.84	0.18 (18 min)
$I3^{ADP}$	103	1.80	1.0	1.80	0.18 (18 min)
$\tau2/I2^{ADP}$	102	2.70	1.0	2.70	0.05 (5 min)
$\tau2/I3^{ADP}$	101	2.05	1.0	2.05	0.11 (11 min)
$I3/C3^{ADP}$	102	1.83	1.0	1.83	0.18 (18 min)
$I2/I3^{ADP}$	102	2.00	1.0	2.00	0.13 (13 min)

[a] *Relative cell mass was measured by flow cytometry, using forward scattered light of mutant strain relative to the wild type strain.*
[b] *Ai (age at initiation), as a fraction of the cell cycle generation time, was calculated using the formula Ai = -ln(1-F/2)/ln2 where F is the fraction of cells in a population that have not initiated chromosome replication, measured from flow cytometric histograms of cells treated with rifampicin and cephalexin (**Figures 2, 3**). The calculated Ai was multiplied by the generation time to give age of initiation in minutes.*

TABLE 2 | Doubling times, number of origins, and cell mass of strains carrying mutated sites growing in minimal media supplemented with glucose and casamino acids.

Site mutated	Doubling time (min)	Number of origins/cell	Relative cell mass[a]	Origins/cell mass	Ai[b]
None	33	4.20	1.0	4.20	0.1 (3 min)
$\tau2^{ADP}$	33	4.23	1.0	4.23	0.1 (3 min)
$I1^{ADP}$	34	4.21	1.0	4.21	0.1 (3 min)
$I2^{ADP}$	33	4.20	1.0	4.20	0.1 (3 min)
$C2^{ADP}$	34	4.19	1.0	4.19	0.1 (3 min)
$C3^{ADP}$	32	4.20	1.0	4.20	0.1 (3 min)
$I3^{ADP}$	33	4.20	1.0	4.20	0.1 (3 min)
$\tau2/I2^{ADP}$	34	4.68	1.0	4.68	–
$\tau2/I3^{ADP}$	34	4.55	1.0	4.55	–
$I3/C3^{ADP}$	33	3.40	1.0	3.40	–
$I2/I3^{ADP}$	33	4.20	1.0	4.20	0.1 (3 min)

[a] *Relative cell mass was measured by flow cytometry, using forward scattered light of mutant strain relative to the wild type strain*
[b] *Ai (age at initiation), as a fraction of the cell cycle generation time, was calculated from cultures with synchronous initiations using the formula Ai = -ln(1-F/2)/ln2 where F is the fraction of cells in a population that have not initiated chromosome replication, measured from flow cytometric histograms of cells treated with rifampicin and cephalexin (**Figures 4, 5**). The calculated Ai was multiplied by the generation time to give age of initiation in minutes.*

times seen in slower growing cells. While most of the cells in the population had 4 origins, odd numbers of origins were also detected, indicating that initiations in these cells were no longer synchronous (**Figure 5B**). The number of origins per cell was also slightly increased (**Table 2**). A similar pattern of asynchronous initiations was seen when τ2 was mutated in the left sub-assembly and I3 was mutated in the right sub-assembly (**Figure 5C** and **Table 2**). However, origins carrying both the $I2^{ADP}$ (left) and $I3^{ADP}$ (right) mutations showed essentially normal initiation timing (**Figure 5E**), as was previously reported (Riber et al., 2009). These results support idea that τ2 plays a greater role in initiation timing than the other sites.

Unlike the slight over-initiation seen in cells with origins carrying $\tau2^{ADP}/I2^{ADP}$ and $\tau2^{ADP}/I3^{ADP}$ mutations, cells with $oriC^{I3ADP/C3ADP}$ contained fewer origin copies per cell, with a larger percentage of cells containing 2 or 3 origins, and a lower number of cells containing 8 origins (**Figure 5D** and **Table 2**). These results reveal a clear difference in the response to changes in the left and right halves of *oriC*. One possible cause of this difference is the role of the right DnaA sub-assembly in displacing Fis (Ryan et al., 2004). While further studies are needed to verify this idea, it is intriguing to note that the pattern of initiation in cells lacking Fis is similar to the $I3^{ADP}/C3^{ADP}$ mutant (**Figure 5F**).

DISCUSSION

An important question about the regulation of DNA replication during the *E. coli* cell cycle is whether low affinity DnaA recognition sites in *oriC* are components of the timing mechanism. There are several possible roles for recognition sites in a clock model based on the availability of DnaA-ATP. In one simple scenario, each low affinity recognition site would require an equivalent amount of time to become occupied. Alternatively, the timer might be based on site occupation, but sites could play a non-equivalent role, i.e., some sites would require more or less time to become occupied than others. A third possibility is that despite a requirement to fill sites with DnaA-ATP during orisome assembly, some other factor acts as the initiation timer by

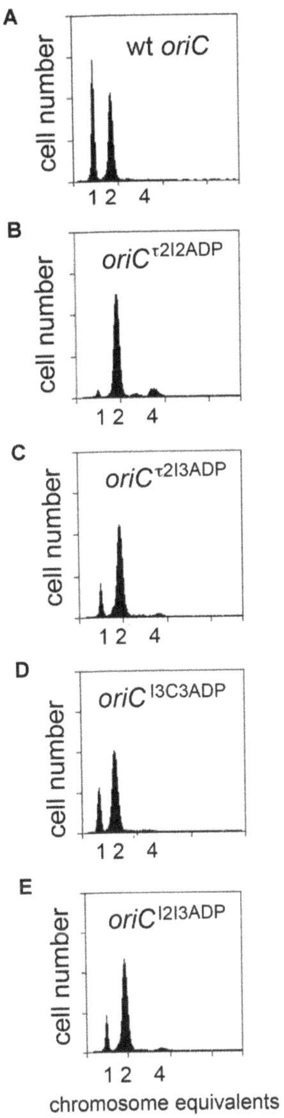

FIGURE 3 | Initiation timing of slowly growing *E. coli* cells carrying two DnaA-ATP site mutations in *oriC*. **(A–E)** DNA histograms of cells with wild type *oriC* or with *oriC*s carrying the indicated two site mutations, growing in minimal media supplemented with succinate, were treated with cephalexin and rifampicin, and analyzed by flow cytometry. The number of chromosome equivalents, corresponding to the number of origins at the time of drug treatment, is shown.

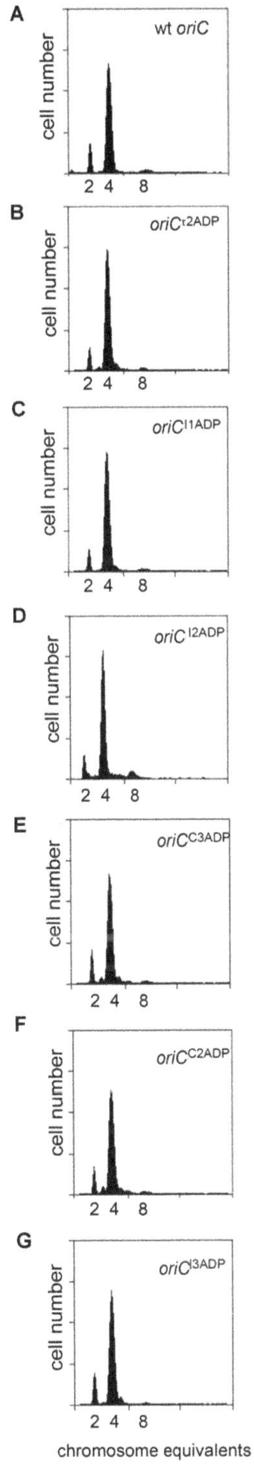

FIGURE 4 | Mutation of individual DnaA-ATP sites into sequences that bind DnaA-ADP equivalently does not change initiation timing in rapidly growing *E. coli* cells. **(A–G)** DNA histograms of cells with wild type *oriC* or with *oriC*s carrying the indicated single site mutations, growing in minimal media supplemented with glucose and casamino acids were treated with cephalexin and rifampicin, and analyzed by flow cytometry. The number of chromosome equivalents, corresponding to the number of origins at the time of drug treatment, is shown.

restricting or facilitating DnaA-ATP access to *oriC*, thus setting the DnaA-ATP level for site occupation and origin activation. Since previous DnaA over-expression studies both support and oppose DnaA's role in initiation timing (Churchward et al., 1983; Atlung et al., 1987; Pierucci et al., 1987; Xu and Bremer, 1988; Løbner-Olesen et al., 1989; Skarstad et al., 1989; Atlung and Hansen, 1993; Flåtten et al., 2015), our approach was to examine minimally disturbed cells with altered chromosomal *oriC* low affinity DnaA recognition sites that bind to DnaA-ATP or DnaA-ADP equivalently. Any differences in DNA replication during

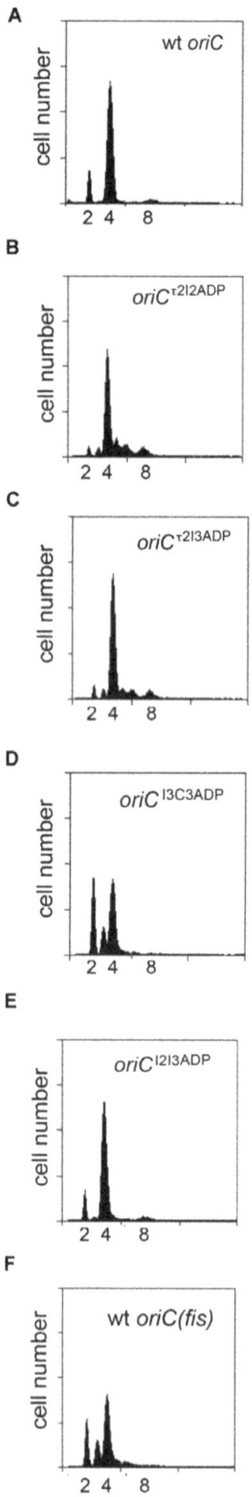

FIGURE 5 | Initiation timing of rapidly growing *E. coli* cells carrying two DnaA-ATP site mutations in *oriC*. **(A–E)** DNA histograms of cells with wild type *oriC* or with *oriC*s carrying the two indicated site mutations **(F)** growing in minimal media supplemented with glucose and casamino acids were treated with cephalexin and rifampicin, and analyzed by flow cytometry. The number of chromosome equivalents, corresponding to the number origins at the time of drug treatment, is shown.

the cell cycle compared to the wild-type version of *oriC* should implicate the altered site as a timing component.

We obtained evidence that loss of DnaA-ATP preference at *oriC* low affinity recognition sites can shift initiation timing during the cell cycle. However, the timing role for recognition sites was more complex than expected, and we identify a dependence on both recognition site location and cellular growth rate. During slow growth in minimal medium with a succinate carbon source, we found that removing DnaA-ATP preference at any of the low affinity recognition sites caused a shift in initiation timing, supporting a site filling-dependent clock mechanism where sites fill when a molecule of DnaA-ATP becomes available to them. But, the unexpectedly larger change in initiation time caused by mutations at τ2 suggests that occupation of this site acts as a bottleneck step during orisome assembly. Although we can only speculate about the mechanism responsible, we note that filling the arrayed low affinity sites in *oriC*'s left half requires IHF-stabilized DNA bending to position the DnaA occupying R1 close enough to R5(M) and τ2 to allow cross-strand DnaA donation (Rozgaja et al., 2011; Ozaki and Katayama, 2012; Kaur et al., 2014; Shimizu et al., 2016; **Figure 6**). It is possible that some structural aspect of this orisome assembly step could impede DnaA-ATP access to τ2. It is also possible that cross-strand interactions between DnaA-ATP bound to τ2 and R1 are required to stabilize an origin configuration that promotes DnaA binding to the unwound DUE (Ozaki and Katayama, 2012; Sakiyama et al., 2017), and achieving this configuration may be more time consuming than lateral cooperative DnaA-ATP binding between arrayed sites.

Any bottleneck in orisome assembly functioning during slow growth would also be expected to function during the synchronous initiations of multiple *oriC* copies during rapid growth. However, none of the single site loss of preference mutations changed initiation timing in fast growing *E. coli*, suggesting that there could be a fast growth-specific mechanism that controls DnaA-ATP availability or its access to *oriC*. One likely component of this mechanism is the growth rate-regulated protein, Fis. We previously reported that binding of purified Fis protein to *oriC* increases the amount of DnaA-ATP required to assemble orisomes *in vitro* (Ryan et al., 2004), with displacement of Fis allowing IHF binding and immediate filling of unoccupied recognition sites by the excess DnaA (Ryan et al., 2004; Leonard and Grimwade, 2011). It is reasonable to expect an analogous situation exists *in vivo*. If the level of DnaA-ATP required to displace Fis from *oriC* over-shoots the amount needed to fill all remaining low affinity DnaA-ATP recognition sites, initiation timing would be determined by how long it takes to accumulate enough DnaA-ATP to displace Fis (see **Figure 6**). In this model, Fis serves as both a negative and positive regulator during orisome assembly, negatively regulating IHF-dependent orisome assembly steps, but positively regulating synchronous initiation by ensuring enough DnaA-ATP accumulates to fire all *oriC* copies (Ryan et al., 2004; Flåtten and Skarstad, 2013). In support of this model, we observed that mutations allowing DnaA-ADP binding at two low affinity recognition sites in the right half of *oriC* caused under-initiation as well as initiation asynchrony (**Figure 5D**). This is the expected result if the mutations resulted in Fis

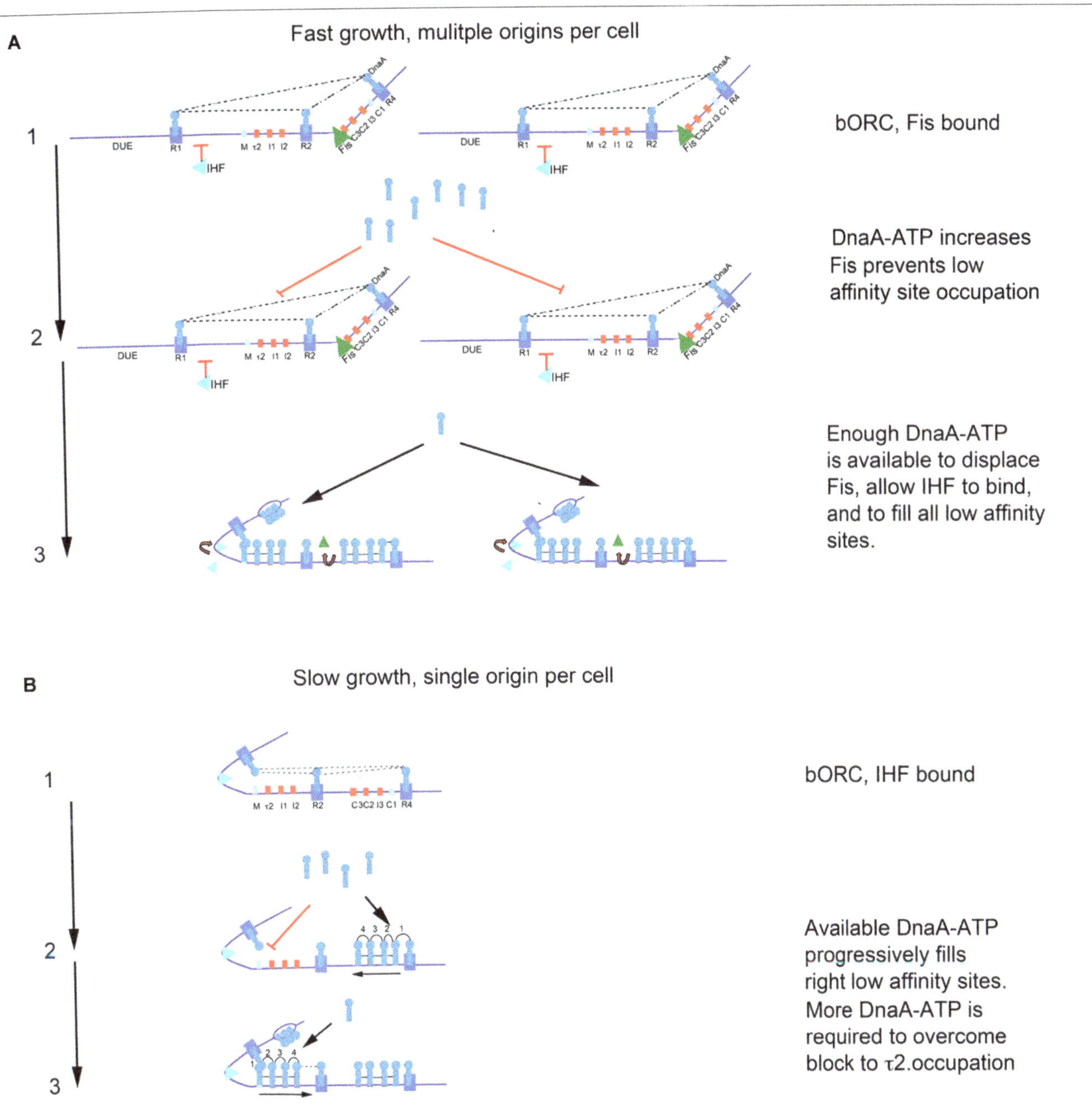

FIGURE 6 | Model of growth-rate dependent regulation of DnaA-ATP site occupation during orisome assembly in *E. coli*. **(A)** In rapidly growing cells, DnaA-ATP site occupation is regulated by Fis. Stage 1 (bacterial Origin Recognition Complex; bORC): Cells contain more than one copy of *oriC*, and DnaA is bound to high affinity R1, R2, and R4 sites throughout most of the cell cycle. Fis is also bound at this stage, and inhibits IHF from occupying its site. Dotted lines indicate putative interactions between R1, R2, and R4. Low affinity sites are unoccupied. Stage 2: Bound Fis prevents accumulating DnaA-ATP from occupying low affinity sites. Stage 3: When enough DnaA-ATP accumulates and becomes available, Fis is displaced. Loss of Fis allows IHF to bind to *oriC*, and also allows all low affinity sites on all *oriC* copies in the cell to become occupied by DnaA. The sharp bend induced by IHF binding allows DnaA bound to R1 to assist loading of R5M and the left array of DnaA-ATP sites. In this configuration, *oriC* DNA is unwound in the DUE, and DnaA-ATP in the form of a compact filament binds to the ssDNA. **(B)** In slowly growing cells, DnaA-ATP site occupation is determined by the availability of DnaA-ATP, and the ability to occupy the τ2 site. Stage 1: Cells contain only one copy of *oriC*, and DnaA is bound to high affinity sites throughout most of the cell cycle. IHF is also bound at this stage. Dotted lines indicate possible interaction between DnaA bound at R1, R2, and R4. Stage 2:As DnaA-ATP become available, it progressively binds right-side low affinity sites, nucleated by DnaA bound to R4. Occupation of the low affinity sites in the left half of *oriC* is delayed by a bottleneck in filling the τ2 site. When enough DnaA-ATP has accumulated to overcome this bottleneck, *oriC* DNA is unwound in the DUE, and DnaA-ATP in the form of a compact filament binds to the ssDNA.

displacement at lower DnaA-ATP levels than would be required for synchronous initiations at all *oriC* copies. Interestingly, there is more than one Fis recognition site in the right half of *oriC* (Filutowicz et al., 1992; Hengen et al., 2003), and it is possible that the amount of accumulated DnaA needed to displace Fis might change over a range of growth rates. There may also be additional mechanisms, such as DnaA acetylation (Zhang et al., 2016; Li et al., 2017), that regulate DnaA access to *oriC* and initiation timing at very slow growth rates, where Fis is absent. Support for this idea comes from reports that *E. coli* cells using acetate as the sole carbon source have higher DnaA/*oriC* ratios than expected (Flåtten et al., 2009). It should also be noted that other DNA binding proteins can contribute to initiation timing by blocking DNA access, changing topology, or changing the availability of DnaA-ATP, reviewed in Riber et al. (2016), and it is possible that some of these could also be growth rate-dependent.

Taken together, our results suggest that *E. coli* may not use a single initiation timing mechanism at all growth rates. Instead, it appears that there are multiple timing mechanisms, each optimized for particular growth conditions, with the ability to shift between them if nutrient availability changes. Timing mechanisms might also vary among bacterial types, depending on their particular lifestyles, and this may be one reason for the large variety in the number and positioning of DnaA recognition sites seen in different bacterial replication origins (Leonard and Méchali, 2013). Although it remains to be determined whether low affinity recognition sites with DnaA-ATP preference are ubiquitous among *oriC*s (Charbon and Lobner-Olesen, 2011), variations in high affinity recognition site number and position may also result in bottlenecks and/or growth rate regulators similar to those that we suggest for *E. coli*.

Finally, it is worth noting that the change in initiation age that we observe for our mutant *oriC*s during slow growth suggests that while there must be free DnaA-ADP available to interact with recognition sites during the majority of the cell cycle, supporting previous findings for rapidly growing cells (Kurokawa et al., 1999; Grimwade et al., 2018), the availability of DnaA-ATP must be stringently restricted even near the time of initiation, such that each low affinity DnaA-ATP site fills only when a DnaA-ATP molecule becomes available. Our ability to remove the preference for DnaA-ATP from low affinity recognition sites in *oriC* and retain orisome activity also demonstrates the functional equivalence of the active and inactive (DnaA-ADP) form of the initiator for most of the steps required for orisome assembly, as long as DnaA-ADP can access the appropriate recognition sites.

AUTHOR CONTRIBUTIONS

PR, JG, and AL planned the experiments. PR, JG, AA, and TR performed the experiments. PR, JG, and AL analyzed data. JG and AL wrote the manuscript.

FUNDING

The work in our laboratories was supported by Public Health Service grant GM54042. Publication of this article was funded in part by the Open Access Subvention Fund and the Florida Tech Libraries.

ACKNOWLEDGMENTS

We thank Charlotte Chen and Jessica Blanco for technical assistance, and Reid Johnson and Elliot Crooke for gifts of anti-Fis and anti-DnaA antibodies, respectively.

REFERENCES

Atlung, T., and Hansen, F. G. (1993). Three distinct chromosome replication states are induced by increasing concentrations of DnaA protein in *Escherichia coli*. *J. Bacteriol.* 175, 6537–6545. doi: 10.1128/jb.175.20.6537-6545.1993

Atlung, T., Løbner-Olesen, A., and Hansen, F. G. (1987). Overproduction of DnaA protein stimulates initiation of chromosome and minichromosome replication in *Escherichia coli*. *Mol. Gen. Genet.* 206, 51–59. doi: 10.1007/BF00326535

Bell, S. P., and Kaguni, J. M. (2013). Helicase loading at chromosomal origins of replication. *Cold Spring Harb. Perspect. Biol.* 5:a010124. doi: 10.1101/cshperspect.a010124

Bramhill, D., and Kornberg, A. (1988). Duplex opening by dnaA protein at novel sequences in initiation of replication at the origin of the *E. coli* chromosome. *Cell* 52, 743–755. doi: 10.1016/0092-8674(88)90412-6

Browning, S. T., Castellanos, M., and Shuler, M. L. (2004). Robust control of initiation of prokaryotic chromosome replication: essential considerations for a minimal cell. *Biotechnol. Bioeng.* 88, 575–584. doi: 10.1002/bit.20223

Charbon, G., and Lobner-Olesen, A. (2011). A role for the weak DnaA binding sites in bacterial replication origins. *Mol. Microbiol.* 82, 272–274. doi: 10.1111/j.1365-2958.2011.07840.x

Chodavarapu, S., and Kaguni, J. M. (2016). Replication initiation in bacteria. *Enzymes* 39, 1–30. doi: 10.1016/bs.enz.2016.03.001

Churchward, G., Holmans, P., and Bremer, H. (1983). Increased expression of the dnaA gene has no effect on DNA replication in a dnaA+ strain of *Escherichia coli*. *Mol. Gen. Genet.* 192, 506–508. doi: 10.1007/BF00392197

Cooper, S., and Helmstetter, C. E. (1968). Chromosome replication and the division cycle of *Escherichia coli* B/r. *J. Mol. Biol.* 31, 519–540. doi: 10.1016/0022-2836(68)90425-7

Court, D. L., Sawitzke, J. A., and Thomason, L. C. (2002). Genetic engineering using homologous recombination. *Annu. Rev. Genet.* 36, 361–388. doi: 10.1146/annurev.genet.36.061102.093104

Filutowicz, M., Ross, W., Wild, J., and Gourse, R. L. (1992). Involvement of Fis protein in replication of the *Escherichia coli* chromosome. *J. Bacteriol.* 174, 398–407. doi: 10.1128/jb.174.2.398-407.1992

Flåtten, I., Fossum-Raunehaug, S., Taipale, R., Martinsen, S., and Skarstad, K. (2015). The DnaA protein is not the limiting factor for initiation of replication in *Escherichia coli*. *PLoS Genet.* 11:e1005276. doi: 10.1371/journal.pgen.1005276

Flåtten, I., Morigen, and Skarstad, K. (2009). DnaA protein interacts with RNA polymerase and partially protects it from the effect of rifampicin. *Mol. Microbiol.* 71, 1018–1030. doi: 10.1111/j.1365-2958.2008.06585.x

Flåtten, I., and Skarstad, K. (2013). The Fis protein has a stimulating role in initiation of replication in *Escherichia coli* in vivo. *PLoS One* 8:e83562. doi: 10.1371/journal.pone.0083562

Grant, M. A., Saggioro, C., Ferrari, U., Bassetti, B., Sclavi, B., and Cosentino Lagomarsino, M. (2011). DnaA and the timing of chromosome replication in *Escherichia coli* as a function of growth rate. *BMC Syst. Biol.* 5:201. doi: 10.1186/1752-0509-5-201

Grimwade, J. E., Rozgaja, T. A., Gupta, R., Dyson, K., Rao, P., and Leonard, A. C. (2018). Origin recognition is the predominant role for DnaA-ATP in initiation of chromosome replication. *Nucleic Acids Res.* 46, 6140–6151. doi: 10.1093/nar/gky457

Grimwade, J. E., Ryan, V. T., and Leonard, A. C. (2000). IHF redistributes bound initiator protein, DnaA, on supercoiled *oriC* of *Escherichia coli*. *Mol. Microbiol.* 35, 835–844. doi: 10.1046/j.1365-2958.2000.01755.x

Grimwade, J. E., Torgue, J. J., McGarry, K. C., Rozgaja, T., Enloe, S. T., and Leonard, A. C. (2007). Mutational analysis reveals *Escherichia coli oriC* interacts with both DnaA-ATP and DnaA-ADP during pre-RC assembly. *Mol. Microbiol.* 66, 428–439. doi: 10.1111/j.1365-2958.2007.05930.x

Hansen, F. G., Atlung, T., Braun, R. E., Wright, A., Hughes, P., and Kohiyama, M. (1991a). Initiator (DnaA) protein concentration as a function of growth rate in *Escherichia coli* and *Salmonella typhimurium*. *J. Bacteriol.* 173, 5194–5199.

Hansen, F. G., Christensen, B. B., and Atlung, T. (1991b). The initiator titration model: computer simulation of chromosome and minichromosome control. *Res. Microbiol.* 142, 161–167.

Helmstetter, C. E., and Krajewski, C. A. (1982). Initiation of chromosome replication in dnaA and dnaC mutants of *Escherichia coli* B/r F. *J. Bacteriol.* 149, 685–693.

Helmstetter, C. E., and Leonard, A. C. (1987). Coordinate initiation of chromosome and minichromosome replication in *Escherichia coli*. *J. Bacteriol.* 169, 3489–3494. doi: 10.1128/jb.169.8.3489-3494.1987

Hengen, P. N., Lyakhov, I. G., Stewart, L. E., and Schneider, T. D. (2003). Molecular flip-flops formed by overlapping Fis sites. *Nucleic Acids Res.* 31, 6663–6673. doi: 10.1093/nar/gkg877

Katayama, T., Kasho, K., and Kawakami, H. (2017) . The DnaA Cycle in *Escherichia coli*: activation, function and inactivation of the initiator protein. *Front. Microbiol.* 8:2496. doi: 10.3389/fmicb.2017.02496

Kaur, G., Vora, M. P., Czerwonka, C. A., Rozgaja, T. A., Grimwade, J. E., and Leonard, A. C. (2014). Building the bacterial orisome: high-affinity DnaA recognition plays a role in setting the conformation of *oriC* DNA. *Mol. Microbiol.* 91, 1148–1163. doi: 10.1111/mmi.12525

Koppes, L. J. (1987). *OriC* plasmids do not affect timing of chromosome replication in *Escherichia coli* K12. *Mol. Gen. Genet.* 209, 188–192. doi: 10.1007/BF00329857

Kowalski, D., and Eddy, M. J. (1989). The DNA unwinding element: a novel, cis-acting component that facilitates opening of the *Escherichia coli* replication origin. *EMBO J.* 8, 4335–4344.

Kurokawa, K., Nishida, S., Emoto, A., Sekimizu, K., and Katayama, T. (1999). Replication cycle-coordinated change of the adenine nucleotide-bound forms of DnaA protein in *Escherichia coli*. *EMBO J.* 18, 6642–6652. doi: 10.1093/emboj/18.23.6642

Langer, U., Richter, S., Roth, A., Weigel, C., and Messer, W. (1996). A comprehensive set of DnaA-box mutations in the replication origin, *oriC*, of *Escherichia coli*. *Mol. Microbiol.* 21, 301–311. doi: 10.1046/j.1365-2958.1996.6481362.x

Leonard, A. C., and Grimwade, J. E. (2011). Regulation of DnaA assembly and activity: taking directions from the genome. *Annu. Rev. Microbiol.* 65, 19–35. doi: 10.1146/annurev-micro-090110-102934

Leonard, A. C., and Grimwade, J. E. (2015). The orisome: structure and function. *Front. Microbiol.* 6:545. doi: 10.3389/fmicb.2015.00545

Leonard, A. C., and Méchali, M. (2013). DNA replication origins. *Cold Spring Harb. Perspect. Biol.* 5:a010116. doi: 10.1101/cshperspect.a010116

Li, S., Zhang, Q., Xu, Z., and Yao, Y. F. (2017). Acetylation of lysine 243 inhibits the *oriC* binding ability of DnaA in *Escherichia coli*. *Front. Microbiol.* 8:699. doi: 10.3389/fmicb.2017.00699

Li, Z., and Crooke, E. (1999). Functional analysis of affinity-purified polyhistidine-tagged DnaA protein. *Protein Expr. Purif.* 17, 41–48. doi: 10.1006/prep.1999.1094

Link, A. J., Phillips, D., and Church, G. M. (1997). Methods for generating precise deletions and insertions in the genome of wild-type *Escherichia coli*: application to open reading frame characterization. *J. Bacteriol.* 179, 6228–6237. doi: 10.1128/jb.179.20.6228-6237.1997

Løbner-Olesen, A. (1999). Distribution of minichromosomes in individual *Escherichia coli* cells: implications for replication control. *EMBO J.* 18, 1712–1721. doi: 10.1093/emboj/18.6.1712

Løbner-Olesen, A., Skarstad, K., Hansen, F. G., von Meyenburg, K., and Boye, E. (1989). The DnaA protein determines the initiation mass of *Escherichia coli* K-12. *Cell* 57, 881–889. doi: 10.1016/0092-8674(89)90802-7

Mahaffy, J. M., and Zyskind, J. W. (1989). A model for the initiation of replication in *Escherichia coli*. *J. Theor. Biol.* 140, 453–477. doi: 10.1016/S0022-5193(89)80109-2

Mallik, P., Paul, B. J., Rutherford, S. T., Gourse, R. L., and Osuna, R. (2006). DksA is required for growth phase-dependent regulation, growth rate-dependent control, and stringent control of fis expression in *Escherichia coli*. *J. Bacteriol.* 188, 5775–5782. doi: 10.1128/JB.00276-06

McGarry, K. C., Ryan, V. T., Grimwade, J. E., and Leonard, A. C. (2004). Two discriminatory binding sites in the *Escherichia coli* replication origin are required for DNA strand opening by initiator DnaA-ATP. *Proc. Natl. Acad. Sci. U.S.A.* 101, 2811–2816. doi: 10.1073/pnas.0400340101

Miller, D. T., Grimwade, J. E., Betteridge, T., Rozgaja, T., Torgue, J. J., and Leonard, A. C. (2009). Bacterial origin recognition complexes direct assembly of higher-order DnaA oligomeric structures. *Proc. Natl. Acad. Sci. U.S.A.* 106, 18479–18484. doi: 10.1073/pnas.0909472106

Nievera, C., Torgue, J. J., Grimwade, J. E., and Leonard, A. C. (2006). SeqA blocking of DnaA-*oriC* interactions ensures staged assembly of the *E. coli* pre-RC. *Mol. Cell* 24, 581–592. doi: 10.1016/j.molcel.2006.09.016

Nilsson, L., Verbeek, H., Vijgenboom, E., van Drunen, C., Vanet, A., and Bosch, L. (1992). FIS-dependent trans activation of stable RNA operons of *Escherichia coli* under various growth conditions. *J. Bacteriol.* 174, 921–929. doi: 10.1128/jb.174.3.921-929.1992

Ozaki, S., and Katayama, T. (2012). Highly organized DnaA-*oriC* complexes recruit the single-stranded DNA for replication initiation. *Nucleic Acids Res.* 40, 1648–1665. doi: 10.1093/nar/gkr832

Ozaki, S., Noguchi, Y., Hayashi, Y., Miyazaki, E., and Katayama, T. (2012). Differentiation of the DnaA-*oriC* subcomplex for DNA unwinding in a replication initiation complex. *J. Biol. Chem.* 287, 37458–37471. doi: 10.1074/jbc.M112.372052

Pierucci, O., Helmstetter, C. E., Rickert, M., Weinberger, M., and Leonard, A. C. (1987). Overexpression of the dnaA gene in *Escherichia coli* B/r: chromosome and minichromosome replication in the presence of rifampin. *J. Bacteriol.* 169, 1871–1877. doi: 10.1128/jb.169.5.1871-1877.1987

Riber, L., Frimodt-Møller, J., Charbon, G., and Løbner-Olesen, A. (2016). Multiple DNA binding proteins contribute to timing of chromosome replication in *E. coli*. *Front. Mol. Biosci.* 3:29. doi: 10.3389/fmolb.2016.00029

Riber, L., Fujimitsu, K., Katayama, T., and Løbner-Olesen, A. (2009). Loss of Hda activity stimulates replication initiation from I-box, but not R4 mutant origins in *Escherichia coli*. *Mol. Microbiol.* 71, 107–122. doi: 10.1111/j.1365-2958.2008.06516.x

Rozgaja, T. A., Grimwade, J. E., Iqbal, M., Czerwonka, C., Vora, M., and Leonard, A. C. (2011). Two oppositely oriented arrays of low-affinity recognition sites in *oriC* guide progressive binding of DnaA during *Escherichia coli* pre-RC assembly. *Mol. Microbiol.* 82, 475–488. doi: 10.1111/j.1365-2958.2011.07827.x

Ryan, V. T., Grimwade, J. E., Camara, J. E., Crooke, E., and Leonard, A. C. (2004). *Escherichia coli* prereplication complex assembly is regulated by dynamic interplay among Fis, IHF and DnaA. *Mol. Microbiol.* 51, 1347–1359. doi: 10.1046/j.1365-2958.2003.03906.x

Sakiyama, Y., Kasho, K., Noguchi, Y., Kawakami, H., and Katayama, T. (2017). Regulatory dynamics in the ternary DnaA complex for initiation of chromosomal replication in *Escherichia coli*. *Nucleic Acids Res.* 45, 12354–12373. doi: 10.1093/nar/gkx914

Samitt, C. E., Hansen, F. G., Miller, J. F., and Schaechter, M. (1989). In vivo studies of DnaA binding to the origin of replication of *Escherichia coli*. *EMBO J.* 8, 989–993.

Schaper, S., and Messer, W. (1995). Interaction of the initiator protein DnaA of *Escherichia coli* with its DNA target. *J. Biol. Chem.* 270, 17622–17626. doi: 10.1074/jbc.270.29.17622

Schenk, K., Hervás, A. B., Rösch, T. C., Eisemann, M., Schmitt, B. A., Dahlke, S., et al. (2017). Rapid turnover of DnaA at replication origin regions contributes to initiation control of DNA replication. *PLoS Genet.* 13:e1006561. doi: 10.1371/journal.pgen.1006561

Sekimizu, K., Bramhill, D., and Kornberg, A. (1987). ATP activates dnaA protein in initiating replication of plasmids bearing the origin of the *E. coli* chromosome. *Cell* 50, 259–265. doi: 10.1016/0092-8674(87)90221-2

Shimizu, M., Noguchi, Y., Sakiyama, Y., Kawakami, H., Katayama, T., and Takada, S. (2016). Near-atomic structural model for bacterial DNA replication initiation complex and its functional insights. *Proc. Natl. Acad. Sci. U.S.A.* 113, E8021–E8030. doi: 10.1073/pnas.1609649113

Simmons, L. A., Felczak, M., and Kaguni, J. M. (2003). DnaA protein of *Escherichia coli*: oligomerization at the *E. coli* chromosomal origin is required for initiation and involves specific N-terminal amino acids. *Mol. Microbiol.* 49, 849–858. doi: 10.1046/j.1365-2958.2003.03603.x

Skarstad, K., Boye, E., and Steen, H. B. (1986). Timing of initiation of chromosome replication in individual *Escherichia coli* cells. *EMBO J.* 5, 1711–1717.

Skarstad, K., and Katayama, T. (2013). Regulating DNA replication in bacteria. *Cold Spring Harb. Perspect. Biol.* 5:a012922. doi: 10.1101/cshperspect.a012922

Skarstad, K., Løbner-Olesen, A., Atlung, T., von Meyenburg, K., and Boye, E. (1989). Initiation of DNA replication in *Escherichia coli* after overproduction of the DnaA protein. *Mol. Gen. Genet.* 218, 50–56. doi: 10.1007/BF00330564

Weigel, C., Messer, W., Preiss, S., Welzeck, M., and Morigenand Boye, E. (2001). The sequence requirements for a functional *Escherichia coli* replication origin are different for the chromosome and a minichromosome. *Mol. Microbiol.* 40, 498–507. doi: 10.1046/j.1365-2958.2001.02409.x

Wolański, M., Donczew, R., Zawilak-Pawlik, A., and Zakrzewska-Czerwińska, J. (2014). *oriC*-encoded instructions for the initiation of bacterial chromosome replication. *Front. Microbiol.* 5:735. doi: 10.3389/fmicb.2014.00735

Xu, Y. C., and Bremer, H. (1988). Chromosome replication in *Escherichia coli* induced by oversupply of DnaA. *Mol. Gen. Genet.* 211, 138–142. doi: 10.1007/BF00338404

Yung, B. Y., Crooke, E., and Kornberg, A. (1990). Fate of the DnaA initiator protein in replication at the origin of the *Escherichia coli* chromosome in vitro. *J. Biol. Chem.* 265, 1282–1285.

Zhang, Q., and Shi, H. (2012). Coupling chromosomal replication to cell growth by the initiator protein DnaA in *Escherichia coli*. *J. Theor. Biol.* 314, 164–172. doi: 10.1016/j.jtbi.2012.08.045

Zhang, Q., Zhou, A., Li, S., Ni, J., Tao, J., Lu, J., et al. (2016). Reversible lysine acetylation is involved in DNA replication initiation by regulating activities of initiator DnaA in *Escherichia coli*. *Sci. Rep.* 6:30837. doi: 10.1038/srep30837

17

The membrane: Transertion as an organizing principle in membrane heterogeneity

Kouji Matsumoto[1], Hiroshi Hara[1], Itzhak Fishov[2], Eugenia Mileykovskaya[3] and Vic Norris[4]*

[1] Department of Biochemistry and Molecular Biology, Graduate School of Science and Engineering, Saitama University, Saitama, Japan, [2] Department of Life Sciences, Ben-Gurion University of the Negev, Beer-Sheva, Israel, [3] Department of Biochemistry and Molecular Biology, University of Texas Medical School at Houston, Houston, TX, USA, [4] Laboratory of Microbiology Signals and Microenvironment EA 4312, Department of Science, University of Rouen, Mont-Saint-Aignan, France

Correspondence:
Kouji Matsumoto,
Department of Biochemistry and Molecular Biology, Graduate School of Science and Engineering, Saitama University,
255 Shimo-ohkubo, Sakura-ku, Saitama-shi, Saitama 338-8570, Japan
koumatsu@mail.saitama-u.ac.jp

The bacterial membrane exhibits a significantly heterogeneous distribution of lipids and proteins. This heterogeneity results mainly from lipid–lipid, protein–protein, and lipid–protein associations which are orchestrated by the coupled transcription, translation and insertion of nascent proteins into and through membrane (transertion). Transertion is central not only to the individual assembly and disassembly of large physically linked groups of macromolecules (*alias* hyperstructures) but also to the interactions between these hyperstructures. We review here these interactions in the context of the processes in *Bacillus subtilis* and *Escherichia coli* of nutrient sensing, membrane synthesis, cytoskeletal dynamics, DNA replication, chromosome segregation, and cell division.

Keywords: membrane, transertion, hyperstructures, lipid domain, *Escherichia coli, Bacillus subtilis*

Introduction

The laterally heterogeneous distribution of the lipid and the protein components of bacterial cell membranes is the current paradigm, having replaced the homogeneous distribution of these components that is assumed in the fluid mosaic membrane model. This heterogeneity is involved in producing the specific environments needed for membrane proteins to participate in the many important processes that are associated with cell membranes. This raises the question of the nature of the mechanisms responsible for membrane heterogeneity.

One of the main causes of large-scale heterogeneity in bacterial membranes is the coupled transcription, translation and insertion of nascent proteins into and through membrane, *alias* transertion (Norris and Madsen, 1995; Binenbaum et al., 1999; Bakshi et al., 2014; **Figure 1**). The tethering of nascent proteins orchestrates interactions that result in the formation of membrane domains. These interactions include protein–protein interactions and lipid–protein interactions – and therefore also include lipid–lipid interactions (Vereb et al., 2003). The multitude of processes in which transertion is implicated has led to it being proposed as a powerful force in maintaining the nucleoid in expanded state, in the integrative sensing of metabolic activity and in coupling chromosome segregation to cell division (Norris and Amar, 2012; Fishov and Norris, 2012b).

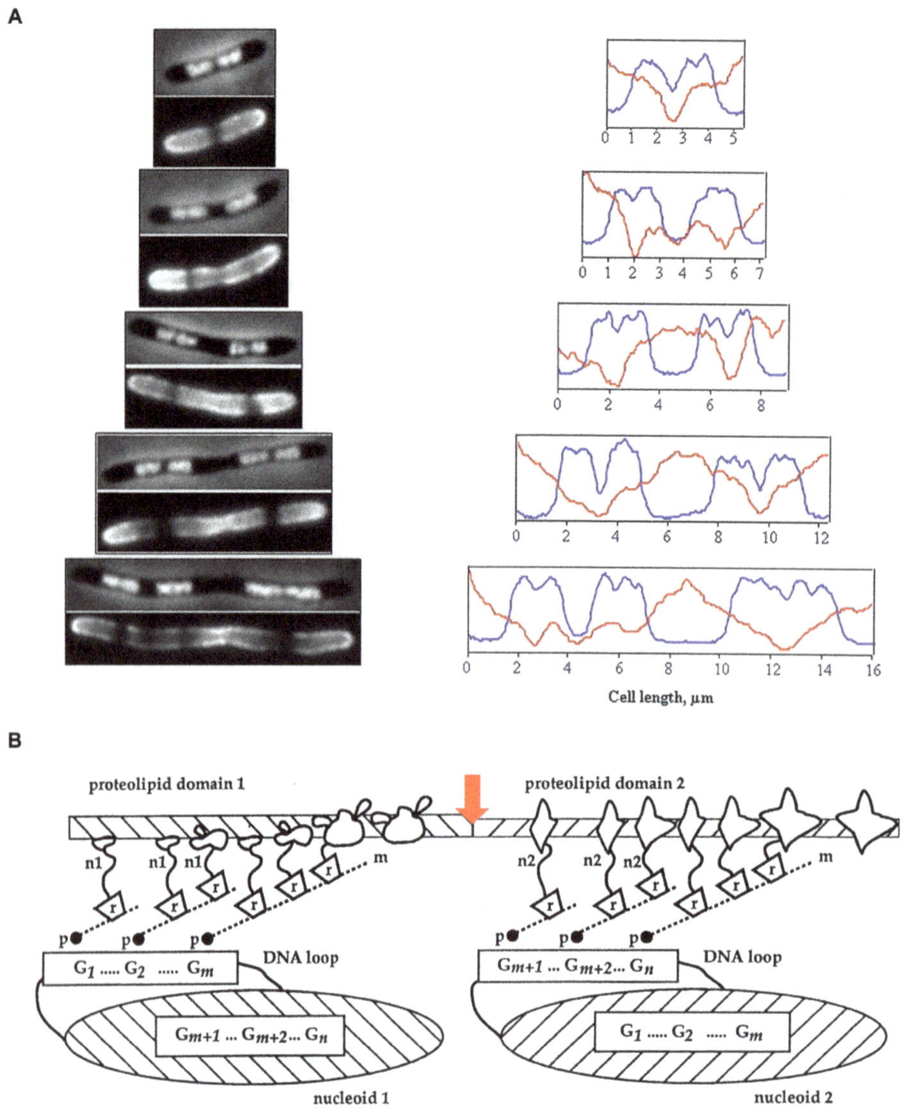

FIGURE 1 | (A) Visualization of membrane heterogeneity in *Escherichia coli* cells by fluorescent microscopy. Left: image of an *E. coli* cell division mutant *pbpB*(ts) transferred to the non-permissive temperature (time zero, bacterium at the top), double-stained with DAPI (top half of image) and FM4-64 (bottom half of image) for visualization of nucleoids and membrane, respectively; adapted from Fishov and Woldringh (1999), © John Wiley and Sons. On the right, plots of fluorescence intensity profiles along filaments (DAPI, blue line; FM4-64, red). **(B)** The relationship between proteolipid domains, transertion, and clusters of genes either "looped out" from or buried within the nucleoid. n1, n2, nascent proteins; m, mRNA; r, ribosomes; p, RNA polymerase; G, gene clusters; red arrow indicates division site; adapted from Norris and Madsen (1995), © Elsevier.

It has long been evident that bacteria are highly structured (for references see Norris et al., 1994). It is now emerging that this structuring is in the form of *hyperstructures* which are large, physically linked, assemblies of macromolecules that serve specific functions (Norris et al., 2007a). Transertion plays a central role in the formation of one class of hyperstructures and also in the dialog between these hyperstructures that determines the behavior of the cell itself. Here, we review membrane heterogeneity focussing on transertion hyperstructures and their relationships with the processes of nutrient sensing, membrane synthesis, cytoskeletal dynamics, DNA replication, chromosome segregation, and cell division.

Lipid Domains

Cardiolipin and Other Anionic Phospholipid Domains

Although the lipids present in the inner membrane of *Escherichia coli* will spontaneously separate into domains *in vitro* (Zerrouk et al., 2008), until fairly recently, the lipid molecules in the membranes of bacterial cells were assumed to be homogeneously distributed, since the fluidity of biological membranes had become generally accepted following the publication of the fluid mosaic model (Singer and Nicolson, 1972). However, cell membranes must be laterally polarized to produce specific environments for certain membrane proteins, in particular, chemoreceptor proteins and proteins that promote polymerization of actin in eukaryotic hosts, as well as proteins involved in cell division at mid-cell and at asymmetric positions (Shapiro et al., 2002). Microscopic visualization of membrane lipids in cells has reinforced the view that bacterial membranes do possess structural heterogeneity. Uneven distribution of fluorescent lipophilic dyes and selective staining of septal regions has been observed in mycobacteria and an uneven distribution of fluorescence has been observed in lipophilic dye (FM4-64)-stained *E. coli* cells (Christensen et al., 1999; Fishov and Woldringh, 1999; **Figure 1**).

Unequivocal visualization of cardiolipin (CL) domains in the septal region and in the poles in the membranes of *E. coli* cells has been accomplished by means of the CL-specific fluorescent dye 10-*N*-nonyl-3,6-bis(dimethylamino)acridine (10-*N*-nonyl-acridine orange: NAO; Mileykovskaya and Dowhan, 2000, 2009; Mileykovskaya, 2007; **Figures 2A–C**). Time-lapse microscopy of *E. coli* cells showed positioning of NAO stained domains at nascent division sites and their gradual development into septal domains (Mileykovskaya and Dowhan, 2009; Mileykovskaya and Margolin, 2012; **Figure 2D**). NAO only binds to anionic phospholipids owing to an interaction between its quaternary amine and the phosphate residue of phospholipids and an intercalation of the hydrophobic acridine moiety into the membrane bilayer (Petit et al., 1992). The stoichiometry between NAO and monoacidic phospholipids is 1:1 and the emitted fluorescence is green whilst with CL, which contains two phosphate groups per molecule, NAO forms a dimer and the emitted fluorescence of the CL complex shifts to red due to the metachromatic effect of acridine molecules (Petit et al., 1994; Mileykovskaya et al., 2001; **Figure 2G**). CL domains have also been observed in *Bacillus subtilis* cells, both during exponential growth in the septal region and the poles and during sporulation in the engulfment and the forespore membranes (Kawai et al., 2004, 2006; **Figures 2E,F**). Note that in *B. subtilis* cells PE is also localized in polar and septal membranes (Nishibori et al., 2005). The CL localization to the polar membranes in *E. coli* is consistent with the increase in the content of CL in the membranes of minicells, which are rich in polar membranes (Koppelman et al., 2001).

An *E. coli* Δ*pgsA* mutant lacks phosphatidylglycerophosphate synthase, which catalyzes the committed step of biosynthesis of the major acidic phospholipids, and thus lacks phosphatidylglycerol (PG) and CL (both $< 0.01\%$ of total phospholipids); this mutant, which is viable if it also has an *lpp* mutation, accumulates the anionic biosynthetic precursors, phosphatidic acid (PA) and CDP-diacylglycerol (4.0 and 3.2%, respectively; Kikuchi et al., 2000; Shiba et al., 2004). PG is required for modification of prolipoprotein (the precursor of Braun's lipoprotein, the product of *lpp*); cells lacking PG in an lpp^+ background are not viable because the accumulation of unmodified prolipoprotein in the inner membrane causes a tight membrane fusion (Suzuki et al., 2002). The viability of a Δ*pgsA* lpp^- mutant suggests that the anionic phospholipids can substitute for one another in essential biological functions that include (1) the initiation of DNA replication, which depends on the rejuvenation of the DnaA protein by CL, and (2) the selection of the division site, which depends on the inhibition of inappropriate FtsZ polymerization in the polar regions by a complex, MinCD, with a high affinity for anionic phospholipids (Matsumoto, 2001; Szeto et al., 2002; Mileykovskaya et al., 2003; Norris et al., 2004; Mileykovskaya and Dowhan, 2005; Mazor et al., 2008; Vecchiarelli et al., 2014). In the Δ*pgsA* mutant cells, *N*-acyl-phosphatidylethanolamine (*N*-acyl-PE) and PA have been localized to polar and septal membrane domains (Mileykovskaya et al., 2009), indicating that these normally minor anionic phospholipids are segregated, like CL, into similar anionic lipid domains. Thus, *E. coli* has a mechanism for preferential segregation of anionic phospholipids to the polar and septal regions. One possible segregation mechanism depends on the shape of lipid molecules. 'Cone-shaped' lipid molecules, in which the cross-sectional area of the polar head group is less than the cross-sectional area of the hydrophobic domain, prefer regions of membranes that have a negative curvature; hence CL and PA are concentrated in negatively curved regions of the inner leaflet of the bacterial membranes (Cullis et al., 1983; Gennis, 1989; Seddon, 1990; Renner and Weibel, 2011; Jouhet, 2013; Renner et al., 2013). Using a mathematical model, this sensing of membrane curvature – and consequent positioning in the poles – has been attributed to stable, conical clusters of CL molecules forming due to the membrane being pinned to the cell wall, which is itself the result of the balance between the osmotic-pressure difference across the membrane and the inward pressure of the cell wall (Huang et al., 2006; Huang and Ramamurthi, 2010; **Figure 2F**).

A recent report from Weibel's lab has claimed that there is no clear preference of NAO for binding to CL compared to PG and other anionic phospholipids, and that NAO produces an intense red-shifted fluorescence emission with PG, PA, and phosphatidylserine (PS) that is comparable to that with CL (Oliver et al., 2014). They suggested that not only CL but also PG is concentrated in the polar regions of *E. coli* cell membranes. However, they used much higher concentrations of NAO than previous authors used for bacterial staining (Mileykovskaya and Dowhan, 2000; Kawai et al., 2004; Romantsov et al., 2007, 2008; Rosch et al., 2007; Lobasso et al., 2009, 2013; Maloney et al., 2011; Muchová et al., 2011; Kicia et al., 2012; Tran et al., 2013). Moreover, PG microdomains, in contrast to CL domains, cannot sense membrane curvature, since the cross-sectional area of PG molecule is cylindrical and lamellar structures are preferred (Cullis et al., 1983; Gennis, 1989; Seddon, 1990). As Weibel's lab had already demonstrated that the location of NAO-stained domains depends directly on membrane curvature (Renner and Weibel, 2011; Renner et al., 2013), this group did not propose a specific physical mechanism responsible for localizing anionic phospholipid domains stained by NAO (Oliver et al., 2014). In *B. subtilis*, phospholipid-specific dyes including FM4-64 are localized in a helical pattern extending along the long axis (Barák et al., 2008). Helical FM4-64 domains are missing in cells depleted of MurG, an enzyme involved in peptidoglycan synthesis, indicating a link between the helical

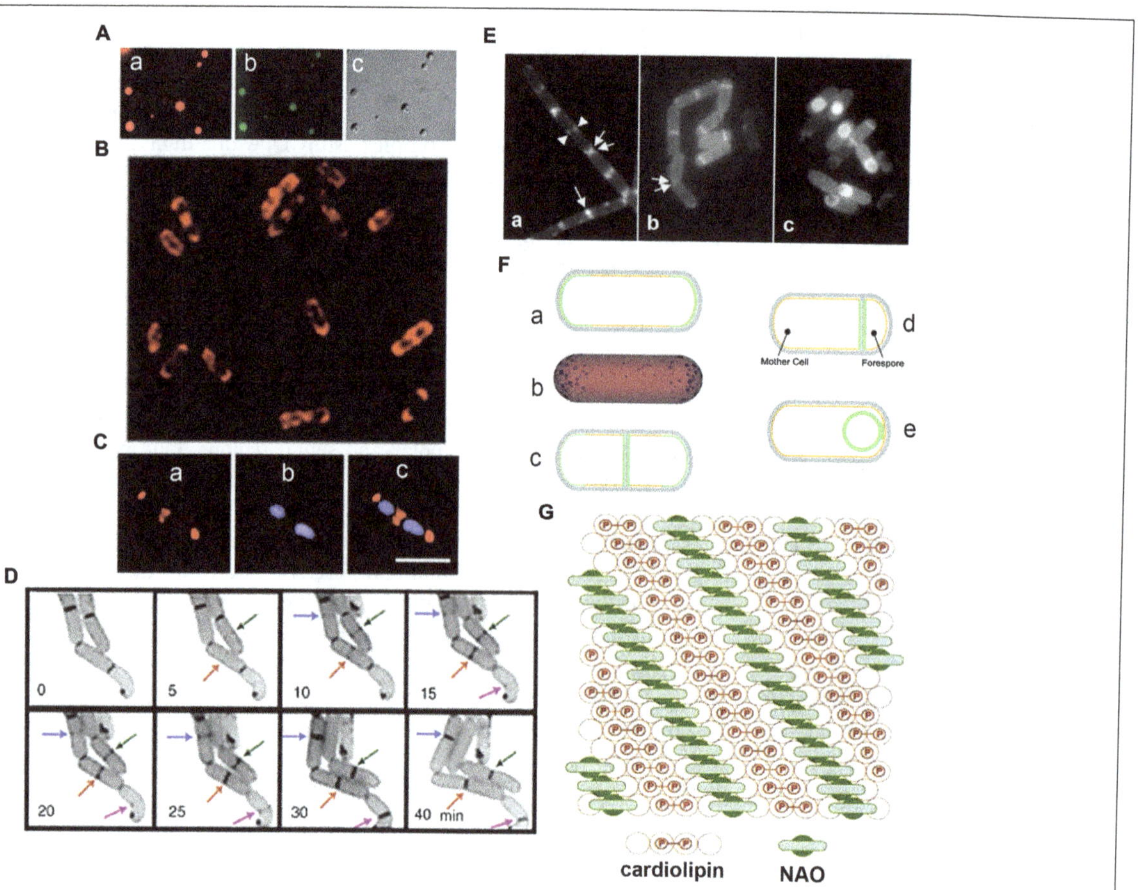

FIGURE 2 | Anionic phospholipid domains in *E. coli* and *B. subtilis* cells revealed by NAO staining. (A) Fluorescence microscopy of cardiolipin (CL)-containing liposomes stained with NAO. Liposomes composed of CL and phosphatidylcholine (60 mol%/40 mol%). Excitation: 490 nm, emission: 617 nm (a) and 528 (b), DIC (c); modified from Mileykovskaya (2007), © John Wiley and Sons. **(B)** Deconvolved images of an optical section of *E. coli* W3899 stained with NAO. Cells were grown in LB media in the presence of 200 nM NAO to OD_{600} 1.5, and then immobilized on a microscope slide cover glass with poly-L-lysine; modified from Mileykovskaya and Dowhan (2009), © Elsevier. Excitation is 490 nm and emission is 617 nm. **(C)** *E. coli* cells stained with NAO (red) for CL (a) and DAPI (blue) for the nucleoid (b). Overlay of images (c) demonstrates localization of CL at the cell poles and at the division site; scale bar, 2.5 μM; modified from Mileykovskaya and Dowhan (2000), © American Society for Microbiology. **(D)** The dynamics of CL domain formation during the cell cycle of growing *E. coli*. Arrows of the same color indicate the progression of the corresponding NAO-stained structures with time; adapted from Mileykovskaya and Dowhan (2009). **(E)** Staining of wild type *B. subtilis* cells with NAO. Wild type *B. subtilis* 168 cells were harvested in exponential growth and sporulation phase (at T2 and T4) and stained with 100 nM NAO for 20 min. Fluorescence images of exponential growth (a) and sporulation phase cells at T2 (b) and T4 (c). An arrow indicates a sharp fluorescence band in the center of the cell. Two fluorescence dot structures in the cell center are indicated by a pair of arrowheads. Regions of NAO stained nascent poles in the cells that are separating are indicated with a pair of arrows. Panels are adapted from Kawai et al. (2004), © American Society for Microbiology. **(F)** Osmotic pressure and turgor-induced localization of lipid clusters during cell division and sporulation. During exponential growth, high-intrinsic-curvature clusters of CL localize to the poles of the inner leaflet [green curve (a) or blue clusters in (b)], driven by differences in membrane curvature. A lower osmotic-pressure differential across the septal/forespore-engulfing membrane [green area, (c)/(d), respectively] induces relocalization of the lipid clusters to the septal membrane; (d) CL clusters migrate along the continuous leaflet consisting of the inner leaflet of the mother cell and the outer leaflet of the forespore-engulfing membrane to localize around the spore due to low osmotic-pressure differential (green circle); modified from Huang et al. (2006) and Huang and Ramamurthi (2010), © John Wiley and Sons with permission. **(G)** The proposed arrangement of CL in the presence of NAO. A top view of the bilayer in which the hexagonal array of large circles represents the fatty acid chains is shown. The small internal circles containing P represent the phosphate groups, hydrogen-bonded tightly by the hydroxyl of the connecting glycerol, above the two central circles of the four fatty acid chains of CL (red). This tight array provides room for the NAO molecules (green) to stack in between the rows of CL head groups; adapted from Mileykovskaya et al. (2001), © Elsevier.

domains and peptidoglycan synthesis (Muchová et al., 2011). As the helical structures are absent from the cells with repressed *pgsA* expression (in which the levels of PG and CL are much reduced; Hashimoto et al., 2009), the dyes are considered to be associated with the anionic phospholipids (Barák et al., 2008). As the CL-specific NAO dye is located in the septal regions and the poles of *B. subtilis* cells (Kawai et al., 2004), the principal component of the FM4-64 helical structures is likely to be PG (Barák et al., 2008). It has, however, been suggested that experiments showing changes in FM 4-64 staining of *B. subtilis* cells with repressed *pgsA* expression cannot serve as unambiguous evidence of FM 4-64 specificity for PG since, first, changes in the distribution of FM 4-64 resulted from the dissipation of membrane potential by the uncoupler CCCP and, second, the membrane potential was much lower in *B. subtilis* cells with repressed *pgsA* expression (Strahl et al., 2014). In *E. coli* no discernible helical structures were observed with FM4-64 staining (Barák et al., 2008) as reported previously (Fishov and Woldringh, 1999).

Microdomains that are functionally similar to the lipid rafts of eukaryotic cells have recently been found in *B. subtilis* membranes (Lopez and Kolter, 2010). These microdomains contain homologues of eukaryotic flotillins, YuaG and YqfA, referred to as FloT and FloA, respectively (see Heterogeneous Distribution of Envelope and Envelope-Associated Proteins). The distribution of these microdomains was in a punctuated pattern along the cytoplasmic membrane (Donovan and Bramkamp, 2009; Lopez and Kolter, 2010), which differs from the patterns of FM4-64-stained helices and of CL-domains in the septal region and the poles. YisP is involved in biofilm formation in *B. subtilis* and has been predicted to produce C30 isoprenoids; the enzyme acts as a phosphatase, catalyzing formation of farnesol from farnesyl diphosphate (Feng et al., 2014). Inhibition of formation of these microdomains using zaragozic acid impaired biofilm formation and protein secretion but not cell viability and it has been suggested that polyisoprenoids are constituents of the microdomains (Lopez and Kolter, 2010; Feng et al., 2014; Bramkamp and Lopez, 2015).

Lipid Synthases and Membrane Domains

The septal and polar locations of specific phospholipids may be related to the subcellular location of the enzymes involved in their synthesis. In *E. coli*, a relationship between the location of CL (Mileykovskaya and Dowhan, 2009; Mileykovskaya and Margolin, 2012; see Cardiolipin and Other Anionic Phospholipid Domains) and that of CL synthase (ClsA) and phosphatidylglycerolphosphate synthase (PgsA; Tan et al., 2012; Dowhan, 2013) has yet to be reported. In *B. subtilis*, green fluorescent protein (GFP) fusions to ClsA and PssA, a PS synthase that catalyzes the committed step of biosynthesis of PE, are located in the septal membranes even when the corresponding genes are expressed at a low level from their natural promoters (Nishibori et al., 2005). Thus, ClsA is probably concentrated in the septal membranes under natural conditions thereby playing an important role in the septal localization of CL in *B. subtilis* (**Figure 3**). Localizing the enzymes involved in the synthesis of other lipids has yielded unexpected, interesting results. GFP fusions to several phospholipid synthases were localized to the septum in a thick, bright fluorescence band. These synthases include PgsA, PssA, Psd, which converts PS into PE, CdsA, which produces CDP-diacylglycerol, MprF, which transfers lysine to PG to produce lysylPG, and UgtP, which is responsible for glucolipid synthesis. The dot-pair distribution of UgtP corresponds to its role in regulating FtsZ assembly and hence differs from the distribution of the phospholipid synthases

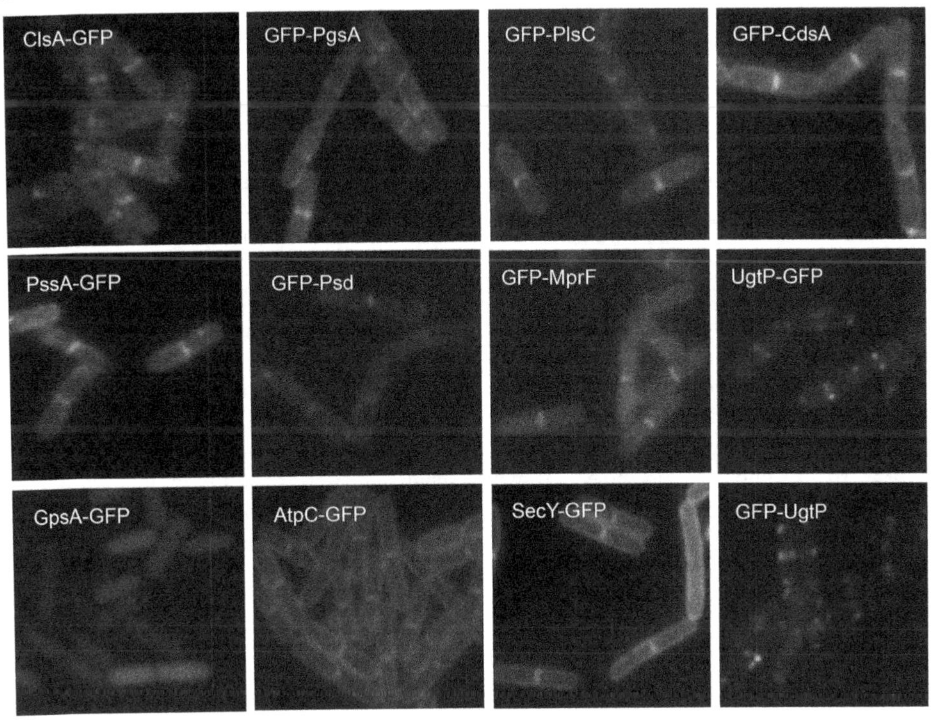

FIGURE 3 | Septal localization of lipid synthases in *B. subtilis* cells. Typical florescence images from cells harboring *gfp* fusions in the *amyE* locus cultivated in sporulation medium (DSM) up to late logarithmic growth phase are shown. Green fluorescence from the GFP-fusions was detected by using a standard GFP(R)-BP filter unit. For the name of the enzymes and their catalytic activities refer to the figure of the biosynthetic pathway (**Figure 5B**). Panels are from Figures 4 and 5 of Nishibori et al. (2005), © American Society for Microbiology. GFP-fusions of ClsA and PssA are located in the septal membranes even when the corresponding genes are expressed from their natural promoters (see Figure 4 of Nishibori et al., 2005).

(see Sensing). The locations of these enzymes thus differ from the cytoplasmic location of GpsA, which catalyzes the production of G3P (Nishibori et al., 2005; **Figure 3**), and the uniformly distributed locations of the membrane proteins AtpC, a subunit of ATP synthase, and SecY (Matsumoto et al., 2012). It has also been shown that the enzymes for lipoteichoic acid (LTA) synthesis, LtaS and YfnI, are septally localized (Schirner et al., 2009; Matsuoka et al., 2011b), but DG kinase (DgkB), which converts DG that is produced in LTA synthesis into PA, is not localized (Matsuoka et al., 2011b).

How are the lipid synthases targeted to the septal membranes? These enzymes presumably have specific regions responsible either directly for their localization or indirectly via interaction with certain cell division proteins. Selective inactivation of putative targeting regions has shown that ClsA has regions of amphipathic α-helices at its COOH-terminus that are responsible for its septal localization (Kusaka et al., manuscript in preparation). PgsA has no such amphipathic α-helices at its COOH-terminus, suggesting that the system for its septal localization is different from that of ClsA. Many other proteins with amphipathic α-helices, including MinD (see Lipid–Protein Interactions), have membrane binding properties that depend on electrostatic interactions, *via* their net positive charges, and on hydrophobicity. In addition, the catalytic domains (or consensus sequences) of *B. subtilis* PssA (Saha et al., 1996b) and *E. coli* and *B. subtilis* PgsA (Usui et al., 1994; Matsumoto, 1997) have an amphipathic α-helix structure which allows them to access their hydrophilic substrates, serine and G3P, in the cytoplasm (Matsumoto et al., 2012). The septal location of the phospholipid synthases in *B. subtilis* cells implies that most phospholipids are synthesized at the septal membranes, and, given the septal location of PE and CL, it is likely that these newly synthesized lipids are somehow prevented from diffusing into the lateral membranes (Matsumoto et al., 2006).

Chemical Basis for Generation of Lipid Domains

What then is the chemical basis for the generation of lipid domains? How do the lipid molecules form domains in membranes that are fluid? Both lipid–lipid interactions and lipid–membrane protein interactions are suspected to induce the formation of microdomains, which comprise at most some tens of molecules of a specific lipid (for a review, see Edidin, 1997). The following properties of phospholipid molecules may account for the formation of microdomains through lipid–lipid interactions.

Cardiolipin, which is a diphosphatidylglycerol, has a unique head group, with a tightly locked, surprisingly small configuration and only one negative charge (Haines and Dencher, 2002); this result contradicts the previous view reported in textbooks. The two phosphates in the head group trap a proton and are locked in a bicyclic array held together by the hydroxyl residue of the glycerol which connects the two halves. This small polar head group makes for a tighter packing of hydrophobic acyl chains between CL molecules by van der Waals force interactions than is found in lipids with larger polar head groups. It has also been suggested (Haines and Dencher, 2002) that interaction between adjacent head-groups creates a compact array of CL molecules, which becomes manifest in the presence of associated NAO arrays (Mileykovskaya et al., 2001).

Recently, a microtechnology for manipulating bacterial membrane curvature and quantitatively measuring its effect on the localization of CL in spheroplasts has been developed (Renner and Weibel, 2011). CL domains were localized to regions of high intrinsic negative curvature and MinD was found to be associated to the region of large negative curvature (see below). This localization of CL to membranes of highly negative curvature is ascribable to its cone-like molecular configuration (Gennis, 1989; Jouhet, 2013; see Cardiolipin and Other Anionic Phospholipid Domains). Polar localization of PA domains found in *E. coli*Δ*pgsA* cells lacking both CL and PG can also be ascribable to the cone-like configuration of PA molecule (Norris et al., 2002; Mileykovskaya et al., 2009).

Polar head groups of PG molecules interact by an extensive network of hydrogen bonds, ionic bonds, and coordination bonds between glycerol hydroxyls and the unesterified phosphate oxygen both in the anhydrous crystal and the hydrated gel state (Boggs, 1987; Pascher et al., 1987; Seddon, 1990). In fact, PG is probably segregated into distinct domains, in both *B. subtilis* and *E. coli* membranes, according to studies using pyrene-labeled phospholipids (Vanounou et al., 2003). This extensive and tight network of PG molecules may cause exclusion of CL molecules to produce patches of CL in the membranes (Mileykovskaya et al., 2001), since the small and tightly locked head group of CL cannot interact with PG.

The head group of the PE molecule has both a cationic amine residue and an anionic phosphate residue. Each amine and unesterified phosphate oxygen can participate in short distance intermolecular hydrogen bonds. The ethanolamine groups thus form a linkage between phosphorus groups of adjacent PE molecules and the tendency to be hydrated must be less than that of phosphatidylcholine, so producing a very compact, rigid head-group network of PE molecules (Elder et al., 1977; Hauser et al., 1981; Boggs, 1987; Seddon, 1990), and giving PE substantially higher T_m values than phosphatidylcholine which has an identical acyl chain structure. This compact head-group network of PE may well suffice to explain the formation of PE microdomains. Interactions with certain membrane proteins might then modify the lipid organization and further stabilize the head-group network, as suggested (Edidin, 1997). Polar and septal localization of PE in *B. subtilis* cells (Nishibori et al., 2005) is likely ascribable to its cone-like molecular shape (see Cardiolipin and Other Anionic Phospholipid Domains). For details refer to the previous review (Matsumoto et al., 2006).

The potential role for polyamines in lipid domain formation is largely unexplored although it has been suggested that polyamine binding to the acidic groups of phospholipids in membranes could lead to clusters forming (Schuber, 1989). Ionic interactions between polyamines and acidic phospholipids are strongest between the polyamine with the highest positive charges and the phospholipid with the highest content of negative groups (Yung and Green, 1986). Interaction with polyamines is believed to reduce the repulsive forces between negatively charged membrane components by bridging proteins and lipids and by shielding the surface charges (Schuber, 1989). The binding of ions such as calcium to anionic phospholipids can also result in domain formation (Haverstick and Glaser, 1987) and for references see Cannon et al. (2003), but, again, this is relatively unexplored.

Transertion and Membrane Heterogeneity

Heterogeneous Distribution of Envelope and Envelope-Associated Proteins

A heterogeneous distribution of proteins in the cell envelope of bacteria, despite their size, has been widely demonstrated (see reviews, Shapiro et al., 2002; Fishov and Norris, 2012b; Govindarajan et al., 2012). During the course of investigation of specific biological functions and processes, this heterogeneity has been revealed mainly by visualization with GFP-fusions and immunofluorescence to take the form of polar, patchy and helix-like distributions in the membranes. Examples of proteins localized to the septal regions include lipid synthases in *B. subtilis* (Nishibori et al., 2005; Matsumoto et al., 2012; Takada et al., 2014; described in Lipid Synthases and Membrane Domains). Reciprocally, ATP synthase and succinate dehydrogenase are at low levels in (or absent from) the mid-cell region at the onset of cell division in *B. subtilis*, which may reflect an association with lipid domains elsewhere that are rich in PG or other lipids rather than with lipid domains in the mid-cell region (i.e., the division site) where these lipids may be at low levels (Meredith et al., 2008). Recently, FloT and FloA, homologues of eukaryotic flotillin proteins found exclusively in lipid rafts along with proteins involved in signaling and transport have been localized to discrete microdomains in the membrane of *B. subtilis* (see Cardiolipin and Other Anionic Phospholipid Domains); significantly, these microdomains, which are likely to be present in many other bacteria, also contain other proteins involved in signal transduction and cell–cell communication such as the sensor kinase, KinC, and protein secretion such as SecY in *B. subtilis* (Donovan and Bramkamp, 2009; Lopez and Kolter, 2010; Bach and Bramkamp, 2013; Bramkamp and Lopez, 2015). Flotillins are believed to play a large part in maintaining the overall physical heterogeneity of the membrane since, in their absence, lipid-ordered domains coalesce (Bach and Bramkamp, 2013; Bramkamp and Lopez, 2015). Cytoplasmic membrane proteins located in the polar regions of *E. coli* cells include ProP, LacY, and MscS (Romantsov et al., 2010) and the MCPs (methyl-accepting chemotaxis proteins; Alley et al., 1992; Sourjik and Armitage, 2010) whilst proteins located at the sites of cell constriction in *E. coli* include the components of *trans*-envelope Tol-Pal complex, TolA, TolQ, and TolR (in the cytoplasmic membrane), the peptidoglycan-associated lipoprotein Pal, (anchored to the outer membrane), and TolB (a soluble periplasmic protein; Gerding et al., 2007).

Two systematic approaches have been taken to find proteins that are preferentially located in the poles. Using two-dimensional gel electrophoresis, Lai et al. (2004) identified proteins preferentially located in *E. coli* minicells and created a catalog of polar proteins; these included the inner membrane proteins MCPs, AtpA, AtpB, YiaF, and AcrA, the membrane-associated protein FtsX, and the outer membrane proteins YbhC, OmpW, Tsx, Pal, FadL, OmpT, and BtuB. In a complementary, cataloging approach, Li and Young (2012) used FLAG peptide-tagged Tar, a known polar MCP, to isolate by affinity capture those inner membrane vesicles that originated from the poles. These vesicles were enriched in over 30 proteins. Most were in or associated with the inner membrane and included Aer, PBP1B, TnaA, DcuA, PutP, TrxA, FliP, AccD, CpxA, FliC, RpsD, YcbC, GlnP, GroEL, MlaF, NarZ, YqjD, YniB, MCPs, CheA, YijP, PBP1A, SppA, LepB, NupC, and YiaF, although a few exceptions, TolC, Pal, and BamA, were in the outer membrane. In following up the latter catalog by fusing proteins to GFP and expressing the gene from their own promoters, Aer, TnaA, GroES (rather than GroEL), YqjD, and YiaF were found to form polar foci that were distinct from inclusion bodies, thereby suggesting that these proteins are located in polar membranes *in vivo*. Moreover, the MCPs and Pal figure in the polar membrane class in both catalogues, thus lending some confidence to the results. Extensions of approaches like these should identify the remaining proteins with polar preference using the full list of *E. coli* proteins, which includes as many as 1,133 predicted integral inner membrane proteins (Bernsel and Daley, 2009) and 503 peripheral, inner membrane-associated ones (Papanastasiou et al., 2013).

GFP-fusions and immunofluorescence have provided many examples of specific proteins in polar and helix-like distributions in the membranes of *E. coli* and *B. subtilis* cells; these proteins include MreB (as well as homologues MreBH and Mbl) and the Sec translocon (for the list and references see Table 1 in Fishov and Norris, 2012a). These results have, however, been questioned due to recent findings showing that artefactual distributions can be generated by high levels of expression. The translocon in *E. coli* includes the SecYEG translocon which can associate with SecDF–YajC–YidC and SecA to enable protein transport in a reaction that is stimulated by CL (Gold et al., 2010; Schulze et al., 2014). The GFP-fusions of SecA and SecY over-expressed using 1% xylose are helically distributed in *B. subtilis* (Campo et al., 2004) whilst expression from the native promoter Psec-*secA* gives a peripherally homogeneous and septal distribution of the fusion products in conditions in which careful high-level expression controls show helices (Carballido-Lopez et al., 2006-Supplementary Figure S6; Halbedel et al., 2014). This peripherally homogeneous distribution of SecA in *B. subtilis* is different from that of *E. coli* GFP-SecE which has a patchy subcellular distribution (Shiomi et al., 2006). Localization of FlAsH-tagged SecY and SecE in *E. coli* also shows quite different patterns from that of *B. subtilis* SecA, with helix-like (patchy), polar dots and a diffuse distribution in cells under conditions of no induction (Gold et al., 2010); that said, the cells still have higher levels of tagged products from the high copy number plasmid pBAD22, even in the absence of the inducer, compared with those from their native promoter on the chromosome, i.e., single copy. The translocon distributions in *B. subtilis* and *E. coli* also differ from that of the *Streptococcus pyogenes* translocon, which forms a single microdomain (Rosch and Caparon, 2005; Rosch et al., 2007) and that of *S. pneumonia* D39, expressed from the native chromosomal loci, which localizes dynamically to different places (Tsui et al., 2011). GFP-fusions of MreB have been shown to move, as fragmented discrete patches, perpendicular to the long axis of *B. subtilis* cells using highly sensitive time-lapse imaging (Dominguez-Escobar et al., 2011; Garner et al., 2011). This movement of MreB did not, as one would have expected, follow a helical pattern. Recent reports and reviews suggest that the distribution pattern reported for *E. coli* MreB mainly resembles that reported for *B. subtilis* MreB: a punctuated pattern with pairs of dots or small bands generally described as helical (van Teeffelen et al., 2011; Chastanet and Carballido-Lopez, 2012; Margolin, 2012; Errington, 2015). Indeed, MreB proteins form elongated filamentous structures when over-produced or when observed in late phases of growth. In virtually all older reports describing MreB localization, inducible, mostly over-expressed, GFP fusions were used, and observations were often made during the late exponential stage, when the structures were easier to visualize. This could be the case for other fusion proteins showing polar and helix-like distributions under highly expressed conditions. Thus, it is difficult at present to decide how many of the proteins do indeed form long-range, helical structures at the membrane, as noted by Margolin (2012).

Lipid–Protein Interactions

The last stage of transertion (see below) involves interaction of nascent proteins with the polar head groups of lipids followed by insertion into non-polar acyl-chain layer of the membrane. Proteins with a preference for a specific polar head group interact to form a specific domain in the membrane. Integral inner membrane proteins with specific lipid preferences could therefore be located in specific membrane domains (for the list of proteins see Table 2 in the review of Fishov and Norris, 2012a). In addition, peripheral membrane proteins that associate weakly with specific lipids could help form specific domains and, if these proteins bind other proteins, could bring them into the domain too. Among the peripheral membrane proteins, there are amphitropic proteins that have two apparent locations: one form of the protein is in the aqueous compartment of the cytosol whilst the other form is weakly associated with the cell membrane. Amphitropic proteins can be classified into three major categories (Johnson and Cornell, 1999), based on the way they associate with membranes. The first class (A) contains motifs with binding pockets for a lipid monomer, called lipid clamps. The second class (B) contains lipid covalent anchors embedded in the lipid bilayer; this class includes lipoproteins such as the major outer membrane lipoprotein (Braun's lipoprotein), LolB, components of Bam complexes and RcsF with an *N*-acyl chain and a diacylglyceryl moiety at the NH_2-terminus (Sankaran and Wu, 1994; Matsumoto, 2001; Suzuki et al., 2002; Okuda and Tokuda, 2011; Shiba et al., 2012; Konovalova et al., 2014). *E. coli* has 96 lipoproteins, 58% of which have completely unknown functions (Brokx et al., 2004). The third class (C) of amphitropic proteins contains amphipathic α-helices, which bind to a membrane by partitioning into the membrane bilayer such that the hydrophobic face of the protein is sequestered away from water, yet its polar face can contact the aqueous phase (Johnson and Cornell, 1999). The helix axis lies parallel to the membrane surface. The weak (reversible) interaction of the proteins with amphipathic α-helices plays an important role in regulation of their functions. Representatives of this class include a family of prokaryotic cytoskeletal proteins such as MreB, MinD, and FtsA, having an amphipathic α-helix at their NH_2− or COOH-terminus (Szeto et al., 2003; Pichoff and Lutkenhaus, 2005; Salje et al., 2011; Szwedziak et al., 2012; **Figure 4**). Membrane anchoring with amphipathic helices has yet to be reported for any eukaryotic filament (Salje et al., 2011). It has been demonstrated recently that the nucleoid occlusion protein of *B. subtilis*, Noc, also associates with the cell membrane via an NH_2-terminal amphipathic α-helix. It occludes assembly of the division machinery by simultaneous binding to DNA and the membrane (Adams et al., 2015). PgsA, *B. subtilis* PssA (Matsumoto, 1997), and the well-understood mammalian CTP: phosphocholine cytidylyltransferase (CCT) have an amphipathic α-helical structure in the middle of the protein (Cornell and Taneva, 2006). The amphipathic α-helices that were found in DnaA (Garner and Crook, 1996; Makise et al., 2001) are not involved in membrane association; instead, the results of membrane retention experiment obtained with various fragments of the protein indicated that the association is through a concerted interaction of distant residues forming a surface (Regev et al., 2012).

MreB is one of the key components of the bacterial cytoskeleton that lies just underneath the membrane and organizes the cell wall synthesis machinery. In *E. coli*, MreB has an amphipathic α-helix (an 11 amino acid residue sequence which includes two methionine, two phenylalanine, one leucine on one face of the helical wheel and three basic residues on the other face) at its NH_2-terminus that is responsible for its location underneath the membrane. In gram-positive bacteria and the thermotrophic archaeum *Thermotoga maritima*, the counterpart of MreB has no such helix (Salje et al., 2011) and the membrane binding of *T. maritima* MreB is mediated by a small insertion loop that contains leucine and phenylalanine (residue 93 and 94), two other hydrophobic residues, the start methionine and a leucine (residues 1 and 2; Salje et al., 2011). MinD, plays a key role in division site selection by protecting the poles from aberrant positioning of FtsZ ring, and has a small amphipathic α-helix of 8–12 residues, termed the membrane targeting sequence or MTS, at its COOH-terminus that is conserved across eubacteria, archaea, and chloroplasts; this MTS is essential for association with anionic phospholipid-enriched membrane from which, in the case of *E. coli*, it can be easily detached by MinE at the appropriate time in the Min oscillation cycle (Raskin and de Boer, 1999; Szeto et al., 2002, 2003; Mileykovskaya et al., 2003; Vecchiarelli et al., 2014). The actin-like protein FtsA has an MTS that comprises a conserved COOH-terminal amphipathic α-helix of 13–14 residues (Pichoff and Lutkenhaus, 2005); FtsA polymerizes to form an "FtsA ring." The tubulin-like protein FtsZ is tethered to FtsA through a COOH-terminal tail of 16 residues to polymerize to form an FtsZ ring underneath the A-ring (Szwedziak et al., 2012; **Figure 4**). Work on liposome division has shown the importance of the MTS in division. ATP

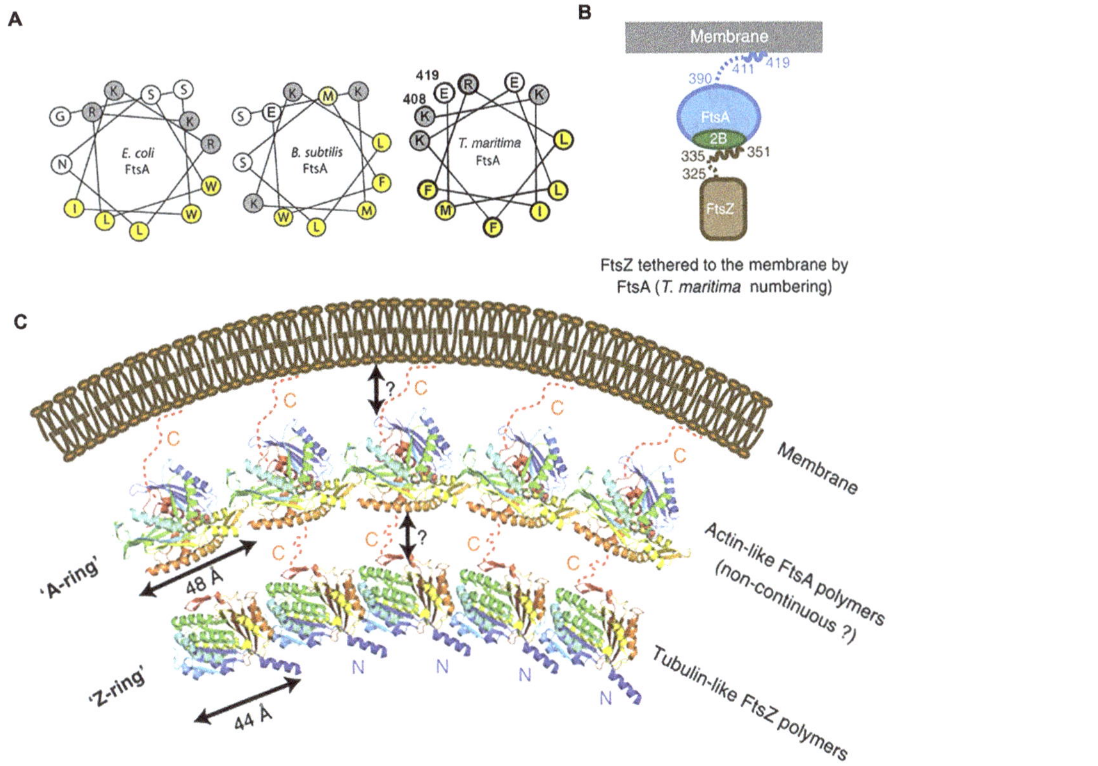

FIGURE 4 | Crucial role of the COOH-terminal amphipathic α-helix of FtsA in tethering FtsZ to the membrane. (A) Helical wheel diagrams of the COOH-terminal domains of FtsA from *E. coli*, *B. subtilis* (Pichoff and Lutkenhaus (2005), © John Wiley and Sons with permission.) and *Thermotoga maritima* (prepared using Helixator-TCDB: www.tcdb.org/progs/helical_wheel.php). Hydrophobic residues are highlighted in yellow and negatively charged residues are highlighted in gray. **(B)** FtsA/FtsZ membrane-bound complex of *T. maritima*. **(C)** The Z-ring made of polymerized tubulin-like FtsZ may be attached to the membrane by an A-ring, made of actin-like FtsA polymers. The two cartoons are adapted from Szwedziak et al. (2012), © John Wiley and Sons with permission.

is needed not only for the polymerization of FtsA and also for its attachment via its MTS to lipid monolayers and to vesicle membranes; this polymerization of FtsA caused vesicle shrinkage, consistent with the protein facilitating division both indirectly by interacting with FtsZ and directly by altering the membrane (Krupka et al., 2014). A fusion of FtsZ-YFP-mts was constructed with an MTS at its COOH-terminus to tether it directly to the membrane; this allowed Z-rings to assemble in multi-lamellar tubular liposomes and to generate a constriction force in the presence of GTP without the need for any other protein (Osawa et al., 2008). Uni-lamellar liposomes incorporating FtsA and FtsZ-YFP produced more natural Z-rings, which constricted liposomes and in some cases appeared to complete the division (Osawa and Erickson, 2013). The co-polymerization of FtsZ and FtsA is proposed to lead to bending, curvature and membrane constriction because the subunit repeat lengths of FtsZ and FtsA are different, being roughly 4 and 5 nm, respectively (Szwedziak et al., 2014; **Figure 4**).

Membrane Curvature and Protein Location

In rod-shaped bacterial cells, negatively curved membranes characterize the poles and parts of the developing septum. In *B. subtilis*, it has been suggested that this curvature is sensed by DivIVA (Lenarcic et al., 2009; Ramamurthi and Losick, 2009), which is located at the septum and at mature poles and which is responsible for the polar location of the division inhibitor MinC/MinD via MinJ, a system different from the Min system of *E. coli* (Marston et al., 1998; Bramkamp et al., 2008; Patrick and Kearns, 2008). In cells with engineered curvature, DivIVA is indeed located preferentially in regions of high negative curvature (Renner et al., 2013; **Figure 2F**). The location of DivIVA depends on SecA although DivIVA does not contain a signal sequence (Halbedel et al., 2014). Instead, SecA is considered to act as a chaperone in the folding of DivIVA, which has essentially an α-helical structure with several coiled-coil helices that contribute to the curvature of the DivIVA tetramer (Halbedel et al., 2014). The crystal structure of this tetramer resembles the crescent shape of eukaryotic BAR domains which bind to curved membranes and also introduce curvature (Lemmon, 2008; Oliva et al., 2010; Mim and Unger, 2012). Enzyme I (EI), which is a part of the phosphoenolpyruvate-phosphotransferase system (PTS) responsible for the sensing and uptake of many extracellular sugars, is located in the poles of *E. coli* and *B. subtilis* due, it is proposed, to its affinity for negatively curved membrane, an affinity shared by DivIVA (Govindarajan et al., 2013). Conversely, SpoVM, a small peripheral membrane protein with an amphipathic α-helix, associates with the positively curved (convex) membrane surfaces of the forespore to form the spore coat complex with SpoVA, in the mother cell during sporulation (Ramamurthi, 2010; **Figure 2F**). Recently, it has been suggested that the actual amphipathic α-helix region of SpoVM (26 aa) is atypical; this region is short with only 13 residues (from 11 to 23), contains only one positively charged residue (Arg) and three Gly residues, and inserts deeply into the membrane, unlike many amphipathic α-helix molecules that only insert shallowly into the membrane (Gill et al., 2015). The authors hypothesize that deep insertion of SpoVM into membrane, which involves extensive interactions with acyl chains to sense packing differences in differently curved membranes and which may involve cooperative interactions with other SpoVM proteins, drives its preferential localization onto slightly convex membranes surface such as the outer surface of the forespore.

Transertion

In the transertion hypothesis, it is proposed that the coupled transcription-translation of genes encoding membrane proteins (1) structures the membranes, both physically and chemically, and (2) positions these genes close to the membrane (Norris and Madsen, 1995; Woldringh, 2002; **Figure 1**). Despite early evidence in its favor (Binenbaum et al., 1999), this hypothesis lacked further evidence to support it for long time. Recently, Libby et al. (2012) have shown that the synthesis of membrane proteins expression affects the position of chromosome loci in *E. coli* cells. They observed that two genetic loci, *lacY* and *tetA*, which encode membrane proteins, were rapidly shifted toward the membrane upon induction whilst the chromosomal locus for a cytoplasmic protein was not shifted; in addition, antibiotics that block transcription and translation prevented these shifts toward the membrane. More recently still, a radial contraction of the *E. coli* nucleoids was observed immediately after the addition of inhibitors of either transcription or translation, again consistent with transertion normally exerting an expanding force on the nucleoid (Bakshi et al., 2014). In the case of inhibition of transcription and hence transertion by Rifampicin, the eventual expansion of the nucleoid has recently been attributed to the penetration of the nucleoid by the ribosomal subunits (Bakshi et al., 2014; Sanamrad et al., 2014). That said, it should be noted that transcription is the essential source of supercoiling (particularly at high growth rates when 50–80% of active RNA polymerases are transcribing *rrn* genes) which is a major cause of nucleoid compaction; hence, inhibition of transcription initiation with Rifampicin results in RNA polymerase run-off, less supercoiling, and nucleoid expansion; this expansion can happen even after nucleoid compaction by chloramphenicol when transertion is absent (for a review, see Jin et al., 2013).

Over 1,100 genes distributed throughout the *E. coli* genome are predicted to encode integral inner membrane proteins (Bernsel and Daley, 2009) and therefore their expression could maintain the chromosome in an expanded and dynamic state, consistent with the transertion hypothesis. Significant numbers of ribosome and RNA polymerase copies have been shown by superresolution imaging to extend from the nucleoid to the cytoplasmic membrane (Bakshi et al., 2012), indicating that transertion exerts a direct radially expanding force on the nucleoid (Woldringh, 2002; Woldringh and Nanninga, 2006). In this study, however, few copies of RNA polymerase were found near the polar caps, indicating that transertion could not occur in the regions away from the nucleoid (polar caps) and therefore it might seem that the radially expanding force on the nucleoid of transertion could not be a direct axially expanding force. Such axial expansion might, however, result from the penetration of the nucleoid by the 30S and 50S subunits released from the 70S ribosomes following treatment with inhibitors of translation (Bakshi et al., 2014; Sanamrad et al., 2014). It should be noted that although less than 15% of *E. coli*'s ribosomes are reported to be available to participate in transertion (Bakshi et al., 2012; Gahlmann and Moerner, 2014), this figure does not necessarily reflect the percentage of the ribosomes that are actually engaged in translation. A higher percentage is likely to be available in *C. crescentus* where genes and mRNA have been found to be close to one another (Llopis et al., 2010).

There are plenty of polar membrane proteins as noted in the previous section (see Heterogeneous Distribution of Envelope and Envelope-Associated Proteins). Transertion is unlikely to be an important factor in directly controlling their location since, firstly, RNA polymerase is relatively scarce near the polar caps and, secondly, the Sec machinery responsible for inserting integral membrane proteins into membranes is not specifically located in the poles in either *B. subtilis* (Carballido-Lopez et al., 2006-Supplementary Figure S6) or *E. coli* (Shiomi et al., 2006; Gold et al., 2010; see Heterogeneous Distribution of Envelope and Envelope-Associated Proteins). Thus, some integral membrane proteins (e.g., the MCPs) are inserted into the cytoplasmic membrane around the nucleoid via the Sec machinery and then freely migrate in the membrane to the polar regions (Shiomi et al., 2006). One possibility is that integral membrane proteins such as the MCPs are excluded from the transertion domains around the chromosome because they have no affinity for the lipids in which those domains are enriched. The reason for this is that transertion may create a problem if it leads to the formation of an array of MCPs of the wrong size in the wrong place. Another, recently proposed, possibility is that polar localization of the diffused MCPs occurs independently of the phospholipid composition of the cytoplasmic membrane and is not be dictated by the curvature of the cell poles, instead, MCPs interact with components of the trans-envelope Tol-Pal complex which restricts the diffusion of MCP arrays (Santos et al., 2014; see Transertion Problems).

In the original transertion hypothesis, the structuring of the membrane and the tethering of genes was restricted to those genes encoding proteins inserted into or through membranes. Although the role of the former class of proteins (i.e., integral membrane proteins) is easy to imagine, the latter class (i.e., secreted and exported proteins) is also of interest. For example, there could be a role in transertion for Braun's lipoprotein if the lipid modification at the NH_2-terminus of the lipoprotein were to occur before the synthesis of the rest of the protein. Moreover, the transertion hypothesis could be extended to genes encoding peripheral membrane proteins such as SeqA (Slater et al., 1995; Shakibai et al., 1998) and DnaA in *E. coli* (Regev et al., 2012) and Noc in *B. subtilis* (Adams et al., 2015).

Finally, it has been proposed recently on entropic grounds that transertion enables ribosomal subunits to penetrate the nucleoids and to initiate the more general process of co-transcriptional translation (Bakshi et al., 2014). Co-transcriptional translation would then protect nascent mRNA (Deana and Belasco, 2005; Deneke et al., 2013) and prevent the backtracking of RNA polymerases (McGary and Nudler, 2013), thereby giving transertion a key role in the optimization of transcription and translation (Bakshi et al., 2014). In the context in which two chemically identical chromosomes are in competition for the transcriptional and translational apparatuses, such optimization constitutes a positive feedback that could underpin differentiation (Norris and Madsen, 1995; Norris and Amar, 2012).

Hyperstructures

A hyperstructure is an assembly of elements (such as genes, RNA, proteins, small molecules and ions) that performs a function and that constitutes a substantial proportion of the cellular material (Norris et al., 2007a; Saier, 2013; Meyer et al., 2014). The accumulating evidence for their existence allows them to be classed in several ways (Norris et al., 2007b). In the case of the membrane, the best known include those involved in motility (such as the MCPs and flagella), lactose metabolism, energy generation, and cell division. It has been further proposed that a dialog between hyperstructures determines the phenotype of the cell (Norris et al., 2014a). Here we focus on some of the hyperstructures involved with the membrane.

Transertion Hyperstructures

The coupled transcription-translation-insertion of membrane proteins (transertion) has been proposed to tether networks of nascent proteins, mRNAs and genes to the cytoplasmic membrane thereby attaching DNA dynamically to the bacterial cell envelope (Norris, 1995; Norris and Madsen, 1995; Woldringh, 2002; **Figure 1**). This means that the high level expression of a gene encoding a membrane protein could lead to the formation of a domain in the membrane enriched by the nascent proteins, possibly specific lipids, and translocon complexes, along with an adjacent, highly structured cytoplasm to give a transertion hyperstructure (Norris, 1995; Norris and Madsen, 1995; Binenbaum et al., 1999; Woldringh, 2002).

The *lac* operon is an obvious candidate for giving rise to a transertion hyperstructure. In *E. coli*, full induction of this operon has been calculated to give 23 RNA polymerases on *lacZ* and 5 or 6 on *lacY* and *lacA* with the ensemble of the nascent, full length and decaying mRNAs being translated by several hundred ribosomes (Kennell and Riezman, 1977). *lacY* encodes the membrane-bound permease with an average of four ribosomes calculated as translating the corresponding full length mRNA (Kennell and Riezman, 1977). LacZ-encoding mRNA has been shown to remain close to the *lacZYA* locus, consistent with both transcription and translation occurring in the same place (Llopis et al., 2010). Expression of the *lac* operon brings the locus closer to the membrane and this movement depends on the presence in the operon of *lacY*, consistent with transertion (Libby et al., 2012). Hence, calculations and experiments suggest that a transertion hyperstructure exists that comprises the *lac* genes dynamically attached to tens of nascent mRNAs and to 100s of ribosomes and the nascent enzymes some of which would be inserted into the membrane. This does not, of course, mean that the Lac proteins once synthesized remain within the transertion hyperstructure. Indeed, a tagged version of LacY – albeit expressed from a plasmid – was localized to the poles (Romantsov et al., 2010).

RNA degradation may also play a part in the dynamics of transertion hyperstructures. Using epifluorescence microscopy and molecular dynamics simulation, RNase E, the backbone of the RNA degradosome, was shown to diffuse over the entire inner membrane of *E. coli* to form, along with other proteins, short-lived hyperstructures (Strahl et al., 2015). The existence of these hyperstructures depends on the presence of the RNA substrate (Strahl et al., 2015) so they may be considered as *functioning-dependent* structures, that is, structures that form due to their activity (Norris et al., 2007a). It should be noted that such separation of the degradosomes from the sites of transcription would favor nascent transcripts being translated by polyribosomes, that is, would favor the coupling of transcription and translation over the separation of these processes (Strahl et al., 2015).

Transertion Problems

Transertion is not only a likely solution to many of the problems that confront cells but also a source of potential problems. One problem would occur if a transertion hyperstructure were to determine, inappropriately, the size, position or composition of another class of hyperstructure. Consider, for example, the importance in signal transduction of the distribution of the MCPs and related proteins into many small clusters or into one giant cluster (Bray et al., 1998). These distribution confer different sensitivities but this relationship could be lost if transertion were to dominate by creating a large chemosignaling hyperstructure. In *E. coli*, part of the solution may be that the genes that encode chemotaxis proteins are located on the chromosome so that their transertion associates them with another major hyperstructure, that of the flagellum; it is then conceivable that differences in the affinities of the chemotaxis and flagellar proteins for the chemical or physical properties of lipids are responsible for the chemotaxis proteins relocating from the flagellar hyperstructure(s) to the poles (Cabin-Flaman et al., 2005). Consistent with this, Tar-GFP does diffuse away from the sites where this MCP is synthesized to the poles (Shiomi et al., 2006). It has been recently found that this polar localization of the diffused MCPs does not depend on either the phospholipid composition of the cytoplasmic membrane or the curvature of the cell poles; instead, MCPs interact with components of the trans-envelope Tol-Pal complex which restricts the diffusion of MCP arrays (Santos et al., 2014). The Tol-Pal complex is also part of the cell division hyperstructure, to which it is recruited by FtsN so as to play a role in the invagination of the outer membrane during division (Gerding et al., 2007).

A second problem would arise if the lipid preferences of the constituents of the transertion hyperstructure were to menace the planar, bilayer structure of the cytoplasmic membrane. The reality of this danger is evident from the proteo-lipid structures that result from the overproduction of peripheral and integral membrane proteins (Weiner et al., 1984; Arechaga et al., 2000; van den Brink-van der Laan et al., 2003; Eriksson et al., 2009). In mitochondria, the bioenergetic supercomplexes are indeed in specific membrane structures whilst in bacteria such as *E. coli*, fluorescent fusions suggest that the complexes are spatially dispersed in mobile 100–200 nm domains containing 10s–100s of complexes (Erhardt et al., 2014; Llorente-Garcia et al., 2014). A possible solution would be for an abundant structure such as the ATP synthase to have subunits with complementary lipid preferences (Arechaga et al., 2000) since even overproduction of all eight subunits results in morphological changes (von Meyenburg et al., 1984). Extending this argument, one might expect different proteins to have preferences for different lipids, which may help explain why cells have so many different lipids. An alternative or complementary solution would be the creation by a large, dynamic hyperstructure of many mobile domains with an affinity for ATP synthase.

Finally, problems due to the formation of an inappropriate transertion hyperstructure might be avoided by reducing the time during its synthesis in which the nascent protein interacts with the membrane; this might be achieved if, for example, evolution were to have selected for the membrane-interacting sequences (such as amphipathic helices) to be located at the COOH-terminus rather than at the NH_2−terminus.

Sensing

The membrane is the final frontier between the cell and its environment so membrane-based hyperstructures are in the right place to sense and to respond to environmental changes. This task is not trivial. Not only do new mRNAs and proteins have to be made but many existing mRNAs and proteins must be degraded. There are, for example, probably over a 1000 membrane proteins in *E. coli* (Bernsel and Daley, 2009) and a similar number in *B. subtilis* (Hahne et al., 2008; Otto et al., 2010; van Dijl et al., 2012) and the functioning of these proteins must be coordinated; this task is particularly difficult when major changes in the protein composition of the membrane must occur as during changes in growth conditions. On the entry of *B. subtilis* to stationary phase, the levels of the many proteins that increase include those involved in the uptake of glycerol, ribose, lactate, nucleoside, succinate, fumarate and zinc whilst the levels of those that decrease include transporters of malate, Fe^{3+} citrate, Fe^{3+} hydroxamate, and hydroxymethylthiamine (HMP)/thiamine transport (Otto et al., 2010). In *E. coli* cells supplemented with glucose, which is the preferred carbon and energy source, the transcriptional levels of the many genes that increase include those involved in the import of polyamines, inorganic phosphate and Mg^{2+} whilst the levels of those that are repressed (in other words, upregulated upon glucose–starvation), are of transporters and periplasmic receptor proteins related to the import of alternative carbon and carbon–nitrogen sources, which include amino acids, carbohydrates, lactate, glycerol, peptides, dipeptides, and nucleotides (Gutierrez-Ríos et al., 2007). This transcriptome pattern is consistent with the consequences of carbon catabolite repression exerted by glucose (Saier et al., 1996). The phosphoenolpyruvate-dependent PTS controls the uptake of a large number of energetically preferred sugars. It comprises proteins, EI and HPr, which are common to all substrates, as well as sugar-specific permeases, enzymes II (EIIs). Interestingly, EI and HPr are mainly located near the poles of *E. coli* cell (Patel et al., 2004; Lopian et al., 2010; Govindarajan et al., 2012). Upon addition of the sugar to the growth medium, HPr is phosphorylated by EI; HPr-P produced is released from the polar membranes and distributes in the cell, though EI remains located near the poles of negatively curved sites (Govindarajan et al., 2013; see Membrane Curvature and Protein Location). A fraction of HPr-P gets near the membrane to phosphorylate the permeases, allowing them to transport the sugars into the cell and to phosphorylate them (Govindarajan et al., 2012).

Common affinities for molecules and ions within and between hyperstructures may help in this task of coordination. These molecules include polyamines, poly-(R)-3-hydroxybutyrate, polyphosphate and, of course, lipids (Norris et al., 2014b). Polyamines, for example, bind to nucleic acids, ribosomes and porins, and stimulate the synthesis of around 300 proteins in *E. coli* that include the sigma factors RpoS, FecI, RpoN, and related RNA polymerase omega subunit RpoZ, Cya, Cra, Fis, H-NS and SpoT (Igarashi and Kashiwagi, 2006). These multiple actions may allow polyamines to play an important, coordinating, role in the survival of *E. coli* in the transition to stationary phase (Terui et al., 2012; Norris et al., 2014b) and during osmotic changes (Munro et al., 1972).

Similar global regulatory roles can be ascribed to CL and an increase in CL levels has been proposed as a general physiological response that protects microorganisms from lysis due to osmotic stress (Catucci et al., 2004) and, in line with this, an increase in osmolality leads to transcriptional activation of the *cls* gene in *E. coli* (Romantsov et al., 2007). This increase in CL and its distribution to the poles was correlated with the polar distribution of the osmosensory transporter ProP which actively transports osmo-protectants into the cell (Mileykovskaya, 2007; Romantsov et al., 2007). In such coordination of lipid metabolism with the environment, the formation of a hyperstructure containing membrane components may help. Acyl carrier protein (ACP), a small protein of 9 kDa, interacts with diverse proteins associated with many biosynthetic pathways, including enzymes involved in synthesis of fatty acid in the cytoplasm as well as enzymes on or in the membrane involved in phospholipid or LPS synthesis; the latter assocations could account for the partial localization of ACP with *E. coli* membranes. Two enzymes involved in phospholipid synthesis, PlsB, a *sn*-glycerol-3-phosphate acyltransferase and PssA, a PS synthase, have been shown to interact with ACP and YbgC, an acyl-CoA thioesterase involved in fatty acid synthesis, using a tandem affinity purification method (Gully and Bouveret, 2006). *E. coli* PssA, which has a preference for acidic phospholipids, exists as both a membrane-associated active form and a cytoplasmic latent form and is at the heart of a cross-feedback regulation of the synthesis of zwitterionic (PE) and acidic phospholipids (Shibuya, 1992; Okada et al., 1994; Saha et al., 1996a; Matsumoto, 1997; Rilfors et al., 1999; **Figure 5A**). Bacterial two-hybrid analysis has shown that the enzymes in the phospholipid synthetic pathway, PlsB-PssA and ACP, form a complex in/on the inner membrane and further that YbgC, a fatty acid synthesis enzyme, and PlsB, form a complex in association with ACP (Gully and Bouveret, 2006). Thus, physical interactions between many of the enzymes responsible for fatty acid synthesis and phospholipid synthesis could form a hyperstructure to help coordinate lipid metabolism.

Sensing changes in nutrient availability entails bacteria transmitting this information to flagella hyperstructures to perform chemotaxis (see Transertion Problems) and to the division apparatus to regulate their size, grow faster and become larger when they are grown in nutrient-rich media. UgtP, a glucosyltransferase responsible for the synthesis of glucosylated diacylglycerols, one of which is the anchor of LTA synthesis, is localized, in a dot-pair structure similar to that of the open ring of FtsZ, to the division site in *B. subtilis* (Nishibori et al., 2005). UgtP is a division inhibitor, which prevents assembly of FtsZ and which results in an increase in the length of the cells under nutrient-rich conditions (Shiomi and Margolin, 2007; Weart et al., 2007; **Figure 5**). It has been proposed that UDP-glucose acts as a proxy for nutrient availability and modulates the equilibrium between the UgtP–UgtP oligomer and the UgtP-FtsZ complex thereby serving as a molecular rheostat to help ensure that cell size is precisely co-ordinated with growth rate and nutrient availability (Chien et al., 2012). OpgH, a glucosyltransferase, is an integral membrane protein in the inner membrane of *E. coli* that is involved in the synthesis of osmoregulated periplasmic glucans, OPG (MDO); recently, OpgH has been localized to the division ring and shown to inhibit FtsZ assembly in the presence of UDP-glucose so as to delay the timing of cell division (Hill et al., 2013; **Figure 5A**). These two, very different, bacteria employ unrelated glucosyltransferases – a peripheral membrane enzyme for LTA synthesis in *B. subtilis* and an integral inner membrane enzyme for OPG synthesis in *E. coli* – that both have the second, "moonlighting," function of coupling nutrient availability to cell division by regulating FtsZ assembly (Hill et al., 2013). Disruptions of the moonlighting function of UgtP not only affect cell length: *ugtP* null mutants, which lack glucolipid products of the UgtP-catalyzed reaction, membrane glucolipids, are rounder (Price et al., 1997) and abnormally bent and distended (Lazarevic et al., 2005; Matsuoka et al., 2011a). Moreover, the extracytoplasmic function (ECF) σ factors, σ^M, σ^V, and σ^X are constitutively activated (Salzberg and Helmann, 2008; Matsuoka et al., 2011a; Hashimoto et al., 2013), consistent with glucolipids directly influencing anti-σ^M and anti-σ^V factors by stabilizing conformations that sequester the respective ECF σ factors (Seki et al., 2015).

A two-hybrid-based investigation of interactions between the components of cell division machinery (27 proteins) and the enzymes involved in lipid synthesis (12 enzymes) in *B. subtilis* has revealed that FtsA interacts with PlsX (Takada et al., 2014). PlsX is an acyl-ACP:phosphate acyltransferase that synthesizes acyl-phosphate, an essential substrate for lysophosphatidic acid (LPA) production by PlsY, an acyltransferase that uses acyl-phosphate for the first acylation of *sn*-glycerol-3-phosphate (G3P); these are novel enzymes in a unique LPA synthesis pathway found in prokaryotic but not in eukaryotic cells (Lu et al., 2006;

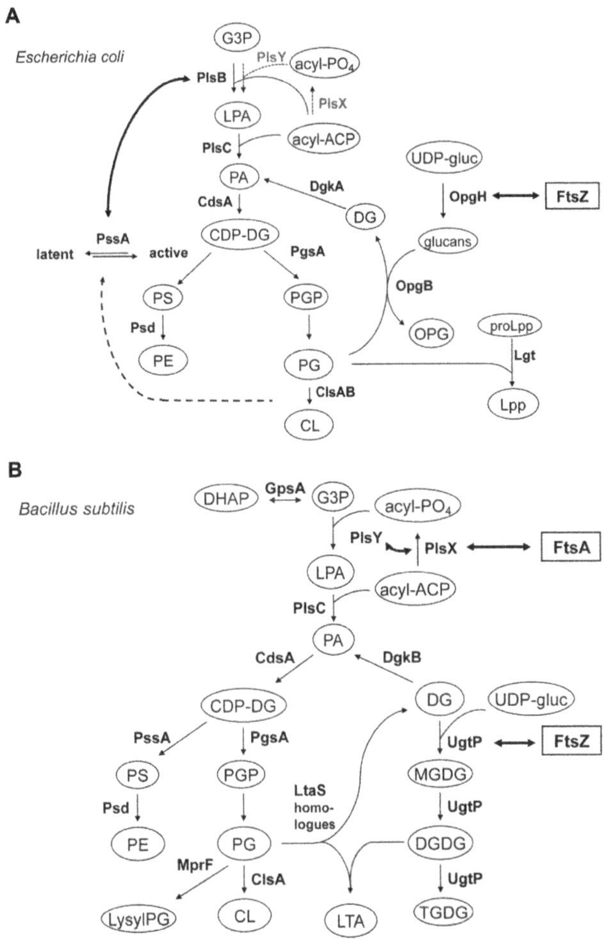

FIGURE 5 | Lipid biosynthetic pathways and interactions of phospholipid synthases and proteins involved in cell division. **(A)** Phospholipid biosynthetic pathway in *E. coli*. The gene product catalyzing each step is indicated. Abbreviations used are: G3P, glycerol-3-phosphate; LPA, lysophosphatidic acid; PA, phosphatidic acid; acyl-PO_4, acylphosphate; acyl-ACP, acyl acyl-carrier protein; CDP-DG, (d)CDP-diacylglycerol; PS, phosphatidylserine; PE, phosphatidylethanolamine; PGP, phosphatidylglycerophosphate; PG, phosphatidylglycerol; CL, cardiolipin; OPG, osmoregulated periplasmic glucans (or MDO); UDP-gluc, UDP-glucose; DG, diacylglycerol; Lpp, major outer membrane lipoprotein; proLpp, prolipoprotein. Trace activities of PlsX-PlsY for LPA synthesis in *E. coli* (Lu et al., 2006/Hara et al., 2008) were indicated with thin letters and dotted arrows, different from those of the major pathway in *B. subtilis*. For Cls homologues, ClsA and ClsB, refer to Tan et al. (2012). Interaction of OpgH (a glucosyltransferase) and FtsZ indicated with thick arrow is from Hill et al. (2013); see Sensing). Activation of phosphatidylserine synthase, PssA, with acidic phospholipids indicated with thick dotted arrow is adapted from Shibuya (1992) and Matsumoto (1997). Interaction of PlsB and PssA is from Gully and Bouveret (2006), but to avoid apparent complexity the interactions between PlsB, ACP and YbgC, an Acyl-CoA thioesterase, shown by the authors, are not depicted. **(B)** Biosynthetic pathways for phospholipids, glucolipids and LTA in *B. subtilis*. Abbreviations used are: lysylPG, lysylphosphatidylglycerol; MGDG, monoglucosyldiacylglycerol; DGDG, diglucosyldiacylglycerol; TGDG, triglucosyldiacylglycerol; LTA, lipoteichoic acid; DHAP, dihydroxyacetone phosphate. For other abbreviations refer to (A) for *E. coli*. The pathway is adapted and compiled from Jerga et al. (2007) and Matsuoka et al. (2011b). For LtaS homologues (YflE, YfnI, YqgS and YvgJ) refer to Gründling and Schneewind (2007), Matsuoka et al. (2011b) and

(Continued)

FIGURE 5 | Continued
Hashimoto et al. (2013). Interactions of FtsA and PlsX, and PlsY, both indicated with thick arrows, are from Hara et al. (2008) and Takada et al. (2014), respectively, but to avoid apparent complexity the interaction of PlsX with EzrA, DivIA and others, shown by the authors, are not depicted. Interaction of UgtP (a glucosyltransferase) and FtsZ indicated with thick arrow is from Weart et al. (2007) and Chien et al. (2012); see Sensing).

Paoletti et al., 2007; Yoshimura et al., 2007, and Hara et al., 2008; **Figure 5B**). *In vivo* cross-linking shows that PlsX interacts with cell division proteins (EzrA, DivIVA), cytoskeletal proteins and several metabolic enzymes, in addition to FtsA. PlsX is in a punctuated pattern in the peripheral membrane and is present at potential division sites independently of FtsA and FtsZ. Inactivation of PlsX leads to aberrant Z-ring formation. Thus, the key enzyme for phospholipid metabolism interacts with the cell division machinery in order to complete septum assembly (Takada et al., 2014). Although PlsX interacts with PlsY, many other enzymes involved in phospholipid synthesis, which localize in the septal region (Nishibori et al., 2005), do not interact with it, except MprF that synthesizes lysylPG (Takada et al., 2014). Depletion of PlsX leads to the cessation of both fatty acid and phospholipid biosynthesis (Paoletti et al., 2007; Yao and Rock, 2013).

Cell Cycle

Transertion may help provide solutions to the problems of coupling growth to the cell cycle, of triggering chromosome replication, and of coupling chromosome segregation spatially and temporally to cell division. The first problem, the way in which growth is coupled to the cell cycle, has been a mystery that has lasted for over half a century (Schaechter et al., 1958; Osella et al., 2014). Many different metabolic processes occur during growth and these processes are needed for the synthesis of RNA, proteins and lipids. These processes come together in transertion, which is therefore well-placed to integrate metabolic information (Fishov and Norris, 2012b). Moreover, transertion is a primary determinant of the structure and composition of the membrane, which is central to the initiation of DNA replication, segregation and cell division, as well as to the structure of the chromosome, which is kept in an expanded state (Woldringh et al., 1995; Libby et al., 2012). Hence transertion is also well-placed to couple growth to the cell cycle. Consistent with this, major effects of transertion on the microviscosity of the membrane have been shown via inhibition of transcription and translation (Binenbaum et al., 1999). The second problem, the nature of the mechanism that initiates chromosome replication, is another long-standing mystery (Eliasson and Nordström, 1997; Wang et al., 2011). This mechanism in *E. coli* is generally attributed to the relative proportions of the DnaA 'initiator' protein in the ATP-DnaA form, which is active in initiation, and the ADP-DnaA, which is inactive (Castuma et al., 1993). The reasons to invoke lipids in a DnaA-based control of initiation include the involvement of unsaturated fatty acids in initiation (Fralick and Lark, 1973), the promotion of formation of ATP-DnaA *in vitro* by anionic phospholipids (Castuma et al., 1993) and the association of 10% of the DnaA in a cell with the membrane (Regev et al., 2012). Given its role in determining membrane dynamics, transertion may well lie at the heart of the initiator mechanism and act via DnaA and, more generally, via promoting strand separation (Norris and Amar, 2012).

The third problem, the coupling between chromosome segregation and cell division, is essentially a membrane problem in the sense that the cell must arrange for the membrane to invaginate and separate segregated chromosomes. One role for transertion may be to create membrane domains in a particular physico-chemical state around the segregating chromosomes such that a different, 'septal' domain forms by default between them (Norris, 1995; Woldringh, 2002; Norris et al., 2004). This septal domain might then generate tubular or other structures that would recruit and activate division proteins. This hypothesis is supported by evidence for cell cycle variations in the lipid composition of *E. coli* (Mozharov et al., 1985) and in the spatial distribution of the lipids of *Micrococcus luteus* (Welby et al., 1996), for the CL-rich domains at the division sites of *E. coli* (Mileykovskaya and Dowhan, 2000; Koppelman et al., 2001) and *B. subtilis* (Kawai et al., 2004), and the septal location of the polyunsaturated fatty acid eicosapentaenoic acid of *Shewanella livingstonensis Ac10* (Sato et al., 2012). Significantly, distinct domains appear around and between the segregating chromosomes at a very early stage of the cell cycle in *E. coli* and these domains disappear when translation – and hence transertion – is abolished (Fishov and Woldringh, 1999). Division is inhibited by the Min system and lipids are again important (Suefuji et al., 2002; Kretschmer and Schwille, 2014). The NH_2-terminal domain of MinC perturbs the interactions between FtsZ monomers within an FtsZ polymer, while the COOH-terminal domain perturbs the lateral association of FtsZ protofilaments as well as interacting with MinD, for references see Kretschmer and Schwille (2014); MinD is a peripheral membrane-binding protein with an ATPase activity that is activated by MinE and that triggers the detachment of MinD from the membrane to prevent MinCD from inhibiting division; MinE may self-assemble on the membrane via an N-terminal helix that acts as an MTS. The idea is that, in *E. coli*, MinD binds cooperatively to the membrane at one pole with MinE forming an "E-ring" on the rim of this MinD zone that induces its disassembly leading to the diffusion of MinD and MinE and repetition of this assembly disassembly process at the other pole; MinC follows these oscillations of MinD and MinE to inhibit division (Rowlett and Margolin, 2013; Zheng et al., 2014 and references therein). *In vitro* and *in vivo*, MinD interacts with anionic lipids, which determine its distribution (Mileykovskaya et al., 2003), as does MinE (Vecchiarelli et al., 2014). In the case of the *E. coli* MinD introduced into *B. subtilis*, MinD makes spirals that coincide with the spirals of what are probably anionic phospholipids (Pavlendova et al., 2010). Moreover, MinD can convert phospholipid vesicles into tubes (Hu et al., 2002), consistent with the possible importance of phospholipid structures in nucleating division (Norris et al., 2004; Li et al., 2011). Finally, in spherical *E. coli* cells, FtsZ is associated with intracellular membraneous structures and MinD accumulates and oscillates between the places where these structures form (Bendezú and de Boer, 2008).

The 2 min or *dcw* cluster includes 16 genes, such as *ftsZ*, involved in peptidoglycan synthesis and cell division, and the promoter at the start of the cluster probably contributes to the transcription of the whole cluster, leading to the synthesis of a quite long mRNA (Hara et al., 1997; Mengin-Lecreulx et al., 1998; Vicente et al., 1998). Transertion of this cluster may therefore be an important factor in the spatio-temporal control of division. Transertion structures not only the membrane but also the cytoplasm thereby raising the possibility that chromosome segregation is accompanied by a phase separation in the cytoplasm that affects membrane dynamics and drives cell division. Hence, it is significant that phase separation of dextran and polyethylene glycol within giant vesicles leads to tubes of membrane forming at their interface (Li et al., 2011).

If transertion from the *dcw* cluster is indeed important, what is the consequence of having transertion from two clusters (one per segregating chromosome)? Are the lipid preferences of the secretion machinery different from those of the proteins encoded by this cluster so that the domain-producing potential of transertion is reduced? In bacterial L-forms, which lack a cell wall, it might be supposed that the requirement for many of the *dcw* genes would be reduced and less transertion would occur. This would be consistent with the findings that, in L-forms of *E. coli*, some of the genes were mutated (Siddiqui et al., 2006) and the levels of FtsZ were fivefold lower (Onoda et al., 2000) whilst in an L-form of *B. subtilis* FtsZ could be eliminated altogether (Leaver et al., 2009).

Discussion

The different ways of making membrane domains include lipid–lipid, lipid–protein, protein–protein, and polyamine–lipid interactions. By integrating these interactions, transertion plays a central role in the organization of the bacterial cell. It might even be argued that transertion helps provide a conceptual context for thinking about a plethora of intracellular events, structures and processes including the regulation and execution of the cell cycle, hyperstructure dynamics and the origins of life. In 1989, the term ‘nucleoid occlusion’ was coined to describe the inhibitory effect of the nucleoids on cell division (Cook et al., 1989) and it was suggested that nucleoid occlusion “may be related to a transcription or translation activity of the nucleoid” (Mulder and Woldringh, 1989). A few years later, it was proposed that the coupled transcription-translation-insertion of nascent proteins into and through membrane (i.e., transertion) is the mechanism responsible for nucleoid occlusion (Norris, 1995; Norris and Madsen, 1995; Woldringh, 2002). In this proposal, transertion creates membrane domains at the right times and in the right places not only to control chromosome replication, segregation and cell division but also to create the positive feedback needed for differentiation (Norris, 1995; Norris and Madsen, 1995; Binenbaum et al., 1999; Norris et al., 2004; Norris and Amar, 2012). Recently, it has been suggested that the role of transertion in nucleoid occlusion is complemented by the recruitment of Noc nucleoprotein complexes and associated DNA to the membrane (Adams et al., 2015).

Transertion can generate hyperstructures (Norris et al., 2007a; Llopis et al., 2010). Such transertion hyperstructures form a large class that, along with the other classes of hyperstructures, constitute the bacterial cell. The exact conformation of transertion hyperstructures is unknown but could be very important in bacterial cell physiology if, as proposed, interactions between hyperstructures were to be the primary determinants of the phenotype (Norris et al., 2007a). For example, interactions based on the condensation and decondensation of ions (Manning, 2007) might enable transertion hyperstructures containing membrane domains and/or linear assemblies of macromolecules to drive the cell cycle (Norris and Amar, 2012) whilst the kinetics of the reactions in the vicinity of a membrane domains are likely to be different from elsewhere, possibly due to the structuring of water (Wichmann et al., 2003).

Transertion may help explain the controversial existence of many spiral hyperstructures (Errington, 2015). Suppose a particular protein has a tendency to form a spiral via protein–protein and protein–lipid interactions. In the absence of transertion, this tendency may result in a spiral whereas, in the presence of transertion, the high concentration of nascent proteins and associated lipids in a circular membrane domain may compete effectively for the mature proteins and thereby prevent a spiral forming. Alternatively, transertion might help generate a protein spiral if the ensemble of gene, nascent RNA (and perhaps mature RNA) were constrained to be essentially linear via RNA–RNA interactions etc. Given the potential of a transertion hyperstructure for altering the morphology of the membrane or for directing a protein to the wrong location, avoidance of these inappropriate results of transertion may act as a powerful factor in evolution that affects both the position in a protein of membrane-interacting sequences and the type of translocon with which the nascent protein interacts.

Finally, several coupled processes in addition to transertion exist in modern bacteria. These include transcription-translation-assembly as in the case of ribosome synthesis (Woldringh and Nanninga, 1985; French and Miller, 1989; Norris et al., 2007a; Cabrera et al., 2009) and transcription-replication as in the case of initiation of DNA replication and DNA repair (Kogoma, 1997; McGlynn, 2010). It might therefore be expected that these processes in their coupled forms would have played a role in the origins of the first cells. This has been suggested for transertion based on its force-generating properties, which are responsible for expanding the nucleoid in modern bacteria (Woldringh et al., 1995; Cabrera et al., 2009; Libby et al., 2012; Bakshi et al., 2014): in this scenario, transertion would have helped maintain membrane integrity in early cells and have been of selective value even without the advantages of coding (Norris et al., 1999).

Acknowledgments

KM and HH would like to thank professors emeriti Isao Shibuya and Yoshito Sadaie for discussion and encouragement. We would also like to thank the editor, Arieh Zaritsky and the reviewers, Marc Bramkamp, Ding J. Jin, and Germán Rivas for helpful comments and guidance.

References

Adams, D. W., Wu, L. J., and Errington, J. (2015). Nucleoid occlusion protein Noc recruit DNA to the bacterial cell membrane. *EMBO J.* 34, 491–501. doi: 10.15252/embj.201490177

Alley, M. R. K., Maddock, J. R., and Shapiro, L. (1992). Polar localization of a bacterial chemoreceptor. *Genes Dev.* 6, 825–836. doi: 10.1101/gad.6.5.825

Arechaga, I., Miroux, B., Karrasch, S., Huijbregts, R., de Kruijff, B., Runswick, M. J., et al. (2000). Characterization of new intracellular membranes in *Escherichia coli* accompanying large scale over-production of the b subunit of F(1)F(o) ATP synthase. *FEBS Lett.* 482, 215–219. doi: 10.1016/S0014-5793(00)02054-8

Bach, J. N., and Bramkamp, M. (2013). Flotillins functionally organize the bacterial membrane. *Mol. Microbiol.* 88, 1205–1217. doi: 10.1111/mmi.12252

Bakshi, S., Choi, H., Mondai, J., and Weisshaar, J. C. (2014). Time-dependent effects of transcription- and translation-halting drugs on the spatial distributions of the *Escherichia coli* chromosome and ribosomes. *Mol. Microbiol.* 94, 871–887. doi: 10.1111/mmi.12805

Bakshi, S., Siryaporn, A., Goulian, M., and Weisshaar, J. C. (2012). Superresolution imaging of ribosomes and RNA polymerase in live *Escherichia coli* cells. *Mol. Microbiol.* 85, 21–38. doi: 10.1111/j.1365-2958.2012.08081.x

Barák, I., Muchová, K., Wilkinson, A. J., O'Toole, P. J., and Pavlendova, N. (2008). Lipid spirals in *Bacillus subtilis* and their role in cell division. *Mol. Microbiol.* 68, 1315–1327. doi: 10.1111/j.1365-2958.2008.06236.x

Bendezú, F. O., and de Boer, P. A. (2008). Conditional lethality, division defects, membrane involution, and endocytosis in mre and mrd shape mutants of *Escherichia coli. J. Bacteriol.* 190, 1792–1811. doi: 10.1128/JB.01322-07

Bernsel, A., and Daley, D. O. (2009). Exploring the inner membrane proteome of *Escherichia coli*: which proteins are eluding detection and why? *Trends Microbiol.* 17, 444–449. doi: 10.1016/j.tim.2009.07.005

Binenbaum, Z., Parola, A. H., Zaritsky, A., and Fishov, I. (1999). Transcription- and translation- dependent changes in membrane dynamics in bacteria: testing the transertion model for domain formation. *Mol. Microbiol.* 32, 1173–1182. doi: 10.1046/j.1365-2958.1999.01426.x

Boggs, J. M. (1987). Lipid intermolecular hydrogen bonding: influence on structural organization and membrane function. *Biochim. Biophys. Acta* 906, 353–404. doi: 10.1016/0304-4157(87)90017-7

Bramkamp, M., Emmins, R., Weston, L., Donovan, C., Daniel, R. A., and Errington, J. (2008). A novel component of the division-site selection system of Bacillus sutilis and a new mode of action for the division inhibitor MinCD. *Mol. Microbiol.* 70, 1556–1569. doi: 10.1111/j.1365-2958.2008.06501.x

Bramkamp, M., and Lopez, D. (2015). Exploring the existence of lipid rafts in bacteria. *Microbiol. Mol. Biol. Rev.* 79, 81–100. doi: 10.1128/MMBR.00036-14

Bray, D., Levin, M. D., and Morton-Firth, C. J. (1998). Receptor clustering as a cellular mechanism to control sensitivity. *Nature* 393, 85–88. doi :10.1038/30018

Brokx, S. J., Ellison, M., Locke, T., Bottorff, D., Frost, L., and Weiner, J. H. (2004). Genome-wide analysis of lipoprotein expression in *Escherichia coli* MG1655. *J. Bacteriol.* 186, 3254–3258. doi: 10.1128/JB.186.10.3254-3258.2004

Cabin-Flaman, A., Ripoll, C., Saier, M. H. Jr., and Norris, V. (2005). Hypothesis: chemotaxis in *Escherichia coli* results from hyperstructure dynamics. *J. Mol. Microbiol. Biotechnol.* 10, 1–14. doi: 10.1159/000090343

Cabrera, J. E., Cagliero, C., Quan, S., Squires, C. L., and Jin, D. J. (2009). Active transcription of rRNA operons condenses the nucleoid in *Escherichia coli*: examining the effect of transcription on nucleoid structure in the absence of transertion. *J. Bacteriol.* 191, 4180–4185. doi: 10.1128/JB.01707-08

Campo, N., Tjalsma, H., Buist, G., Stepniak, D., Meijer, M., Veenhuis, M., et al. (2004). Subcellular sites for bacterial protein export. *Mol. Microbiol.* 53, 1583–1599. doi: 10.1111/j.1365-2958.2004.04278.x

Cannon, B., Hermansson, M., Gyorke, S., Somerharju, P., Virtanen, J. A., and Cheng, K. H. (2003). Regulation of calcium channel activity by lipid domain formation in planar lipid bilayers. *Biophys. J.* 85, 933–942. doi: 10.1016/S0006-3495(03)74532-9

Carballido-Lopez, R., Formstone, A., Li, Y., Ehrlich, S. D., Noirot, P., and Errington, J. (2006). Actin homolog MreBH governs cell morphogenesis by localization of the cell wall hydrolase LytE. *Dev. Cell* 11, 399–409. doi: 10.1016/j.devcel.2006.07.017

Castuma, C. E., Crooke, E., and Kornberg, A. (1993). Fluid membranes with acidic domains activate DnaA, the initiator protein of replication in *Escherichia coli. J. Biol. Chem.* 268, 24665–24668.

Catucci, L., Depalo, N., Lattanzio, V. M. T., Agostiano, A., and Corcelli, A. (2004). Neosynthesis of cardiolipin in Rhodobacter sphaeroides under osmotic stress. *Biochemistry* 43, 15066–15072. doi: 10.1021/bi048802k

Chastanet, A., and Carballido-Lopez, R. (2012). The actin-like MreB proteins in *Bacillus subtilis*: a new turn. *Front. Biosci.* S4:1582–1606. doi: 10.2741/S354

Chien, A. C., Zareh, S. K., Wang, Y. M., and Levin, P. A. (2012). Changes in the oligomerization potential of the division inhibitor UgtP co-ordinate *Bacillus subtilis* cell size with nutrient availability. *Mol. Microbiol.* 86, 594–610. doi: 10.1111/mmi.12007

Christensen, H., Garton, N. J., Horobin, R. W., Minnikin, D. E., and Barer, M. R. (1999). Lipid domains of mycobacteria studied with fluorescent molecular probes. *Mol. Microbiol.* 31, 1561–1572. doi: 10.1046/j.1365-2958.1999.01304.x

Cook, W. R., de Boer, P. A. J., and Rothfield, L. I. (1989). Differentiation of the bacterial cell division site. *Int. Rev. Cytol.* 118, 1–31.

Cornell, R. B., and Taneva, S. G. (2006). Amphipathic helices as mediators of the membrane interaction of amphitropic proteins, and as modulators of bilayer physical properties. *Curr. Protein Pept. Sci* 7, 539–552. doi: 10.2174/138920306779025675

Cullis, P. R., de Kruijff, B., Hope, M. J., Verkleij, A. J., Nayar, R., Farren, S. B., et al. (1983). "Structure properties of lipids and their functional roles in biological membranes," in *Membrane Fluidity in Biology*, ed. R. C. Aloia (New York, NY: Academic Press), 39–81.

Deana, A., and Belasco, J. G. (2005). Lost in translation: the influence of ribosomes on bacterial mRNA decay. *Genes Dev.* 19, 2526–2533. doi: 10.1101/gad.1348805

Deneke, C., Lipowsky, R., and Valleriani, A. (2013). Effect of ribosome shielding on mRNA stability. *Phys. Biol.* 10, 046008. doi: 10.1088/1478-3975/10/4/046008

Dominguez-Escobar, J., Chastanet, A., Crevenna, A. H., Fromion, V., Wedlich-Soldner, R., and Carballido-Lopez, R. (2011). Processive movement of MreB-associated cell wall biosynthetic complexes in bacteria. *Science* 333, 225–228. doi: 10.1126/science.1203466

Donovan, C., and Bramkamp, M. (2009). Characterization and subcellular localization of a bacterial flotillin homologue. *Microbiology* 155, 1786–1799. doi: 10.1099/mic.0.025312-0

Dowhan, W. (2013). A retrospective: use of *Escherichia coli* as a vehicle to study phospholipid synthesis and function. *Biochim. Biophys. Acta* 1831, 471–494. doi: 10.1016/j.bbalip.2012.08.007

Edidin, M. (1997). Lipid microdomains in cell surface membranes. *Curr. Opin. Struct. Biol.* 7, 528–532. doi: 10.1016/S0959-440X(97)80117-0

Elder, M., Hitchcock, P., Mason, F. R. S., and Shipley, G. G. (1977). A refinement analysis of crystallography of the phospholipid, 1,2-dilauroyl-DL-phosphatidylethanolamine, and some remarks on lipid–lipid and lipid–protein interactions. *Proc. R. Soc. London A* 354, 157–170. doi: 10.1098/rspa.1977.0062

Eliasson, A., and Nordström, K. (1997). Replication of minichromosomes in a host in which chromosome replication is random. *Mol. Microbiol.* 23, 1215–1220. doi: 10.1046/j.1365-2958.1997.2981663.x

Erhardt, H., Dempwolff, F., Pfreundschuh, M., Riehle, M., Schäfer, C., Pohl, T., et al. (2014). Organization of the *Escherichia coli* aerobic enzyme complexes of oxidative phosphorylation in dynamic domains within the cytoplasmic membrane. *Microbiol. Open* 3, 316–326. doi: 10.1002/mbo3.163

Eriksson, H. M., Wessman, P., Ge, C., Edwards, K., and Wieslander, Å. (2009). Massive formation of intracellular membrane vesicles in *Escherichia coli* by a monotopic membrane-bound lipid glycosyltransferase. *J. Biol. Chem.* 284, 33904–33914. doi: 10.1074/jbc.M109.021618

Errington, J. (2015). Bacterial morphogenesis and the enigmatic MreB helix. *Nat. Rev. Microbiol.* 13, 241–248. doi: 10.1038/nrmicro3398

Feng, X., Hu, Y., Zheng, Y., Zhu, W., Li, K., Guo, R. T., et al. (2014). Structural and functional analysis of *Bacillus subtilis* YisP reveals a role of its product in biofilm production. *Chem. Biol.* 21, 1557–1563. doi: 10.1016/j.chembiol.2014.08.018

Fishov, I., and Norris, V. (2012a). "Membrane structure and heterogeneity," in *Escherichia coli and Bacillus subtilis*: *The Frontiers of Molecular Microbiology Revisited*, eds Y. Sadaie and K. Matsumoto (Kerara: Research Signpost), 93–113.

Fishov, I., and Norris, V. (2012b). Membrane heterogeneity created by transertion is a global regulator in bacteria. *Curr. Opin. Microbiol.* 15, 724–730. doi: 10.1016/j.mib.2012.11.001

Fishov, I., and Woldringh, C. L. (1999). Visualization of membrane domains in *Escherichia coli*. *Mol. Microbiol.* 32, 1166–1172. doi: 10.1046/j.1365-2958.1999.01425.x

Fralick, J. A., and Lark, K. G. (1973). Evidence for the involvement of unsaturated fatty acids in the initiation of chromosome replication in *Escherichia coli*. *J. Mol. Biol.* 80, 459–475. doi: 10.1016/0022-2836(73)90416-6

French, S. L., and Miller, O. L. Jr. (1989). Transcription mapping of the *Escherichia coli* chromosome by electron microscopy. *J. Bacteriol.* 171, 4207–4216.

Gahlmann, A., and Moerner, W. E. (2014). Exploring bacterial cell biology with single-molecule tracking and super-resolution imaging. *Nat. Rev. Microbiol.* 12, 9–22. doi: 10.1038/nrmicro3154

Garner, E. C., Bernard, R., Wang, W., Zhuang, X., Rudner, D. Z., and Mitchison, T. (2011). Coupled, circumferential motions of the cell wall synthesis machinery and MreB filaments in *B. subtilis*. *Science* 333, 222–225. doi: 10.1126/science.1203285

Garner, J., and Crook, E. (1996). Membrane regulation of the chromosomal replication activity of *E. coli DnaA requires a discrete site on the protein. EMBO J.* 15, 1313–2321.

Gennis, R. (1989). *Biomembranes*. New York: Springer-Verlag. doi: 10.1007/978-1-4757-2065-5

Gerding, M. A., Ogata, Y., Pecora, N. D., Niki, H., and de Boer, P. A. J. (2007). The trans-envelope Tol–Pal complex is part of the cell division machinery and required for proper outer-membrane invagination during cell constriction in *E. coli*. *Mol. Microbiol.* 63, 1008–1025. doi: 10.1111/j.1365-2958.2006.05571.x

Gill, R. L. Jr., Castaing, J.-P., Hsin, J., Tan, I. S., Wang, X., Huang, K. C., et al. (2015). Structural basis for the geometry-driven localization of a small protein. *Proc. Natl. Acad. Sci. U.S.A.* 112, E1908–E1915. doi: 10.1073/pnas.1423868112

Gold, V. A. M., Robson, A., Bao, H., Romantsov, T., Duong, F., and Collinson, I. (2010). The action of cardiolipin on the bacterial translocon. *Proc. Natl. Acad. Sci. U.S.A.* 107, 10044–10049. doi: 10.1073/pnas.0914680107

Govindarajan, S., Elisha, Y., Nevo-Dinur, K., and Amster-Choder, O. (2013). The general phosphotransferase system proteins localize to sites of strong negative curvature in bacterial cells. *mBio* 4:e00443-13. doi: 10.1128/mBio.00443-13

Govindarajan, S., Nevo-Dinur, K., and Amster-Choder, O. (2012). Compartmentalization and spatiotemporal organization of macromolecules in bacteria. *FEMS Microbiol. Rev.* 36, 1005–1022. doi: 10.1111/j.1574-6976.2012.00348.x

Gründling, A., and Schneewind, O. (2007). Synthesis of glycerol phosphate lipoteichoic acid in *Staphylococcus aureus*. *Proc. Natl. Acad. Sci. U.S.A.* 104, 8478–8483. doi: 10.1073/pnas.0701821104

Gully, D., and Bouveret, E. (2006). A protein network for phospholipid synthesis uncovered by a variant of the tandem affinity purification method in *Escherichia coli*. *Proteomis* 6, 282–293. doi: 10.1002/pmic.200500115

Gutierrez-Ríos, R. M., Freyre-Gonzalez, J. A., Resendis, O., Collado-Vides, J., Saier, M., and Gosset, G. (2007). Identification of regulatory network topological units coordinating the genome-wide transcriptional response to glucose in *Escherichia coli*. *BMC Microbiol.* 7:53. doi: 10.1186/1471-2180-7-53

Hahne, H., Wolff, S., Hecker, M., and Becher, D. (2008). From complementarity to comprehensiveness–targeting the membrane proteome of growing Bacillus subtilis by divergent approaches. *Proteomics* 8, 4123–4136. doi: 10.1002/pmic.200800258

Haines, T. H., and Dencher, N. A. (2002). Cardiolipin: a proton trap for oxidative phosphorylation. *FEBS Lett.* 528, 35–39. doi: 10.1016/S0014-5793(02)03292-1

Halbedel, S., Kawai, M., Breitling, R., and Hamoen, L. (2014). SecA is required for membrane targeting of the cell division protein DivIVA in vivo. *Front. Microbiol.* 5:58. doi: 10.3389/fmicb.2014.00058

Hara, H., Yasuda, S., Horiuchi, K., and Park, J. T. (1997). A promoter for the first nine genes of the *Escherichia coli* mra cluster of cell division and cell envelope biosynthesis genes, including ftsI and ftsW. *J. Bacteriol.* 179, 5802–5811.

Hara, Y., Seki, M., Matsuoka, S., Hara, H., Yamashita, A., and Matsumoto, K. (2008). Involvement of PlsX and the acyl-phosphate dependent sn-glycerol-3-phosphate acyltransferase PlsY in the initial stage of glycerolipid synthesis in *Bacillus subtilis*. *Genes Genet. Syst.* 83, 433–442. doi: 10.1266/ggs.83.433

Hashimoto, M., Seki, T., Matsuoka, S., Hara, H., Asai, K., Sadaie, Y., et al. (2013). Induction of extracytoplasmic function sigma factors in *Bacillus subtilis* cells with defects in lipoteichoic acid synthesis. *Microbiology* 159, 23–35. doi. 10.1099/mic.0.063420-0

Hashimoto, M., Takahashi, H., Hara, Y., Hara, H., Asai, K., Sadaie, Y., et al. (2009). Induction of extracytoplasmic function sigma factors in *Bacillus subtilis* cells with membranes of reduced phosphatidylglycerol content. *Genes Genet. Syst.* 84, 191–198. doi: 10.1266/ggs.84.191

Hauser, H., Pascher, I., Pearson, R. H., and Sundell, S. (1981). Preferred conformation and molecular packing of phosphatidylethanolamine and phosphatidylcholine. *Biochim. Biophys. Acta* 650, 21–51. doi: 10.1016/0304-4157(81)90007-1

Haverstick, D. M., and Glaser, M. (1987). Visualization of Ca^{2+}-induced phospholipid domains. *Proc. Natl. Acad. Sci. U.S.A.* 84, 4475–4479. doi: 10.1073/pnas.84.13.4475

Hill, N. S., Buske, P. J., Shi, Y., and Levine, P. A. (2013). A moonlighting enzyme links *Escherichia coli* cell size with central metabolism. *PLoS Genetics* 9:e1003663. doi: 10.1371/journal.pgen.1003663

Hu, Z., Gogol, E. P., and Lutkenhaus, J. (2002). Dynamic assembly of MinD on phospholipid vesicles regulated by ATP and MinE. *Proc. Natl. Acad. Sci. U.S.A.* 99, 6761–6766. doi: 10.1073/pnas.102059099

Huang, K. C., Mukhopadhyay, R., and Wingreen, N. S. (2006). A curvature-mediated mechanism for localization of lipids to bacterial poles. *PLoS Comput. Biol.* 2:e151. doi: 10.1371/journal.pcbi.0020151

Huang, K. C., and Ramamurthi, K. S. (2010). Macromolecules that prefer their membranes curvy. *Mol. Microbiol.* 76, 822–832. doi: 10.1111/j.1365-2958.2010.07168.x

Igarashi, K., and Kashiwagi, K. (2006). Polyamine modulon in *Escherichia coli*: genes involved in the stimulation of cell growth by polyamines. *J. Biochem.* 139, 11–16. doi: 10.1093/jb/mvj020

Jerga, A., Miller, D. J., White, S. W., and Rock, C. O. (2007). Molecular determinants for interfacial binding and conformational change in a soluble diacylglycerol kinase. *J. Biol. Chem.* 284, 7246–7254. doi: 10.1074/jbc.M805962200

Jin, D. J., Cagliero, C., and Zhou, Y. N. (2013). Role of RNA polymerase and transcription in the organization of the bacterial nucleoid. *Chem. Rev.* 113, 8662-8682. doi: 10.1021/cr4001429

Johnson, J. E., and Cornell, R. B. (1999). Amphitropoic proteins: regulation by reversible membrane interactions (review). *Molec. Membr. Biol.* 16, 217–235. doi: 10.1080/096876899294544

Jouhet, J. (2013). Importance of the hexagonal lipid phase in biological membrane organization. *Front. Plant Sci.* 4:494. doi: 10.3389/fpls.2013.00494

Kawai, F., Hara, H., Takamatsu, H., Watabe, K., and Matsumoto, K. (2006). Cardiolipin enrichment in spore membranes and its involvement in germination of *Bacillus subtilis* Marburg. *Genes Genet. Syst.* 81, 69–76. doi: 10.1266/ggs.81.69

Kawai, F., Shoda, M., Harashima, R., Sadaie, Y., Hara, H., and Matsumoto, K. (2004). Cardiolipin domains in *Bacillus subtilis* Marburg membranes. *J. Bacteriol.* 186, 1475–1483. doi: 10.1128/JB.186.5.1475-1483.2004

Kennell, D., and Riezman, H. (1977). Transcription and translation frequencies of the *Escherichia coli* lac operon. *J. Mol. Biol.* 114, 1–21. doi: 10.1016/022-2836(77)90279-0

Kicia, M., Janeczko, N., Lewicka, J., and Hendrich, A. B. (2012). Comparison of the effects of subinhibitory concentrations of ciprofloxacin and colistin on the morphology of cardiolipin domains in *Escherichia coli* membranes. *J. Med. Microbiol.* 61, 520–524. doi: 10.1099/jmm.0.037788-0

Kikuchi, S., Shibuya, I., and Matsumoto, K. (2000). Viability of an *Escherichia coli* pgsA null mutant lacking detectable phosphatidylglycerol and cardiolipin. *J. Bacteriol.* 182, 371–376. doi: 10.1128/JB.182.2.371-376.2000

Kogoma, T. (1997). Stable DNA replication: interplay between DNA replication, homologous recombination and transcription. *Microbiol. Molec. Biol. Rev.* 61, 212–238.

Konovalova, A., Perlman, D. H., Coweles, C. E., and Silhavy, T. J. (2014). Transmembrane domain of surface-exoised outer membrane lipoprotein RcsF is threaded through the lumen of β-barrel proteins. *Proc. Natl. Acad. Sci. U.S.A.* 111, E4350-8. doi: 10.1073/pnas.1417138111

Koppelman, C.-M., den Blaauwen, T., Duursma, M. C., Heeren, R. M. A., and Nanninga, N. (2001). *Escherichia coli* minicell membranes are enriched in cardiolipin. *J. Bacteriol.* 183, 6144–6147. doi: 10.1128/JB.183.20.6144-6147.2001

Kretschmer, S., and Schwille, P. (2014). Toward spatially regulated division of protocells: insights into the *E. coli Min System from in vitro studies. Life* (*Basel*) 4, 915–928. doi: 10.3390/life4040915

Krupka, M., Cabré, E. J., Jiménez, M., Rivas, G., Rico, A. I., and Vicente, M. (2014). Role of the FtsA C terminus as a switch for polymerization and membrane association. *mBio* 5:e02221-14. doi: 10.1128/mBio.02221-14

Lai, E. M., Nair, U., Phadke, N. D., and Maddock, J. R. (2004). Proteomic screening and identification of differentially distributed membrane proteins in *Escherichia coli*. *Mol. Microbiol.* 52, 1029–1044. doi: 10.1111/j.1365-2958.2004.04040.x

Lazarevic, V., Soldo, B., Médico, N., Pooley, H., Bron, S., and Karamata, D. (2005). *Bacillus subtilis* α-phosphoglucomutase is required for normal cell morphology and biofilm formation. *Appl. Environ. Microbiol.* 71, 39–45. doi: 10.1128/AEM.71.1.39-45.2005

Leaver, M., Dominguez-Cuevas, P., Coxhead, J. M., Daniel, R. A., and Errington, J. (2009). Life without a wall or division machine in *Bacillus subtilis*. *Nature* 457, 849–853. doi: 10.1038/nature07742

Lemmon, M. A. (2008). Membrane recognition by phospholipid-binding domains. *Nat. Rev. Molec. Cell Biol.* 9, 99–111. doi: 10.1038/nrm2328

Lenarcic, R., Halbedel, S., Visser, L., Shaw, M., Wu, L. J., and Errington, J. (2009). Localization of DivIVA by targeting to negatively curved membranes. *EMBO J.* 28, 2272–2282. doi: 10.1038/emboj.2009.129

Li, G., and Young, K. (2012). Isolation and identification of new inner membrane-associated proteins that localize to cell poles in *Escherichia coli*. *Mol. Microbiol.* 84, 276–295. doi: 10.1111/j.1365-2958.2012.08021.x

Li, Y., Lipowsky, R., and Dimova, R. (2011). Membrane nanotubes induced by aqueous phase separation and stabilized by spontaneous curvature. *Proc. Natl. Acad. Sci. U.S.A.* 108, 4731–4736. doi: 10.1073/pnas.1015892108

Libby, E., Roggiani, M., and Goulian, M. (2012). Membrane protein expression triggers chromo-somal locus repositioning in bacteria. *Proc. Natl. Acad. Sci. U.S.A.* 109, 7445–7450. doi: 10.1073/pnas.1109479109

Llopis, P. M., Jackson, A. F., Sliusarenko, O., Surovtsev, I., Heinritz, J., Emonet, T., et al. (2010). Spatial organization of the flow of genetic information in bacteria. *Nature* 466, 77–82. doi: 10.1038/nature09152

Llorente-Garcia, I., Lenn, T., Erhardt, H., Harriman, O. L., Liu, L. N., Robson, A., et al. (2014). Single-molecule in vivo imaging of bacterial respiratory complexes indicates delocalized oxidative phosphorylation. *Biochim. Biophys. Acta* 1837, 811–824. doi: 10.1016/j.bbabio.2014.01.020

Lobasso, S., Palese, L. L., Angelini, R., and Corcelli, A. (2013). Relationship between cardiolipin metabolism and oxygen availability in *Bacillus subtilis*. *FEBS Open Bio* 3, 151–155. doi: 10.1016/j.fob.2013.02.002

Lobasso, S., Saponetti, M. S., Polidoro, F., Lopalco, P., Urbanija, J., Kralj-Iglic, V., et al. (2009). Archaebacterial lipid membranes as models to study the interaction of 10-N-nonyl acridine orange with phospholipids. *Chem. Phys. Lipids* 157, 12–20. doi: 10.1016/j.chemphyslip.2008.09.002

Lopez, D., and Kolter, R. (2010). Functional microdomains in bacterial membranes. *Genes Dev.* 24, 1893–1902. doi: 10.1101/gad.1945010

Lopian, L., Elisha, Y., Nussbaum-Shochat, A., and Amster-Choder, O. (2010). Spatial and temporal organization of the *E. coli PTS components. EMBO J.* 29, 3630–3645. doi: 10.1038/emboj.2010.240

Lu, Y. J., Zhang, Y. M., Grimes, K. D., Qi, J., Lee, R. E., and Rock, C. O. (2006). Acyl-phosphates initiate membrane phospholipid synthesis in gram-positive pathogens. *Molec. Cell* 23, 765–772. doi: 10.1016/j.molcel.2006.06.030

Makise, M., Mima, S., Tsuchiya, T., and Mizushima, T. (2001). Molecular mechanism for functional interaction between DnaA protein and acidic phospholipids: identification of important amino acids. *J. Biol. Chem.* 276, 7450–7456. doi: 10.1074/jbc.M009643200

Maloney, E., Madiraju, S. C., Rajagopalan, M., and Madiraju, M. (2011). Localization of acidic phospholipid cardiolipin and DnaA in mycobacteria. *Tuberculosis* (*Edinb*) 1, S150–S155. doi: 10.1016/j.tube.2011.10.025

Manning, G. S. (2007). Counterion condensation on charged spheres, cylinders, and planes. *J. Phys. Chem.* 111, 8554–8559. doi: 10.1021/jp067084417

Margolin, W. (2012). The price of tags in protein localization studies. *J. Bacteriol.*194, 6369–6371. doi: 10.1128/JB.01640-12

Marston, A. L., Thomaides, H. B., Edwards, D. H., Sharpe, M. E., and Errington, J. (1998). Polar localization of the MinD protein of *Bacillus subtilis* and its role in selection of the mid-cell division site. *Genes Dev.* 12, 3419–3430. doi: 10.1101/gad.12.21.3419

Matsumoto, K. (1997). Phosphatidylserine synthase from bacteria. *Biochim. Biophys. Acta* 1348, 214–227. doi: 10.1016/S0005-2760(97)00110-0

Matsumoto, K. (2001). Dispensable nature of phosphatidylglycerol in *Escherichia coli*: dual roles of anionic phospholipids. *Mol. Microbiol.* 39, 1427–1433. doi: 10.1046/j.1365-2958.2001.02320.x

Matsumoto, K., Kusaka, J., Nishibori, A., and Hara, H. (2006). Lipid domains in bacterial membranes. *Mol. Microbiol.* 61, 1110–1117. doi: 10.1111/j.1365-2958.2006.05317.x

Matsumoto, K., Matsuoka, S., and Hara, H. (2012). "Membranes and lipids," in *Escherichia coli and Bacillus subtilis: The Frontiers of Molecular Microbiology Revisited*, eds Y. Sadaie and K. Matsumoto (Kerala: Research Signpost), 61–91.

Matsuoka, S., Chiba, M., Tanimura, Y., Hashimoto, M., Hara, H., and Matsumoto, K. (2011a). Abnormal morphology of *Bacillus subtilis* ugtP mutant cells lacking glucolipids. *Genes Genet. Syst.* 86, 295–304. doi: 10.1266/ggs.86.295

Matsuoka, S., Hashimoto, M., Kamiya, Y., Miyazawa, T., Ishikawa, K., Hara, H., et al. (2011b). The *Bacillus subtilis* essential gene dgkB is dispensable in mutants with defective lipoteichoic acid synthesis. *Genes Genet. Syst.* 86, 365–376. doi: 10.1266/ggs.86.365

Mazor, S., Regev, T., Mileykovskaya, E., Margolin, W., Dowhan, W., and Fishov, I. (2008). Mutual effects of MinD-membrane interaction: II. Domain structure of the membrane enhances MinD binding. *Biochim. Biophys. Acta* 1778, 2505–2511. doi: 10.1016/j.bbamem.2008.08.004

McGary, K., and Nudler, E. (2013). RNA polymerase and the ribosome: the close relationship. *Curr. Opin. Microbiol.* 16, 112–117. doi: 10.1016/j.mib.2013.01.010

McGlynn, P. (2010). Linking transcription with DNA repair, damage tolerance, and genome duplication. *Proc. Natl. Acad. Sci. U.S.A.* 107, 15314–15315. doi: 10.1073/pnas.1010659107

Mengin-Lecreulx, D., Ayala, J., Bouhss, A., van Heijenoort, J., Parquet, C., and Hara, H. (1998). Contribution of the *Pmra* promoter to expression of genes in the *Escherichia coli mra* cluster of cell envelope biosynthesis and cell division genes. *J. Bacteriol.* 180, 4406–4412.

Meredith, D. H., Plank, M., and Lewis, P. J. (2008). Different patterns of integral membrane protein localization during cell division in *Bacillus subtilis*. *Microbiology* 154, 64–71. doi: 10.1099/mic.0.2007/013268-0

Meyer, P., Cecchi, G., and Stolovitzky, G. (2014). Spatial localization of the first and last enzymes effectively connects active metabolic pathways in bacteria. *BMC Syst. Biol.* 8:131. doi: 10.1186/s12918-014-0131-1

Mileykovskaya, E. (2007). Subcellular localization of *Escherichia coli* osmosensory transporter ProP: focus on cardiolipin membrane domains. *Mol. Microbiol.* 64, 1419–1422. doi: 10.1111/j.1365-2958.2007.05766.x

Mileykovskaya, E., and Dowhan, W. (2000). Visualization of phospholipid domains in *Escherichia coli* by using the cardiolipin-specific fluorescent dye 10-N-nonyl acridineorange. *J. Bacteriol.* 182, 1172–1175. doi: 10.1128/JB.182.4.1172-1175.2000

Mileykovskaya, E., and Dowhan, W. (2005). Role of membrane lipids in bacterial division-site selection. *Curr. Opin. Microbiol.* 8, 135–142. doi: 10.1016/j.mib.2005.02.012

Mileykovskaya, E., and Dowhan, W. (2009). Cardiolipin membrane domains in prokaryotes and eukaryotes. *Biochim. Biophys. Acta* 1788, 2084–2091. doi: 10.1016/j.bbamem.2009.04.003

Mileykovskaya, E., Dowhan, W., Birke, R. L., Zheng, D., Lutterodt, L., and Haines, T. H. (2001). Cardiolipin binds nonyl acridine orange by aggregating the dye at exposed hydrophobic domains on bilayer surfaces. *FEBS Lett.* 507, 187–190. doi: 10.1016/S0014-5793(01)02948-9

Mileykovskaya, E., Fishov, I., Fu, X., Corbin, B. D., Margolin, W., and Dowhan, W. (2003). Effects of phospholipids composition on MinD-membrane interaction in vitro and in vivo. *J. Biol. Chem.* 278, 22193–22198. doi: 10.1074/jbc.M302603200

Mileykovskaya, E., and Margolin, W. (2012). "Cell division" in *Escherichia coli and Bacillus subtilis: The frontiers of Molecular Microbiology Revisited*, eds Y. Sadaie and K. Matsumoto (Kerala: Research Signpost), 149–177.

Mileykovskaya, E., Ryan, A. C., Mo, X., Lin, C. C., Khalaf, K. I., Dowhan, W., et al. (2009). Phosphatidic acid and N-acylphosphatidylethanolamine from membrane domains in *Escherichia coli* mutant lacking cardiolipin and phosphatidylglycerol. *J. Biol. Chem.* 284, 2990–3000. doi. 10.1074/jbc.M805189200

Mim, C., and Unger, V. M. (2012). Membrane curvature and its generation by BAR proteins. *Trends Bichem. Sci.* 37, 526–533. doi: 10.1016/j.tibs.2012.09.001

Mozharov, A. D., Shchipakin, V. N., Fishov, I. L., and Evtodienko, Y. V. (1985). Changes in the composition of membrane phospholipids during the cell cycle of *E. coli*. *FEBS Lett.* 186, 103–106. doi: 10.1016/0014-5793(85)81348-X

Muchová, K., Wilkinson, A., and Barak, I. (2011). Change of lipid domains in *Bacillus subtilis* cells with disrupted cell wall peptidoglycan. *FEMS Microbiol. Lett.* 325, 92–98. doi: 10.1111/j.1574-6968.2011.02417

Mulder, E., and Woldringh, C. (1989). Actively replicating nucleoids influence positioning of division sites in *Escherichia coli* filaments forming cells lacking DNA. *J. Bacteriol.* 171, 4303–4314.

Munro, G. F., Hercules, K., Morgan, J., and Sauerbier, W. (1972). Dependence of the putrescine content of *Escherichia coli* on the osmotic strength of the medium. *J. Biol. Chem.* 247, 1272–1280.

Nishibori, A., Kusaka, J., Hara, H., Umeda, M., and Matsumoto, K. (2005). Phosphatidylethanolamine domains and localization of phospholipids synthases in *Bacillus subtilis* membranes. *J. Bacteriol.* 187, 2163–2174. doi: 10.1128/JB.187.6.2163-2174.2005

Norris, V. (1995). Hypothesis: chromosome separation in *Escherichia coli* involves autocatalytic gene expression, transertion and membrane-domain formation. *Mol. Microbiol.* 16, 1051–1057. doi: 10.1111/j.1365-2958.1995.tb02330.x

Norris, V., and Amar, P. (2012). Chromosome replication in *Escherichia coli*: life on the scales. *Life* 2, 286–312. doi: 10.3390/life2040286

Norris, V., Ayala, J. A., Begg, K., Bouché, J. P., Bouloc, P., Boye, E., et al. (1994). Cell cycle control: prokaryotic solutions to eukaryotic problems? *J. Theor. Biol.* 168, 227–230. doi: 10.1006/jtbi.1994.1102

Norris, V., den Blaauwen, T., Doi, R. H., Harshey, R. M., Janniere, L., Jiménez-Sánchez, A., et al. (2007a). Toward a hyperstructure taxonomy. *Annu. Rev. Microbiol.* 61, 309–329. doi: 10.1146/annurev.micro.61.081606.103348

Norris, V., den Blaauwen, T., Cabin-Flaman, A., Doi, R. H., Harshey, R., Janniere, L., et al. (2007b). Functional taxonomy of bacterial hyperstructures. *Microbiol. Mol. Biol. Rev.* 71, 230–253. doi: 10.1128/MMBR.00035-06

Norris, V., and Madsen, M. S. (1995). Autocatalytic gene expression occurs via transertion and membrane domain formation and underlies differentiation in bacteria: a model. *J. Mol. Biol.* 253, 739–748. doi: 10.1006/jmb.1995.0587

Norris, V., Misevic, G., Delosme, J.-M., and Oshima, A. (2002). Hypothesis: a phospholipid translocase couples lateral and transverse bilayer asymmetries in dividing bacteria. *J. Mol. Biol.* 318, 455–462. doi: 10.1016/S0022-2836(02)00098-0

Norris, V., Nana, G. G., and Audinot, J. N. (2014a). New approaches to the problem of generating coherent, reproducible phenotypes. *Theory Biosci.* 133, 47–61. doi: 10.1007/s12064-013-0185-4

Norris, V., Reusch, R. N., Igarashi, K., and Root-Bernstein, R. (2014b). Molecular complementarity between simple, universal molecules and ions limited phenotype space in the precursors of cells. *Biol. Direct.* 10, 28. doi: 10.1186/s13062-014-0028-3

Norris, V., Onoda, T., Pollaert, H., and Grehan, G. (1999). The mechanical advantages of DNA. *BioSystems* 49, 71–78. doi: 10.1016/S0303-2647(98)00031-8

Norris, V., Woldringh, C. L., and Mileykovskaya, E. (2004). Hypothesis: a hypothesis to explain division site selection in *Escherichia coli* by combining nucleoid occlusion and min. *FEBS Lett.* 561, 3–10. doi: 10.1016/S0014-5793(04)00135-8

Okada, M., Matsuzaki, H., Shibuya, I., and Matsumoto, K. (1994). Cloning, sequencing, and expression in *Escherichia coli* of the *Bacillus subtilis* gene for phosphatidylserine synthase. *J. Bacteriol.* 176, 7456–7461.

Okuda, S., and Tokuda, H. (2011). Lipoprotein sorting in bacteria. *Annu. Rev. Microbiol.* 65, 239–259. doi: 10.1146/annurev-micro-090110-102859

Oliva, M. A., Halbedel, S., Freund, S. M., Dutow, P., Leonard, T. A., Veprintsev, D. B., et al. (2010). Features critical for membrane binding reveal ed by DivIVA crystal structure. *EMBO J.* 29, 1988–2001. doi: 10.1038/emboj.2010.99

Oliver, P. M., Crooks, J. A., Leidl, M., Yoon, E. J., Saghatelian, A., and Weibel, D. B. (2014). Localization of anionic phospholipids in *Escherichia coli* cells. *J. Bacteriol.* 196, 3386–3398. doi: 10.1128/JB.01877-14

Onoda, T., Enokizono, J., Kaya, H., Oshima, A., Freestone, P., and Norris, V. (2000). Effects of calcium and calcium chelators on growth and morphology of *Escherichia coli* L-form NC-7. *J. Bacteriol.* 182, 1419–1422. doi: 10.1128/JB.182.5.1419-1422.2000

Osawa, M., Anderson, D. E., and Erickson, H. P. (2008). Reconstitution of contractile FtsZ rings in liposomes. *Science* 320, 792–794. doi: 10.1126/science.1154520

Osawa, M., and Erickson, H. P. (2013). Liposome division by a simple bacterial division machinery. *Proc. Natl. Acad. Sci. U.S.A.* 110, 11000–11004. doi: 10.1073/pnas.1222254110

Osella, M., Nugent, E., and Lagomarsino, M. C. (2014). Concerted control of *Escherichia coli* cell division. *Proc. Natl. Acad. Sci. U.S.A.* 111, 3431–3435. doi: 10.1073/pnas.1313715111

Otto, A., Bernhardt, J., Meyer, H., Schaffer, M., Herbst, F.-A., Siebourg, J., et al. (2010). Systems-wide temporal proteomic profiling in glucose-starved *Bacillus subtilis*. *Nat. Commun.* 1, 137. doi: 10.1038/ncomms1137

Paoletti, L., Lu, Y. J., Schujman, G., de Mendoza, and Rock, C. O. (2007). Coupling of fatty acid and phospholipid synthesis in *Bacillus subtilis*. *J. Bacteriol.* 189, 5816–5842. doi: 10.1128/JB.00602-07

Papanastasiou, M., Orfanoudaki, G., Koukaki, M., Kountourakis, N., Frantzeskos Sardis, M. F., Aivaliotis, M., et al. (2013). The *Escherichia coli* peripheral inner membrane proteome. *Molec. Cell Proteom.* 12, 599–610. doi: 10.1074/mcp.M112.024711

Pascher, I., Sundell, S., Harlos, K., and Eibl, H. (1987). Conformation and packing properties of membrane lipids: the crystal structure of sodium dimyristoylphosphatidylglycerol. *Biochim. Biophys. Acta* 896, 77–88. doi: 10.1016/0005-2736(87)90358-0

Patel, H. V., Vyas, K. A., Li, X., Savtchenko, R., and Roseman, S. (2004). Subcellular distribution of enzyme I of the *Escherichia coli* phosphoenolpyruvate: glucose phosphotransferase system depends on growth conditions. *Proc. Natl. Acad. Sci. U.S.A.* 102, 17486–17491. doi: 10.1073/pnas.0407865101

Patrick, J. E., and Kearns, D. B. (2008). MinJ(YvjD) is a topological determinant of cell division in *Bacillus subtilis*. *Mol. Microbiol.* 70, 1166–1179. doi: 10.1111/j.1365-2958.2008.06469.x

Pavlendova, N., Muchova, K., and Barak, I. (2010). Expression of *Escherichia coli* Min system in *Bacillus subtilis* and its effect on cell division. *FEMS Microbiol. Lett.* 302, 58–68. doi: 10.1111/j.1574-6968.2009.01832.x

Petit, J. M., Huet, O., Gallet, P. F., Maftah, A., Ratinaud, M. H., and Julien, R. (1994). Direct analysis and significance of cardiolipin transverse distribution in mitochondrial inner membranes. *Eur. J. Biochem.* 220, 871–879. doi: 10.1111/j.1432-1033.1994.tb18690.x

Petit, J. M., Maftah, A., Ratinaud, M. H., and Julien, R. (1992). 10-N-nonyl acridine orange interacts with cardiolipin and allows the quantification of this phospholipid in isolated mitochondria. *Eur. J. Biochem.* 209, 267–273. doi: 10.1111/j.1432-1033.1992.tb17285.x

Pichoff, S., and Lutkenhaus, J. (2005). Tethering the Z ring to the membrane through a conserved membrane targeting sequence in FtsA. *Mol. Microbiol.* 55, 1722–1734. doi: 10.1111/j.1365-2958.2005.04522.x

Price, K. D., Roels, S., and Losick, R. (1997). A *Bacillus subtilis* gene encoding a protein similar to nucleotide sugar transferases influences cell shape and viability. *J. Bacteriol.* 179, 4959–4961.

Ramamurthi, K. S. (2010). Protein localization by recognition of membrane curvature. *Curr. Opin. Microbiol.* 13, 753–757. doi: 10.1016/j.mib.2010.09.14

Ramamurthi, K. S., and Losick, R. (2009). Negative membrane curvature as a cue for subcellular localization of a bacterial protein. *Proc. Natl. Acad. Sci. U.S.A.* 106, 13541–13545. doi: 10.1073/pnas.0906851106

Raskin, D. M., and de Boer, P. A. J. (1999). Rapid pole-to-pole oscillation of a protein required for directing division to the middle of *Escherichia coli*. *Proc. Natl. Acad. Sci. U.S.A.* 96, 4971–4976. doi: 10.1073/pnas.96.9.4971

Regev, T., Myers, N., Zarivach, R., and Fishov, I. (2012). Association of the chromosome replication initiator DnaA with the *Escherichia coli* inner membrane in vitro: quantity and mode of binding. *PLoS ONE* 7:e36441. doi: 10.1371/journal.pone.0036441

Renner, L. D., Eswaramoorthy, P., Ramamurthi, K. S., and Weibel, D. B. (2013). Studying biomolecule localization by engineering bacterial cell wall curvature. *PLoS ONE* 8:e84143. doi: 10.1371/journal.pone.0084143

Renner, L. D., and Weibel, D. B. (2011). Cardiolipin microdomains localize to negatively curved regions of *Escherichia coli* membranes. *Proc. Natl. Acad. Sci. U.S.A.* 108, 6264–6269. doi: 10.1073/pnas.1015757108

Rilfors, L., Niemi, A., Haraldsson, S., Edwards, K., Andersson, A.-S., and Dowhan, W. (1999). Reconstituted phosphatidylserine synthase from *Escherichia coli* is activated by anionic phospholipids and micelle-forming amphiphiles. *Biochim. Biophys. Acta* 1438, 281–294. doi: 10.1016/S1388-1981(99)00060-8

Romantsov, T., Battle, A. R., Hendel, J. L., Martinac, B., and Wood, J. M. (2010). Protein localization in *Escherichia coli* cells: comparison of the cytoplasmic membrane protein ProP, LacY, ProW, AqpZ, MscS, and MscL. *J. Bacteriol.* 192, 912–924. doi: 10.1128/JB.00967-09

Romantsov, T., Helbig, S., Culham, D. E., Gill, C., Stalker, L., and Wood, J. (2007). Cardiolipin promotes polar localization of osmosensory transporter ProP in *Escherichia coli*. *Mol. Microbiol.* 64, 1455–1465. doi: 10.1111/j.1365-2958.2007.05727.x

Romantsov, T., Stalker, L., Culham, D. E., and Wood, J. M. (2008). Cardiolipin controls the osmotic stress response and the subcellular location of transporter ProP in *Escherichia coli*. *J. Biol. Chem.* 283, 12314–12323. doi: 10.1074/jbc.M709871200

Rosch, J. W., and Caparon, M. G. (2005). The ExPortal: an organelle dedicated to the biogenesis of secreted proteins in *Streptococcus pyogenes*. *Mol. Microbiol.* 58, 959–968. doi: 10.1111/j.1365-2958.2005.04887.x

Rosch, J. W., Hsu, F. F., and Caparon, M. G. (2007). Anionic lipids enriched at the exportal of *Streptococcus pyogenes*. *J. Bacteriol.* 189, 801–806. doi: 10.1128/JB.01549-06

Rowlett, V. W., and Margolin, W. (2013). The bacterial Min system. *Curr. Biol.* 23, R553–R556. doi: 10.1016/j.cub.2013.05.024

Saha, S. K., Furukawa, Y., Matsuzaki, H., Shibuya, I., and Matsumoto, K. (1996b). Directed mutagenesis, Ser-56 to Pro, of *Bacillus subtilis* phosphatidylserine synthase drastically lowers enzymatic activity and relieves amplification toxicity in *Escherichia coli*. *Biosci. Biotech. Biochem.* 60, 630–633. doi: 10.1271/bbb.60.630

Saha, S. K., Nishijima, S., Matsuzaki, H., Shibuya, I., and Matsumoto, K. (1996a). A regulatory mechanism for the balanced synthesis of membrane phospholipid species in *Escherichia coli*. *Biosci. Biotech. Biochem.* 60, 111–116. doi: 10.1271/bbb.60.111

Saier, M. H. Jr. (2013). Microcompartments and protein machines in prokaryotes. *J. Mol. Microbiol. Biotechnol.* 23, 243–269. doi: 10.1271/bbb.60.630

Saier, M. H. Jr., Ramseier, T. M., and Reizer, J. (1996). "Regulation of carbon utilization," in *Escherichia coli and Salmonella. Cellular and Molecular Biology*, eds F. C. Neidhardt, R. Curtiss, J. L. Ingraham, E.C.C. Lin, K. B. Low, B. Magasanik, et al. (Washington, DC: American Society for Microbiology), 1325–1343.

Salje, J., van den Ent, F., de Boer, P., and Löwe, J. (2011). Direct membrane binding by bacterial actin MreB. *Molec. Cell* 43, 478–487. doi: 10.1016/j.molec.2011.07.008

Salzberg, L. I., and Helmann, J. D. (2008). Phenotypic and transcriptomic characterization of *Bacillus subtilis* mutants with grossly altered membrane composition. *J. Bacteriol.* 190, 7797–7807. doi: 10.1128/JB.00720-08

Sanamrad, A., Persson, F., Lundius, E. G., Fange, D., Gynnå, A. H., and Elf, J. (2014). Single-particle tracking reveals that free ribosomal subunits are not excluded from the *Escherichia coli* nucleoid. *Proc. Natl. Acad. Sci. U.S.A.* 111, 11413–11418. doi: 10.1073/pnas.1411558111

Sankaran, K., and Wu, H. C. (1994). Lipid modification of bacterial prolipoprotein. Transfer of diacylglycerol moiety from phosphatidylglycerol. *J. Biol. Chem.* 269, 19701–19706.

Santos, T. M. A., Lin, T. Y., Rajendran, M., Anderson, S. M., and Weibel, D. B. (2014). Polar localization of *Escherichia coli* chemoreceptors requires an intact Tol-Pal complex. *Moelc. Microbiol.* 92, 985–1004. doi: 10.1111/mm.12609

Sato, S., Kawamoto, J., Sato, S. B., Watanabe, B., Hiratake, J., Esaki, N., et al. (2012). Occurrence of a bacterial membrane microdomain at the cell division site enriched in phospholipids with polyunsaturated hydrocarbon chains. *J. Biol. Chem.* 287, 24113–24121. doi: 10.1074/jbc.M111.318311

Schaechter, M., Maaloe, O., and Kjeldgaard, N. O. (1958). Dependency on medium and temperature of cell size and chemical composition during balanced grown of *Salmonella typhimurium*. *J. Gen. Microbiol.* 19, 592–606. doi: 10.1099/00221287-19-3-592

Schirner, K., Marles-Wright, J., Lewis, R. J., and Errington, J. (2009). Distinct and essential morphogenic functions for wall- and lipoteichoic acids in *Bacillus subtilis*. *EMBO J.* 28, 830–842. doi: 10.1038/emboj.2009.25

Schuber, F. J. (1989). Influence of polyamines on membrane functions. *Biochemistry* 260, 1–10.

Schulze, R. J., Komar, J., Botte, M., Allen, W. J., Whitehouse, S., Gold, V. A., et al. (2014). Membrane protein insertion and proton-motive-force-dependent secretion through the bacterial holo-translocon SecYEG-SecDF-YajC-YidC. *Proc. Natl. Acad. Sci. U.S.A.* 111, 4844–4849. doi: 10.1073/pnas.1315901111

Seddon, J. M. (1990). Structure of the inverted hexagonal (HII) phase, and non-lamellar phase transitions of lipids. *Biochim. Biophys. Acta* 1031, 1–69. doi: 10.1016/0304-4157(90)90002-T

Seki, T., Mineshima, R., Hashimoto, M., Matsumoto, K., Hara, H., and Matsuoka, S. (2015). Repression of the activities of extracytoplasmic function σ factors, σM and σV of *Bacillus subtilis* by glucolipids in *Escherichia coli* cells. *Genes Genet. Syst.* (in press).

Shakibai, N., Ishidate, K., Reshetnyak, E., Gunji, S., Kohiyama, M., and Rothfield, L. (1998). High-affinity binding of hemimethylated oriC by *Escherichia coli* membranes is mediated by a multiprotein system that includes SeqA and a newly identified factor, SeqB. *Proc. Natl. Acad. Sci. U.S.A.* 95, 11117–11121. doi: 10.1073/pnas.95.19.11117

Shapiro, L., McAdams, H. H., and Losick, R. (2002). Generating and exploiting polarity in bacteria. *Science* 298, 1942–1946. doi: 10.1126/science.1072163

Shiba, Y., Miyagawa, H., Nagahama, H., Matsumoto, K., Kondo, D., Matsuoka, S., et al. (2012). Exploring the relationship between lipoprotein mislocalization and activation of the Rcs signal transduction system in *Escherichia coli*. *Microbiology* 158, 1238–1248. doi: 10.1099/mic.0.056945-0

Shiba, Y., Yokoyama, Y., Aono, Y., Kiuchi, T., Kusaka, J., Matsumoto, K., et al. (2004). Activation of the Rcs signal transduction system is responsible for the thermosensitive growth defect of an *Escherichia coli* mutant lacking phosphatidylglycerol and cardiolipin. *J. Bacteriol.* 186, 6526–6535. doi: 10.1128/JB.186.19.6526-6535.2004

Shibuya, I. (1992). Metabolic regulations and biological functions of phospholipids in *Escherichia coli*. *Prog. Lipid Res.* 31, 245–299. doi: 10.1016/0163-7827(92)90010-G

Shiomi, D., and Margolin, W. (2007). Sweet sensor for size-conscious bacteria. *Cell* 130, 21–218. doi: 10.1016/j.cell.2007.07.011

Shiomi, D., Yoshimoto, M., Homma, M., and Kawagishi, I. (2006). Helical distribution of the bacterial chemoreceptor via colocalization with the Sec protein translocation machinery. *Mol. Microbiol.* 60, 894–906. doi: 10.1111/j.1365-2958.2006.05145.x

Siddiqui, R. A., Hoischen, C., Holst, O., Heinze, I., Schlott, B., Gumpert, J., et al. (2006). The analysis of cell division and cell wall synthesis genes reveals mutationally inactivated ftsQ and mraY in a protoplast-type L-form of *Escherichia coli*. *FEMS Microbiol. Lett.* 258, 305–311. doi: 10.1111/j.1574-6968.2006.00237.x

Singer, S. J., and Nicolson, G. L. (1972). The fluid mosaic model of the structure of cell membranes. *Science* 175, 720–731. doi: 10.1126/science.175.4023.720

Slater, S., Wold, S., Lu, M., Boye, E., Skarstad, K., and Kleckner, N. (1995). *E. coli* SeqA protein binds oriC in two different methyl-modulated reactions appropriate to its roles in DNA replication initiation and origin sequestration. *Cell* 82, 927–936. doi: 10.1016/0092-8674(95)90272-4

Sourjik, V., and Armitage, J. P. (2010). Spatial organization in bacterial chemotaxis. *EMBO J.* 29, 2724–2733. doi: 10.1038/emboj.2010.178

Strahl, H., Bürmann, F., and Hamoen, L. W. (2014). The actin homologue MreB organizes the bacterial cell membrane. *Nat. Commun.* 5, 3442. doi: 10.1038/ncomms4442

Strahl, H., Turlan, C., Khalid, S., Bond, P. J., Kebalo, J. M., Peyron, P., et al. (2015). Membrane recognition and dynamics of the RNA degradosome. *PLoS Genet.* 11:e1004961. doi: 10.1371/journal.pgen.1004961

Suefuji, K., Valluzzi, R., and RayChaudhuri, D. (2002). Dynamic assembly of MinD into filament bundles modulated by ATP, phospholipids, and MinE. *Proc. Natl. Acad. Sci. U.S.A.* 99, 16776–16781. doi: 10.1073/pnas.262671699

Suzuki, M., Hara, H., and Matsumoto, K. (2002). Envelope disorder of *Escherichia coli* cells lacking phosphatidylglycerol. *J. Bacteriol.* 184, 5418–5425. doi: 10.1128/JB.184.19.5418-5425.2002

Szeto, T. H., Rowland, S. L., Habrukowich, C. L., and King, G. F. (2003). The MinD membrane targeting sequence is a transdplantable lipid-binding helix. *J. Biol. Chem.* 278, 40050–40056. doi: 10.74/jbc.M306876200

Szeto, T. H., Rowland, S. L., Rothfield, L. I., and King, G. F. (2002). Membrane localization of MinD is mediated by a C-terminal motif that is conserved across eubacteria, archaea, and chloroplasts. *Proc. Natl. Acad. Sci. U.S.A.* 99, 15693–15698. doi: 10.1073/pnas.232590599

Szwedziak, P., Wang, Q., Bharat, T. A. M., Tsim, M., and Löwe, J. (2014). Architecture of the ring formed by the tubulin homologue FtsZ in bacterial cell division. *elife* 3:e04601. doi: 10.7554/eLife.04601

Szwedziak, P., Wang, Q., Freund, S. M. V., and Löwe, J. (2012). FtsA forms actin-like protofilaments. *EMBO J.* 31, 2249–2260. doi: 10.1038/emboj.2012.76

Takada, H., Fukushima-Tanaka, S., Morita, M., Kasahara, Y., Watanabe, S., Chibazakura, T., et al. (2014). An essential enzyme for phospholipid synthesis associates with the *Bacillus subtilis* divisome. *Mol. Microbiol.* 91, 242–255. doi: 10.1111/mmi.12457

Tan, B. K., Bogdanov, M., Zhao, J., Dowhan, W., Raetz, C. R. H., and Guan, Z. (2012). Discovery of a cardiolipin synthase utilizing phosphatidylethanolamine and phosphatidylglycerol as substrates. *Proc. Natl. Acad. Sci. U.SA.* 109, 16504–16509. doi: 10.1073/pnas.1212797109

Terui, Y., Akiyama, M., Sakamoto, A., Tomitori, H., Yamamoto, K., Ishihama, A., et al. (2012). Increase in cell viability by polyamines through stimulation of the synthesis of ppGpp regulatory protein and ω protein of RNA polymerase in *Escherichia coli*. *Int. J. Biochem. Cell Biol.* 44, 412–422. doi: 10.1016/j.biocel.2011.11.017

Tran, T. T., Panesso, D., Mishra, N. N., Mileykovskaya, E., and Arias, C. A. (2013). Daptomycin-resistant *Enterococcus faecalis* diverts the antibiotic molecule from the division septum and remodels cell membrane phospholipids. *mBio* 4:e00281-13. doi: 10.1128/mBio.00281-13

Tsui, H. C. T., Keen, S., Sham, L. T., Wayne, K. J., and Winkler, M. E. (2011). Dynamic distribution of the SecA and SecY translocase subunits and septal localization of the HtrA surfacee chaperon/protease during *Streptococcus pneumoniae* D39 cell division. *mBio* 2:e00202-11. doi: 10.1128/mBio.00202-11

Usui, M., Sembongi, H., Matsuzaki, H., Matsumoto, K., and I Shibuya, I. (1994). Primary structures of the wild-type and mutant alleles encoding the phosphatidylglycerophosphate synthase of *Escherichia coli*. *J. Bacteriol.* 176, 3389–3392.

van den Brink-van der Laan, E., Boots, J. W. P., Spelbrink, R. E. J., Kool, G. M., Breukink, E. J., Killian, A., et al. (2003). Membrane interaction of the glycosyltransferase MurG: a special role for cardiolipin. *J. Bacteriol.* 185, 3773–23779. doi: 10.1128/JB.185.13.3773-3779.2003

van Dijl, J. M., Dreisbach, A., Skwark, M. J., Sibbald, M. J. J. B., Tjalsma, H., Zweers, J. C., et al. (2012). "Ins and outs of the *Bacillus subtilis* membrane proteome," in *Bacillus Cellular and Molecular Biology*, ed. P. Graumann (Norfolk: Caister Academic Press), 253–283.

van Teeffelen, S., Wang, S., Furchtgott, L., Huang, K. H., Wingreen, N. S., Shaevitz, J. W., et al. (2011). The bacterial actin MreB rotates, and rotation depends on cell-wall assembly. *Proc. Natl. Acad. Sci. U.S.A.* 108, 15822–15827. doi: 10.1073/pnas.1108999108

Vanounou, S., Parola, A. H., and Fishov, I. (2003). Phosphatidylethanolamine and phosphatidyl-glycerol are segregated into different domains in bacterial membrane. A study with pyren-labelled phospholipids. *Mol. Microbiol.* 49, 1067–1079. doi: 10.1046/j.1365-2958.2003.03614.x

Vecchiarelli, A. G., Li, M., Mizuuchi, M., and Mizuuchi, K. (2014). Differential affinities of MinD and MinE to anionic phospholipid influence Min pattering dynamics in vitro. *Mol. Microbiol.* 93, 453–463. doi: 10.1111/mmi.12669

Vereb, G., Szöllösi, J., Matkó, J., Nagy, P., Farkas, T., Vigh, L., et al. (2003). Dynamic, yet structure: the cell membrane three decades after the Singer-Nicolson model. *Proc. Natl. Acad. Sci. U.S.A.* 100, 8053–8058. doi: 10.1073/ pnas.1332550100

Vicente, M., Gomez, M. J., and Ayala, J. A. (1998). Regulation of transcription of cell division genes in the *Escherichia coli* dcw cluster. *Cell. Mol. Life Sci.* 54, 317–324. doi: 10.1007/s000180050158

von Meyenburg, K., Jørgsen, B. B., and van Deurs, B. (1984). Physiological and morphological effects of overproduction of membrane-bound ATP synthase in *Escherichia coli* K-12. *EMBO J.* 3, 1791–1797.

Wang, X., Lesterlin, C., Reyes-Lamothe, R., Ball, G., and Sherratt, D. J. (2011). Replication and segregation of an *Escherichia coli* chromosome with two replication origins. *Proc. Natl. Acad. Sci. U.S.A.* 108, E243–E250. doi: 10.1073/pnas.1100874108

Weart, R. B., Lee, A. H., Chien, A. C., Haeusser, D. P., Hill, N. S., and Levin, P. A. (2007). A metabolic sensor governing cell size in bacteria. *Cell* 130, 335–347. doi: 10.1016/j.cell.2007.05.043

Weiner, J. H., Lemire, B. D., Elmes, M. L., Bradley, R. D., and Scraba, D. G. (1984). Overproduction of fumarate reductase in *Escherichia coli* induces a novel intracellular lipid-protein organelle. *J. Bacteriol.* 158, 590–596.

Welby, M., Poquet, Y., and Tocanne, J.-F. (1996). The spatial distribution of phospholipids and glycolipids in the membrane of the bacterium *Micrococcus luteus* varies during the cell cycle. *FEBS Lett.* 384, 107–111. doi. 10.1016/0014-5793(96)00278-5

Wichmann, C., Naumann, P. T., Spangenberg, O., Konrad, M., Mayer, F., and Hoppert, M. (2003). Liposomes for microcompartmentation of enzymes and their influence on catalytic activity. *Biochem. Biophys. Res. Commun.* 310, 1104–1110. doi: 10.1016/j.bbrc.2003.09.128

Woldringh, C. L. (2002). The role of co-transcriptional translation and protein translocation (transertion) in bacterial chromosome segregation. *Mol. Microbiol.* 45, 17–29. doi: 10.1046/j.1365-2958.2002.02993.x

Woldringh, C. L., Jensen, P. R., and Westerhoff, H. V. (1995). Structure and partitioning of bacterial DNA: determined by a balance of compaction and expansion forces? *FEMS Microbiol. Lett.* 131, 235–242. doi: 10.1111/j.1574-6968.1995.tb07782.x

Woldringh, C. L., and Nanninga, N. (1985). "Structure of the nucleoid and cytoplasm in the intact cell," in *Molecular Cytology of Escherichia coli*. ed. N. Nanninga (London: Academic Press), 161–197.

Woldringh, C. L., and Nanninga, N. (2006). Structural and physical aspects of bacterial chromosome segregation. *J. Struct. Biol.* 156, 273–283. doi: 10.1016/j.jsb.2006.04.013

Yao, J., and Rock, C. O. (2013). Phosphatidic acid synthesis in bacteria. *Biochim. Biophys. Acta* 1831, 495–502. doi: 10.1016/j.bbalip.2012.08.018

Yoshimura, M., Oshima, T., and Ogasawara, N. (2007). Involvement of the YneS/YgiH and PlsX proteins in phospholipid biosynthesis in both *Bacillus subtilis* and *Escherichia coli*. *BMC Microbiol.* 7:69. doi: 10.1186/1471-2180-7-69

Yung, M. W., and Green, C. (1986). The binding of polyamines to phospholipid bilayers. *Biochem. Pharmacol.* 35, 4037–4041. doi: 10.1016/0006-2952(86)90024-9

Zerrouk, Z., Alexandre, S., Lafontaine, C., Norris, V., and Valleton, J. M. (2008). Inner membrane lipids of *Escherichia coli* form domains. *Colloids and Surfaces B Biointerfaces* 63, 306–310. doi: 10.1016/j.colsurfb.2007.12.016

Zheng, M., Chiang, Y. L., Lee, H. L., Kong, L. R., Hsu, S.-T. D., Hwang, I.-S., et al. (2014). Self-assembly of MinE on the membrane underlies formation of the MinE ring to sustain function of the *Escherichia coli* Min system. *J. Biol. Chem.* 289, 21252–21266. doi: 10.1074/jbc.M114.571976

18

Chromosome replication, cell growth, division and shape: A personal perspective

Arieh Zaritsky [1*] *and Conrad L. Woldringh* [2*]

[1] *Faculty of Natural Sciences, Ben-Gurion University of the Negev, Be'er-Sheva, Israel,* [2] *Swammerdam Institute for Life Sciences, University of Amsterdam, Amsterdam, Netherlands*

Correspondence:
Arieh Zaritsky,
Faculty of Natural Sciences,
Ben-Gurion University of the Negev,
Kiryat Bergman N2 (2110), PO Box 653, Be'er-Sheva 85105, Israel
ariehzar@gmail.com;
Conrad L. Woldringh,
Swammerdam Institute for Life Sciences, University of Amsterdam,
Science Park 904, NL-1098 XH Amsterdam, Netherlands
c.woldringh@gmail.com

The origins of Molecular Biology and Bacterial Physiology are reviewed, from our personal standpoints, emphasizing the coupling between bacterial growth, chromosome replication and cell division, dimensions and shape. Current knowledge is discussed with historical perspective, summarizing past and present achievements and enlightening ideas for future studies. An interactive simulation program of the bacterial cell division cycle (BCD), described as "The Central Dogma in Bacteriology," is briefly represented. The coupled process of transcription/translation of genes encoding membrane proteins and insertion into the membrane (so-called transertion) is invoked as the functional relationship between the only two unique macromolecules in the cell, DNA and peptidoglycan embodying the nucleoid and the sacculus respectively. We envision that the total amount of DNA associated with the replication terminus, so called "nucleoid complexity," is directly related to cell size and shape through the transertion process. Accordingly, the primary signal for cell division transmitted by DNA dynamics (replication, transcription and segregation) to the peptidoglycan biosynthetic machinery is of a physico-chemical nature, e.g., stress in the plasma membrane, relieving nucleoid occlusion in the cell's center hence enabling the divisome to assemble and function between segregated daughter nucleoids.

Keywords: bacterial cell division cycle, nucleoid complexity and segregation, size and shape determination, transertion, peptidoglycan biosynthesis

Bacteriology and the Molecular Biology Revolution

Bacteriology was conceived by the Dutch Scientist Antony van Leeuwenhoek in the 17th Century (Porter, 1976), but considered "The Last Stronghold of Lamarckism" until 1943, when the ingenious Fluctuation Test was performed (Luria and Delbrück, 1943). The Phage Group of reductionists led by Max Delbrück (Cairns et al., 1966) revolutionized Basic Genetics to explain the flow of genetic information from Mendelian genes to proteins in molecular terms. This transformation was preceded by the era of protein biochemistry that could not easily pass the concept hurdle of enzyme-cannot-make-enzyme paradox (Stent and Calendar, 1978). Pure logic supported by simple, clear-cut experiments forced them to conclude that the long, seemingly monotonous DNA macromolecule is the storehouse of genetic information.

Molecular Biology developed quickly by clarifying that the transforming principle (Avery et al., 1944) was DNA: its structure was deciphered (Watson and Crick, 1953), semi-conservative replication demonstrated (Measelson and Stahl, 1958), functions in transcription/translation into proteins disclosed (Nirenberg, 2004), and manipulations crossed species barriers (Balbás et al., 1986).

Two mutually-exclusive groups that hardly exchanged information were responsible for the revolution: those mentioned above led by Physicist Max Delbrück and Chemists led by Arthur Kornberg (Kornberg and Baker, 1992). Exposing the DNA's symmetrical beauty and crucial role required seminal studies by persistent scientists such as Erwin Chargaff and Rosalind Franklin, who were in the frontier's cutting edge but individualistic and less lucky (Watson, 1996).

There were unavoidable diversions: some excellent scientists considered penicillin resistance to be an adaptive response, others described enzyme induction in terms of kinetics, still others thought of bacterial conjugation as zygote formation, but these and many more failed attempts were indispensable for the main thrust of advance. Furthermore, the absolute acceptance of the operon model (Jacob and Monod, 1961) for example, misled others to consider *lon* mutants as regulatory (Markovitz, 1964).

Merging molecular biology with general bacteriology, basic genetics and sophisticated microscopic and physical techniques discovered the sexuality and circularity of the bacterial chromosome (Jacob and Wollman, 1956; Cairns, 1963; Hayes, 1968), its replication schedule (Helmstetter et al., 1968), and the nucleoid structure (Kellenberger et al., 1958; Woldringh and Odijk, 1999).

The Origins of Bacterial Physiology

Until the late 1920's, bacterial cultures were thought to be composed of cells that constantly change size, form and structure in a meaningless fashion. In his book, Henrici (1928) noted that these changes during a single growth cycle *"occur with great regularity and are governed by simple laws which,... may probably be very precisely formulated."* It took 30 years to achieve this goal in descriptive terms, and additional decades to begin deciphering the fundamental laws anticipated by Henrici (1928) in robust, molecular terms. The multitude of forms and sizes in a random, single-species pure culture could only be explained when age distribution (Powell, 1956) and balanced growth (Campbell, 1957) were defined, and the Copenhagen School (Maalϕe and Kjeldgaard, 1966; Schaechter, 2006) described how cell size and composition change with the medium (Schaechter et al., 1958) and during transitions between growth rates (Kjeldgaard et al., 1958).

Ole Maalϕe was working at The State Serum Institute (Cooper, 1993) until he was named a Professor and started, late in 1958, The Institute of Microbiology. It seems to some of us that Ole entertained the idea to imitate Niels Bohr's Physics Institute, likely because he held Bohr in the highest admiration and was a good friend of his son Aage, also a Nobel laureate in Physics. This Institute and Ole's strong personality influenced dramatically several generations of scientists involved in investigating physiological aspects of the bacterial cell, nicknaming it The Copenhagen School. The numerous scientists who passed through it during their careers (Anderson et al., 2006), mostly young, promising and subsequently influential, demonstrate that it was a success.

The seminal series of experiments with *Salmonella typhimurium* published in 1958 in two back-to-back articles (Kjeldgaard et al., 1958; Schaechter et al., 1958), established the field of Bacterial Physiology and turned into its main hallmark. The stream of articles stemming from the Institute became a flood of crucial information published in the most prestigious periodicals of the time. One major motto of Ole in understanding the cell was "Look–Do Not Touch" hence studies were performed with minimal perturbations of the so-called steady-state of exponential growth (Fishov et al., 1995). After physiological manipulations were seemingly exhausted, the use of drugs and mutants became common when the mechanisms of their actions were, or thought to be deciphered. The multi-faceted phenotypes exerted by these (lack of specificity and pleiotropism, respectively) occasionally remind us to stick to this rule-of-thumb in order to keep interpretations of results as crystal-clear as possible.

This first leg of the journey to understand the logic behind the duplication of a bacterial cell, which took place in the 1950s, is described in this collection by Schaechter (2015), and the other two, partially overlapping legs in the 1960s–by Hanawalt (2015) and Helmstetter (2015). Phil studied the phenomenon of thymineless-death (TLD) in thymine-starved populations of *thyA* mutants (Cohen and Barner, 1954) employing it to better understand the connection between chromosome replication and cell growth and viability (Hanawalt et al., 1961), and Charles exploited the neat, so-called "baby-machine" that he devised (Helmstetter and Cummings, 1964) to derive the temporal aspects of the bacterial cell cycle (Helmstetter et al., 1968).

Being students during the early 1970's, here we try to fill-in the development in a perspective of half a century and in line with our view-points. To this effect, we acknowledge with admiration the ingenuity of Noboru Sueoka and Hirosho Yoshikawa, whose results with *Bacillus subtilis* (Yoshikawa and Sueoka, 1963) revealed Ole's prediction (Maalϕe, 1961) that replication initiates from a single point (later defined as *oriC*) and is sequential and multi-forked at fast growth rates (Oishi et al., 1964). Thinking rigorously, they derived marker frequency equations (Sueoka and Yoshikawa, 1965) that survived the test of time. Bidirectionality of the replication has later been demonstrated by various genetic, physiological and microscopic means (e.g., Masters and Broda, 1971; Bird et al., 1972; Prescott and Kuempel, 1972; Wake, 1972).

Experiments that investigated the fractional increase of DNA (ΔG) in amino acids-starved cultures of *Escherichia coli* 15T^{-} (so-called "runout") using dense and radioactive isotopes of thymine (Lark et al., 1963) led to the discovery of the so-called premature initiation (Pritchard and Lark, 1964), distinguishing between the two independent processes of replication, initiation and elongation. This distinction had clearly been indicated by Phil's classical experiments (Hanawalt et al., 1961), and was later supported by isolating two groups of conditional-lethal replication mutants (Hirota et al., 1968) that either stopped replication immediately upon transfer to the restrictive temperature (elongation) or allowed completion of the ongoing cycle but not new initiations.

Growth, Chromosome Replication and Cell Division; the BCD

Two essential, unique macromolecules (structures) exist in a bacterium: DNA (nucleoid) that stores the genetic information, and the shape-maintaining peptidoglycan (sacculus), which also protects the cell from rupture by its osmotic pressure (turgor). To survive, the cell must divide after its genome doubles and in a plane between the two emerging sets, hence duplications of the two are coupled, temporally and spatially. Much effort is expended to discover the mechanism responsible for this coupling, which raises the efficacy of competition among species. To study this coupling, reproducible steady-state conditions and well-defined perturbations (Maalϕe and Kjeldgaard, 1966) have been exploited.

Wild-type *E. coli* can synthesize all of its component macromolecules necessary for duplication from aqueous salts solution. Multiplication rate is carbon source-dependent, the most efficient of which is glucose, supporting doubling time τ of about 40 min at 37°C. Slower rates are obtained on poorer sources, whereas adding organic building blocks result in faster rates, the maximum achievable being about 3 h^{-1} (i.e., $\tau_{min} \approx 20$ min). Irrespectively, the time *C* taken to duplicate the chromosome (of ~4.6 Mb) is constant, *ca.* 40 min (Helmstetter et al., 1968). A cell divides into two morphologically-identical daughters (Trueba and Woldringh, 1980) about 20 min (designated *D*) after termination of replication hence division follows replication-initiation by about 1 h. This model was experimentally confirmed for cells growing at τ ranging 20–70 min (growth rate μ of 3–0.9 h^{-1}, respectively). Situations with $\tau < C$ are achieved by initiating new replication rounds before completing the previous ones. Under slow growth rates, on the other hand, the cycle includes a period $B\ [= \tau - (C+D)]$ in which cells have not initiated yet hence they continue to grow—much like in the G_1 period of the eukaryotic cell division cycle. This (B, C, D, τ) model has survived over 40 years with minor modifications of parameter values (e.g., Bipatnath et al., 1998; Michelsen et al., 2003), and many of its conclusions have been confirmed in other eubacteria (Helmstetter, 1996; Toro and Shapiro, 2010). It can thus be termed (Zaritsky et al., 2011, 2012) as "*The Central Dogma of The Bacterial Cell Division Cycle*" (two meanings for *BCD*). A cell cycle is divided in 3 (or 4) periods by two major events between successive fissions, initiation and termination of replication that can occur in reverse order depending on the values of *C*, *D*, and τ (Jiménez Sánchez, 2015).

Combining the noted constancy of *C* and *D* values (Helmstetter et al., 1968) with the way mean cell mass change with τ (Schaechter et al., 1958) resulted in an important insight: cell mass M_i at the time of replication-initiation is roughly constant per replication origin *oriC* (Donachie, 1968; Pritchard, 1968; Pritchard et al., 1969). The molecular mechanism regulating initiation of replication, occurring synchronously from all existing *oriC* copies and once per cell cycle, is under investigation (e.g., Leonard and Grimwade, 2010), but the apparent constancy of the M_i/*oriC* ratio is very useful, conferring a quantitative description of the bacterial cell. The cycle ends *C*+*D* min after initiation, when cell mass reaches $M_i \times 2^{(C+D)/\tau}$. The changing exponential rate of mass growth in different media is not matched by the linear, constant DNA elongation rate (1/*C*), but the faster increase of cell mass in richer media leads to increased initiation frequency as prescribed by the constant M_i/*oriC*. BCD thus explains changes in cell composition and size with τ and predicts the consequences of perturbations such as nutritional shifts (Kjeldgaard et al., 1958). These basic features and other examples are illustrated and can be followed by the user-friendly Cell Cycle Simulation program (CCSim) at https://sils.fnwi.uva.nl/bcb/ that was partially described before (Zaritsky et al., 2006, 2007, 2011, 2012) and will be re-mentioned below. It must be noted that the values of these constants do change slightly with τ—more so at longer values, can be manipulated experimentally by various means (e.g., Meacock and Pritchard, 1975; Zaritsky and Zabrovitz, 1981; Wold et al., 1994; Bipatnath et al., 1998), and inserted in the CCSim program to confirm or reject working hypotheses.

Dissociating Rates of Replication and Growth

Capitalizing on Helmstetter's "baby machine" (Helmstetter and Cummings, 1964) and just before the description of BCD (Helmstetter et al., 1968), Clark and Maalϕe (1967) demonstrated a constant rate of replication along the chromosome, with distinct discontinuities in DNA synthesis rate during the cell cycle interpreted as occurring due to initiation and completion of replication cycles. Chromosomes with multiple replication forks (also termed dichotomously replicating) is the reason for bigger ΔG added DNA in amino acids-starved, faster growing cells (Schaechter, 1961). This was the current knowledge at the end of 1968, upon the arrival of one of us (AZ) at Leicester University for graduate studies, supervised by Robert Pritchard, who had established the Genetics Department there merely 4 years earlier[1].

Digressing to some personal involvements, one of us (AZ) was very lucky to enter the atmosphere inspired by Bob and at the right time to be assigned a project in the just-opened BCD field, about which I had no clue. During 6-years of previous studies (1962–1968) at the Hebrew University of Jerusalem, my M.Sc. (with distinction but no publication) in Bacterial Genetics was supervised by Amiram Ronen, I finished 4 full years of Pre-Medical studies and attended several courses in Mathematics (my ever-lasting love). The latter was helpful to sharpen rigorous thinking, to derive the equation relating DNA concentration to the number of replication positions (Sueoka and Yoshikawa, 1965) $n\ (= C/\tau)$ irrespective of the value of *D* (Pritchard and Zaritsky, 1970; Zaritsky, 1971) and to program the huge computer at Leicester University (using card-punching). It may have been important for my active participation in developing CCSim, as described below. Bob and his large team of students were instrumental for my learning both, proper English and the BCD, mainly in the tea/coffee/seminar room that was inhabited during many hours, days and nights.

Simultaneously, the other (CLW) extended his biological and microscopic skills at the University of Amsterdam. There are at least three at that time commonly-accepted ideas that I ruled out

[1] http://www2.le.ac.uk/news/blog/2014-archive-1/october/50th-anniversary-of-department-of-genetics-celebrates-world-changing-research-1

during my Ph.D. studies and beyond namely, existence of direct DNA-membrane attachments (Woldringh, 1974), of peri-septal annuli (Woldringh, 1994) and rapid nucleoid displacement (van Helvoort and Woldringh, 1994), all has meanwhile disappeared from our knowledge-base, justifiably so. My close association with Nanne Nanninga (e.g., Woldringh and Nanninga, 1985), who in the late 1960's demonstrated the artifactual origin of mesosomes (Nanninga, 1971), enabled the establishment of a department that attracted distinguished students and scientists from all over the world, microbiologists as well as physicists and engineers. In their search to define the structural changes occurring during fixation and dehydration necessary for visualizing the bacterial nucleoid in the electron microscope, the possibilities to study live cells were improved with the reinvention and development of the confocal scanning light microscope (CSLM) by Brakenhoff (see Valkenburg et al., 1985).

Back to the main subject, at Leicester, Bob realized existence of literature-recorded contradictory results, the common feature of most is that they were obtained in thymine-requiring strains. These observations (e.g., Maalϕe and Rasmussen, 1963; Friesen and Maaloe, 1965; Lark and Lark, 1965; Beacham et al., 1968) led him to hypothesize that the replication time of the chromosome in *thyA* strains depends on the external concentration of thymine [T] present in their growth medium (Pritchard, 1974). This hypothesis could explain all discrepancies and is consistent with lack of active thymine-transport, in *E. coli* (Itsko and Schaaper, 2011) and other bacterial species (Carmody and Herriott, 1970; Reinhart and Copeland, 1973). It was strongly confirmed by four physiological methods, more or less independent of each other (Pritchard and Zaritsky, 1970; Zaritsky, 1971), and later supported by various means in other laboratories (reviewed in Zaritsky et al., 2006).

Thus, the dissociation between syntheses rates of mass and DNA, originally observed by changing the former alone (Helmstetter et al., 1968), was confirmed by exclusively manipulating *C* by limiting [T] in *thyA* strains (Pritchard and Zaritsky, 1970), affected through the intracellular [dTTP] (Beacham et al., 1971). This method is more amenable to analysis than nutritional shifts because modulating [dTTP] by changing [T] occurs abruptly, without affecting the multitude of metabolic pathways and interactions between them that accompany nutritional shifts (Scott and Hwa, 2011).

Dissociating Cell Growth and Division; the Eclipse

In a steady-state exponentially growing culture, concentrations of all cell components increase in parallel to each other and in pace with divisions (Campbell, 1957; Fishov et al., 1995). The puzzling phenomenon of division rate-maintenance after a nutritional shift-up (Kjeldgaard et al., 1958) was instantly explained by the BCD model (Helmstetter et al., 1968): a cell divides a constant time, $C+D$ min after initiation of chromosome replication, which in turn follows mass growth. The division-rate therefore changes $C+D$ (*ca.* 65) min after the change in growth rate is affected by enriching the medium. Most perturbations, by chemical/physical agents or under restrictive conditions of *ts* mutants, cause immediate block of division (Slater and Schaechter, 1974)—one that is usually restored upon transfer back to permissive conditions. Specific inhibition of protein or DNA synthesis, however, allows divisions to continue during the *D* period; these so-called residual divisions cause a decrease in average cell length (cf. entry into stationary phase) and enable estimation of the *D* period (Dix and Helmstetter, 1973; Kubitschek, 1974; Woldringh et al., 1977).

Determination of *C* and *D* periods for batch cultures of *E. coli* cells have also been performed by flow cytometry (Michelsen et al., 2003) or by image cytometry (cf. Huls et al., 1999). From these studies it becomes clear how these cell cycle periods can vary with different strains and growth conditions. The measurements indicate that the *D* period is especially variable, making it difficult to generalize the *E. coli* cell cycle.

When thymine-limited *thyA* mutants grow at fast growth rates, another puzzling phenomenon appears, namely dissociation between growth and division that is related to replication. Under these conditions, the inter-division time is longer than mass doubling time (i.e., $\tau_d > \tau_m$) thus cell size increases continuously (Zaritsky and Pritchard, 1973), and seemingly indefinitely. The 40 years-old observation (Zaritsky, 1975a) that indicated existence of a minimal possible distance l_{min} between two successive replisomes, promptly explains this phenomenon (Zaritsky et al., 2007). The question whether the mechanism involved is structural (replisome size; Norris et al., 2007) or chemical (sequestration of membrane-attached hemi-methylated DNA; Olsson et al., 2002) remains moot, but breaching this distance would extend the inter-initiation time I ($= \tau_i$) beyond the mass doubling time (τ_m) thus delay initiations, and cumulatively so (Zaritsky et al., 2007). Such a breach can be achieved by enhanced initiation frequency (Simmons et al., 2004) or slowed replication rate (Zaritsky and Pritchard, 1973). This distance is estimated to be about half of the chromosome length (l_{chr}), termed the Eclipse (l_{min}/l_{chr}) and can be expressed in units of time depending on the rate of replication (l_{min}/l_{chr}) $\times$ C (e.g., how long it takes to reach this fraction of chromosome at a given, constant rate C^{-1}). Release from this situation by restoring the permissive conditions causes a transient increase in the frequency of divisions (Zaritsky et al., 2011) thus substantiating this concept and facilitating its investigation.

The Cell Cycle Simulation Program

Our fortuitous encounter at the Luntern Conference in November 1974 was very fortunate. We had apparently met 3 years earlier in a previous meeting there, but being students it hadn't engendered significant mutual impressions. In 1974, both of us had already acquired results related to morphometric variations of *E. coli* cells under different growth conditions, theoretical (Zaritsky, 1975b) and experimental (Woldringh, 1974), and ideas about joint research sprang in the air during a long night of extensive discussions. It was just 7 months later that EMBO financed a 3-month visit for CLW in Be'er-Sheva (**Figure 1**), followed by another short-term fellowship for AZ to visit Amsterdam a couple of years later. These and follow-up visits culminated in detailed descriptions of cell dimensional rearrangements during nutritional shift-up experiments (Grover et al., 1980; Woldringh

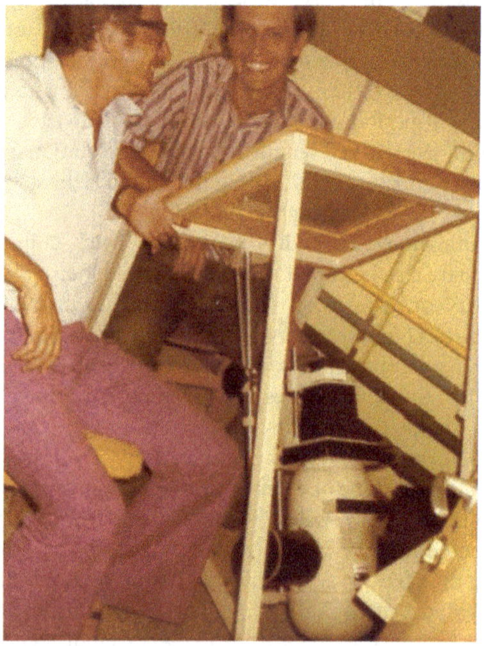

FIGURE 1 | Conrad (R) and Arieh (L) at the unique "Chezi" 30 mm-film projector, manually measuring cell dimensions and constrictions, Be'er-Sheva, Summer 1976. This primitive, bulky "machine" was designed and constructed by the Workshop of the Natural Sciences Faculty at Ben-Gurion University, led by Mr. Yechezkel Tahori in the "pre-history" of computer visualization, initiated in the University of Amsterdam by (Trueba and Woldringh, 1980) and developed further into a versatile measuring plugin "ObjectL," which runs under ImageJ (see Vischer et al., 2015).

et al., 1980; Zaritsky et al., 1982), organization of two EMBO Workshops on Duplication of Bacteria (1980 in Holland; 1984 in Israel[2]), and 40 years of continuous cooperation. One notable outcome of our interactions was implementation of an interactive simulation program (Zaritsky et al., 2011) that integrates all quantitative knowledge about the BCD (Helmstetter et al., 1968), including the anticipated behavior of various existing and prospective mutants. This program implementation was enabled by the recruitment of Norbert Vischer, a computer engineer, by the Amsterdam department chair and faculty dean Nanne Nanninga. The lab in Swammerdam Institute is thus frequently referred to as The Amsterdam School (*à la* the Copenhagen School mentioned above).

All considerations described so far and by the CCSim (**Figure 2**) do not relate to cell dimensions and shape nor to nucleoid segregation. Future versions of CCSim may be extended to incorporate these aspects.

Cell Size and Dimensions

An exponentially growing bacillary cell elongates with unnoticeable change in width, and divides evenly at a perpendicular plane (Trueba and Woldringh, 1980). The seminal observation (Schaechter et al., 1958) that larger cells at faster growth rates in richer media are both longer *and* wider led to the proposal (Zaritsky and Pritchard, 1973; Pritchard, 1974; Zaritsky, 1975b) that cell dimensions and cell shape could be directly coupled to the process of DNA replication and segregation. It was initially interpreted to involve active regulation of length L (Grover et al., 1977) or surface area S (Rosenberger et al., 1978a,b) extension, and passive response of width W to the changes of volume V and L (or S), the so-called linear/log model. Cell elongation was assumed to proceed at a constant rate (either dependent on μ or not) that is proportional to the number of *oriCs*, *terCs* (replication termini) or replisomes (Zaritsky and Pritchard, 1973). This view was later abandoned when peptidoglycan synthesis was demonstrated to be diffuse throughout the cylindrical periphery and only localized during the division process (Woldringh et al., 1987).

With such models in mind, we measured (**Figure 1**) the dimensions of *E. coli* cells cultured under steady-state of exponential growth in different media supporting various rates, prepared for electron microscopy by the agar filtration method (Woldringh et al., 1977; **Figure 3**), and compared the results with the various models (Zaritsky et al., 1982). Our nutritional-upshift experiment (Woldringh et al., 1980) revealed that the increase in cell diameter was slow and occurred mainly during the division process in the vicinity of the deepening constriction site, forming transiently tapered cells (**Figure 4**). Consequent to this slow adaptation and almost immediate change in the rate of mass synthesis, cell length overshoots, but the mechanism governing this diameter change is still enigmatic. A diameter increase during the constriction process has also been implied in populations growing in steady state where the cells showed a diameter decrease during elongation (see Figure 4 in Trueba and Woldringh, 1980). It should be noted that in all these preparations the cells had been fixed with osmium tetroxide and were air-dried, causing their flattening (Vardi and Grover, 1993). Nevertheless, the measurements compared well with those obtained from hydrated cells with phase-contrast microscopy (cf. Table 3 in Trueba and Woldringh, 1980).

Associated with cell widening, the nucleoids (bright areas in **Figure 4**) start replicating in planes tilted to the long cell axis (**Figure 4**), rather than parallel to it as during slow growth conditions. The differences in cell dimensions and nucleoids replication-planes are pronounced when *thyA* cells grow under identical conditions but with limiting [T] that impose slow replication rate (compare, e.g., panels A and B of Figure 6 of Zaritsky et al., 2006; and see Figure 1 in Woldringh et al., 1994).

Homeostasis of Cell Size and Shape

In the 1970's, the period of DNA replication during a division cycle was determined by pulse-labeling cells with ^{3}[H]-thymidine and measuring size distributions of cells prepared for radio-autographic electron microscopy (Koppes et al., 1978). These studies clarified that individual cells elongate exponentially (i.e., at a rate proportional to their length) and provided information about length variations at different events in the cycle as well as size and time correlations between these events (Koppes and Nanninga, 1980). The results led Koppes et al. (1978) in The

[2]http://ariehz.weebly.com/uploads/2/9/6/1/29618953/emboworkshop1984.pdf

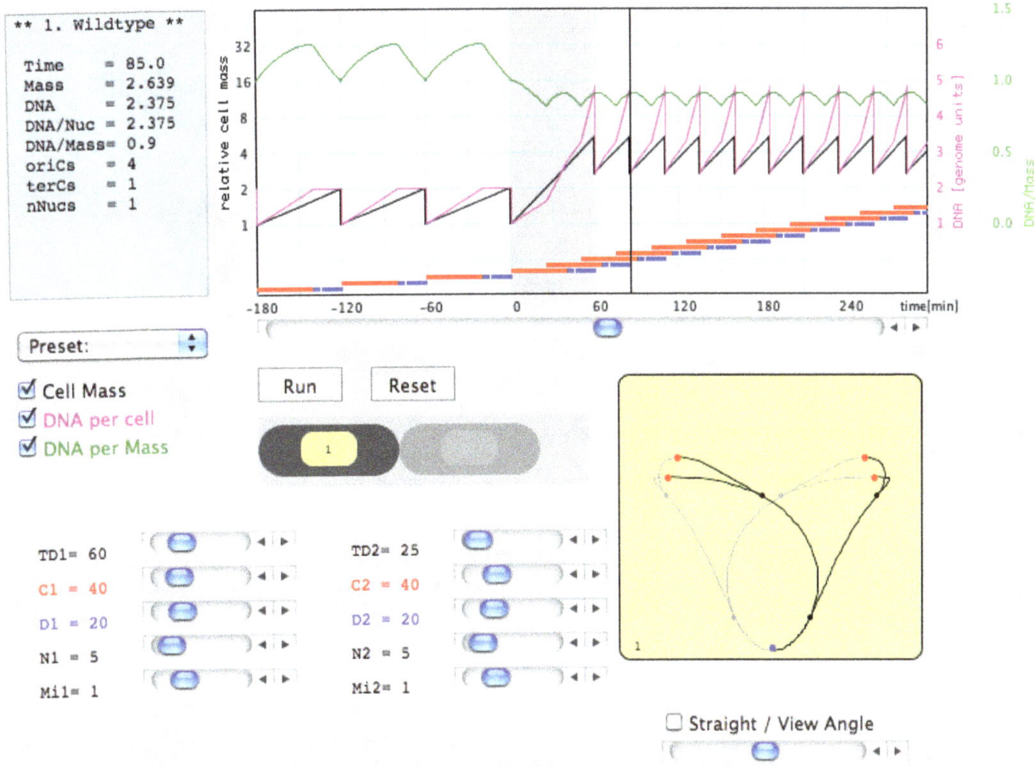

FIGURE 2 | Example of a work window of the Cell Cycle Simulation program (CCSim) for a nutritional shift-up from a doubling time $\tau_1 = 60$ min to $\tau_2 = 25$ min. The interactive program can be downloaded from: https://sils.fnwi.uva.nl/bcb/.

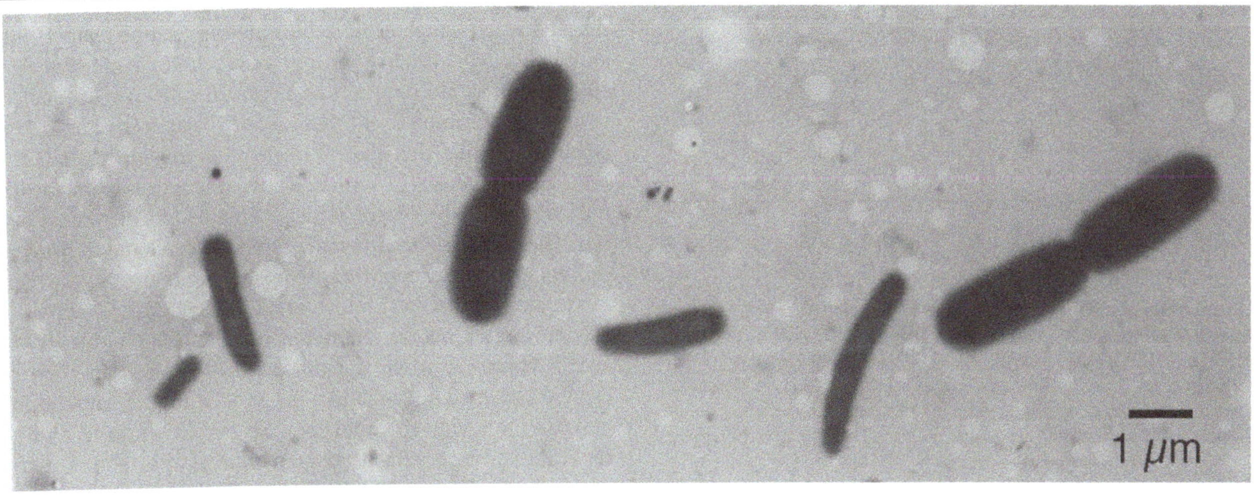

FIGURE 3 | Electron micrograph of a mixture of two *E. coli* B/r cultures prepared by agar filtration. The big cells were grown in trypton broth with a doubling time of 22 min; the small cells were grown in synthetic alanine-medium with a doubling time of 150 min. Compare with a similar preparation of mixed populations in Figure 2 of Nanninga and Woldringh (1985).

Amsterdam School to propose that cells initiate constriction after a constant length increment ΔL following initiation of DNA replication (**Figure 5**) thus establishing a correlation between cell sizes at replication initiation and at initiation of visible constriction C min later. This model of constant ΔL was recently revived (Amir, 2014) and supported by measurements of live cells (Campos et al., 2014; Iyer-Biswas et al., 2014; Taheri-Araghi et al., 2015) confirming that a growing bacterium maintains stable size by adding a constant incremental length ΔL each generation irrespective of its size at birth. This automatically leads to size

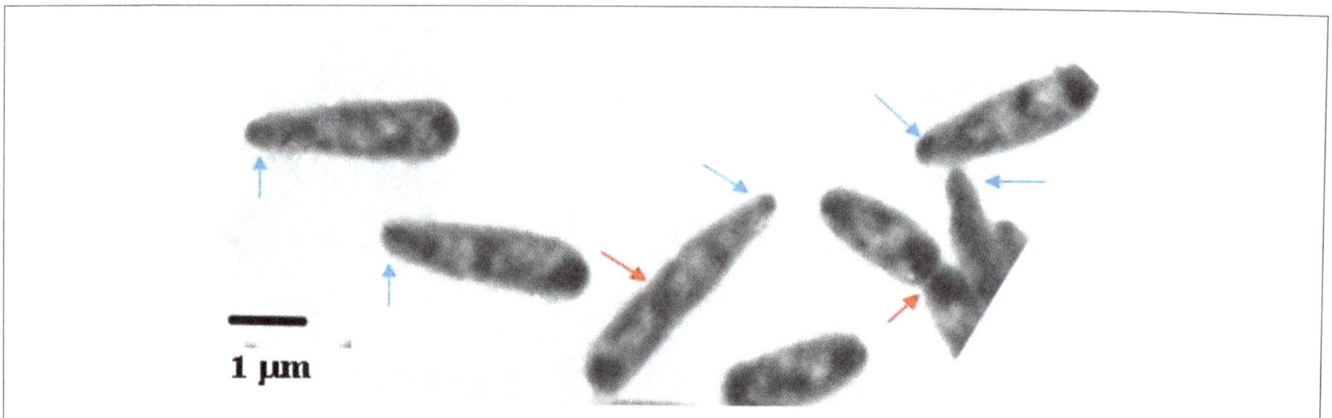

FIGURE 4 | *Escherichia coli* B/r cells prepared for electron microscopy by agar filtration, 60 min after a nutritional shift-up from $\tau_1 = 72$ to $\tau_2 = 24$ min (cf. Figure 3 in Nanninga and Woldringh, 1985). The nucleoids show up as electron-transparent regions in the air-dried cells, flattened by surface tension (cf. Woldringh et al., 1977). Red arrows indicate constriction sites, blue arrows, tapered tips.

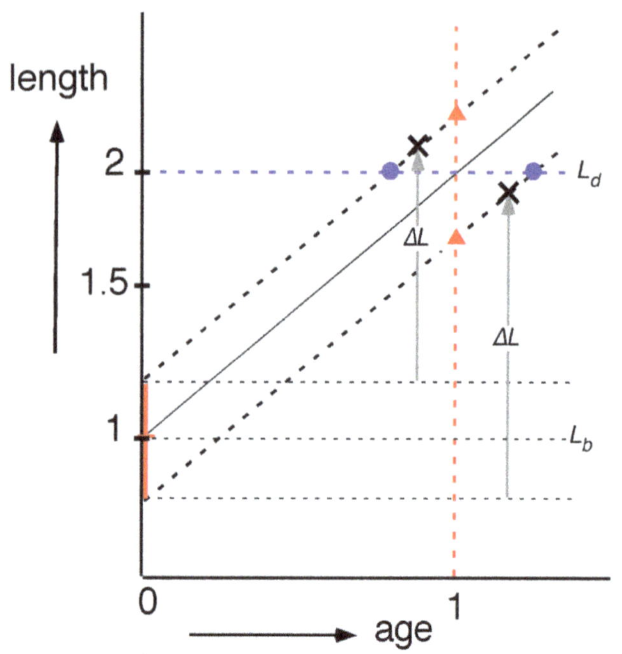

FIGURE 5 | Semi-log plot of cell length as a function of cell age (cf. Figure 6 in Koppes et al., 1978). Irrespective of newborn cell size (vertical, red bar; $L_b \pm$ SD) cells elongate exponentially (same specific elongation rate). According to the "timer" model (red triangles), newborn cells divide (at age 1) after a constant period; according to the "sizer" model (blue circles, L_d), newborn cells divide after reaching a critical size (at length 2); according to the "adder" model (black crosses) newborn cells divide after elongating with a constant length increment ΔL.

homeostasis that is valid at all growth rates obtained in different media, and since faster growing cells are longer, ΔL changes accordingly.

Jun and colleagues (Jun and Taheri-Araghi, 2014; Taheri-Araghi et al., 2015) proposed that the molecular mechanism underlying the size homeostasis by the so-called "adder" model (**Figure 5**) is related to the P-sector proteins of the *E. coli* proteome of which the total number per cell is relatively constant at different growth conditions. According to this hypothesis, accumulation of these proteins to a fixed threshold each generation would serve as a trigger for cell division. This proposal, however, does not relate mass growth to the DNA replication cycle, as suggested four decades ago (Zaritsky, 1975b). If P-sector proteins are at a fixed number per cell, then they would become diluted during the interdivision time (molecules fixed, but cell volume increases). Therefore, it is not clear how it could result in their accumulation to trigger division. Other aspects of this idea have recently been rebutted in more details (Zaritsky, 2015).

Coupling between DNA replication and cell elongation could be obtained by the nucleoid occlusion mechanism that is being relieved when daughter nucleoids are segregating apart (Mulder and Woldringh, 1989; Nanninga et al., 1990; Woldringh et al., 1990). This would require that newborn cells contain nucleoids with the same amount of DNA (*G/terC*) irrespective of their size at birth and that the state of nucleoid segregation parallels the cell's length increase. In other words, a length increment of the nucleoid would be sensed rather than a length increment of the cell. That DNA replication and segregation go hand in hand with cell elongation is supported by observations on the movement of duplicated *oriC's* (Elmore et al., 2005) and of segregating chromosome arms (cf. Youngren et al., 2014; Woldringh et al., 2015). However, while during slow growth all newborn cells can be assumed to contain nucleoids with the same amount of DNA, this will not hold for fast growth showing multifork replication. Here, stochastic premature or postponed division of mother cells will produce small and large daughter cells, respectively, with different amounts of DNA per nucleoid and thus different stages of segregation. Such cells will not signal division after a constant length increment as predicted by the "adder" model.

Another proposal (Ho and Amir, 2015; see also Robert, 2015) couples DNA replication and cell elongation to the time of initiation of DNA replication. Here, sensing of a constant length increment is starting at the last initiation of DNA replication. How a size increment rather than a critical size is monitored and whether nucleoid segregation is involved in such a model remains to be seen. Presently, information is lacking on the size of the nucleoids in newborn cells at different growth rates at the individual cell level. Better DNA staining techniques are required to observe nucleoid growth and segregation in individual cells growing in microfluidic systems.

Whatever property a cell is sensing to enable it to divide after a constant size increment irrespective of its size at birth, some communication will be necessary between the dynamics of DNA (transcription, replication and segregation) and the biosynthetic activities of peptidoglycan (elongation and constriction at perpendicular angles). It has been proposed (Rabinovitch et al., 2003) that DNA could exert stress on the membrane through the transertion mechanism (Woldringh, 2002): coupled transcription/translation of genes encoding membrane proteins and inserting these proteins into the membrane. The strength of this interaction varies along cell length with a minimum in between the segregating nucleoids. By a yet-unknown mechanism, this stress-change signal that is relayed to initiate division is proposed to be sensed by the peptidoglycan-synthetic machinery. As described by Typas et al. (2012), this may involve stretching of the peptidoglycan network hence influencing the activity of outer membrane-anchored lipoproteins. These proteins reach through the pores of the peptidoglycan network to interact with peptidoglycan synthases (penicillin binding proteins) as required for constriction (Woldringh et al., 1987). Proteins interfering with FtsZ-ring formation were recently also related to the NO phenomenon (reviewed by Wu and Errington, 2012).

The notion that a functional relationship exists between DNA dynamics and peptidoglycan biosynthesis is supported by the high correlations found between cell dimensions and the amount of DNA per nucleoid (*G*/*terC*) over a wide range of conditions (Zaritsky, 2015). Moreover, the constant aspect ratio (cell length/width ratio) supports the view that the expansion of the nucleoid during replication and segregation (and cell mass growth) occurs equally in three dimensions.

Concluding Remarks

It is well known that the formulas describing cell mass and DNA content, as well as nucleoid complexity (amount of DNA per nucleoid), can only be applied in populations that grow under steady-state conditions (Campbell, 1957; Fishov et al., 1995). However, confirmation of steady state is seldom mentioned or documented. In many studies, bacterial batch cultures growing in rich media are used after a 100- to 1000-fold dilution of an overnight culture. In such populations the steady state has probably not been reached as it requires unperturbed, exponential growth at the same rate for some 20 generations (e.g., Maalϕe and Kjeldgaard, 1966).

How do single-cell growth studies in microfluidic channels measure up to the requirements for steady state growth? It appears that constancy of growth rate and length distributions of newborn cells dividing in the channels can accurately be monitored (Wang et al., 2010; Campos et al., 2014; Osella et al., 2014; Taheri-Araghi et al., 2015). If in addition the growth experiments could include observations on nucleoid extension and segregation after labeling with, for instance, fluorescent DNA binding proteins (e.g., Männik et al., 2012; Pelletier et al., 2012), it would be possible to test the present proposal, that DNA replication and cell growth are coupled via a segregation signal for cell division. If the presumed segregation signal could be related to forces exerted by the nucleoid on the plasma membrane (Rabinovitch et al., 2003) and on the peptidoglycan network (Typas et al., 2012), it would support a belief expressed by Bob Pritchard more than 50 years ago: "...that an understanding of the determination of cell size and shape will not be possible without taking into account the physical forces to which the cell boundary is exposed." (Pritchard, 1974). We believe that the task of Physicists in expanding and deepening understanding of Cell Biology, bacteria included of course, is as critical as it was for Molecular Biology during the last Century, and similar, tight cooperation with Biologists is as crucial. The novel technologies continuously developed to enhance this end, as exemplified in the whole series of articles of this *Research Topic*, facilitate the study on both levels, single cells and single molecules in real-time.

In this memoir-style review, we try to bridge between past achievements and future prospects in the relatively-young field of Bacterial Physiology through present knowledge; scientists and students who are involved can exploit the information, which by no means is exhaustive, for the benefit of their current investigations, in the never-ending endeavor to understand Nature.

Acknowledgments

This review is a personal tribute to our mentors and inspiring colleagues Robert H Pritchard, Nanne Nanninga, Ole Maalϕe and Charles E Helmstetter. Appropriate remarks by the Editor and Reviewers, particularly lain Duggin, have improved the article.

References

Amir, A. (2014). Cell size regulation in bacteria. *Phys. Rev. Lett.* 112, 208102. doi: 10.1103/PhysRevLett.112.208102

Anderson, K. B., Atlung, T., Bennett, P. M., Cooper, S., Dennis, P., Diderichsen, B., et al. (2006). Honoring Ole Maalϕe. *Microbe* 1, 210–211.

Avery, O. T., MacLeod, C. M., and McCarty, M. (1944). Studies on the chemical nature of the substance inducing transformation of pneumococcal types. Induction of transformation by a desoxyribonucleic acid fraction isolated from pneumococcus type III. *J. Cell Biol.* 79, 137–158. doi: 10.1084/jem.79.2.137

Balbás, P., Soberón, X., Merino, E., Zurita, M., Lomeli, H., Fernando, V., et al. (1986). Plasmid vector pBR322 and its special-purpose derivatives—a review. *Gene* 50, 3–40. doi: 10.1016/0378-1119(86)90307-0

Beacham, I. R., Barth, P. T., and Pritchard, R. H. (1968). Constitutivity of thymidine phosphorylase in deoxyriboaldolase negative strains: dependence on thymine requirement and concentration. *Biochim. Biophys. Acta* 166, 589–592. doi: 10.1016/0005-2787(68)90251-7

Beacham, I. R., Beacham, K., Zaritsky, A., and Pritchard, R. H. (1971). Intracellular thymidine triphosphate concent-ration in wild-type and thymine-requiring mutants of *Escherichia coli* 15 and K12. *J. Mol. Biol.* 60, 75–86. doi: 10.1016/0022-2836(71)90448-7

Bipatnath, M., Dennis, P. P., and Bremer, H. (1998). Initiation and velocity of chromosome replication in *Escherichia coli* B/r and K-12. *J. Bacteriol.* 180, 265–273.

Bird, R. E., Louarn, J., Martuscelli, J., and Caro, L. (1972). Origin and sequence of chromosome replication in *Escherichia coli*. *J. Mol. Biol.* 70, 549–566. doi: 10.1016/0022-2836(72)90559-1

Cairns, J. (1963). The chromosome of *Escherichia coli*. *Cold Spring Harbor Symp. Quant. Biol.* 33, 43–45. doi: 10.1101/SQB.1963.028.01.011

Cairns, J., Stent, G. S., and Watson, J. D. (1966). *Phage and the Origins of Molecular Biology*. New York: Cold Spring Harbor Laboratory Press.

Campbell, A. (1957). Synchronization of cell division. *Bacteriol. Rev.* 21, 263–272.

Campos, M., Surovtsev, I. V., Kato, S., Paintdakhi, A., Beltran, B., Ebmeier, S. E., et al. (2014). A constant size extension drives bacterial cell size homeostasis. *Cell* 159, 1433–1446. doi: 10.1016/j.cell.2014.11.022

Carmody, J. M., and Herriott, R. M. (1970). Thymine and thymidine uptake by *Haemophilus influenzae* and the labeling od deoxyribonucleic acid. *J. Bacteriol.* 101, 525–530.

Clark, D. J., and Maalϕe, O. (1967). DNA replication and the division cycle in *Escherichia coli*. *J. Mol. Biol.* 23, 99–112. doi: 10.1016/S0022-2836(67)80070-6

Cohen, S. S., and Barner, H. D. (1954). Studies on unbalanced growth in *Escherichia coli*. *Proc. Natl. Acad. Sci. U.S.A.* 40, 885–893. doi: 10.1073/pnas.40.10.885

Cooper, S. (1993). The origins and meaning of the Schaechter-Maalϕe-Kjeldgaard experiments. *J. Gen. Microbiol.* 139, 1117–1124. doi: 10.1099/00221287-139-6-1117

Dix, D. E., and Helmstetter, C. E. (1973). Coupling between chromosome completion and cell division in *Escherichia coli*. *J. Bacteriol.* 115, 786–795.

Donachie, W. (1968). Relationships between cell size and time of initiation of DNA replication. *Nature* 219, 1077–1079. doi: 10.1038/2191077a0

Elmore, S., Müller, M., Vischer, N., Odijk, T., and Woldringh, C. L. (2005). Single-particle tracking of oriC-GFP fluorescent spots during chromosome segregation in *Escherichia coli*. *J. Struct. Biol.* 151, 275–287. doi: 10.1016/j.jsb.2005.06.004

Fishov, I., Grover, N. B., and Zaritsky, A. (1995). On microbial states of growth. *Mol. Microbiol.* 15, 789–794. doi: 10.1111/j.1365-2958.1995.tb02349.x

Friesen, J. D., and Maalϕe, O. (1965). On the control of DNA synthesis by amino acids in *E. coli*. *Biochim. Biophys. Acta* 95, 436–445. doi: 10.1016/0005-2787(65)90190-5

Grover, N. B., Woldringh, C. L., Zaritsky, A., and Rosenberger, R. F. (1977). Elongation of rod-shaped bacteria. *J. Theor. Biol.* 67, 181–193. doi: 10.1016/0022-5193(77)90192-8

Grover, N. B., Zaritsky, A., Woldringh, C. L., and Rosenberger, R. F. (1980). Dimensional rearrangement of rod-shaped bacteria following nutritional shift-up. I. Theory. *J. Theor. Biol.* 86, 421–439. doi: 10.1016/0022-5193(80)90343-4

Hanawalt, P. C. (2015). A balanced perspective on unbalanced growth and thymineless death. *Front. Microbiol.* 6:504. doi: 10.3389/fmicb.2015.00504

Hanawalt, P. C., Maalϕe, O., Cummings, D. J., and Schaechter, M. (1961). The normal DNA replication cycle. II. *J Mol. Biol.* 3, 156–165. doi: 10.1016/S0022-2836(61)80042-9

Hayes, W. (1968). *The Genetics of Bacteria and their Viruses*, 2nd Edn. Oxford: Blackwell Scientific Publications. 925.

Helmstetter, C. E. (1996). "Timing and synthetic activities in the cell cycle," in *Escherichia coli and Salmonella Typhimurium: Cellular and Molecular Biology*, eds F. C. Neidhardt, R. III, Curtiss, J. L. Ingraham, K. B. Low, B. Magasanik, M. Schaechter, and H. E. Umbarger (Washington, DC: ASM Press), 1627–1649.

Helmstetter, C. E. (2015). A ten-year search for synchronous cells: obstacles, solutions and practical applications. *Front. Microbiol.* 6:238. doi: 10.3389/fmicb.2015.00238

Helmstetter, C. E., and Cummings, D. J. (1964). An improved method for the selection of bacterial cells at division. *Biochim. Biophys. Acta* 82, 608–610. doi: 10.1016/0304-4165(64)90453-2

Helmstetter, H. E., Cooper, S., Pierucci, O., and Revelas, E. (1968). On the bacterial life sequence. *Cold Spring Harb. Symp. Quant. Biol.* 33, 809–822. doi: 10.1101/SQB.1968.033.01.093

Henrici, A. T. (1928). *Morphological Variations and the Rate of Growth of Bacteria*. London: Bailliere, Tindall & Cox. doi: 10.5962/bhl.title.7269

Ho, P.-Y., and Amir, A. (2015). Simultaneous regulation of cell size and chromosome replication in bacteria. *Front. Microbiol.* 6:662. doi: 10.3389/fmicb.2015.00662

Hirota, Y., Ryter, A., and Jacob, F. (1968). Thermosensitive mutants of *E. coli* affected in the processes of DNA synthesis and cellular division. *Cold Spring Harb. Symp. Quant. Biol.* 33, 677–694. doi: 10.1101/SQB.1968.033.01.077

Huls, P. G., Vischer, N. O. E., and Woldringh, C. L. (1999). Delayed nucleoid segregation in *Escherichia coli*. *Mol. Microbiol.* 33, 959–970. doi: 10.1046/j.1365-2958.1999.01535.x

Itsko, M., and Schaaper, R. M. (2011). The dgt gene of *E. coli* facilitates thymine utilization in thymine-requiring strains. *Mol. Microbiol.* 81, 1221–1232. doi: 10.1111/j.1365-2958.2011.07756.x

Iyer-Biswas, S., Wright, C. S., Henry, J. T., Lo, K., Burov, S., Lin, Y., et al. (2014). Scaling laws governing stochastic growth and division of single bacterial cells. *Proc. Natl. Acad. Sci. U.S.A.* 111, 15912–15917. doi: 10.1073/pnas.1403232111

Jacob, F., and Monod, J. (1961). Genetic regulatory mechanisms in the synthesis of proteins. *J. Mol. Biol.* 3, 318–356. doi: 10.1016/S0022-2836(61)80072-7

Jacob, F., and Wollman, E. L. (1956). Processes of conjugation and recombination in *Escherichia coli*. I. Induction by conjugation or zygotic induction. *Ann. Inst. Pasteur* 91, 486–510.

Jiménez Sánchez, A. (2015). Chromosome replication status and DNA content at any cell age in a bacterial cell cycle. *J. Theor. Biol.* doi: 10.1016/j.jtbi.2015.06.008 [Epub ahead of print].

Jun, S., and Taheri-Araghi, S. (2014). Cell-size maintenance: universal strategy revealed. *Trends Microbiol.* 23, 4–6. doi: 10.1016/j.tim.2014.12.001

Kellenberger, E., Ryter, A., and Séchaud, J. (1958). Electron microscope study of DNA-containing plasms. II. Vegetative and mature phage DNA as compared with normal bacterial nucleoids in different physiological states. *J. Cell Biol.* 4, 671–678. doi: 10.1083/jcb.4.6.671

Kjeldgaard, N. O., Schaechter, M., and Maalϕe, O. (1958). The transition between different physiological states during balanced growth of *Salmonella typhimurium*. *J. Gen. Microbiol.* 19, 607–616. doi: 10.1099/00221287-19-3-607

Koppes, L. J. H., Woldringh, C. L., and Nanninga, N. (1978). Size variations and correlation on different cell cycle events in slow-growing *Escherichia coli*. *J. Bacteriol.* 134, 423–433.

Koppes, L. J. H., and Nanninga, N. (1980). Positive correlation between size at initiation of chromosome replication in *Escherichia coli* and cell size at initiation of cell constriction. *J. Bacteriol.* 143, 89–99.

Kornberg, A., and Baker, T. A. (1992). *DNA Replication*, 2nd Edn. New York: WH Freeman and Company. 952.

Kubitschek, H. E. (1974). Estimation of the D period from residual division after exposure of exponential phase bacteria to chloramphenicol. *Mol. Gen. Genet.* 135, 123–130. doi: 10.1007/BF00264780

Lark, K. G., Repko, T., and Hoffman, E. J. (1963). The effect of amino acid deprivation on subsequent DNA replication. *Biochim. Biophys. Acta* 79, 9–24. doi: 10.1016/0926-6550(63)90003-3

Lark, K. G., and Lark, C. (1965). Regulation of chromosome replication in *Escherichia coli*: alternate replication of two chromosomes at slow growth rates. *J. Mol. Biol.* 13, 105–126. doi: 10.1016/S0022-2836(65)80083-3

Leonard, A. C., and Grimwade, J. E. (2010). "Initiation of DNA replication," in *EcoSal—Escherichia coli and Salmonella: Cellular and Molecular Biology*, eds A. Böck, R. III. Curtiss, J. B. Kaper, P. D. Karp, F. C. Neidhardt, T. Nyström, et al. (Washington, DC: ASM Press).

Luria, S. E., and Delbrück, M. (1943). Mutations of bacteria from virus sensitivity to virus resistance. *Genetics* 28, 491–511.

Maalϕe, O. (1961). The control of normal DNA replication in bacteria. *Cold Spring Harb. Symp. Quant. Biol.* 26, 45–52. doi: 10.1101/SQB.1961.026.01.010

Maalϕe, O., and Kjeldgaard, N. O. (1966). *Control of Macromolecular Synthesis*. New York: WA Benjamin Inc. 284.

Maalϕe, O., and Rasmussen, K. V. (1963). On the *in vivo* replication of bacterial DNA. *Colloq. Intern. Centre Nat. Rech. Sci.* 124, 165–168.

Männik, J., Wua, F., Hol, F. J. H., Bisicchia, P., Sherratt, D. J., Keymer, J. E., et al. (2012). Robustness and accuracy of cell division in *Escherichia coli* in diverse cell shapes. *Proc. Natl. Acad. Sci. U.S.A.* 109, 6957–6962. doi: 10.1073/pnas.1120854109

Markovitz, A. (1964). Regulatory mechanisms for synthesis of capsular polysaccharide in mucoid mutants of *Escherichia coli* K-12. *Proc. Natl. Acad. Sci. U.S.A.* 51, 239–246. doi: 10.1073/pnas.51.2.239

Masters, M., and Broda, P. (1971). Evidence for the bidirectional replication of the *Escherichia coli* chromosome. *Nat. New Biol.* 232, 137–140. doi: 10.1038/newbio232137a0

Measelson, M., and Stahl, F. W. (1958). The replication of DNA in *Escherichia coli*. *Proc. Natl. Acad. Sci. U.S.A.* 44, 671–682. doi: 10.1073/pnas.44.7.671

Meacock, P. A., and Pritchard, R. H. (1975). Relationship between chromosome replication and cell division in a thymineless mutant of *Escherichia coli* B/r. *J. Bacteriol.* 122, 931–942.

Michelsen, O., de Mattos, M. J. T., Jensen, P. R., and Hansen, F. G. (2003). Precise determinations of C and D periods by flow cytometry in *E. coli K*-12 and B/r. *Microbiology* 149, 1001–1010. doi: 10.1099/mic.0.26058-0

Mulder, E., and Woldringh, C. L. (1989). Actively replicating nucleoids influence the positioning of division sites in DNA-less cell forming filaments of *Escherichia coli*. *J. Bacteriol.* 171, 4303–4314.

Nanninga, N. (1971). The mesosome of *Bacillus subtilis* as affected by chemical and physical fixation. *J. Cell Biol.* 48, 219–224. doi: 10.1083/jcb.48.1.219

Nanninga, N., and Woldringh, C. L. (1985). "Cell growth, genome duplication and cell division in *Escherichia coli*," in *Molecular Cytology of Escherichia coli*, ed. N. Nanninga (London: Academic Press Inc.), 259–318.

Nanninga, N., Wientjes, F. B., de Jonge, B. L. M., and Woldringh, C. L. (1990). Polar cap formation during cell division in *Escherichia coli*. *Res. Microbiol.* 141, 103–118. doi: 10.1016/0923-2508(90)90102-V

Nirenberg, M. (2004). Historical review: deciphering the genetic code—a personal account. *Trends Biochem. Sci.* 29, 46–54. doi: 10.1016/j.tibs.2003.11.009

Norris, V., den Blaauwen, T., Cabin-Flaman, A., Doi, R. H., Harshey, R., Janniere, L., et al. (2007). Functional taxonomy of bacterial hyperstructures. *Microbiol. Mol. Biol. Rev.* 71, 230–253. doi: 10.1128/MMBR.00035-06

Oishi, M., Yoshikawa, H., and Sueoka, N. (1964). Synchronous and dichotomous replication of the *Bacillus subtilis* chromosome during spore germination. *Nature* 204, 1069–1073. doi: 10.1038/2041069a0

Olsson, J., Dasgupta, S., Berg, O. G., and Nordström, K. (2002). Eclipse period without sequestration in *Escherichia coli*. *Mol. Microbiol.* 44, 1429–1440. doi: 10.1046/j.1365-2958.2002.02954.x

Osella, M., Nugent, E., and Cosentino Lagomarsino, M. (2014). Concerted control of *Escherichia coli* cell division. *Proc. Natl. Acad. Sci. U.S.A.* 111, 3431–3435. doi: 10.1073/pnas.1313715111

Pelletier, J., Halvorsen, K., Ha, B.-Y., Paparcone, R., Sandler, S. J., Woldringh, C. L., et al. (2012). Physical manipulation of the *Escherichia coli* chromosome reveals its soft nature. *Proc. Natl. Acad. Sci. U.S.A.* 109, E2649–E2656. doi: 10.1073/pnas.1208689109

Porter, J. R. (1976). Antony van Leeuwenhoek: tercentenary of his discovery of bacteria. *Bacteriol. Rev.* 40, 260–269.

Powell, E. O. (1956). Growth rate and generation time in bacteria with special reference to continuous culture. *J. Gen. Microbiol.* 15, 492–511. doi: 10.1099/00221287-15-3-492

Prescott, D. M., and Kuempel, P. L. (1972). Bidirectional replication of the chromosome of *Escherichia coli*. *Proc. Natl. Acad. Sci. U.S.A.* 69, 2842–2845. doi: 10.1073/pnas.69.10.2842

Pritchard, R. H. (1968). Control of DNA synthesis in bacteria. *Heredity* 23, 472–473.

Pritchard, R. H. (1974). On the growth and form of a bacterial cell. *Phil. Trans. R. Soc. London Ser B* 267, 303–336. doi: 10.1098/rstb.1974.0003

Pritchard, R. H., Barth, P. T., and Collins, J. (1969). Control of DNA synthesis in bacteria. *Microbial Growth. Symp. Soc. Gen. Microbiol.* 19, 263–297.

Pritchard, R. H., and Lark, K. G. (1964). Induction of replication by thymine starvation at the chromosome origin in *Escherichia coli*. *J. Mol. Biol.* 9, 288–307. doi: 10.1016/S0022-2836(64)80208-4

Pritchard, R. H., and Zaritsky, A. (1970). Effect of thymine concentration on the replication velocity of DNA in a thymineless mutant of *Escherichia coli*. *Nature* 226, 126–131. doi: 10.1038/226126a0

Rabinovitch, A., Zaritsky, A., and Feingold, M. (2003). DNA-membrane interactions can localize bacterial cell center. *J. Theor. Biol.* 225, 393–396. doi: 10.1016/S0022-5193(03)00292-3

Reinhart, K. V., and Copeland, J. C. (1973). Evidence that thymine is not a normal metabolite in wild-type *Bacillus subtilis*. *Biochim. Biophys. Acta* 294, 1–7. doi: 10.1016/0005-2787(73)90308-0

Robert, L. (2015). Size sensors in bacteria, cell cycle control, and size control. *Front. Microbiol.* 6:515. doi: 10.3389/fmicb.2015.00515

Rosenberger, R. F., Grover, N. B., Zaritsky, A., and Woldringh, C. L. (1978a). Surface growth in rod-shaped bacteria. *J. Theor. Biol.* 73, 711–721. doi: 10.1016/0022-5193(78)90132-7

Rosenberger, R. F., Grover, N. B., Zaritsky, A., and Woldringh, C. L. (1978b). Control of microbial surface growth by density. *Nature* 271, 244–245. doi: 10.1038/271244a0

Schaechter, M. (1961). Pattern of cellular control during unbalanced growth. *Cold Spring Harb. Symp. Quant. Biol.* 26, 53–62. doi: 10.1101/SQB.1961.026.01.011

Schaechter, M. (2006). From growth physiology to systems biology. *Int. Microbiol.* 9, 157–161.

Schaechter, M. (2015). A brief history of bacterial growth physiology. *Front. Microbiol.* 6:289. doi: 10.3389/fmicb.2015.00289

Schaechter, M., Maalφe, O., and Kjeldgaard, N. O. (1958). Dependency on medium and temperature of cell size and chemical composition during balanced growth of *Salmonella typhimurium*. *J. Gen. Microbiol.* 19, 592–606. doi: 10.1099/00221287-19-3-592

Scott, M., and Hwa, T. (2011). Bacterial growth laws and their applications. *Curr. Opin. Biotechnol.* 22, 559–565. doi: 10.1016/j.copbio.2011.04.014

Simmons, L. A., Breier, A. M., Cozzarelli, N. R., and Kaguni, J. M. (2004). Hyperinitiation of DNA replication in *Escherichia coli* leads to replication forks collapse and invariability. *Mol. Microbiol.* 51, 349–358. doi: 10.1046/j.1365-2958.2003.03842.x

Slater, M., and Schaechter, M. (1974). Control of cell division in bacteria. *Bacteriol. Rev.* 38, 199–221.

Stent, G. S., and Calendar, R. (1978). *Molecular Genetics—An Introductory Narrative*. 2nd Edn. San Francisco: W. H. Freeman and Company. 773.

Sueoka, N., and Yoshikawa, H. (1965). The chromosome of *Bacillus subtilis*. I. Theory of marker frequency analysis. *Genetics* 52, 747–757.

Taheri-Araghi, S., Bradde, S., Sauls, J. T., Hill, N. S., Levin, P. A., Paulsson, J., et al. (2015). Size-control and homeostasis in bacteria. *Curr. Biol.* 25, 385–391. doi: 10.1016/j.cub.2014.12.009

Toro, E., and Shapiro, L. (2010). Bacterial chromosome organization and segregation. *Cold Spring Harb. Perspect. Biol.* 2, a000349. doi: 10.1101/cshperspect.a000349

Trueba, F. J., and Woldringh, C. L. (1980). Changes in cell diameter during the division cycle of *Escherichia coli*. *J. Bacteriol.* 142, 869–878.

Typas, A., Banzhaf, M., Gross, C. A., and Vollmer, W. (2012). From the regulation of peptidoglycan synthesis to bacterial growth and morphology. *Nat. Rev. Microbiol.* 10, 123–135. doi: 10.1038/nrmicro2677

Valkenburg, J. A. C., Woldringh, C. L., Brakenhoff, G. J., van der Voort, H. T. M., and Nanninga, N. (1985). Confocal scanning light microscopy of the *Escherichia coli* nucleoid: comparison with phase-contrast and electron microscope images. *J. Bacteriol.* 161, 478–483.

van Helvoort, J. M. L. M., and Woldringh, C. L. (1994). Nucleoid partitioning in *Escherichia coli* during steady state growth and upon recovery from chloramphenicol treatment. *Mol. Microbiol.* 13, 577–583. doi: 10.1111/j.1365-2958.1994.tb00452.x

Vardi, E., and Grover, N. B. (1993). Shape changes in *Escherichia coli* B/rA during agar filtration. *Cytometry* 14, 173–178. doi: 10.1002/cyto.990140209

Vischer, N. O. E., Verheul, J., Postma, M., Bart van den Berg van Saparoea, B., Galli, E., Natale, P., et al. (2015). Cell age dependent concentration of *Escherichia coli* divisome proteins analyzed with ImageJ and ObjectJ. *Front. Microbiol.* 6:586. doi: 10.3389/fmicb.2015.00586

Wake, R. G. (1972). Visualization of reinitiated chromosomes in *Bacillus subtilis*. *J. Mol. Biol.* 68, 501–509. doi: 10.1016/0022-2836(72)90102-7

Wang, P., Robert, L., Pelletier, J., Dang, W.-L., Taddei, F., Wright, A., et al. (2010). Robust growth of *Escherichia coli*. *Curr. Biol.* 20, 1099–1103. doi: 10.1016/j.cub.2010.04.045

Watson, J. D. (1996). *The Double Helix—A Personal Account of the Discovery of the Structure of DNA*. New York: Scribner.

Watson, J., and Crick, F. H. C. (1953). Molecular structure of nucleic acids. *Nature* 171, 737–738. doi: 10.1038/171737a0

Wold, S., Skarstad, K., Steen, H. B., Stokke, T., and Boye, E. (1994). The initiation mass for DNA replication in *Escherichia coli* K-12 is dependent on growth rate. *EMBO J.* 13, 2097–2102.

Woldringh, C. L. (1974). *The Nucleoplasm of Escherichia coli*. Ph.D. thesis, University of Amsterdam (NL), Amsterdam, 129.

Woldringh, C. L. (1994). Significance of plasmolysis spaces as markers for periseptal annuli and adhesion sites. *Mol. Microbiol.* 14, 597–607. doi: 10.1111/j.1365-2958.1994.tb01299.x

Woldringh, C. L. (2002). The role of co-transcriptional translation and protein translocation (transertion) in bacterial chromosome segregation. *Mol. Microbiol.* 45, 17–29. doi: 10.1046/j.1365-2958.2002.02993.x

Woldringh, C. L., de Jong, M. A., van den Berg, W., and Koppes, L. (1977). Morphological analysis of the division cycle of two *Escherichia coli* substrains during slow growth. *J. Bacteriol.* 131, 270–279.

Woldringh, C. L., Grover, N. B., Rosenberger, R. F., and Zaritsky, A. (1980). Dimensional rearrangement of rod-shaped bacteria following nutritional shift-up. II. Experiments with *Escherichia coli* B/r. *J. Theor. Biol.* 86, 441–454. doi: 10.1016/0022-5193(80)90344-6

Woldringh, C. L., Hansen, F. G., Vischer, N. O. E., and Atlung, T. (2015). Segregation of chromosome arms in growing and non-growing *Escherichia coli* cells. *Front. Microbiol.* 6:448. doi: 10.3389/fmicb.2015.00448

Woldringh, C. L., Huls, P., Pas, E., Brakenhoff, G. J., and Nanninga, N. (1987). Topography of peptidoglycan synthesis during elongation and polar cap formation in a cell division mutant of *Escherichia coli* MC4100. *J. Gen. Microbiol.* 133, 575–586. doi: 10.1099/00221287-133-3-575

Woldringh, C. L., Mulder, E., Valkenburg, J. A. C., Wientjes, F. B., Zaritsky, A., and Nanninga, N. (1990). Role of nucleoid in toporegulation of division. *Res. Microbiol.* 141, 39–49. doi: 10.1016/0923-2508(90)90096-9

Woldringh, C. L., and Nanninga, N. (1985). "Structure of nucleoid and cytoplasm in the intact cell," in *Molecular Cytology of Escherichia coli*, ed. N. Nanninga (London: Academic Press), 161–197.

Woldringh, C. L., and Odijk, T. (1999). "Structure of DNA within the bacterial cell: physics and physiology," in *Organization of the Prokaryotic Genome*, ed. R. L. Charlebois (Washington, DC: ASM Press), 171–187.

Woldringh, C. L., Zaritsky, A., and Grover, N. B. (1994). Nucleoid partitioning and the division plane in *Escherichia coli*. *J. Bacteriol.* 176, 6030–6038.

Wu, L. J., and Errington, J. (2012). Nucleoid occlusion and bacterial cell division. *Nat. Rev. Microbiol.* 10, 8–12. doi: 10.1038/nrmicro2671

Yoshikawa, H., and Sueoka, N. (1963). Sequential replication of the *Bacillus subtilis* chromosome. I. Comparison of marker frequencies in exponential and stationary growth phases. *Proc. Natl. Acad. Sci. U.S.A.* 49, 559–566. doi: 10.1073/pnas.49.4.559

Youngren, B., Nielsen, H. J., Jun, S., and Austin, S. (2014). The multifork *Escherichia coli* chromosome is a self-duplicating and self-segregating thermodynamic ring polymer. *Genes Dev.* 28, 71–84. doi: 10.1101/gad.231050.113

Zaritsky, A. (1971). *Studies on DNA Replication and Cell Division in Bacteria*. Ph.D. thesis, The University of Leicester (UK), Leicester, 97.

Zaritsky, A. (1975a). Rate stimulation of DNA synthesis after inhibition. *J. Bacteriol.* 122, 841–846.

Zaritsky, A. (1975b). On dimensional determination of rod-shaped bacteria. *J. Theor. Biol.* 54, 243–248. doi: 10.1016/S0022-5193(75)80129-9

Zaritsky, A. (2015). Cell shape homeostasis in *Escherichia coli* is driven by growth, division, and nucleoid complexity. *Biophys. J.* 109, 178–181. doi: 10.1016/j.bpj.2015.06.026

Zaritsky, A., and Pritchard, R. H. (1973). Changes in cell size and shape associated with changes in the replication time of the chromosome of *Escherichia coli*. *J. Bacteriol.* 114, 824–837.

Zaritsky, A., Vischer, N., and Rabinovitch, A. (2007). Changes of initiation mass and cell dimensions by the 'eclipse'. *Mol. Microbiol.* 63, 15–21. doi: 10.1111/j.1365-2958.2006.05501.x

Zaritsky, A., Wang, P., and Vischer, N. O. E. (2011). Instructive simulation of the bacterial cell cycle. *Microbiology* 157, 1876–1885. doi: 10.1099/mic.0.049403-0

Zaritsky, A., Woldringh, C. L., Einav, M., and Alexeeva, S. (2006). Thymine limitation and thymine starvation to study bacterial physiology and cytology. *J. Bacteriol.* 188, 1667–1679. doi: 10.1128/JB.188.5.1667-1679.2006

Zaritsky, A., Woldringh, C. L., Grover, N. B., Naaman, J., and Rosenberger, R. F. (1982). Growth and form in bacteria. *Comments Mol. Cell. Biophys.* 1, 237–260.

Zaritsky, A., Woldringh, C. L., Vischer, N. O. E., Wang, P., and Helmstetter, C. E. (2012). Simulation of the prokaryotic cell cycle at http://simon.bio.uva.nl/cellcycle/. *J. Sytem. Cybernet. Informat.* 10, 18–23.

Zaritsky, A., and Zabrovitz, S. (1981). DNA synthesis in *Escherichia coli* during a nutritional shiftup. *Mol. Gen. Genet.* 181, 564–565. doi: 10.1007/BF00428756

19

Molecular mechanisms for the evolution of bacterial morphologies and growth modes

*Amelia M. Randich and Yves V. Brun**

Department of Biology, Indiana University, Bloomington, IN, USA

***Correspondence:**
Yves V. Brun,
Department of Biology, Indiana University, 1001 E 3rd Street, Bloomington, IN 47405-7005, USA
ybrun@indiana.edu

Bacteria exhibit a rich diversity of morphologies. Within this diversity, there is a uniformity of shape for each species that is replicated faithfully each generation, suggesting that bacterial shape is as selectable as any other biochemical adaptation. We describe the spatiotemporal mechanisms that target peptidoglycan synthesis to different subcellular zones to generate the rod-shape of model organisms *Escherichia coli* and *Bacillus subtilis*. We then demonstrate, using the related genera *Caulobacter* and *Asticcacaulis* as examples, how the modularity of the core components of the peptidoglycan synthesis machinery permits repositioning of the machinery to achieve different growth modes and morphologies. Finally, we highlight cases in which the mechanisms that underlie morphological evolution are beginning to be understood, and how they depend upon the expansion and diversification of the core components of the peptidoglycan synthesis machinery.

Keywords: bacterial shape, bacterial morphology, peptidoglycan synthesis, *Caulobacter*, *Asticcacaulis*, FtsZ, MreB

Introduction

The typical of bacterial shapes as simply rods, variations of rods, or cocci belies the great diversity of bacterial morphologies. Within these simple classifications, bacteria exhibit a broad range of morphologies—helical or vibrioid twists, filaments, hyphae or branched filaments, prosthecae or stalks (thin cylindrical extensions of the cell envelope)—as well as distinct growth modes, forms of cellular differentiation, and life cycles (**Figure 1**, Young, 2006). The mechanisms by which bacteria achieve these morphologies, much less evolve them, has remained unclear. Historically, the murein sacculus, or peptidoglycan layer, has been posited as the underlying determinant of bacterial shape (Vollmer et al., 2008a). A heteropolymer of peptide cross-linked glycan strands, the sacculus confers strength to the cell wall and maintains cell shape and size. It is becoming clear that the sacculus in and of itself is not the only determinant of cell shape; rather it is the spatiotemporal regulation of the enzymes that build the sacculus and modify it in response to environmental changes that drives morphogenesis (Young, 2010; Cava and de Pedro, 2014).

In bacteria, morphologies and growth modes are interconnected (Brown et al., 2011). In many taxa, maintenance of the rod shape itself requires at least two well-studied modes of growth: lateral elongation (**Figure 2A**), or incorporation of peptidoglycan along the sidewalls; and septation (**Figure 2B**), the generation of nascent poles, usually at the cell center. In these canonical cases, which are exemplified by model organisms *Escherichia coli* and *Bacillus subtilis*, material at the poles remains inert, with no evidence of new peptidoglycan incorporation or turnover (Mobley et al., 1984; Schlaeppi et al., 1985; de Pedro et al., 1997; Janakiraman and Goldberg, 2004). However, other, less understood approaches to achieving rod-shaped morphologies exist. Actinobacteria exhibit polar

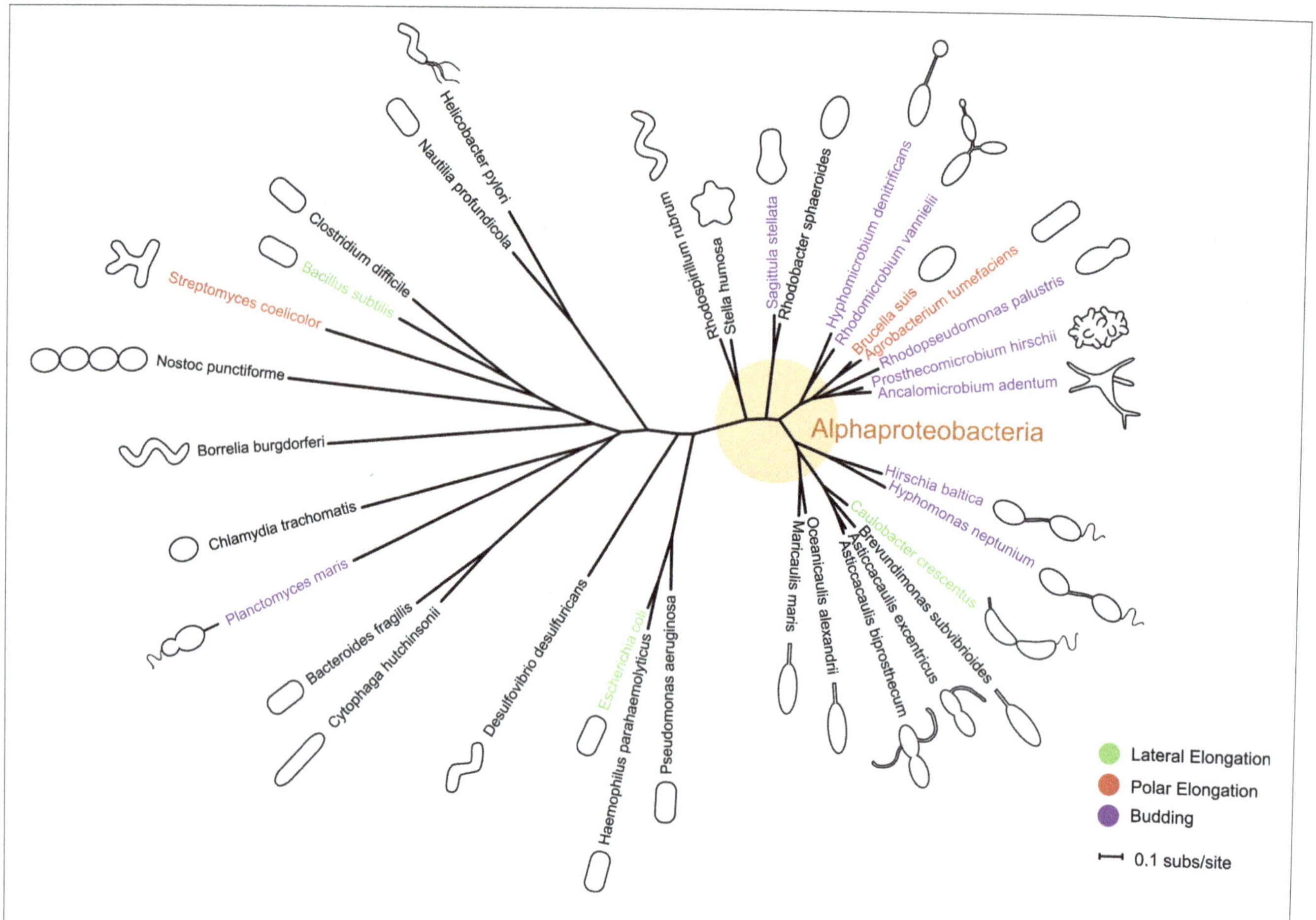

FIGURE 1 | Phylogeny and morphology of bacteria exhibiting different modes of growth. *rpoC* sequences were aligned using MUSCLE v3.8.31 (Edgar, 2004). RAxML v7.0.4 (Stamatakis, 2006) reconstructed the maximum likelihood phylogeny using the JTT amino acid substitution matrix with rate variation among sites modeled by a four-category discrete gamma distribution and an additional invariant category. Note that the stalk of *Planctomyces maris* is composed of several parallel fibrils (Bauld and Staley, 1976), and is not bound by cell envelop layers like that of the Caulobacterales. Bacteria are not drawn to scale. Scale bar reflects numbers of substitutions per site.

growth (**Figure 2C**), where elongation occurs strictly at one or both poles, leaving the sidewalls, instead of the poles, inert (Umeda and Amako, 1983; Daniel and Errington, 2003; Chauhan et al., 2006; Flärdh and Buttner, 2009; Flärdh, 2010). Alphaproteobacteria (the superfamily highlighted in **Figure 1**), exhibit a diverse mix of growth strategies including lateral elongation and polar growth, but also the unknown mechanisms of zonal growth to produce buds and/or stalks (**Figures 2D,E**; Brown et al., 2011). In a striking example, bacteria in the family *Hyphomicrobiaceae* not only produce a stalk, but also bud daughter cells from the end of the stalk (**Figure 2D**) via an unknown mechanism (Whittenbury and Dow, 1977; Moore, 1981). The life cycle of these cells involves at least three separate modes of growth: stalk elongation at the junction of the cell body and stalk, daughter cell elongation at the tip of the stalk, and septum formation within the stalk to complete division (Moore and Hirsch, 1973; Whittenbury and Dow, 1977; Moore, 1981).

In comparison to lateral elongation and septation, much less is known genetically or mechanistically about polar growth or budding, making it hard to rigorously distinguish between these growth modes. The next section will describe how, although lateral elongation and septation differ at the genetic level, they ultimately drive distinct growth modes by positioning functionally similar peptidoglycan synthesis machineries at specific locations in the cell. It is likely that polar growth, budding, and stalk growth also result from the repositioning of yet to be described peptidoglycan synthesis machineries. In this way, all growth modes can be described as zonal growth (Brown et al., 2011). Already with these few examples, a great plurality of growth modes clearly has evolved to generate a "common" rod shape, as well as to achieve novel morphologies and life cycles. This review aims to summarize the spatiotemporal mechanisms that target peptidoglycan synthesis to different subcellular zones to generate the rod-shape of model organisms *E. coli* and *B. subtilis* and to then describe how the modularity of the core components of the peptidoglycan synthesis machinery permits repositioning of the growth machinery to achieve different growth modes and morphologies. Because of space limitations, we were unable to cover the septal elongation mechanisms that generate round and ovoid cells and we refer the reader to excellent

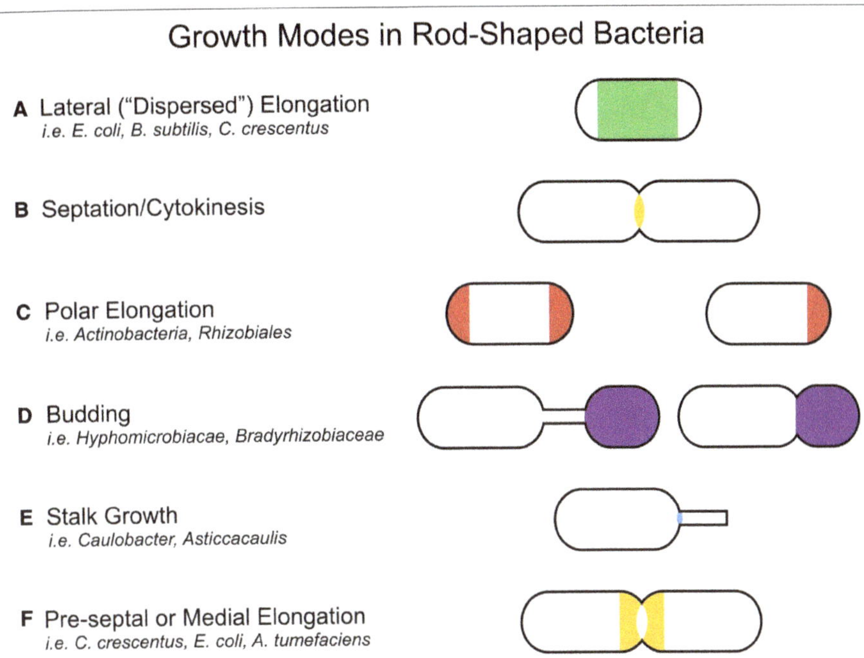

FIGURE 2 | Growth modes in rod-shaped bacteria. Several growth modes with the regions of active peptidoglycan synthesis are schematized. Colors indicate regions of active peptidoglycan synthesis. **(A)** Lateral elongation is well-studied in *E. coli*, *B. subtilis*, and *C. crescentus*, and often assumed for many rod-shaped bacteria. **(B)** A majority of bacterial families divide in a FtsZ-dependent manner. Notable exceptions are members of the PVC (Planctomycetes, Verrucomicrobia, and Chlamydiae) superfamily. **(C)** Polar elongation is characteristic of Actinobacteria, where it is bipolar. It has recently been demonstrated in Rhizobiales (Brown et al., 2012) where it is unipolar. **(D)** Budding occurs both at the ends of stalks and off the cell body in Alphaproteobacteria. **(E)** Stalks are wide-spread in Alphaproteobacteria. While the exact mechanism for growth may differ among families, all depend on zonal growth of the peptidoglycan at the cell-stalk junction. **(F)** Some species exhibit medial (pre-septal) growth, which appears to be peptidoglycan synthesis near the division plane before full assembly of the Z-ring.

recent reviews on this topic (Sham et al., 2012; Pinho et al., 2013).

Molecular Machinery of Bacterial Growth and Division in Rod-shaped Bacteria

The molecular underpinnings of how bacteria grow, divide, and maintain shape are slowly emerging, especially in model rod-shaped bacteria such as *E. coli* and *B. subtilis*. In these species, two protein assemblies direct the modification and synthesis of the sacculus at specific times and locations during the cell cycle: the elongasome inserts new peptidoglycan along the length of the rod during growth, and the divisome completes the steps of constriction and new peptidoglycan synthesis at the cell center during cell division (Typas et al., 2012). Both of these assemblies utilize similar components and likely share a common evolutionary history (Szwedziak and Löwe, 2013). The preponderance of conserved protein classes for peptidoglycan synthesis machinery across gram-negative and positive bacteria, as well as between the elongasome and divisome assemblies, suggests a general strategy for shaping the bacterial cell as well as molecular mechanisms for the evolution of novel morphologies.

The divisome and elongasome have long been hypothesized to form large complexes (Höltje, 1998). Both consist of cytoplasmic scaffolding cytoskeletal-like proteins; inner membrane-spanning elements; and a host of periplasmic enzymes including peptidoglycan synthases and hydrolases (**Figure 3**; Margolin, 2009; Typas et al., 2012; Egan and Vollmer, 2013). This machinery works in concert to create the sacculus, a meshwork consisting of chains of two alternating sugar types, *N*-acetyl glucosamine and *N*-acetyl muramic acid, joined through beta-(1,4)-glycosidic bonds that are cross-linked together via peptide chains. The steps in peptidoglycan synthesis proceed as follows: a series of enzymes generates a pool of nucleotide precursors in the cytoplasm, a flippase then transfers the lipidated monomers over the inner membrane, and finally, peptidoglycan synthases polymerize the monomers into glycan chains and cross-link them through peptide bonds, thus forming the sacculus. Other peptidoglycan-modifying enzymes further diversify the periplasmic/extracellular components, such as the carboxypeptidases that trim peptide chains and the lytic transglycosylases that reduce the length of glycan strands. Moreover, peptidoglycan hydrolases, which cleave either the glycosidic or amide bonds of peptidoglycan, play important roles in turning over old peptidoglycan to allow insertion of new material in the growing cell and help shape the new poles (Priyadarshini et al., 2007; Vollmer et al., 2008b; Frirdich and Gaynor, 2013; Lee and Huang, 2013).

The cytoplasmic scaffold FtsZ organizes and regulates the activity and localization of various divisome components. Much progress has been made in elucidating how the divisome is spatiotemporally regulated in canonical rod-shaped bacteria.

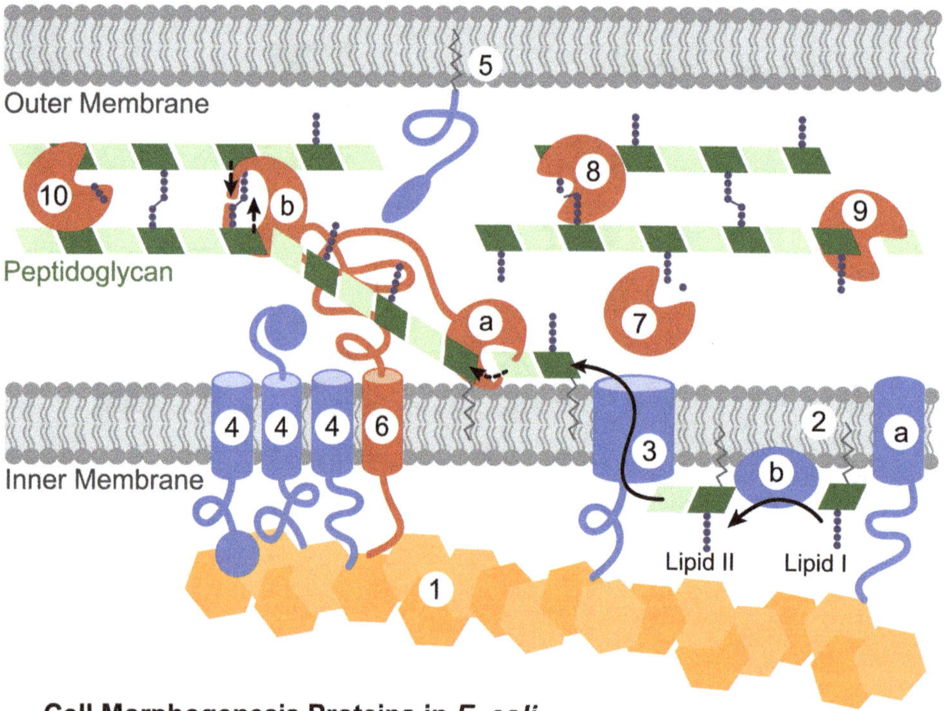

Cell Morphogenesis Proteins in *E. coli*

		Divisome	Elongasome
1	**Cytoplasmic Scaffolds**	FtsZ, FtsA	MreB
	Membrane-Spanning Elements		
2	Pecursor Synthesis	(a) MraY, (b) MurG	
3	Flippase	FtsW/MurJ?	RodA/MurJ?
4	Associated Proteins	FtsEX, FtsK, FtsN	MreCD, RodZ
5	(OM) PG Synthase Activator	LpoB	LpoA
	PG-Modifying Enzymes		
6	PG-Synthase (Bifunctional)	PBP1B	PBP1A
	a Glycosyltransferase	MtgA	
	b Transpeptidase	PBP3	PBP2
7	Carboxypeptidase	PBP5, PBP4B, PBP6, PBP6B	
8	Endopeptidase	PBP4, PBP7	Spr, YdhO, YebA
9	Lytic Transglycosylase	Slt70, MltABCDEF	?
10	Amidase	AmiABCD	?

FIGURE 3 | Peptidoglycan synthesis machinery schematic. The divisome and elongasome both consist of cytoplasmic scaffolds (orange), membrane-spanning elements that include regulatory proteins as well as peptidoglycan precursor synthesis machinery (blue), and peptidoglycan-modifying enzymes (red). Each assembly has genetic components that are biochemically distinct despite retaining similar functions (Table). In general, (1) cytoplasmic scaffolds such as FtsZ or MreB direct the location of peptidoglycan synthesis and recruit various cytoplasmic and inner-membrane components. (2) Cytoplasmic and inner membrane enzymes synthesize peptidoglycan monomers (lipid II) in the cytoplasm [only (a) MraY and (b) MurG shown] and (3) a flippase flips them across the inner membrane. (4, 5) Various membrane proteins function to regulate or organize the divisome or elongasome protein assemblies. (6) Bifunctional or monofunctional peptidoglycan synthases (a) polymerize lipid II into glycan strands and (b) crosslink the peptide chains to form the sacculus. Various enzymes modify the peptidoglycan after synthesis: (7) carboxypeptidases trim peptide chains, (8) endopeptidases cleave crosslinks, (9) lytic transglycosylases cleave the glycan strand, (10) and amidases remove peptide chains from the glycan strand. Question marks in the table indicate that elongasome proteins responsible for the indicated activities remain to be identified. Some proteins essential to the divisome have been omitted: ZipA, ZapABCD, FtsQ, FtsL, and FtsB.

During division, the GTP-dependent polymerization of the tubulin-like FtsZ creates a ring-shaped structure called the Z ring at the cytoplasmic face of the membrane at the center of the cell (Bi and Lutkenhaus, 1991; Löwe and Amos, 1998; Mukherjee and Lutkenhaus, 1998). The exact architecture of the Z-ring has recently been the subject of intense study (Li et al., 2007; Fu et al., 2010; Strauss et al., 2012; Szwedziak et al., 2014). Once assembled, the Z-ring recruits over 10 essential divisome components (Egan and Vollmer, 2013). In *E. coli*, the mechanism for targeting the Z-ring to the center of the cell involves the well-characterized Min system, which prevents Z-ring assembly at the cell poles (Raskin and de Boer, 1999; Hale et al., 2001), and nucleoid occlusion protein SlmA (Bernhardt and de Boer, 2005). *B. subtilis* appears to maintain many of the same essential division proteins with the exception of utilizing DivIVA to regulate the septation site by positioning and stabilizing Min proteins at the cell poles (Edwards and Errington, 1997; Marston et al., 1998; Hamoen and Errington, 2003) and using a nucleoid occlusion mechanism mediated by a different protein, Noc (Wu and Errington, 2004). Other species also utilize different systems for positioning FtsZ, such as the MipZ system in *Caulobacter crescentus* (Thanbichler and Shapiro, 2006), the MapZ/LocZ system in *Streptococcus pneumoniae* (Fleurie et al., 2014; Holečková et al., 2015), and the SsgB system in *Streptomyces* (Willemse et al., 2011). Each of these systems (with the exception of SsgB) faithfully localizes the Z ring to the division plane with high precision at onset of division (Trueba, 1982; Yu and Margolin, 1999; Männik et al., 2012).

Much less is known about the spatiotemporal regulation of the elongasome, for which MreB appears to be the major scaffold for coordinating peptidoglycan precursor synthesis and peptidoglycan polymerization. MreB is an actin-like protein that forms membrane-associated filaments (Esue et al., 2005, 2006; Salje et al., 2011; Ozyamak et al., 2013). Once thought to form a cytoskeletal meshwork, MreB has since been shown to form discrete, motile patches that move, independently of MreB polymerization or treadmilling, near-perpendicularly to the long cell axis in *E. coli*, *B. subtilis*, and *C. crescentus* (Dominguez-Escobar et al., 2011; Garner et al., 2011; van Teeffelen et al., 2011). MreB interacts with inner membrane proteins MreC, MreD, RodZ (Kruse et al., 2004; van den Ent et al., 2006; Shiomi et al., 2008; Alyahya et al., 2009; Bendezú et al., 2009), as well as lipid II synthesis enzymes MraY and MurG (Mohammadi et al., 2007), and its movement during elongation depends on both the synthesis of essential peptidoglycan components and the activity of peptidoglycan synthases (Dominguez-Escobar et al., 2011; Garner et al., 2011; van Teeffelen et al., 2011). Other components of the elongasome, such as MreB-associated proteins MreCD and certain PBPs, have been shown to exhibit the same motile, spatiotemporal localization as MreB (Dominguez-Escobar et al., 2011; Garner et al., 2011), suggesting that parts of the elongasome indeed travel as a complex. MreB could function to coordinate elongasome complexes and perhaps restrict their mobility to ensure a uniform distribution of peptidoglycan insertion. Recent work combining time-lapse and 3D-imaging with computational analysis indicates that MreB preferentially localizes to and directs peptidoglycan synthesis at regions of negative curvature in *E. coli* cells (Ursell et al., 2014). Therefore MreB could selectively drive peptidoglycan growth away from the positive curvature of the cell poles to straighten the cell and create the rod shape. Possibly, other factors remain to be discovered that could regulate the localization and dynamics of MreB to in turn direct the activity of the elongasome. For example, as will be covered in the next sections, MreB interacts with *Caulobacter crescentus*-specific genes TipN and CreS.

While both the divisome and elongasome depend on similar molecular components, they clearly drive disparate morphological changes: cutting the cell in half versus lengthening the cell. These particular growth modes are achieved by the distinct localization patterns of their respective cytoplasmic scaffolds. During division, FtsZ forms a ring at the site of septation and thereby directs peptidoglycan growth and hydrolysis to shape the new poles of the daughter cells. In contrast, MreB distributes along the sidewalls of the cell to drive homogeneous insertion of new peptidoglycan in the lengthening cell. While these assemblies appear to share a similar strategies for growth, their respective components make them biochemically distinguishable (Table in **Figure 3**) and allow for independent regulation and incorporation into specific cell events. This modularity—evident here in the cytoplasmic scaffolds—permits wholesale repositioning of the growth machinery to achieve different growth modes and morphologies.

Mechanisms for Evolving Novel Morphologies and Growth Modes

One can imagine multiple ways in which complexes such as the elongasome and divisome could be repurposed or retooled to evolve new morphologies and modes of growth. Theoretically, expansion and diversification of any of the core elements of the peptidoglycan synthesis assembly—cytoskeletal-like proteins, inner membrane-spanning elements, periplasmic or extracellular enzymes and outer membrane proteins—present opportunities for such evolution. The spatiotemporal interdependency of peptidoglycan synthesis machinery components in both the elongasome and divisome suggests that new relationships can evolve between different components. Moreover, the addition of new regulatory or recruitment components to the complexes may drive species-specific morphologies. Recent studies in non-canonical rod-shaped bacteria indicate that different modes of growth depend upon the same core machinery, but that these are organized or regulated differently by species-specific components.

A large diversity of bacterial cytoskeletal elements appears to underlie species-specific lifestyles and morphologies. Besides FtsZ and MreB, other self-oligomerizing protein scaffolds have been identified and shown to be widespread in bacteria (Bagchi et al., 2008; Kühn et al., 2010). In many cases, these scaffolds have roles in driving morphology. For example, intermediate filament-like crescentin (CreS) forms a single filamentous structure in *Caulobacter crescentus* that produces the characteristic vibrioid shape of the bacterium by inducing differential growth of the sides of the rod-shaped sacculus (Ausmees et al., 2003; Cabeen et al., 2009). In contrast, *Helicobacter pylori* utilizes a family of coiled-coil rich proteins (Ccrp) that form extended filamentous structures to maintain a helical rod shape (Waidner et al., 2009; Specht et al., 2011). These examples demonstrate how two scaffolding proteins, unrelated in sequence, have arisen in two different species to create twists in the rod shape. In contrast to the eukaryotic cytoskeleton, in which classes of scaffolding proteins such as actin or microtubules have been adapted to suit multiple processes by using adaptor proteins, it appears bacteria may have developed larger numbers of species- and process-specific scaffolding proteins (Ozyamak et al., 2013).

Although FtsZ and MreB comprise the core of the canonical divisome and elongasome, there are many cases in which their roles have changed or they have been eliminated from genomes completely. For example, MreB plays a role in chromosome segregation but not growth in *H. pylori* (Waidner et al., 2009), and is absent in the majority of Actinobacteria, some Firmicutes, and a minority of Proteobacteria (Margolin, 2009; Brown et al., 2011; Jiang et al., 2015). It is tantalizing to infer that MreB delineates specific growth modes since bacteria lacking MreB, such as Actinobacteria and many Alphaproteobacteria—*Agrobacterium*, *Hyphomicrobium*, and *Rhizobium*—utilize polar growth mechanisms (Brown et al., 2012, 2011). However, the story appears to be much more complex than associating MreB with lateral elongation of rod-shaped bacteria. Members of the *Chlamydiae* family represent an interesting family of bacteria that lack FtsZ while maintaining a nearly complete suite of other peptidoglycan synthesis machinery components, including MreB (Stephens et al., 1998; McCoy and Maurelli, 2006; Pilhofer et al., 2008; Bertelli et al., 2010). This is in contrast to the great majority of bacterial phyla that utilize FtsZ for cell division (Erickson et al., 2010). In fact, the presence of peptidoglycan in the *Chlamydiae* cell wall was long debated and has only recently been demonstrated in several members of the family (Pilhofer et al., 2013; Liechti et al., 2014). Localization studies in *Waddlia chondrophila* suggest that MreB acts in lieu of FtsZ to facilitate division in this species (Jacquier et al., 2014). Other studies have demonstrated that various components of the peptidoglycan synthesis pathway from FtsZ-independent species function *in vitro* and in complementation studies (Henrichfreise et al., 2009; Frandi et al., 2014). Therefore, even in genera that have diverged greatly, the overarching mechanism of peptidoglycan synthesis remains.

The expansion of various classes of peptidoglycan-modifying enzymes is becoming a common theme in studies of alternative morphologies and growth modes as less studied genera come to the forefront. *H. pylori* appears to employ a diverse cast of endo- and carboxypeptidases that likely shape the helical sacculus through alternative cross-linking (Bonis et al., 2010; Sycuro et al., 2010, 2012, 2013). In *Agrobacterium tumefaciens*, which has been recently shown to grow polarly (Brown et al., 2012), polar growth appears to depend on a class of alternative L,D-transpeptidases (Cameron et al., 2014). Overall, the modularity of the common core components of the peptidoglycan synthesis machinery provides ample flexibility for new growth modes and morphologies. Although some bacterial families have clearly shifted their dependence from central components such as MreB or FtsZ to other scaffolds, the general mechanism for assembling core peptidoglycan synthesis machineries appears to remain intact.

Cell Growth and Morphology of *Caulobacter crescentus*

The dimorphic alphaproteobacterium *Caulobacter crescentus* represents an excellent case study in understanding how new morphologies arise from the diversification of the common core peptidoglycan synthesis machinery. *C. crescentus* divides asymmetrically, producing a swarmer cell with a polar flagellum and a DNA replication-competent cell with a polar stalk. This process of division requires strict coordination in time and space with other cell cycle events such as cell growth, chromosome segregation, and differentiation. Therefore it is unsurprising that although *C. crescentus* maintains similar core peptidoglycan synthesis machinery components in its divisome and elongasome, it has evolved discrete scaffolds, regulatory proteins, and cell cycle control mechanisms to adapt the machinery to its specific lifestyle. Determining exactly how *C. crescentus* has repurposed conserved peptidoglycan synthesis components for elongation, division, and stalk synthesis will help to expand our understanding of bacterial morphology.

The diversification of proteins interacting with and regulating FtsZ and MreB allows for strict control of asymmetric division and morphological development in *C. crescentus*. One significant difference between *C. crescentus* and the canonical rod-shaped bacteria *E. coli* and *B. subtilis* is that *C. crescentus* does not utilize a Min system or nucleoid occlusion proteins. Instead, *C. crescentus* uses MipZ gradients to position the FtsZ ring at the division plane (Thanbichler and Shapiro, 2006). MipZ directly interacts with FtsZ to inhibit ring polymerization. When MipZ associates with ParB at the stalk pole prior to S phase, it drives FtsZ monomers to the new, non-stalked pole (**Figure 4B**). During initiation of chromosome replication and origin duplication, some of the MipZ-ParB complex binds the new origin and migrates with it to the new pole, displacing FtsZ from the new pole to the division plane. Although MipZ shares some distant domain similarity with MinD, it likely evolved from ParA-like DNA partitioning proteins and is conserved amongst all Alphaproteobacteria without MinCD orthologues (Thanbichler and Shapiro, 2006).

Other polar regulators have been shown to influence cell morphology and interact with peptidoglycan synthesis machinery (**Figure 4B**). TipN, a polytopic membrane protein with a large coiled-coil cytoplasmic domain, ensures transmission of cell polarity to the daughter cell by marking the new pole and regulating MreB dynamics at the division plane (Lam et al., 2006). PopZ forms oligomeric networks at the poles to form polar ribosome exclusion zones and serves two roles during the cell cycle, switching function during the swarmer to stalk transition: it first interacts with the parS/ParB centromere at the stalked pole prior to the initiation of DNA replication, and later, during polar maturation, it recruits remodeling factors, such as peptidoglycan synthesis machinery, to transform the flagellated pole into a stalked pole (Bowman et al., 2010; Laloux and Jacobs-Wagner, 2013; Ptacin et al., 2014). All of these regulatory polar protein systems serve as examples of alternative networks that regulate the otherwise conserved core peptidoglycan synthesis machinery.

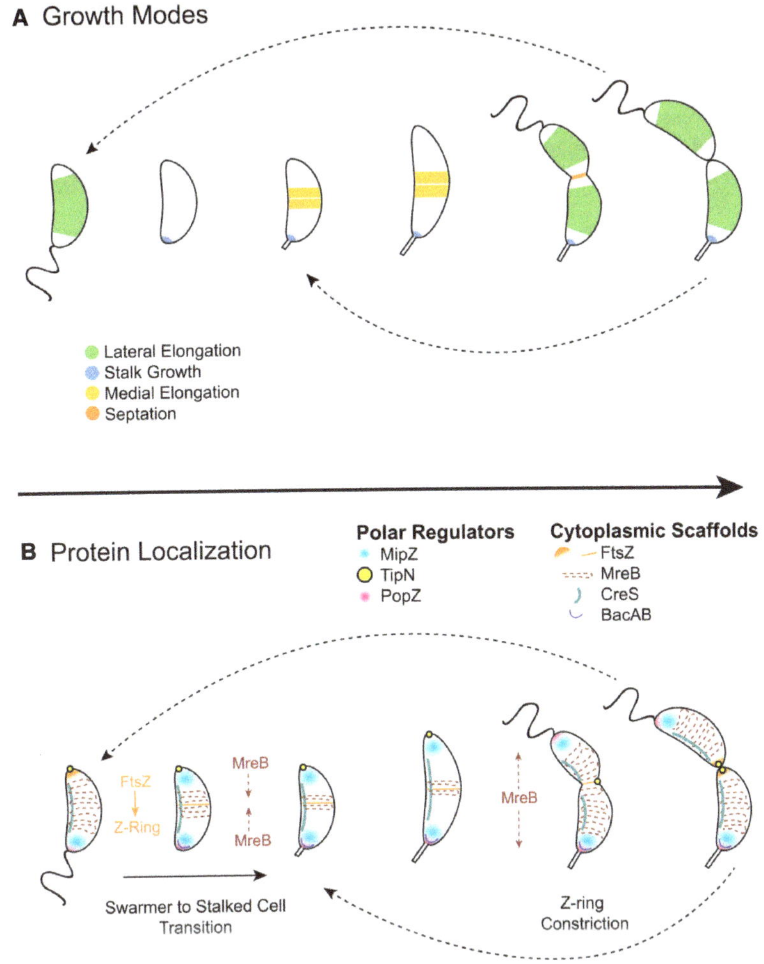

FIGURE 4 | Cytoplasmic scaffold localization dynamics and growth modes in *Caulobacter crescentus*. (A) *C. crescentus* exhibits four growth modes over the course of its dimorphic cell cycle: lateral elongation (swarmer and stalked cells), stalk synthesis and elongation (stalked cells), medial elongation (stalked cells), and septation. Colors indicate regions of active peptidoglycan synthesis. **(B)** The activity and dynamics of various peptidoglycan synthesis machineries, which are regulated by polar regulators, underlie the different growth modes. In swarmer cells, MipZ localizes at the flagellar pole, driving FtsZ to the opposite pole, and MreB is dispersed along the length of the cell. During the swarmer to stalked cell transition, the flagellar pole undergoes remodeling to become the stalked pole. This transition involves the recruitment of bactofilin and peptidoglycan synthesis machinery to the stalked pole by the PopZ network to initiate stalk synthesis. During initiation of chromosome replication and origin duplication, some of the MipZ-ParB complex binds the new origin and migrates with it to the new pole, displacing FtsZ from the new pole to the division plane, where it forms the Z-ring. MreB condenses at the division plane before dispersing again during constriction due to regulation by TipN. After septation, FtsZ reassembles into the Z-ring quickly in stalked cells, making medial elongation the primary mode of elongation in this morphology.

Another difference between *C. crescentus* and canonical rod-shaped bacteria is that it exhibits at least four peptidoglycan synthesis modes over its cell cycle: lateral elongation, elongation at the division plane before division (also called "pre-septal" or "medial" elongation, **Figure 2F**), septation, and stalk synthesis/elongation (**Figure 4A**). As in canonical rod-shaped cells, cell elongation in *C. crescentus* is MreB-dependent. Early studies showed that depletion of MreB disrupts cell morphology and results in lemon-shaped cells (Figge et al., 2004; Gitai et al., 2004). However, the role of MreB has expanded in *C. crescentus* to accommodate the cell cycle and additional growth modes. For example, MreB appears to influence cell polarity and division, as its depletion disturbs the localization of origins of replication as well as the localization of polar regulatory proteins PleC, DivJ, CckA, and DivK (Gitai et al., 2004). Moreover, MreB exhibits dynamic localization throughout the cell cycle, cycling between dispersed helical patterns in swarmer cells (similar to that seen in *E. coli* and *B. subtilis*) and an intense band at the division plane in pre-divisional cells (**Figure 4B**).

The dynamic localization of MreB drives a shift in growth mode from dispersed, lateral elongation mediated by MreB to zonal, medial elongation mediated by both MreB and FtsZ. The condensation of MreB at the division plane requires assembly of the FtsZ ring (Figge et al., 2004; Aaron et al., 2007), and high-resolution temporal studies have shown that MreB recruitment to the FtsZ ring coincides with medial elongation (Goley et al., 2011). *E. coli* also exhibits a transient pre-septal elongation mode (de Pedro et al., 1997; Aarsman et al., 2005;

Potluri et al., 2012), in which FtsZ and MreB interact directly to transfer peptidoglycan synthesis machinery from the elongasome to the divisome (Fenton and Gerdes, 2013). However, in *C. crescentus,* this growth mode predominates a portion of the cell cycle. Because the FtsZ quickly reassembles after division in stalked cells, stalked cells likely elongate primarily via medial elongation (Aaron et al., 2007). Therefore, a similar medial elongation growth mode has been extended in *C. crescentus* to accommodate the asymmetric cell cycle.

MreB appears to have species-specific functions in *C. crescentus.* For example, it appears to interact directly or indirectly with CreS, and is required for proper attachment of the CreS filament to the membrane (Ausmees et al., 2003; Cabeen et al., 2009). Mutations along one surface of MreB alter cell curvature and disrupt the association of the CreS filament with the membrane (Charbon et al., 2009; Dye et al., 2011). Therefore MreB may organize CreS during cell growth, facilitating the attachment of the filament in a stretched conformation (Charbon et al., 2009). Another set of MreB mutations, which occur within the nucleotide-binding pocket, result in morphological defects and affect the localization pattern and dynamics of MreB (Dye et al., 2011). The nucleotide binding site mutants suggest that the ATP hydrolysis cycle of MreB could be coupled to the *C. crescentus* cell cycle to spatiotemporally regulate specific growth modes. The regulation of MreB by ATP could drive its relocation from the sides of the elongating cell to the division plane during the cell cycle.

Some MreB mutants have interesting implications for understanding how *C. crescentus* may remodel polar peptidoglycan. For a subset of the nucleotide binding site MreB mutants (E213G, D16G, N21D, and A325P), cells exhibit a variable width phenotype in which the ends of the cells become extremely tapered and pointed (Dye et al., 2011; Harris et al., 2014). In these mutants, although MreB still exhibits wild type function, it localizes to the cell poles instead of dispersing or condensing at the division plane (Harris et al., 2014). This behavior presumably drives aberrant peptidoglycan synthesis to create elongated cell poles. These MreB mutants may be trapped in a phase of the ATP cycle in which MreB localizes to the poles for a specific polar growth mode in wild type *C. crescentus* cells (Dye et al., 2011). Previous observations that the tapered shape of the wild type *C. crescentus* cell pole develops during the next cell cycle, not during septation or medial elongation (Aaron et al., 2007), corroborates the idea that MreB could participate in remodeling the cell wall at the poles. That the polar cell wall in *C. crescentus* undergoes remodeling instead of remaining entirely inert contradicts what is typically assumed for rod-shaped bacteria. Given that *C. crescentus* belongs to the Alphaproteobacteria class, many members of which exhibit polar growth (Brown et al., 2011, 2012), it may not be surprising if its poles do not follow the same rules as *E. coli* and *B. subtilis.*

In addition to its diversified polar regulators and various growth modes, *C. crescentus* hosts an expanded repertoire of bifunctional peptidoglycan synthases (penicillin-binding proteins, or PBPs). *E. coli* has three bifunctional PBPs. Although PBP1A and PBP1B exhibit semi-redundancy in function, PBP1A appears to preferentially associate with the elongasome (Banzhaf et al., 2012) and PBP1B with the divisome (Bertsche et al., 2006; Müller et al., 2007). The function of PBP1C remains unknown (Schiffer and Höltje, 1999). In contrast, *C. crescentus* has five predicted bifunctional PBPs (PBP1A, PbpC, PbpX, PbpY, PbpZ) which appear to be largely redundant, with any of them capable of functioning as the sole bifunctional PBP with the exception of PbpZ (Yakhnina and Gitai, 2013; Strobel et al., 2014). The same four enzymes were shown to be capable of interacting with divisome components FtsN, FtsL, and DipM (Strobel et al., 2014). Despite this redundancy, each PBP has a specific localization pattern during the cell cycle, suggesting specific cellular functions and roles: Pbp1A and PbpZ localize to the cell periphery; PbpY and PbpX to the periphery, division plane, and stalk; and PbpC to the stalk. Therefore, with the exception of PbpZ, these bifunctional PBPs have retained the ability to interact with both the elongasome and divisome while also evolving distinct affinities for specific localization factors.

Caulobacter crescentus exemplifies how the core peptidoglycan synthesis machinery can be adapted to create its specific morphology and to fit its lifestyle. While it retains an MreB-dependent elongasome and an FtsZ-dependent divisome, it has diversified its growth modes by adding what appears to be a dual MreB- and FtsZ-mediated zonal growth mode at the division plane. These growth modes are strictly regulated in order to synchronize them with the cell cycle and to accommodate asymmetric division. The vibrioid shape arises from diversifying cytoplasmic scaffolds (CreS) and changing the interacting partners of MreB to achieve cell cycle-dependent dynamics. In addition, *C. crescentus* has expanded its suite of bifunctional PBPs and may have additional peptidoglycan-modifying enzymes. Finally, *C. crescentus* has evolved a specialized growth mode, stalk synthesis, which is described in the next section.

The Stalk as a Model System for Understanding Zonal Growth

Stalk synthesis and elongation constitutes one growth mode unique to *C. crescentus* among model rod-shaped bacteria (although not among Alphaproteobacteria). A thin extension of the inner membrane, peptidoglycan, and outer membrane layers (Poindexter, 1964), the stalk is a morphologically distinct organelle thought to improve the nutrient scavenging ability of the cell (Poindexter, 1978; Ireland et al., 2002; Wagner et al., 2006). The narrow cytoplasm of the stalk is free of DNA, ribosomes and most cytoplasmic proteins (Poindexter and Cohen-Bazire, 1964; Ireland et al., 2002; Wagner et al., 2006) and compartmentalized by cross-bands, disk-like, proteinaceous structures that intersect the width of the stalk perpendicular to the long axis of the cell (Jones and Schmidt, 1973; Schlimpert et al., 2012). Cross-bands prevent exchange of membrane and soluble proteins between the stalk and cell body (Schlimpert et al., 2012). As a nonessential organelle, the stalk serves as a convenient model for zonal, or targeted, peptidoglycan growth (Wagner and Brun, 2007).

Stalk synthesis initiates at the cell pole during the swarmer to stalked cell transition phase of the cell cycle, and elongation occurs at the cell-stalk junction during pre-divisional cell elongation (Schmidt and Stanier, 1966; Aaron et al., 2007). MreB and RodA (an MreB-associated protein) depletion, or treatment with the PBP2 inhibitor mecillinam (Seitz and Brun, 1998), result in the loss or shortening of stalks, and recovery from depletion leads to the growth of ectopic stalked poles, suggesting that MreB, in concert with RodA and a PBP2 homolog, plays a role in recruiting peptidoglycan synthesis machinery for stalk synthesis (Wagner et al., 2005). Overexpression of RodZ (another MreB-associated protein) resulted in multiple stalks forming at the same pole, opposite pole, or on the cell body (Alyahya et al., 2009). RodZ localization dynamics coincided with FtsZ during the cell cycle but appeared to depend on MreB. Overall, these data suggest that stalk synthesis and elongation in *C. crescentus* potentially depends on a similar core apparatus as elongation, utilizing MreB as a scaffold. The mechanism for cross-band formation may have interesting intersections with the processes of stalk synthesis and elongation. Early on, electron micrographs of *C. crescentus* stalks showed that cross-bands consisted of concentric circular striations, possibly of peptidoglycan and membranes (Jones and Schmidt, 1973). Because the stalk continues to elongate as the stalked cell participates in increasing rounds of division, it was postulated that stalk length and number of cross-bands could indicate cell age. Quantitative analysis suggested that one cross-band is added to the stalk toward the end of each reproductive cycle, possibly accompanying division (Poindexter and Staley, 1996; Schlimpert et al., 2012). The idea that cross-bands consist of peptidoglycan has been challenged by the discovery that cross-bands are the product of the macromolecular assembly of least four proteins, StpABCD (Schlimpert et al., 2012). However, a relationship between cross-band assembly and the peptidoglycan synthesis machinery may still exist.

Like septation, lateral, or medial elongation in *C. crescentus*, stalk synthesis and elongation appear to be highly tuned to cell cycle regulation and polar positioning mechanisms. Moreover, the mechanism for stalk synthesis appears to depend on the expansion and diversification of the core machinery. Bactofilins BacA and BacB constitute a class of scaffolding proteins with proline-rich terminal regions that localize to the stalked pole of the cell during the swarmer to stalked cell transition (Kühn et al., 2010). Deletion of these genes led to a 45% reduction in stalk length, suggesting a recruitment role in stalk assembly. Indeed, the localization of the stalk-specific bifunctional peptidoglycan synthase PbpC to the stalk was shown to be dependent on BacA and BacB. PbpC potentially acts in conjunction with PbpX in stalk elongation (Yakhnina and Gitai, 2013; Strobel et al., 2014), however, it has also been shown to have a role in recruiting and modifying a stalk-specific protein. StpX is a stalk-specific membrane protein that promotes the elongation of the stalk in nutrient-limiting conditions (Hughes et al., 2010). PbpC is not only required for localizing StpX to the stalk-cell junction at onset of stalk synthesis, but appears to be directly or indirectly involved in the process of tethering StpX to the outer membrane or to an outer membrane protein (Hughes et al., 2013). That this phenomenon takes place regardless of PbpC's transglycosylase or transpeptidase activities suggests alternative roles for various PBPs and other enzyme classes in regulating the peptidoglycan synthesis machinery, opening up the diversity of mechanisms for the evolution of novel functions.

Phylogenic studies indicate that stalk positioning evolved from an ancestral, single polar stalk in the order *Caulobacteraceae*, to a single sub-polar stalk in *Asticcacaulis excentricus*, and subsequently to bilateral stalks in *Asticcacaulis biprosthecum* (Jiang et al., 2014). The stalk structure appears identical in *Caulobacter* and *Asticcacaulis* (Pate and Ordal, 1965) and all three species appear to share the same stalk synthesis mechanism of inserting peptidoglycan at the base of the stalk (Jiang et al., 2014). The natural variation and evolution of stalk positioning in these species correlates with the localization of SpmX, a developmental regulator (Radhakrishnan et al., 2008) that has been shown to be necessary for stalk synthesis in *Asticcacaulis* but not in *Caulobacter* (Radhakrishnan et al., 2008; Jiang et al., 2014). Therefore, in *Asticcacaulis* SpmX has been co-opted as a stalk-positioning factor. SpmX consists of three defined regions: a highly conserved N-terminal muramidase domain, an unstructured intermediate region that has been expanded by over 370 amino acids in the *Asticcacaulis* genus, and a two-pass transmembrane domain. Recent work has demonstrated that either the intermediate region or the transmembrane domain contain the residues responsible for SpmX's new role in sub-polar and bilateral stalk positioning (Jiang et al., 2014). The muramidase domain, while necessary for overall function and localization, is interchangeable amongst species. How a murein hydrolase participates in SpmX's function and localization will be an interesting question for future studies.

The recent discovery that SpmX coordinates stalk placement in the *Asticcacaulis* genus underscores the utility of not only studying the stalk as a model system for zonal growth, but of exploiting the natural diversity of stalked morphologies observed in Alphaproteobacteria. In marked contrast to division and elongation genes, which require depletion strains, stalk-less mutants are far easier to identify in a genetic screen. *Asticcacaulis* provides an opportunity for screening stalkless mutants with fewer chances of impacting elements critical for cell viability, since stalk synthesis no longer overlaps with polar regulation. Finally, altering the location of the stalk may prove to be a far more advantageous phenotype than losing the stalk entirely in terms of discovering localization and recruitment factors for stalk initiation and maintenance. Therefore studying stalk localization, initiation, and maintenance in *Asticcacaulis* will likely make significant inroads into understanding how bacteria can redirect the core peptidoglycan machinery to create novel morphologies.

Conclusion and Summary

Critical advances in studying bacterial growth and division in model bacteria such as *E. coli*, *B. subtilis*, and *C. crescentus* have identified and characterized the fundamental components of the molecular machinery responsible for creating and shaping the sacculus. This expanding body of work reveals a general strategy for growing and shaping the bacterial cell wall. Moreover, it suggests ways in which the core can be adapted for various bacterial lifestyles as well as molecular mechanisms for the evolution of novel morphologies. The peptidoglycan synthesis machinery consists of a central core of cytoskeletal-like scaffolding proteins, inner membrane-spanning elements, and periplasmic or extracellular peptidoglycan-modifying elements.

The modularity and evolvability of this system is inherent in how different elements have been retuned or repurposed in each genus for species-specific growth modes and morphologies. As different species of bacteria become genetically tractable, it is becoming clear that the diversification and expansion of any of the core elements of the peptidoglycan synthesis assembly underlie morphological diversity and alternate growth modes.

As studies in *C. crescentus* and the closely related genus *Asticcacaulis* demonstrate, much can be learned about general growth strategies using comparative approaches with related families. The great diversity of morphologies and growth modes in Alphaproteobacteria (**Figure 4**) offers an opportunity to exploit natural diversity to dissect mechanisms for stalk localization and synthesis, as well as polar growth. The recent advances in expressing fluorescent fusions in *Hyphomonas neptunium* open up yet another opportunity for discovering genes responsible for polar growth and stalk development (Jung et al., 2014). The increasing ease and affordability of genome sequencing allows further investigation into the suites of genes underlying uncharacterized growth modes in other genera. New reagents that fluorescently label sites of peptidoglycan synthesis are providing simple methods to determine growth modes, even in understudied species and environmental samples (Kuru et al., 2012; Pilhofer et al., 2013; Liechti et al., 2014). Determining how bacteria actively restructure their morphologies, or how they have evolved various morphologies over time, remains a major goal in bacteriology. Expanding investigations into understudied genera with novel morphologies and growth modes will complement and enrich our understanding of how bacteria grow and proliferate.

Acknowledgments

Our research on morphogenesis is supported by National Institutes of Health Grants GM051986 and GM113172 to YB AR was supported by National Institutes of Health National Research Service Award F32GM112362. We thank David Kysela for the phylogenetic tree, and Kevin Young and members of the Brun laboratory for critical reading of the manuscript.

References

Aaron, M., Charbon, G., Lam, H., Schwarz, H., Vollmer, W., and Jacobs-Wagner, C. (2007). The tubulin homologue FtsZ contributes to cell elongation by guiding cell wall precursor synthesis in *Caulobacter crescentus*. *Mol. Microbiol.* 64, 938–952. doi: 10.1111/j.1365-2958.2007.05720.x

Aarsman, M. E. G., Piette, A., Fraipont, C., Vinkenvleugel, T. M. F., Nguyen-Distèche, M., and den Blaauwen, T. (2005). Maturation of the *Escherichia coli* divisome occurs in two steps. *Mol. Microbiol.* 55, 1631–1645. doi: 10.1111/j.1365-2958.2005.04502.x

Alyahya, S. A., Alexander, R., Costa, T., Henriques, A. O., Emonet, T., and Jacobs-Wagner, C. (2009). RodZ, a component of the bacterial core morphogenic apparatus. *Proc. Natl. Acad. Sci. U.S.A.* 106, 1239–1244. doi: 10.1073/pnas.0810794106

Ausmees, N., Kuhn, J. R., and Jacobs-Wagner, C. (2003). The bacterial cytoskeleton: an intermediate filament-like function in cell shape. *Cell* 115, 705–713. doi: 10.1016/S0092-8674(03)00935-8

Bagchi, S., Tomenius, H., Belova, L. M., and Ausmees, N. (2008). Intermediate filament-like proteins in bacteria and a cytoskeletal function in *Streptomyces*. *Mol. Microbiol.* 70, 1037–1050. doi: 10.1111/j.1365-2958.2008.06473.x

Banzhaf, M., van den Berg van Saparoea, B., Terrak, M., Fraipont, C., Egan, A., Philippe, J., et al. (2012). Cooperativity of peptidoglycan synthases active in bacterial cell elongation. *Mol. Microbiol.* 85, 179–194. doi: 10.1111/j.1365-2958.2012.08103.x

Bauld, J., and Staley, J. T. (1976). Planctomyces maris sp. nov: a marine isolate of the planctomyces-blastocaulis group of budding bacteria. *J. Gen. Microbiol.* 97, 45–55. doi: 10.1099/00221287-97-1-45

Bendezú, F. O., Hale, C. A., Bernhardt, T. G., and de Boer, P. A. J. (2009). RodZ (YfgA) is required for proper assembly of the MreB actin cytoskeleton and cell shape in *E. coli*. *EMBO J.* 28, 193–204. doi: 10.1038/emboj.2008.264

Bernhardt, T. G., and de Boer, P. A. J. (2005). SlmA, a nucleoid-associated, FtsZ binding protein required for blocking septal ring assembly over chromosomes in *E. coli*. *Mol. Cell* 18, 555–564. doi: 10.1016/j.molcel.2005.04.012

Bertelli, C., Collyn, F., Croxatto, A., Rückert, C., Polkinghorne, A., Kebbi-Beghdadi, C., et al. (2010). The waddlia genome: a window into chlamydial biology. *PLoS ONE* 5:e10890. doi: 10.1371/journal.pone.0010890

Bertsche, U., Kast, T., Wolf, B., Fraipont, C., Aarsman, M. E. G., Kannenberg, K., et al. (2006). Interaction between two murein (peptidoglycan) synthases, PBP3 and PBP1B, in *Escherichia coli*. *Mol. Microbiol.* 61, 675–690. doi: 10.1111/j.1365-2958.2006.05280.x

Bi, E. F., and Lutkenhaus, J. (1991). FtsZ ring structure associated with division in *Escherichia coli*. *Nature* 354, 161–164. doi: 10.1038/354161a0

Bonis, M., Ecobichon, C., Guadagnini, S., Prévost, M.-C., and Boneca, I. G. (2010). A M23B family metallopeptidase of *Helicobacter pylori* required for cell shape, pole formation and virulence. *Mol. Microbiol.* 78, 809–819. doi: 10.1111/j.1365-2958.2010.07383.x

Bowman, G. R., Comolli, L. R., Gaietta, G. M., Fero, M., Hong, S.-H., Jones, Y., et al. (2010). *Caulobacter* PopZ forms a polar subdomain dictating sequential changes in pole composition and function. *Mol. Microbiol.* 76, 173–189. doi: 10.1111/j.1365-2958.2010.07088.x

Brown, P. J. B., de Pedro, M. A., Kysela, D. T., Van der Henst, C., Kim, J., De Bolle, X., et al. (2012). Polar growth in the Alphaproteobacterial order Rhizobiales. *Proc. Natl. Acad. Sci. U.S.A.* 109, 1697–1701. doi: 10.1073/pnas.1114476109

Brown, P. J. B., Kysela, D. T., and Brun, Y. V. (2011). Polarity and the diversity of growth mechanisms in bacteria. *Semin. Cell Dev. Biol.* 22, 790–798. doi: 10.1016/j.semcdb.2011.06.006

Cabeen, M. T., Charbon, G., Vollmer, W., Born, P., Ausmees, N., Weibel, D. B., et al. (2009). Bacterial cell curvature through mechanical control of cell growth. *EMBO J.* 28, 1208–1219. doi: 10.1038/emboj.2009.61

Cameron, T. A., Anderson-Furgeson, J., Zupan, J. R., Zik, J. J., and Zambryski, P. C. (2014). Peptidoglycan synthesis machinery in *Agrobacterium tumefaciens* during unipolar growth and cell division. *MBio* 5, e01219–14. doi: 10.1128/mBio.01219-14

Cava, F., and de Pedro, M. A. (2014). Peptidoglycan plasticity in bacteria: emerging variability of the murein sacculus and their associated biological functions. *Curr. Opin. Microbiol.* 18, 46–53. doi: 10.1016/j.mib.2014.01.004

Charbon, G., Cabeen, M. T., and Jacobs-Wagner, C. (2009). Bacterial intermediate filaments: *in vivo* assembly, organization, and dynamics of crescentin. *Genes Dev.* 23, 1131–1144. doi: 10.1101/gad.1795509

Chauhan, A., Lofton, H., Maloney, E., Moore, J., Fol, M., Madiraju, M. V. V. S., et al. (2006). Interference of Mycobacterium tuberculosis cell division by Rv2719c, a cell wall hydrolase. *Mol. Microbiol.* 62, 132–147. doi: 10.1111/j.1365-2958.2006.05333.x

Daniel, R. A., and Errington, J. (2003). Control of cell morphogenesis in bacteria. *Cell* 113, 767–776. doi: 10.1016/S0092-8674(03)00421-5

de Pedro, M. A. Quintela, J. C., Höltje, J. V., and Schwarz, H. (1997). Murein segregation in *Escherichia coli*. *J. Bacteriol.* 179, 2823–2834.

Dominguez-Escobar, J., Chastanet, A., Crevenna, A. H., Fromion, V., Wedlich-Soldner, R., and Carballido-Lopez, R. (2011). Processive movement of MreB-associated cell wall biosynthetic complexes in bacteria. *Science* 333, 225–228. doi: 10.1126/science.1203466

Dye, N. A., Pincus, Z., Fisher, I. C., Shapiro, L., and Theriot, J. A. (2011). Mutations in the nucleotide binding pocket of MreB can alter cell curvature and polar morphology in *Caulobacter*: modifying morphology with mutations in mreB. *Mol. Microbiol.* 81, 368–394. doi: 10.1111/j.1365-2958.2011.07698.x

Edgar, R. C. (2004). MUSCLE: multiple sequence alignment with high accuracy and high throughput. *Nucleic Acids Res.* 32, 1792–1797. doi: 10.1093/nar/gkh340

Edwards, D. H., and Errington, J. (1997). The *Bacillus subtilis* DivIVA protein targets to the division septum and controls the site specificity of cell division. *Mol. Microbiol.* 24, 905–915.

Egan, A. J. F., and Vollmer, W. (2013). The physiology of bacterial cell division. *Ann. N. Y. Acad. Sci.* 1277, 8–28. doi: 10.1111/j.1749-6632.2012.06818.x

Erickson, H. P., Anderson, D. E., and Osawa, M. (2010). FtsZ in bacterial cytokinesis: cytoskeleton and force generator all in one. *Microbiol. Mol. Biol. Rev.* 74, 504–528. doi: 10.1128/MMBR.00021-10

Esue, O., Cordero, M., Wirtz, D., and Tseng, Y. (2005). The assembly of MreB, a prokaryotic homolog of actin. *J. Biol. Chem.* 280, 2628–2635. doi: 10.1074/jbc.M410298200

Esue, O., Wirtz, D., and Tseng, Y. (2006). GTPase activity, structure, and mechanical properties of filaments assembled from bacterial cytoskeleton protein MreB. *J. Bacteriol.* 188, 968–976. doi: 10.1128/JB.188.3.968-976.2006

Fenton, A. K., and Gerdes, K. (2013). Direct interaction of FtsZ and MreB is required for septum synthesis and cell division in *Escherichia coli*. *EMBO J.* 32, 1953–1965. doi: 10.1038/emboj.2013.129

Figge, R. M., Divakaruni, A. V., and Gober, J. W. (2004). MreB, the cell shape-determining bacterial actin homologue, co-ordinates cell wall morphogenesis in *Caulobacter crescentus*: MreB-dependent regulation of cell shape. *Mol. Microbiol.* 51, 1321–1332. doi: 10.1111/j.1365-2958.2003.03936.x

Flärdh, K. (2010). Cell polarity and the control of apical growth in *Streptomyces*. *Curr. Opin. Microbiol.* 13, 758–765. doi: 10.1016/j.mib.2010.10.002

Flärdh, K., and Buttner, M. J. (2009). *Streptomyces* morphogenetics: dissecting differentiation in a filamentous bacterium. *Nat. Rev. Microbiol.* 7, 36–49. doi: 10.1038/nrmicro1968

Fleurie, A., Lesterlin, C., Manuse, S., Zhao, C., Cluzel, C., Lavergne, J.-P., et al. (2014). MapZ marks the division sites and positions FtsZ rings in *Streptococcus pneumoniae*. *Nature* 516, 259–262. doi: 10.1038/nature13966

Frandi, A., Jacquier, N., Théraulaz, L., Greub, G., and Viollier, P. H. (2014). FtsZ-independent septal recruitment and function of cell wall remodelling enzymes in chlamydial pathogens. *Nat. Commun.* 5, 4200. doi: 10.1038/ncomms5200

Frirdich, E., and Gaynor, E. C. (2013). Peptidoglycan hydrolases, bacterial shape, and pathogenesis. *Curr. Opin. Microbiol.* 16, 767–778. doi: 10.1016/j.mib.2013.09.005

Fu, G., Huang, T., Buss, J., Coltharp, C., Hensel, Z., and Xiao, J. (2010). *In vivo* structure of the *E. coli* FtsZ-ring revealed by photoactivated localization microscopy (PALM). *PLoS ONE* 5:e12682. doi: 10.1371/journal.pone.0012680

Garner, E. C., Bernard, R., Wang, W., Zhuang, X., Rudner, D. Z., and Mitchison, T. (2011). Coupled, circumferential motions of the cell wall synthesis machinery and MreB filaments in *B. subtilis*. *Science* 333, 222–225. doi: 10.1126/science.1203285

Gitai, Z., Dye, N., and Shapiro, L. (2004). An actin-like gene can determine cell polarity in bacteria. *Proc. Natl. Acad. Sci. U.S.A.* 101, 8643–8648. doi: 10.1073/pnas.0402638101

Goley, E. D., Yeh, Y.-C., Hong, S.-H., Fero, M. J., Abeliuk, E., McAdams, H. H., et al. (2011). Assembly of the *Caulobacter* cell division machine. *Mol. Microbiol.* 80, 1680–1698. doi: 10.1111/j.1365-2958.2011.07677.x

Hale, C. A., Meinhardt, H., and de Boer, P. A. J. (2001). Dynamic localization cycle of the cell division regulator MinE in *Escherichia coli*. *EMBO J.* 20, 1563–1572. doi: 10.1093/emboj/20.7.1563

Hamoen, L. W., and Errington, J. (2003). Polar targeting of DivIVA in *Bacillus subtilis* is not directly dependent on FtsZ or PBP 2B. *J. Bacteriol.* 185, 693–697. doi: 10.1128/JB.185.2.693-697.2003

Harris, L. K., Dye, N. A., and Theriot, J. A. (2014). A *Caulobacter* MreB mutant with irregular cell shape exhibits compensatory widening to maintain a preferred surface area to volume ratio. *Mol. Microbiol.* 94, 988–1005. doi: 10.1111/mmi.12811

Henrichfreise, B., Schiefer, A., Schneider, T., Nzukou, E., Poellinger, C., Hoffmann, T.-J., et al. (2009). Functional conservation of the lipid II biosynthesis pathway in the cell wall-less bacteria Chlamydia and Wolbachia: why is lipid II needed? *Mol. Microbiol.* 73, 913–923. doi: 10.1111/j.1365-2958.2009.06815.x

Holečková, N., Doubravová, L., Massidda, O., Molle, V., Buriánková, K., Benada, O., et al. (2015). LocZ is a new cell division protein involved in proper septum placement in *Streptococcus pneumoniae*. *mBio* 6, e01700–14. doi: 10.1128/mBio.01700-14

Höltje, J.-V. (1998). Growth of the stress-bearing and shape-maintaining murein sacculus of *Escherichia coli*. *Microbiol. Mol. Biol. Rev.* 62, 181–203.

Hughes, H. V., Huitema, E., Pritchard, S., Keiler, K. C., Brun, Y. V., and Viollier, P. H. (2010). Protein localization and dynamics within a bacterial organelle. *Proc. Natl. Acad. Sci. U.S.A.* 107, 5599–5604. doi: 10.1073/pnas.0909119107

Hughes, H. V., Lisher, J. P., Hardy, G. G., Kysela, D. T., Arnold, R. J., Giedroc, D. P., et al. (2013). Co-ordinate synthesis and protein localization in a bacterial organelle by the action of a penicillin-binding-protein. *Mol. Microbiol.* 90, 1162–1177. doi: 10.1111/mmi.12422

Ireland, M. M. E., Karty, J. A., Quardokus, E. M., Reilly, J. P., and Brun, Y. V. (2002). Proteomic analysis of the *Caulobacter crescentus* stalk indicates competence for nutrient uptake. *Mol. Microbiol.* 45, 1029–1041. doi: 10.1046/j.1365-2958.2002.03071.x

Jacquier, N., Frandi, A., Pillonel, T., Viollier, P. H., and Greub, G. (2014). Cell wall precursors are required to organize the chlamydial division septum. *Nat. Commun.* 5, 3578. doi: 10.1038/ncomms4578

Janakiraman, A., and Goldberg, M. B. (2004). Recent advances on the development of bacterial poles. *Trends Microbiol.* 12, 518–525. doi: 10.1016/j.tim.2004.09.003

Jiang, C., Brown, P. J. B., Ducret, A., and Brun, Y. V. (2014). Sequential evolution of bacterial morphology by co-option of a developmental regulator. *Nature* 506, 489–493. doi: 10.1038/nature12900

Jiang, C., Caccamo, P. D., and Brun, Y. V. (2015). Mechanisms of bacterial morphogenesis: evolutionary cell biology approaches provide new insights. *Bioessays* 37, 413–425. doi: 10.1002/bies.201400098

Jones, H. C., and Schmidt, J. M. (1973). Ultrastructural study of crossbands occurring in the stalks of *Caulobacter crescentus*. *J. Bacteriol.* 116, 466–470.

Jung, A., Eisheuer, S., Cserti, E., Leicht, O., Strobel, W., Möll, A., et al. (2014). A molecular toolbox for the genetic manipulation of the stalked budding bacterium *Hyphomonas neptunium*. *Appl. Environ. Microbiol.* 81, 736–744. doi: 10.1128/AEM.03104-14

Kruse, T., Bork-Jensen, J., and Gerdes, K. (2004). The morphogenetic MreBCD proteins of *Escherichia coli* form an essential membrane-bound complex: MreBCD proteins form an essential membrane complex. *Mol. Microbiol.* 55, 78–89. doi: 10.1111/j.1365-2958.2004.04367.x

Kühn, J., Briegel, A., Mörschel, E., Kahnt, J., Leser, K., Wick, S., et al. (2010). Bactofilins, a ubiquitous class of cytoskeletal proteins mediating polar localization of a cell wall synthase in *Caulobacter crescentus*. *EMBO J.* 29, 327–339. doi: 10.1038/emboj.2009.358

Kuru, E., Hughes, H. V., Brown, P. J., Hall, E., Tekkam, S., Cava, F., et al. (2012). *In situ* probing of newly synthesized peptidoglycan in live bacteria with fluorescent D-amino acids. *Angew. Chem. Int. Ed.* 51, 12519–12523. doi: 10.1002/anie.201206749

Laloux, G., and Jacobs-Wagner, C. (2013). Spatiotemporal control of PopZ localization through cell cycle–coupled multimerization. *J. Cell Biol.* 201, 827–841. doi: 10.1083/jcb.201303036

Lam, H., Schofield, W. B., and Jacobs-Wagner, C. (2006). A landmark protein essential for establishing and perpetuating the polarity of a bacterial cell. *Cell* 124, 1011–1023. doi: 10.1016/j.cell.2005.12.040

Lee, T. K., and Huang, K. C. (2013). The role of hydrolases in bacterial cell-wall growth. *Curr. Opin. Microbiol.* 16, 760–766. doi: 10.1016/j.mib.2013.08.005

Li, Z., Trimble, M. J., Brun, Y. V., and Jensen, G. J. (2007). The structure of FtsZ filaments *in vivo* suggests a force-generating role in cell division. *EMBO J.* 26, 4694–4708. doi: 10.1038/sj.emboj.7601895

Liechti, G. W., Kuru, E., Hall, E., Kalinda, A., Brun, Y. V., VanNieuwenhze, M., et al. (2014). A new metabolic cell-wall labelling method reveals peptidoglycan in *Chlamydia trachomatis*. *Nature* 506, 507–510. doi: 10.1038/nature12892

Löwe, J., and Amos, L. A. (1998). Crystal structure of the bacterial cell-division protein FtsZ. *Nature* 391, 203–206. doi: 10.1038/34472

Männik, J., Wu, F., Hol, F. J. H., Bisicchia, P., Sherratt, D. J., Keymer, J. E., et al. (2012). Robustness and accuracy of cell division in *Escherichia coli* in diverse cell shapes. *Proc. Natl. Acad. Sci. U.S.A.* 109, 6957–6962. doi: 10.1073/pnas.1120854109

Margolin, W. (2009). Sculpting the bacterial cell. *Curr. Biol.* 19, R812–R822. doi: 10.1016/j.cub.2009.06.033

Marston, A. L., Thomaides, H. B., Edwards, D. H., Sharpe, M. E., and Errington, J. (1998). Polar localization of the MinD protein of *Bacillus subtilis* and its role in selection of the mid-cell division site. *Genes Dev.* 12, 3419–3430.

McCoy, A. J., and Maurelli, A. T. (2006). Building the invisible wall: updating the chlamydial peptidoglycan anomaly. *Trends Microbiol.* 14, 70–77. doi: 10.1016/j.tim.2005.12.004

Mobley, H. L., Koch, A. L., Doyle, R. J., and Streips, U. N. (1984). Insertion and fate of the cell wall in *Bacillus subtilis*. *J. Bacteriol.* 158, 169–179.

Mohammadi, T., Karczmarek, A., Crouvoisier, M., Bouhss, A., Mengin-Lecreulx, D., and den Blaauwen, T. (2007). The essential peptidoglycan glycosyltransferase MurG forms a complex with proteins involved in lateral envelope growth as well as with proteins involved in cell division in *Escherichia coli*. *Mol. Microbiol.* 65, 1106–1121. doi: 10.1111/j.1365-2958.2007.05851.x

Moore, R. L. (1981). The biology of *Hyphomicrobium* and other prosthecate, budding bacteria. *Annu. Rev. Microbiol.* 35, 567–594. doi: 10.1146/annurev.mi.35.100181.003031

Moore, R., and Hirsch, P. (1973). First generation synchrony of isolated *Hyphomicrobium* swarmer populations. *J. Bacteriol.* 116, 418–423.

Mukherjee, A., and Lutkenhaus, J. (1998). Dynamic assembly of FtsZ regulated by GTP hydrolysis. *EMBO J.* 17, 462–469. doi: 10.1093/emboj/17.2.462

Müller, P., Ewers, C., Bertsche, U., Anstett, M., Kallis, T., Breukink, E., et al. (2007). The essential cell division protein FtsN interacts with the murein (peptidoglycan) synthase PBP1B in *Escherichia coli*. *J. Biol. Chem.* 282, 36394–36402. doi: 10.1074/jbc.M706390200

Ozyamak, E., Kollman, J. M., and Komeili, A. (2013). Bacterial actins and their diversity. *Biochemistry (Mosc.)* 52, 6928–6939. doi: 10.1021/bi4010792

Pate, J. L., and Ordal, E. J. (1965). The fine structure of two unusual stalked bacteria. *J. Cell Biol.* 27, 133–150.

Pilhofer, M., Aistleitner, K., Biboy, J., Gray, J., Kuru, E., Hall, E., et al. (2013). Discovery of chlamydial peptidoglycan reveals bacteria with murein sacculi but without FtsZ. *Nat. Commun.* 4:2856. doi: 10.1038/ncomms3856

Pilhofer, M., Rappl, K., Eckl, C., Bauer, A. P., Ludwig, W., Schleifer, K. -H., et al. (2008). Characterization and evolution of cell division and cell wall synthesis genes in the bacterial phyla verrucomicrobia, lentisphaerae, chlamydiae, and planctomycetes and phylogenetic comparison with rRNA genes. *J. Bacteriol.* 190, 3192–3202. doi: 10.1128/JB.01797-07

Pinho, M. G., Kjos, M., and Veening, J. -W. (2013). How to get (a)round: mechanisms controlling growth and division of coccoid bacteria. *Nat. Rev. Microbiol.* 11, 601–614. doi: 10.1038/nrmicro3088

Poindexter, J. L. S., and Cohen-Bazire, G. (1964). The fine structure of stalked bacteria belonging to the family *Caulobacteraceae*. *J. Cell Biol.* 23, 587–607. doi: 10.1083/jcb.23.3.587

Poindexter, J. S. (1964). Biological properties and classification of the *Caulobacter* group1. *Bacteriol. Rev.* 28, 231–295.

Poindexter, J. S. (1978). Selection for nonbuoyant morphological mutants of *Caulobacter crescentus*. *J. Bacteriol.* 135, 1141–1145.

Poindexter, J. S., and Staley, J. T. (1996). *Caulobacter* and *Asticcacaulis* stalk bands as indicators of stalk age. *J. Bacteriol.* 178, 3939–3948.

Potluri, L. -P., Kannan, S., and Young, K. D. (2012). ZipA is required for FtsZ-dependent preseptal peptidoglycan Synthesis prior to Invagination during cell division. *J. Bacteriol.* 194, 5334–5342. doi: 10.1128/JB.00859-12

Priyadarshini, R., de Pedro, M. A., and Young, K. D. (2007). Role of peptidoglycan amidases in the development and morphology of the division septum in *Escherichia coli*. *J. Bacteriol.* 189, 5334–5347. doi: 10.1128/JB.00415-07

Ptacin, J. L., Gahlmann, A., Bowman, G. R., Perez, A. M., Diezmann, A. R. S., von, Eckart, M. R., et al. (2014). Bacterial scaffold directs pole-specific centromere segregation. *Proc. Natl. Acad. Sci. U.S.A.* 111, E2046–E2055. doi: 10.1073/pnas.1405188111

Radhakrishnan, S. K., Thanbichler, M., and Viollier, P. H. (2008). The dynamic interplay between a cell fate determinant and a lysozyme homolog drives the asymmetric division cycle of *Caulobacter crescentus*. *Genes Dev.* 22, 212–225. doi: 10.1101/gad.1601808

Raskin, D. M., and de Boer, P. A. J. (1999). Rapid pole-to-pole oscillation of a protein required for directing division to the middle of *Escherichia coli*. *Proc. Natl. Acad. Sci. U.S.A.* 96, 4971–4976.

Salje, J., van den Ent, F., de Boer, P., and Löwe, J. (2011). Direct membrane binding by bacterial actin MreB. *Mol. Cell* 43, 478–487. doi: 10.1016/j.molcel.2011.07.008

Schiffer, G., and Höltje, J. V. (1999). Cloning and characterization of PBP 1C, a third member of the multimodular class A penicillin-binding proteins of *Escherichia coli*. *J. Biol. Chem.* 274, 32031–32039.

Schlaeppi, J. M., Schaefer, O., and Karamata, D. (1985). Cell wall and DNA cosegregation in *Bacillus subtilis* studied by electron microscope autoradiography. *J. Bacteriol.* 164, 130–135.

Schlimpert, S., Klein, E. A., Briegel, A., Hughes, V., Kahnt, J., Bolte, K., et al. (2012). General protein diffusion barriers create compartments within bacterial cells. *Cell* 151, 1270–1282. doi: 10.1016/j.cell.2012.10.046

Schmidt, J. M., and Stanier, R. Y. (1966). The development of cellular stalks in bacteria. *J. Cell Biol.* 28, 423–436.

Seitz, L. C., and Brun, Y. V. (1998). Genetic analysis of mecillinam-resistant mutants of *Caulobacter crescentus* deficient in stalk biosynthesis. *J. Bacteriol.* 180, 5235–5239.

Sham, L.-T., Tsui, H.-C. T., Land, A. D., Barendt, S. M., and Winkler, M. E. (2012). Recent advances in pneumococcal peptidoglycan biosynthesis suggest new vaccine and antimicrobial targets. *Curr. Opin. Microbiol.* 15, 194–203. doi: 10.1016/j.mib.2011.12.013

Shiomi, D., Sakai, M., and Niki, H. (2008). Determination of bacterial rod shape by a novel cytoskeletal membrane protein. *EMBO J.* 27, 3081–3091. doi: 10.1038/emboj.2008.234

Specht, M., Schätzle, S., Graumann, P. L., and Waidner, B. (2011). *Helicobacter pylori* possesses four coiled-coil-rich proteins that form extended filamentous structures and control cell shape and motility. *J. Bacteriol.* 193, 4523–4530. doi: 10.1128/JB.00231-11

Stamatakis, A. (2006). RAxML-VI-HPC: maximum likelihood-based phylogenetic analyses with thousands of taxa and mixed models. *Bioinforma. Oxf. Engl.* 22, 2688–2690. doi: 10.1093/bioinformatics/btl446

Stephens, R. S., Kalman, S., Lammel, C., Fan, J., Marathe, R., Aravind, L., et al. (1998). Genome sequence of an obligate intracellular pathogen of humans: *Chlamydia trachomatis*. *Science* 282, 754–759.

Strauss, M. P., Liew, A. T. F., Turnbull, L., Whitchurch, C. B., Monahan, L. G., and Harry, E. J. (2012). 3D-SIM super resolution microscopy reveals a bead-like arrangement for FtsZ and the division machinery: implications for triggering cytokinesis. *PLoS Biol.* 10:e1001389. doi: 10.1371/journal.pbio.1001389

Strobel, W., Möll, A., Kiekebusch, D., Klein, K. E., and Thanbichler, M. (2014). Function and localization dynamics of bifunctional penicillin-binding proteins in *Caulobacter crescentus*. *J. Bacteriol.* 196, 1627–1639. doi: 10.1128/JB.01194-13

Sycuro, L. K., Pincus, Z., Gutierrez, K. D., Biboy, J., Stern, C. A., Vollmer, W., et al. (2010). Peptidoglycan crosslinking relaxation promotes *Helicobacter pylori*'s helical shape and stomach colonization. *Cell* 141, 822–833. doi: 10.1016/j.cell.2010.03.046

Sycuro, L. K., Rule, C. S., Petersen, T. W., Wyckoff, T. J., Sessler, T., Nagarkar, D. B., et al. (2013). Flow cytometry-based enrichment for cell shape mutants identifies multiple genes that influence *Helicobacter pylori* morphology. *Mol. Microbiol.* 90, 869–883. doi: 10.1111/mmi.12405

Sycuro, L. K., Wyckoff, T. J., Biboy, J., Born, P., Pincus, Z., Vollmer, W., et al. (2012). Multiple peptidoglycan modification networks modulate *Helicobacter pylori*'s cell shape, motility, and colonization potential. *PLoS Pathog.* 8:e1002603. doi: 10.1371/journal.ppat.1002603

Szwedziak, P., and Löwe, J. (2013). Do the divisome and elongasome share a common evolutionary past? *Curr. Opin. Microbiol.* 16, 745–751. doi: 10.1016/j.mib.2013.09.003

Szwedziak, P., Wang, Q., Bharat, T. A. M., Tsim, M., and Löwe, J. (2014). Architecture of the ring formed by the tubulin homologue FtsZ in bacterial cell division. *eLife* 3, e04601. doi: 10.7554/eLife.04601

Thanbichler, M., and Shapiro, L. (2006). MipZ, a spatial regulator coordinating chromosome segregation with cell division in *Caulobacter*. *Cell* 126, 147–162. doi: 10.1016/j.cell.2006.05.038

Trueba, F. J. (1982). On the precision and accuracy achieved by *Escherichia coli* cells at fission about their middle. *Arch. Microbiol.* 131, 55–59. doi: 10.1007/BF00451499

Typas, A., Banzhaf, M., Gross, C. A., and Vollmer, W. (2012). From the regulation of peptidoglycan synthesis to bacterial growth and morphology. *Nat. Rev. Microbiol.* 10, 123–136. doi: 10.1038/nrmicro2677

Umeda, A., and Amako, K. (1983). Growth of the surface of *Corynebacterium diphtheriae*. *Microbiol. Immunol.* 27, 663–671. doi: 10.1111/j.1348-0421.1983.tb00629.x

Ursell, T. S., Nguyen, J., Monds, R. D., Colavin, A., Billings, G., Ouzounov, N., et al. (2014). Rod-like bacterial shape is maintained by feedback between cell curvature and cytoskeletal localization. *Proc. Natl. Acad. Sci. U.S.A.* 111, E1025–E1034. doi: 10.1073/pnas.1317174111

van den Ent, F., Leaver, M., Bendezu, F., Errington, J., de Boer, P., and Löwe, J. (2006). Dimeric structure of the cell shape protein MreC and its functional implications. *Mol. Microbiol.* 62, 1631–1642. doi: 10.1111/j.1365-2958.2006.05485.x van Teeffelen, S., Wang, S., Furchtgott, L., Huang, K. C., Wingreen, N. S., Shaevitz, J. W., et al. (2011). The bacterial actin MreB rotates, and rotation depends on cell-wall assembly. *Proc. Natl. Acad. Sci. U.S.A.* 108, 15822–15827. doi: 10.1073/pnas.1108999108

Vollmer, W., Blanot, D., and de Pedro, M. A. (2008a). Peptidoglycan structure and architecture. *FEMS Microbiol. Rev.* 32, 149–167. doi: 10.1111/j.1574-6976.2007.00094.x

Vollmer, W., Joris, B., Charlier, P., and Foster, S. (2008b). Bacterial peptidoglycan (murein) hydrolases. *FEMS Microbiol. Rev.* 32, 259–286. doi: 10.1111/j.1574-6976.2007.00099.x

Wagner, J. K., and Brun, Y. V. (2007). Out on a limb: how the *Caulobacter* stalk can boost the study of bacterial cell shape. *Mol. Microbiol.* 64, 28–33. doi: 10.1111/j.1365-2958.2007.05633.x

Wagner, J. K., Galvani, C. D., and Brun, Y. V. (2005). *Caulobacter crescentus* requires RodA and MreB for stalk synthesis and prevention of ectopic pole formation. *J. Bacteriol.* 187, 544–553. doi: 10.1128/JB.187.2.544-553.2005

Wagner, J. K., Setayeshgar, S., Sharon, L. A., Reilly, J. P., and Brun, Y. V. (2006). A nutrient uptake role for bacterial cell envelope extensions. *Proc. Natl. Acad. Sci. U.S.A.* 103, 11772–11777. doi: 10.1073/pnas.0602047103

Waidner, B., Specht, M., Dempwolff, F., Haeberer, K., Schaetzle, S., Speth, V., et al. (2009). A novel system of cytoskeletal elements in the human pathogen *Helicobacter pylori*. *PLoS Pathog.* 5:e1000669. doi: 10.1371/journal.ppat.1000669

Whittenbury, R., and Dow, C. S. (1977). Morphogenesis and differentiation in Rhodomicrobium vannielii and other budding and prosthecate bacteria. *Bacteriol. Rev.* 41, 754–808.

Willemse, J., Borst, J. W., de Waal, E., Bisseling, T., and van Wezel, G. P. (2011). Positive control of cell division: FtsZ is recruited by SsgB during sporulation of *Streptomyces*. *Genes Dev.* 25, 89–99. doi: 10.1101/gad.600211

Wu, L. J., and Errington, J. (2004). Coordination of cell division and chromosome segregation by a nucleoid occlusion protein in *Bacillus subtilis*. *Cell* 117, 915–925. doi: 10.1016/j.cell.2004.06.002

Yakhnina, A. A., and Gitai, Z. (2013). Diverse functions for six glycosyltransferases in *Caulobacter crescentus* cell wall assembly. *J. Bacteriol.* 195, 4527–4535. doi: 10.1128/JB.00600-13

Young, K. D. (2006). The selective value of bacterial shape. *Microbiol. Mol. Biol. Rev.* 70, 660–703. doi: 10.1128/MMBR.00001-06

Young, K. D. (2010). Bacterial shape: two-dimensional questions and possibilities. *Annu. Rev. Microbiol.* 64, 223–240. doi: 10.1146/annurev.micro.112408.134102

Yu, X. C., and Margolin, W. (1999). FtsZ ring clusters in min and partition mutants: role of both the Min system and the nucleoid in regulating FtsZ ring localization. *Mol. Microbiol.* 32, 315–326.

Permissions

All chapters in this book were first published by Frontiers; hereby published with permission under the Creative Commons Attribution License or equivalent. Every chapter published in this book has been scrutinized by our experts. Their significance has been extensively debated. The topics covered herein carry significant findings which will fuel the growth of the discipline. They may even be implemented as practical applications or may be referred to as a beginning point for another development.

The contributors of this book come from diverse backgrounds, making this book a truly international effort. This book will bring forth new frontiers with its revolutionizing research information and detailed analysis of the nascent developments around the world.

We would like to thank all the contributing authors for lending their expertise to make the book truly unique. They have played a crucial role in the development of this book. Without their invaluable contributions this book wouldn't have been possible. They have made vital efforts to compile up to date information on the varied aspects of this subject to make this book a valuable addition to the collection of many professionals and students.

This book was conceptualized with the vision of imparting up-to-date information and advanced data in this field. To ensure the same, a matchless editorial board was set up. Every individual on the board went through rigorous rounds of assessment to prove their worth. After which they invested a large part of their time researching and compiling the most relevant data for our readers.

The editorial board has been involved in producing this book since its inception. They have spent rigorous hours researching and exploring the diverse topics which have resulted in the successful publishing of this book. They have passed on their knowledge of decades through this book. To expedite this challenging task, the publisher supported the team at every step. A small team of assistant editors was also appointed to further simplify the editing procedure and attain best results for the readers.

Apart from the editorial board, the designing team has also invested a significant amount of their time in understanding the subject and creating the most relevant covers. They scrutinized every image to scout for the most suitable representation of the subject and create an appropriate cover for the book.

The publishing team has been an ardent support to the editorial, designing and production team. Their endless efforts to recruit the best for this project, has resulted in the accomplishment of this book. They are a veteran in the field of academics and their pool of knowledge is as vast as their experience in printing. Their expertise and guidance has proved useful at every step. Their uncompromising quality standards have made this book an exceptional effort. Their encouragement from time to time has been an inspiration for everyone.

The publisher and the editorial board hope that this book will prove to be a valuable piece of knowledge for researchers, students, practitioners and scholars across the globe.

List of Contributors

Thomas R. Beattie and Rodrigo Reyes-Lamothe
Department of Biology, McGill University, Montreal, QC, Canada

Sue A. Flores, Matthew Howell, Jeremy J. Daniel, Rebecca Piccolo and Pamela J. B. Brown
Division of Biological Sciences, University of Missouri, Columbia, MO, United States

Conrad L. Woldringh and Norbert O. E. Vischer
Bacterial Cell Biology, Faculty of Science, Swammerdam Institute for Life Sciences, University of Amsterdam, Amsterdam, Netherlands

Flemming G. Hansen
Department of Systems Biology, Technical University of Denmark, Lyngby, Denmark

Tove Atlung
Department of Science, Systems and Models, Roskilde University, Roskilde, Denmark

Michelle M. Logsdon
Department of Molecular Biology and Microbiology, Tufts University School of Medicine, Boston, MA, United States
Department of Molecular Microbiology, Sackler School of Graduate Biomedical Sciences, Tufts University, Boston, MA, United States

Bree B. Aldridge
Department of Molecular Biology and Microbiology, Tufts University School of Medicine, Boston, MA, United States
Department of Molecular Microbiology, Sackler School of Graduate Biomedical Sciences, Tufts University, Boston, MA, United States
Department of Biomedical Engineering, Tufts University School of Engineering, Medford, MA, United States

Ding J. Jin, Cedric Cagliero, Carmen M. Martin, Jerome Izard and Yan N. Zhou
transcription Control Section, Gene Regulation and Chromosome Biology Laboratory, National Cancer Institute, National Institutes of Health, Frederick, MD, USA

Da Yang, Anna D. Jennings and Jaan Männik
Department of Physics and Astronomy, The University of Tennessee, Knoxville, TN, United States

Evalynn Borrego
Department of Mechanical, Aerospace and Biomedical Engineering, University of Tennessee, Knoxville, TN, United States

Scott T. Retterer
Oak Ridge National Laboratory, Center for Nanophase Materials Sciences, Oak Ridge, TN, United States

Beronda L. Montgomery
Department of Energy-Plant Research Laboratory, Michigan State University, East Lansing, MI, USA, 2 Department of Biochemistry and Molecular Biology, Michigan State University, East Lansing, MI, USA

Flemming G. Hansen
Department of Bioengineering, Technical University of Denmark, Lyngby, Denmark

Tove Atlung
Department of Science and Environment, Roskilde University, Roskilde, Denmark

Elena C. Guzmán and Carmen M. Martín
Departamento de Bioquímica Biología Molecular y Genética, Facultad de Ciencias, Universidad de Extremadura, Badajoz, Spain

Jacopo Grilli
Santa Fe Institute, Santa Fe, NM, United States

Clotilde Cadart
Centre National de la Recherche Scientifique, Institut Curie, PSL Research University, UMR 144, Paris, France
Institut Pierre-Gilles de Gennes, PSL Research University, Paris, France

Gabriele Micali
Department of Environmental Microbiology, Eawag, Dübendorf, Switzerland
Department of Environmental Systems Science, ETH Zurich, Zurich, Switzerland

Matteo Osella
Physics Department, University of Turin, Turin, Italy
Istituto Nazionale di Fisica Nucleare Sezione di Torino, Turin, Italy

Marco Cosentino Lagomarsino
Sorbonne Université, Paris, France
Centre National de la Recherche Scientifique, UMR 7238, Paris, France
IFOM, FIRC Institute of Molecular Oncology, Milan, Italy

Somenath Bakshi, Heejun Choi and James C. Weisshaar
Department of Chemistry and Molecular Biophysics Program, University of Wisconsin-Madison, Madison, WI, USA

Tsutomu Katayama, Kazutoshi Kasho and Hironori Kawakami
Department of Molecular Biology, Graduate School of Pharmaceutical Sciences, Kyushu University, Fukuoka, Japan

Philip C. Hanawalt
Department of Biology, Stanford University, Stanford, CA, USA

Masaki Osawa and Harold P. Erickson
Department of Cell Biology, Duke University Medical Center, Durham, NC, United States

Lev Ostrer, Bree L. Hamann and Arkady Khodursky
Department of Biochemistry, Molecular Biology and Biophysics, Biotechnology Institute, University of Minnesota, St. Paul, MN, USA

Prassanna Rao, Abdulaziz Alqahtani, Julia E. Grimwade and Alan C. Leonard
Department of Biological Sciences, Florida Institute of Technology, Melbourne, FL, United States

Tania A. Rozgaja
AREVA Med, Plano, TX, United States

Kouji Matsumoto and Hiroshi Hara
Department of Biochemistry and Molecular Biology, Graduate School of Science and Engineering, Saitama University, Saitama, Japan

Itzhak Fishov
Department of Life Sciences, Ben-Gurion University of the Negev, Beer-Sheva, Israel

Eugenia Mileykovskaya
Department of Biochemistry and Molecular Biology, University of Texas Medical School at Houston, Houston, TX, USA

Vic Norris
Laboratory of Microbiology Signals and Microenvironment EA 4312, Department of Science, University of Rouen, Mont-Saint-Aignan, France

Arieh Zaritsky
Faculty of Natural Sciences, Ben-Gurion University of the Negev, Be'er-Sheva, Israel

Conrad L. Woldringh
Swammerdam Institute for Life Sciences, University of Amsterdam, Amsterdam, Netherlands

Amelia M. Randich and Yves V. Brun
Department of Biology, Indiana University, Bloomington, IN, USA

Index

www.ingramcontent.com/pod-product-compliance
Lightning Source LLC
LaVergne TN
LVHW060516210926
848485LV00005B/1090
* 9 7 8 1 6 3 9 2 7 9 1 1 1 *